Praise for *A Commotion in the Blood*

"A terrific read about a subject of great interest."

—Joost Oppenheim, National Cancer Institute

"The concepts and the logic of discovery carry the reader from the first crude experiments to the thrilling and complex modern attempts to eliminate cancers using modern molecular biological tools, and along the way one absorbs important ideas about both cancer and immunology. Most importantly, the non-scientist begins to see what makes these driven men tick. Their egos, their compassion, and their humanity are on display as the story unfolds.

"There is another important reason to read this book: it shows the development of dramatic and sometimes dangerous new treatments that must be tested and refined in human subjects. . . . It is a book that gives lay readers real insight into the way things work in 'Big Science,' and the slow and painful way some of our medical advances have been won."

—C. J. Peters, chief, Viral Epidemiology, Centers for Disease Control

"A mesmerizing read . . . Remarkable . . . A rapacious industry, a voracious press, and a raging desire for share options and more especially for that handshake with the king of Sweden make an explosively rich mixture."

—Walter Gratzer, editor, *The Literary Companion to Science*

"Steve Hall has produced a meticulously researched account of the evolution of the field of immunotherapy of cancer. . . . It should be an inspiration both for those who have suffered from cancer as well as for the next generation of researchers."

—Kendall A. Smith, chief of immunology, Cornell University Medical College

"Thrilling to read . . . Rich in rendering the human drama, behind-the-scenes intrigue and occasional gossip, this is at the same time a scholarly, thoroughly researched work . . . [written] in language that pleases the soul as well as the mind."

—Jan Vilcek, NYU Medical Center

"Hall tells the exciting, involved, century-long story of cancer therapy and the immune system excellently . . . Many current and future medical news items will be clearer to readers of this fascinating, detailed, well-documented book."

—*Booklist*

"As gripping as a spy thriller . . . In the end, Hall leaves the reader optimistic that the 'commotion in the blood' is not random noise but a 'beautifully scored piece of music' that future researchers will be able to read."

—*Kirkus Reviews*

"A recommended science book of the year."

—*Science Friday*, National Public Radio

Also by Stephen S. Hall

Invisible Frontiers

Mapping the Next Millennium

A

COMMOTION

IN THE BLOOD

For M & M
The two sweetest things in my life

CONTENTS

III · THE RISE OF THE T CELL CHAUVINISTS

IV · IN VIVO VERITAS

Blood is the human being's finest juice:
family remedy, household drug, aliment of life.

—Levinus Lennius, *Della complessione del corpo humano,*
libri due: Nuovamente di latino in volgare tradotti
e stampati, Venice, Domenico Nicolini, 1564

Scientific theories . . . begin as imaginative constructions.
They begin, if you like, as stories, and the purpose
of the critical or rectifying episode in scientific
reasoning is precisely to find out whether or not
these stories are stories about real life.

—Peter Medawar, *Pluto's Republic*

Every beginning
is only a sequel, after all,
and the book of events
is always open halfway through.

—Wislawa Szymborska,
"Love at First Sight"

"SHOOTING RUBBER BANDS
AT THE STARS"
A Cautionary Prologue

↝ *Wednesday, August 4, 1993.* In Bethesda, Maryland, the day dawned misty and cool—blessedly so after a long, sodden stretch of humid, 90-degree weather. Thunderstorms had rumbled through overnight, scrubbing the air and leaving behind pockets of coolness in the hollows, as rare and fresh in a Washington summer as they were welcome. When he was asked later, Claude Excell Raiford III said he couldn't recall the thunderstorms, didn't in fact remember Wednesday at all. He went to bed Tuesday night and woke up sometime Thursday, spending most of that Wednesday pinned pale-skinned, still, and Christlike on an operating table, his arms outstretched on either side, his brown beard neatly trimmed and his eyes taped shut, his torso shaved and daubed an antiseptic orange, sheathed in an adhesive antibiotic wrap snug as Saran, as naked as any mortal soul can be in the company of many well-intentioned strangers holding knives.

At exactly 9:17 A.M. that Wednesday, in an operating room of the Clinical Center at the National Institutes of Health in Bethesda, surgeon Douglas Fraker said, "Okay, we're going to start," and made an incision from the center of Claude Raiford's chest to the lower right abdomen; a song called "Shooting Rubber Bands at the Stars" by Edie Brickell and the New Bohemians poured out of a portable CD player in the corner, and given the gravity of Mr. Raiford's illness and the unproven track record of this particular operation—only the fourth time it had been performed anywhere in the world—the musical selection was not inappropriate. Fraker, a tall and broad-shouldered young physician who combines the surgeon's confidence with an easygoing midwestern manner (no one asks for the Iowa-born "Dr. Fraker" in this room, only "Doug"), manipulated his cauterizing needle so swiftly, sealing off blood vessels with so deft and sizzling a touch, that the initial ten-inch incision seemed to produce hardly more blood than a nasty cut. After a few minutes, he was sticking his hand almost to the elbow into Mr. Raiford's torso, trying not to find something while groping blindly as if he were. He was feeling for hard, rocklike knots of tissue—tumors. If he found any in unexpected places, they would have sewn Raiford back up before Edie Brickell finished the third cut of Side A and sent him home to North Carolina to spend time with his daughter, make his final arrangements, make his final peace.

Having satisfied himself that there was no detectable cancer elsewhere, however, Doug Fraker proceeded to expose the liver, a large and odd-shaped organ, looking like a fleshy triangle, neither muscular like the heart nor mushy like the brain (though as essential to life as both), about as big as a woman's medium-sized purse, glistening and regal and dark-hued, the color of a proud bruise. This particular liver did not look regal. Fraker lifted it up and inspected its backside. "Tumor, tumor everywhere," he said wearily, shaking his head. We usually associate cancer with ominous shadows on a CAT scan or x ray; here in the flesh, numerous tumors pocked the surface of the liver, chalky white circles of diseased tissue (doctors prefer the neutral and generic term *lesions*), some the size of a nickel, others as large as a half dollar. The entire organ had a lumpy, irregular appearance, like thick clots of powdery flour that bob up in a poorly mixed pancake batter. Probably 40 percent of the liver, Fraker estimated, had been overtaken by tumor.

At this point, as he steadily snipped away at the thin, tenacious curtains of tissue sheathing parts of the liver, Fraker changed subjects. "We're getting our new basement rug in today," he announced to the operating team. Someone at the table inquired as to the color. "Mauve," he replied, continuing to snip. "We wanted a Berber, but the guy said it makes no sense to have a Berber in the basement. . . ."

You would not immediately surmise from the conversation that ensued, about rugs and *Saturday Night Live* skits and Dr. Seuss (including a spirited recitation by Dr. Fraker of *Green Eggs and Ham*), that these highly skilled physicians and nurses were in the midst of a highly experimental form of cancer treatment, a form of immunotherapy so delicate and complex that it would require about seven hours of surgery before they could even get to the main point of the exercise, which was to inject a syringe full of drug into Claude Raiford's liver;

not any drug, but a biological molecule so potent that Fraker claimed
he could sometimes detect subtle changes in tumors, feel them softening up, after only sixty minutes of treatment, and yet one so rampag-
ingly toxic that a mere thousandth of a gram, if it leaked back into the
general circulation, could plunge an adult male body into fatal shock
within minutes.

It was all part of an experimental procedure adapted from an approach pioneered in Europe against another form of cancer, melanoma; European surgeons reported that nearly 90 percent of patients
undergoing the combined surgical-therapeutic procedure had a complete response, and those promising results made it seem worthwhile
to go through such an extensive surgical prologue simply to administer a drug. What they set out to do—and it would take hours of tedious
dissection, as Fraker called it, allowing time to talk about even mundane things like rugs and TV—was to replumb Claude Raiford's circulation so that his diseased and lumpy liver was temporarily isolated
from the rest of his vasculature; they would then attach the arteries
and veins of his liver to the equivalent of a heart-lung machine, which
would supply to the liver blood warmed as by fever to 42 degrees
centigrade, and into this temporary loop of plastic tubing ferrying
blood to the isolated liver they would introduce a single, powerful,
onetime dose of a drug known as tumor necrosis factor, or TNF. They
would bathe the diseased liver with TNF for exactly sixty minutes,
flush it out, reattach Mr. Raiford's circulation to its accustomed network of arteries and veins, and sew him up before sending him back to
the intensive care unit, where, they hoped, his tumors would begin to
melt away.

The sequence of events that brought Claude Raiford to this juncture
is enough to justify hypochondria in just about anyone. Only four
months previous to this August morning, Mr. Raiford suffered no
more debilitating a symptom than a vague sense of feeling tired; he
felt no pain, had no headaches, complained of no fever, offered not a
single sign of serious illness except a touch of rectal bleeding, which
he attributed to a twenty-year history of hemorrhoids. Thirty-six years
old, with a wife and four-year-old child, Mr. Raiford had recently embarked upon a career change, abandoning his work as an insurance
agent to go back to school and become a psychiatric nurse. The only
hint that something might be wrong was the constant fatigue he felt,

which could easily be attributed to the combined demands of schooling and child-rearing. He was an otherwise handsome and vigorous man, active and outgoing.

When the fatigue persisted, he consulted his family doctor, who discovered traces of blood in his stool; in that ever-accelerating river of bad news known only too well to cancer patients, the first test led to many more tests, which revealed an advanced colon cancer, which led to surgery in June 1993 to remove the diseased segment of intestine with its four-inch tumor. But as frequently occurs with colon cancer, the malignancy had already hopped the surgical fire line and spread to the liver, and CAT scans of Mr. Raiford's liver showed such extensive tumor that his prognosis was extremely grim (in one scan, a major vein, the vena cava, even appeared kinked, shoved to one side by the thrust of a tumor pressing against it). One look at the x ray, and Fraker had muttered, "It's going to be a long day."

As often happens at a major research center like the NIH, other doctors popped into the operating room to watch the experimental procedure, and one of them asked Fraker what made Raiford such a good candidate for the TNF procedure. "He's dying of cancer, he's failed chemotherapy, it's limited to the liver, and he's young," Fraker replied. "It's good, too, because in a lot of colon cancer patients, the metastases are limited to the liver." Experimental cancer treatments require the compliance of desperate patients; Raiford, unfortunately, qualified.

Prior to scrubbing down in the operating room, Fraker had shown me CAT scans of one of the previous three patients to undergo the operation at the NIH, scans taken prior to receiving TNF. "These are all tumors, all these cannonballs," he said, indicating large, round, grapeshot shadows that dotted one man's liver. Next, he pulled out a CAT scan showing the same liver one month after the treatment. "*This* is, like, an incredibly exciting x ray," he said, pointing out where the shadows had shrunk and disappeared, "to take a tumor and have this much shrinkage so soon after a onetime dose of TNF."

Fraker, who at the time headed the Metabolism Branch at the National Cancer Institute (he has since moved to the University of Pennsylvania), and his team began to perform this type of surgery in May 1993. Assisting him were Dr. Arleen Thom, who spent two years testing the procedure in pigs before it was attempted on a human, and Dr. H. Richard Alexander. Fraker not only pushed for approval of the procedure at the National Cancer Institute but rigged from scratch some sophisticated monitoring technology to assure that the operating team could detect any leaks in their anatomical plumbing before injecting the highly toxic TNF. As he pulled on his surgical gown and adjusted his Coaxial 6000 Fiberoptic Headlight, a forehead-mounted surgical flashlight connected by an umbilical cord to a power supply behind him, Rich Alexander paused to say, "I've got to tell you, I've seen thousands of these things in my ten, twelve years of academic surgery, and this—*this* is a real breakthrough."

Here, a momentary intermission to acknowledge an essential component in any account of medical innovation: the chronicler. To anyone who writes about science or medicine for a living, the word *breakthrough* has a kind of transcendent power. The procedure or molecule or discovery under discussion may or may not in fact be a breakthrough, but if anyone in a position of expertise or authority utters the word, the writer possesses a ticket to journalism's most precious real estate: the front page of the newspaper, the magazine cover, prominent play on the evening news. So when you are fortunate enough to be invited to observe a pioneering operation at a place like the National Cancer Institute, you feel a bit blessed and a bit cowed, like the war correspondent smuggled to the front line, grateful for the access but not inclined to say anything that would antagonize the generals. In retrospect, my experience observing this procedure not only documents a promising, cutting-edge approach to medicine but also serves as a cautionary introduction for lay readers to medical reporting in general, and to immunotherapeutic approaches in particular.

What exactly is immunotherapy? Broadly speaking, it is a form of medicine that uses the immune system—its cells, its molecules, and its rules of engagement—to tip the balance in favor of the body in its battle against disease. In a sense, vaccination is the oldest and most glorious form of immune modulation, dating back at least three centuries as a medical intervention and probably back much further as a folk cure. Vaccination involves the direct and specific stimulation of the immune system to elicit protection, traditionally against infectious pathogens and, in recent years, against malignancies as well; other approaches involve isolation of soloists from the immune ensemble, such as T cells or antibodies or the potent molecules known as cytokines (interferon, tumor necrosis factor, and the interleukins are examples), which are then given alone or in combination to bolster the body's natural responses. A long-standing focus of biological approaches, dating back more than a century, is cancer; and it is this effort to attack cancers that informs the crux of this narrative.

Just as the worth of vaccination was proved first by human experimentation, many subsequent developments in the history of immune therapy have been won first in the clinic, only later in the laboratory. This is a pattern not without inherent tensions. The historical disjunction between experience and knowledge, clinical success and scientific understanding, was true of Edward Jenner's smallpox vaccine, of Jonas Salk's polio vaccine, and of many other attempts at immune modulation that have reached the public consciousness before the science explaining their action has made its way into textbooks. Immunotherapy in particular has experienced a long history of booms and busts. The use of bacterial vaccines at the turn of the century, the use of what are called "nonspecific" immune stimulants like BCG in the 1950s and 1960s, the use of cytokines like interferon in the 1970s, the use of so-called LAK cells in the 1980s—their introduction was greeted by excessive optimism, and the disappointment that later settled in has probably been excessive as well. These approaches were never as good as they were initially cracked up to be, but probably not nearly as bad as the odor lingering around their use might suggest. For the better part of a century, the practice of immunotherapy has tended to be ahead of the science, but finally, in the 1990s, the science of immunology has provided an unusually firm foundation for the optimism that currently enlivens its practical applications. Lloyd Old, one of the most thoughtful and global of thinkers in the field, likes to invoke the memory of the British immunologist Peter Gorer, whose work in the 1930s laid the groundwork for the revolution in organ transplantation decades later. "You can't rush science," Gorer said on many occasions, and Old almost sighs with relief when he adds, "Science is *finally* catching up." As a result of more precise knowledge about immunology and cancer, researchers are exploring much more precise manipulations of the immune system.

If there has been one lesson in the history of immunotherapy, it is that reductionism, the single most productive approach in twentieth-century biology, has brought us full circle, leading to a global, almost holistic understanding of an incredibly complex system, namely immune function; and such understanding perforce requires armies of researchers toiling over many years and many scientific generations. But as I write this, more than a year after the TNF perfusion operation in Bethesda, I feel obliged to return to an essential and nettlesome paradox: journalism (medical or otherwise) is defined by its timeliness, while the careful assessment of new medical technologies requires a great deal of time, years usually. This is true of any new medical treatment, but especially true of cancer, against which most efforts in immunotherapy have been directed. Blessed with unusual access, anointed with exclusivity, benefiting from the kind of fly-on-the-wall detail that conveys the genuine excitement (and the equally genuine ennui) of experimental medicine as it unfolds, I had been handed the ball and expected to run with it. And on more than one occasion I rued not publishing an account of this procedure more speedily. For a variety of reasons, I did not.

But one of the luxuries of a book project—indeed, to me its surpassing advantage—is to turn one's back on the deadline clock, to step back from the increasingly hyped hurly-burly of the research milieu and take a somewhat longer view. As much of this book's narrative will suggest, the public's perception of medical innovation—and the role medical journalism plays in that perception—strongly influences the funding, direction, philosophy, and fashions of research and has played a not-peripheral role in a recurrent and unfortunate pattern to which many medical innovations succumb: unrealistic optimism, followed soon thereafter by the sound of a crash, which in turn is greeted soon thereafter by self-satisfied grunts from those to whom unrealistic pessimism is the more comfortable extreme. At a time when anti-tumor experiments in a few mice can land on the front page of the *New York Times*, years before even a first tentative transfer of the technology to the treatment of humans (who in any event are notoriously bad mimickers of mice when it comes to cancer treatments), it bears reiterating that it takes on the order of five to ten years to determine the quality of a new cancer treatment in humans—and then only if the experiments and clinical studies have been well designed and rigorously executed, conditions that pertain much less often than the public might suspect.

All of which is to say a special burden attaches to medical reporting when it discusses new treatments for otherwise grim maladies like cancer and AIDS, and no amount of learned qualification and obfuscatory caution in the description will dilute the hope such reporting invariably inflames among those with the most at stake: patients who do not have five to ten years to wait for an answer. Journalism does not of course reward the impulse to sit on stories, but as I look back now on this particular episode, I realize that the thrust of any account

I wrote about the TNF perfusion technique could easily have changed every few months, depending on trends in the emerging but still incomplete data from the lab. I had observed the procedure firsthand, a respected physician had called it a breakthrough, and there were preliminary data to suggest remarkable responses. But I wanted to wait and see how long these impressive responses lasted, and I wanted to talk with the people who deserved credit for coming up with the idea in the first place. "The Europeans," Richard Alexander had said, "were able to take that leap that we weren't able to."

The platform for that great leap was tumor necrosis factor. Discovered in the 1970s in the laboratory of Lloyd Old at Memorial Sloan-Kettering Cancer Center in New York, TNF is indeed a perfect example of the manic-depressive winds that howl through immunotherapy. The gene for TNF was cloned in 1984 to enormous expectation, and researchers rushed the resulting genetically engineered drug into clinical trials by 1985; with enormous disappointment, clinicians halted testing almost immediately. In Fraker's words, the trials were all "miserable failures." The drug fell between two chairs: the toxicity was extraordinary at the higher doses that worked against cancer in mice, the effectiveness nil at lower doses safe for human consumption. Toward the end of 1988, well after TNF had become the black sheep of the cytokine family, a surgeon in Brussels named Ferdinand ("Ferdy") Lejeune and his colleagues, based now in Lausanne, Switzerland, pioneered a new way to treat melanoma in the rare instances when the tumors were isolated to an arm or a leg. In a procedure called "isolated limb perfusion," they essentially created a surgical tourniquet around the leg, for example, cutting off circulation from the rest of the body, and then pumped in a combination of tumor necrosis factor, another cytokine called interferon gamma, and melphalan, a chemotherapy agent. Writing up their results (which have been confirmed in several follow-up studies) in 1989 in the *European Journal of Cancer*, the Lejeune team claimed that every single one of their first fifty-one melanoma patients responded to the treatment with a diminution of their tumors; an astonishingly high 90 percent had complete responses, meaning that all visible melanoma tumors—dozens of tumors in some cases—totally disappeared. By the end of 1993 Ferdy Lejeune and his team had treated 160 patients with their so-called triple therapy, claiming an overall complete response rate of approximately 85 percent.

Here, the evidence can mislead if not weighed carefully, and these spectacular regressions of tumors must be viewed within a rather more sobering context of relapses. The median duration of these responses is eleven months (in other words, half of all responses last less than a year), and the median survival is two years. The tumors go away, but in many patients the disease comes right back, often because of invisible seeds of malignancy that had already spread to other parts of the body when the procedure was performed. The treatment is promising, tantalizing, even spectacular. But five years after the treatment was first attempted, it is still unclear just how much long-term benefit it provides (the technique is discussed more fully in part IV of this book). This perfectly hews to a general theme in immunotherapy—much has been accomplished, but there is still a long way to go. Against cancer, alas, 99.9 percent success is not good enough.

From the moment of the initial incision, it required the better part of seven hours for Doug Fraker and his colleagues to prepare Claude Raiford for his onetime hit of TNF. They tied off the artery leading from the heart to the liver, removed the gall bladder (the B-52s' "Love Shack" throbbing in the background), and rerouted circulation of the legs around the liver, through which it would normally pass, and back to the heart by way of a shunt of plastic tubing from the left leg to a vein in the left armpit. At ten minutes to four, Arleen Thom walked over to a small table in the corner of the operating room.

There, sitting on the table throughout the surgery, were two brown plastic pouches, looking like containers of freeze-dried camp food. Like the gun in a Chekhov play, they had been onstage the whole time, but nearly forgotten for the bulk of the drama. Inside the pouches were three vials, each containing a tiny white pile of dried tumor necrosis factor. As Thom injected sterile water into each of the vials, the powdered TNF dissolved into a colorless, slightly milky fluid; when it was all mixed and loaded into a single syringe, it amounted to three cubic centimeters of fluid—less than a single teaspoon of medicine, in which floated 0.6 milligram, or slightly less than a thousandth of a gram. Injected directly into the bloodstream, the TNF would kick off an episode of shock that, if not managed aggressively, could prove fatal hours later. All this—a fifteen-inch, omega-shaped incision across Claude Raiford's chest, seven hours of excruciating microscopic dissection, two bypass machines (each accompanied by a personal perfusionist from Walter Reed Army Hospital), three anesthesiologists, three surgeons, three nurses, decades of collective surgical expertise—for a teaspoon of highly toxic and possibly beneficial medicine. After one final check to make sure there was no leakage, one of the perfusionists sent the TNF flooding into the island that was Claude Raiford's liver. "Okay, we're on," Fraker said. A timer on the wall began counting off the hour.

As the TNF began to permeate the diseased organ, Fraker cradled the liver in his left hand, a lobe nestling gently between his little and ring fingers, to prevent any kinks in the cannula, the plastic tube that delivered the drug to the organ. Every once in a while, he gingerly poked at one of the tumors on the surface to see if he could detect any softening. Wishful thinking, probably—the look of friendly exasperation on Arleen Thom's face as she watched Fraker across the operating table suggested that skepticism was the only appropriate response to such optimistic palpations. But Fraker insisted he had felt a difference in tumors only minutes after the TNF had gone to work.

"We had a change in one of the livers we did," he said, "a big lesion in the middle. It wasn't on the surface, but it was a very big lump, and then the [part of the] liver over it turned dark. There was a demarcation over it, as if you'd cut off the blood to the lesion." During the hour, Fraker bathed the liver in warm water to keep it warm, and occasionally the surgeon probed several of the tumors to see if they had softened. When the clock on the wall read 57:07, they began to unplug catheters and commenced flushing out the liver; for five minutes, it received absolutely no blood (or oxygen), and it soon turned faintly rosy and pale, like the pink granite of a Tuscan duomo. Distinct bright red borders formed around each tumor. Once the drug had been flushed out, they hooked up all the arteries and veins to their customary intersections, and the organ slowly blushed back to life, a deep, purplish stain creeping through the tissue. Just prior to giving TNF, and just after the hour of treatment, little bits of the tumor had been excised as biopsy samples. "It doesn't look all that much different," the pathologist said, examining the "after TNF" tumor at a table in the corner. Fraker looked disappointed. "But it may be a little softer," the pathologist added as an afterthought.

"There you go—" Fraker said.

Around 5:40 P.M., they began to close the incision. By now there were fewer nurses, fewer outside observers, no music. Fraker and Alexander had been on their feet, without a break, for about ten hours. While a nurse counted, and accounted for, each of the 240 forceps, hemostats, criles, scissors, clamps, scalpels, and other utensils that had been arrayed on the surgical tray when the day began, each of the gauze pads that had been used, each and every item brought to the edge of that canyon where the operating room stops and the cavity of human flesh begins, Fraker and Alexander sewed Raiford's chest back together. As they inserted the final sutures, a voice came from the corner of the room. "Doug, I take that back," said the pathologist, who had been quietly studying the biopsy sample. "It *is* a little different. When you cool it down, it tends to fall much easier out of its capsule." At 6:29 P.M., the last surgical staple went in. Mr. Raiford was wheeled off to the intensive care unit, and Doug Fraker left to talk to the patient's family.

The following day, when we went into the ICU to visit Claude Raiford, he looked remarkably . . . *okay,* given what he'd gone through the previous day—the color had returned to his face, the breathing tube had been removed, and, in a soft Carolina accent not the least bit frayed by ten hours of surgery, he said he felt pretty comfortable, except for a little pain when he breathed. Predictably, he remembered nothing about the operation. "I went to bed on Tuesday," he said with a slow shake of the head, "and woke up on Thursday." Fraker explained to him that everything had gone well, that several technical problems he had anticipated had not in fact materialized, and that it had simply taken a good deal of time, as these operations tended to do. "You got the full treatment," he continued. "You got 0.6 milligram of TNF, the biggest dose we've given so far."

"Good." Raiford nodded, looking pleased. He managed to add, "That's what we came for."

As in many accounts of this sort, a few things have been left unsaid. Isolated organ perfusions are not a cure for cancer; only a few patients qualify for the treatment; too few patients have been treated for us to know if it has enduring benefits; and the limitations are many and sobering. The surgical aspect alone, so long and arduous, can usually be attempted only once; the drug itself is potentially fatal if for any reason the surgical plumbing fails. If microscopic pods of cancer have spread from the original site, which is virtually impossible to determine ahead of time, the entire effort may be for naught (although Fraker expressed enthusiasm about the liver perfusions because many colon cancers, in one of those quirks of malignancy, spread only to the liver). And no one knows how durable the remissions will prove to be, because the treatment is too new to allow any conclusions to be drawn with statistical reliability.

It is a last-ditch treatment for last-ditch patients, and there is, finally, an underlying irony to the entire approach: it requires exactly the kind of onerous, complicated, massively invasive, and heavy-handed medical intervention that the elegance of immunotherapy, with its genetically engineered molecules and its customized cells and its sophisticated manipulations of our ultimate healing tool, the immune system, was intended to replace. What justifies the effort, in the eyes of surgeons like Fraker, is the remarkable transformations that tumor necrosis factor induces in tumors. Despite its toxicity, and for reasons that remain unclear to this day (though they are the subject of intense laboratory investigation), TNF has the remarkable ability to distinguish between normal blood vessels and the blood vessels that nourish growing tumors; the molecule seems to slide blithely through healthy vasculature but initiates an attack on vessels feeding a tumor in such a way that the walls of these vessels become leaky and porous, entrapping red blood cells that pile up and form a plug, and within days or hours causes a hemorrhage that is lethal to the tumor itself though it spares healthy tissue.

Popular accounts of new medical treatments like this bedevil the medical landscape, because they are long on promise and short on data, especially with regard to the single most critical consideration when it comes to cancer: how long did the response last? The post-surgical saga of Claude Raiford, and the entire evolution of the TNF perfusion story, since my August 1993 visit vividly demonstrate the perils of reaching premature conclusions, an affliction known not only to good medical journalists but to good physicians as well.

If I had done a story soon after the procedure, it would have been quite upbeat. Claude Raiford, like the other patients before him, responded extremely well to the TNF perfusion treatment at first. One month after surgery, the tumors in his liver had shrunk an estimated 85 percent, and a blood test measuring a protein associated with colon cancer called CEA (for "carcinoembryonic antigen") plummeted from 165 before the operation to 16 afterward. His next checkup, at three months, also showed the cancer under control. Back home in North Carolina, he graduated with his nursing degree, found a new job, and began to allow himself a secret smidgen of hope. An article in a local newspaper described his dramatic treatment and quoted him as saying, "Right now it doesn't look too good for the tumors. I think the disease is in serious trouble."

If I had done a story eight months after the operation, it would have been more measured. As happens so often in new treatments of this sort, the cancer had merely paused to regroup. Eight months after the procedure, Raiford learned that the malignancy had grown back to its former extent prior to the operation. By the spring of 1994, his condition was almost as bad as it had been before. By then, he had checked into another hospital in Tennessee to try yet another experimental treatment known as chemo-embolization, in which chemotherapy is delivered by catheterization directly into the liver.

If I had done a story in the fall of 1994, the ultimate efficacy of the TNF perfusions would have looked a bit cloudier. In October of that year, at a meeting of the Society of Biological Therapy in Napa, California, Richard Alexander, one of the surgeons who had operated on Claude Raiford, gave a progress report on the TNF trials at the National Cancer Institute, including preliminary results from a controlled, randomized trial of isolated limb perfusions comparing the chemotherapy drug melphalan alone against the combination of melphalan and TNF. Although he did not say so in so many words, the unpublished data Alexander presented at the meeting left the final outcome uncertain: the numbers suggested to at least some people in the audience that there was little or no difference at all between tumor necrosis factor and melphalan alone in these cases. In less than a year, the TNF "breakthrough" had gone from miraculous and fantastic to marginal and perhaps even nonexistent. A few months later, when I spoke with Fraker, the story seemed to have changed again; he explained that a subsequent study, still underway, suggested that the combination of melphalan and TNF appeared to work better against bulkier disease than melphalan alone. It was, he said, too early to tell.

If I had done a story in the summer of 1995, when I spoke to Mr. Raiford on the phone, it would have changed again. He was undergoing monthly chemotherapy treatments, but had been spared additional chemo-embolization since the previous summer. He'd been back working full-time for a year and a half, and seemed quite happy with his progress. "Considering the disease and where I was," he said, "most people would have to say I'm doing extremely well." Like other patients, he felt the immunotherapeutic treatment had truly stimulated his immune system. "The thing that to me seems so ironic is that since I've had the TNF perfusion, I have only had one to two colds," he said. "I used to get them a lot. Here I've got incurable cancer, but I don't get colds anymore, and that's pretty strange. I seem kind of strangely healthy, given all that's wrong with me." At the end of our conversation, he added that if I wanted to have a follow-up chat "in six months or a year, I'll be here."

And if I had done a story in the spring of 1996, when researchers gathered in Hilton Head, South Carolina, for a meeting, the message would again have been mixed. While several European groups continued to report considerable success using TNF perfusions against melanoma and sarcoma, Fraker told colleagues that liver perfusions of the sort Claude Raiford had undergone seemed to have limited efficacy. The combination of TNF and melphalan proved "feasible" in preliminary testing, and had "significant anti-tumor activity" despite the fact that patients had extensive metastatic disease and had been heavily pretreated.

The story changed, in a nonscientific but profound way, when I planned to give Mr. Raiford that anniversary follow-up call in August 1996. While checking a reference on the Internet, I discovered his obituary and learned that he had died at home on November 24, 1995. He had lived a little more than two years after the TNF perfusion and, like all cancer patients, had waged a courageous battle against the disease. Was the TNF procedure worth it? "Him more than anyone else, I think we helped," Fraker said. And in that elusive way that no machine can measure, of survival and of stolen moments spent with loved ones, forestalling the inevitable, the procedure may have added months if not years to Claude Raiford's time on earth, the value of which is so personal, so precious, and so incalculable that it is far beyond the reach of scientific discussion.

Clearly, something remarkable happens with TNF. Large tumors do not of their own accord liquefy and disappear. Just as clearly, that effect isn't enough. Like molecular magicians, doctors have made sizable tumors shrink and disappear with the use of powerful, immune-derived medicines, and these responses cannot fail to astonish and hearten anyone who has seen a shadow disappear on a CAT scan or a lump palpably soften to the touch. They may even buy a little time. But too often these spectacular responses do not last. Is the glass half empty or half full? If you ponder the enormous resources that go into making a tumor shrink temporarily (a gift conferred with as much brevity by radiation or chemotherapy), the glass looks pretty empty. If it's your life that is extended by such treatment, however, it might look reasonably full. And if you happen to be one of those fortunate souls who respond completely and permanently, the cup runneth over.

As we walked back to his office one day, I asked Doug Fraker to describe exactly what he saw when he noticed changes in a tumor treated with TNF. "You usually see it in two or three days," he said, "although sometimes you see it on the table or by the next day. The tumors feel softer and they look, particularly in the center, necrotic, meaning they have a kind of bull's-eye appearance. They're not always neatly encapsulated. Sometimes they turn into black, dead tissue. . . ."

His words reminded me of passages first published in the medical literature one hundred years ago, when a New York surgeon named William Bradley Coley began treating cancer patients with a crude vaccine composed of killed bacteria. Benefiting in part from the effects of TNF (although quite unaware of it at the time, the molecule not having been identified by biologists until 1975), Coley observed remarkably similar results in his patients, which he described in remarkably similar language. I asked Fraker if he was familiar with Coley's work. Still in his surgical scrubs, he rushed across his small office and yanked open a file cabinet. "I have a whole file, what I call my 'Historical TNF File,' " he explained over his shoulder, pulling out a manila folder containing a number of Coley's papers, along with later papers by Murray Shear of the NIH and Lloyd Old of Sloan-Kettering that were important landmarks in the TNF story.

Did Fraker think the effects described by Coley, much disputed and ultimately discredited by the grandees of medicine during Coley's lifetime, were reliable? "I think it's real," he said, without hesitation. "It's exactly what we see."

Indeed, in order to find "complete and durable responses" (medicine's euphemism for "cure") to cancer, you have to look elsewhere. You have to travel to New York City, to that city's venerable Memorial Hospital. And you have to travel back to the nineteenth century, because that is when William Coley obtained some of the most impressive results ever won by immunological medicine against malignant disease.

I

THE
OCCASIONAL
MIRACLE

*When [John Hunter's] old pupil, Edward Jenner, came
to him and presented his reasons for believing that
smallpox and cowpox were closely related, and wanted
his opinion, Hunter replied, "I think your solution just;
but, why think, why not try the experiment?"*
—William B. Coley, "The Idea of Progress," 1920

1

"LAUDABLE PUS"

*"The first hope of therapeutic success comes with the observation
of the efficiency of unaided nature to accomplish cure. . . .
These cases, rare though they be, are the sum of our hope."*
—Pearce A. Gould, "The Bradshaw Lecture on Cancer," 1910

⚮ *There survives from the beginning* of this century an intriguing
bit of paper ephemera, notable less for the celebrated names marked
down upon it than for a medical future it unknowingly foretells. As
the year 1903 drew to a close, a rather well known New York busi-
nessman went to a Manhattan doctor of his acquaintance for a routine
blood test, which was sent for analysis to an increasingly prominent
pathologist at the Cornell University Medical College. The results of
these tests, typed out on a half-sheet of paper and dated December 7,
1903, are quaintly minimalist and read as follows:

> Hb   95%
> Red cells   4496000
> Leucocytes   5000
>> Mononuclear   44%
>> Polynuclear   55%
>> Eosinophile   1%

"In the stained specimens," the pathologist noted, "I find nothing abnormal."

There is nothing abnormal about the blood chemistry, either. In
quick translation, the results indicate a healthy number of red blood
cells, the cells that carry oxygen to the tissues; a low but acceptable
number of white blood cells (or leukocytes), which are now known to
comprise the main cellular components of immune activity; and a
reasonable proportion of the various cells that make up the "white

blood." What at first seems so remarkable about this lab report is the confluence of names, interests, and destinies attached to it. The pathological analysis came from James Ewing, who would go on to head Memorial Hospital in New York and become one of the most influential cancer authorities of the twentieth century. The doctor ordering the tests was William B. Coley, who in the early days of the century claimed to have cured a number of inoperable cancer patients with a crude antitumor vaccine. And the blood itself belonged to John D. Rockefeller, even then in the process of establishing the great medical institute that would bear his name and whose philanthropic largesse advanced the interests and causes, to greater or lesser extent, of the other two men whose names appear on this otherwise forgettable bit of biochemical trivia.

All three men were dedicated, in their own way, to a cure for cancer, little realizing that certain clues lay in the very cells whose mundane numbers were dutifully recorded in the blood report. And it bespeaks the twists and turns of nearly a hundred years of research into cancer and immunology that the most important word on that fragile piece of paper is not the celebrated names that have come to symbolize wealth or medical authority but rather the word *blood*. All of Mr. Rockefeller's millions, all of Dr. Ewing's encyclopedic knowledge, and all of Dr. Coley's clinical intuition could not together purchase the knowledge we have painstakingly acquired in a century of research into the remarkable qualities of blood—its leukocytes and other cells, its proteins, its molecules and factors then unimaginable—all of which are now understood to function, like sections of an orchestra, in the daily biological symphony that is the immune system. In telling the story of how much of that knowledge has come to be won in the treatment of patients, it is most appropriate to begin with that crucial intersection of Coley and the Rockefellers and a disease whose diagnosis, then even more than now, was tantamount to a death sentence: cancer.

The destinies of the Rockefellers and Dr. Coley first crossed more than a decade earlier in a tragedy that, as the tabloids might put it, left one young woman dead and the lives of two men changed forever. It began in the fall of 1890, shortly after twenty-eight-year-old William Coley had finished his residency as a surgeon at New York Hospital and entered private practice. He was still a green and impressionable physician, still learning his trade, when a young woman named Elizabeth Dashiell came to see him, complaining about a nagging pain in her hand.

She was seventeen at the time, thoughtful and self-possessed, with a face so firmly etched with determination and good humor that she appears in surviving photographs almost preternaturally poised and mature. Born in Minneapolis, Dashiell had moved with her family to Lakewood, New Jersey, when she was two years old; her father, a minister named Mason Dashiell, had died when she was very young. She might have been one more young woman from small-town New Jersey save for the fact that she claimed as a close friend and confidant one of the most famous, and upon his majority one of the wealthiest, Americans ever produced by this nation: John D. Rockefeller Jr. Painfully shy and awkward as a youth, the customary solitudes and discomforts of adolescence no doubt heightened by the attention drawn by his family's notoriety, the only son of the founder of Standard Oil befriended Bessie Dashiell through her older brother Lefferts, who attended the same private school in New York as Rockefeller; one of Rockefeller's three sisters was also named Bessie, and he came to think of Bessie Dashiell as his "adopted sister." They took carriage rides together, rode horses along the Hudson Palisades, and exchanged long, thoughtful letters. There survives a formal portrait of the two of them: "Johnny Rock," as he would later be known to classmates at Brown University, looking less granitic than Fauntleresque, a soft-featured, frail-looking tulip of a youth, all four buttons of his pin-striped suit firmly fixed, a derby in his gloved right hand, and Bessie Dashiell, seeming to tower over him though she sits to his left, clear of eye and with a slight, knowing smile, gussied up in an overcoat with a fur collar, firm of jaw and beguiling in her self-aware sense of humor. That the stars brought this young woman and her notable friend into the orbit of Dr. William Coley ultimately had a profound effect on American philanthropy and cancer research, to say nothing of Coley's career.

Dashiell had a taste for daring. In the summer of 1890, she had undertaken what in retrospect qualifies as Victorian America's equivalent of adventure travel—a railroad trip across the continent, followed by a jaunt up to Alaska. During the trip, as she later wrote young Rockefeller, she had hurt her hand and thought it had become infected. By the time she had returned to New Jersey in August, the hand was still swollen and painful. As it turns out, the injury stemmed from the most trivial of accidents. Dashiell's hand had at some point become caught and pinched between two seats of the Pullman car in which she was riding. That "slight blow," as her doctor would later describe it, left an ordinary bruise upon the back of her right hand.

After some initial swelling and pain, the injury appeared to improve, but after a week, both pain and swelling grew more severe. To all appearances, it was a typical bruise, and a local physician recommended nothing more than "the usual local applications," probably ice. Nothing about so minor a bump seemed ominous, and hardly anything about this young woman suggested frailty—she had been in excellent health her entire life. Little more than a month later, back in New Jersey, Bessie Dashiell went to her own doctor complaining of continued pain; it seemed more pronounced with motion, she reported. In an attempt to alleviate the mysterious and persistent pain, the hand and wrist were placed in a splint, but that brought only temporary relief. During September, the pain grew so persistent, with occasional sharp and shooting flashes of discomfort, that Dashiell had trouble sleeping. Finally, her family decided to seek medical help in New York City. The man to whom they were referred was William Coley.

Coley, barely a year out of medical school, was already a rising star in New York surgical circles, at a time when surgery itself was *the* ascendant medical art. He examined her for the first time on October 1, and in the manner of the day, characterized the swollen area according to the scientifically imprecise but useful idiom of the farmer's market; he noted a small, spindly swelling "about the size of half an olive" on the back of her right hand, just above the large joint leading to her small finger. When he pressed down hard to determine if the swelling was mobile, his young patient complained of the pain. He checked for inflammation of the lymph glands under the armpit, which might indicate an infection, and found none.

Several days after this initial consultation, Coley made a more detailed examination. He suspected the problem was caused by a low-grade inflammation of the membrane, or periosteum, covering the bone, a condition known then as subacute periostitis. To make sure, he ventured a closer look; for a shy country boy who'd dithered over the choice of a profession scarcely five years earlier, Coley had quickly learned to wield the knife without conspicuous indecision. Applying cocaine as a local anesthetic, as was common at the time, he made a small incision through the center of the swollen area down to the bone. He noted a few drops of thin pus, but nothing in an amount suggestive of infection. Puzzled, he probed the tender area and observed that the tissue through which he'd cut "seemed abnormally hard and more of a grayish color than normal." He closed the wound with antiseptic dressings and placed the hand once again in a splint.

Coley probably expected to find telltale traces of a discharging infection when he later changed the dressing, but he did not, and when Dashiell reported that the pain and swelling had increased yet again, he must have felt uncertain about what to do next, because the following week he sought the advice of his mentor at New York Hospital, the celebrated surgeon William Bull. Bull, too, thought it was periostitis. He advised "waiting for further developments." Up to this point, Bessie Dashiell may well have viewed the problem as a nuisance, exceedingly painful but hardly more than a lingering, inconvenient bruise. Indeed, her friend John D. Rockefeller may have been more concerned than she; unaware at first that she had been hospitalized, he wrote her a nine-page letter on October 19 detailing his worry ("I cannot tell you how I shall miss my adopted sister on such occasions") and concluding, "I shall hope to receive just a word from you telling where you are and how your hand is doing before many days."

Her hand was not doing well, and sometime toward the end of October Dashiell must have read considerably greater concern in the face of the young surgeon. As her symptoms slowly but steadily worsened, Coley decided on yet a more detailed look. This time he administered ether and expanded his previous incision to expose a three-quarters-inch length of bone. The membrane covering the bone was thick, but the bone itself appeared clean and free of pus, thus seemingly free of infection. The young surgeon scraped away what he referred to as "grayish granulations" and closed the wound once again. Temporary relief ensued for a day or two, but the pain returned "as severe as before" with redness and swelling, and then Dashiell's condition became ever more frightening: she reported losing sensation in several fingers of her right hand. "The pain soon became so severe," Coley remarked, "that I was obliged to give morphine to secure relief."

Clearly, some process other than routine inflammation seemed to be causing the excruciating pain. Coley wracked his neophyte's brain for an answer. There was one other remote possibility to explain what was going on, and Coley reluctantly began to entertain it. The gradual increase in pain, swelling, loss of sensation, and impaired motion, absent an obvious infection or other inflammation, suggested something much more serious: cancer. Specifically, it suggested sarcoma, a malignant disease of the body's connective tissue, such as muscle, bone, and the miscellaneous gristle in between that holds us together (carcinoma, the other main form of cancer, arises in the epithelial, or surface, layer of tissues, and typically affects major organs like the lung, breast, liver, ovaries, and colon). After securing the consent of Bull and Robert F. Weir, chief surgeon at the hospital, Coley obtained a biopsy sample in early November by cutting away a small, wedge-shaped bit of tissue, which was given to a pathologist at the New York Hospital for microscopic examination. On November 6, the worst-case diagnosis came back: the cells on the slide bore the typical signature of cancer, and the pathologist's report read "Round cell sarcoma."

It is not hard to imagine the consternation with which this news might have been received by Bessie Dashiell, *if* she was told; doctors who practiced nineteenth-century medicine (by which is implied its concealing discretions as well as its often fruitless interventions) were famously circumspect about delivering bad news to patients, to say nothing of delivering it bluntly. In circumstances as dire as these, etiquette hardly mattered. In early July Dashiell had caught her hand between two train seats; now, barely five months later, three of the finest doctors in New York were telling her that, as Coley would later record, "amputation at the middle of the forearm offered the best chance of saving the patient's life." What Coley probably didn't tell her was that even that "best chance" made her odds of surviving at best only about one in ten; and what Coley himself almost certainly did not realize is that his initial exploratory incisions to diagnose the problem may have not only promoted the spread of cancer but accelerated what would become a harrowingly rapid decline, one that would haunt him for many years to come as "one of the most malignant tumors I had ever seen."

Many patients refused such radical treatment, but in this case the family had no choice. On November 8, 1890, little more than a month after he first examined Bessie Dashiell and shortly before her eighteenth birthday, Will Coley amputated her right arm below the elbow.

There is no record of Coley's personal reflections on the operation, whether he considered it a reasonable intervention or thought it a last-ditch, well-intentioned act of barbarism by professionals who had nothing better to offer. His not-quite-matter-of-fact recitation of

Dashiell's ensuing decline suggests he expected a better outcome. "The appetite remained very poor and she did not regain strength as rapidly as I had hoped," he admitted to colleagues later.

Three weeks after the operation, Dashiell experienced extreme abdominal pain lasting several days before it went away. There was "no indiscretion in diet," as Coley put it, to account for the sudden pain. A more seasoned physician might have recognized it as a sign that seeds of malignancy, hatched in one part of Dashiell's body, had spread and nestled in other hospitable niches of the body, there to expand and proliferate as satellite malignancies known as metastases. Indeed, like some cruel, time-lapse movie of malignancy, tumors began to sprout up on Bessie Dashiell's body with alarming speed. On December 11, her doctors discovered a small lump, about the size of "a small almond," in her right breast; the following day, two smaller nodules appeared on her left breast where none had been the day before. A week later, the lymph glands under one armpit became swollen and painful. More lumps appeared in the breasts. Severe aches developed in the left thigh, and she experienced "almost complete anorexia," too weak to go out walking by Christmas. "The pain was so severe," Coley noted, "that the patient had to be kept under the influence of opiates."

The course of Dashiell's illness became a grim seminar for Coley on the speed with which an aggressive cancer can sweep through and conquer the human territory. She began to lose sensation in her lower lip and chin, then in some lower teeth. Dull pain returned to her abdomen. By the first of the year came jaundice; now the liver was failing. More small nodules cropped up on her chest and trunk; more lymph nodes hardened. "From this time," Coley noted, "the loss of strength and flesh was very rapid. She could take almost no nourishment, even liquid causing severe pain in the abdomen." Two weeks after Christmas, Coley located by touch a well-defined and enormous tumor occupying the whole of the abdomen above the stomach; he estimated it to be the size of "a child's head." The liver was enlarged, the heart began to fail. Like street sweepers tidying up after a nuclear holocaust, doctors plied her with large amounts of digitalis and as much brandy as she could stomach. By January 20, the end was in sight.

The endgame in cancer is never pretty, less so in an era where doctors chased rather than managed the last ghastly symptoms. The breast tumors had become the size of goose eggs, the abdominal tumor even larger; the length of her body from head to toe was stippled by small tumors that Coley likened to buckshot or split peas. Last came the vomiting, several times a day, though she had had no solid food; soon, she was regurgitating copious amounts of blood. "The attacks occurred almost hourly," Coley noted, "and were very exhausting to the patient in her extremely weak condition." Elizabeth Dashiell remained conscious of this horrific piracy of her eighteen-year-old body until very nearly the end, when finally, mercifully, she died at home in New Jersey at 7 A.M. on January 23, 1891. Coley was at her bedside and signed the death certificate.

How much meaning can such a short and tragic life embrace? John D. Rockefeller Jr., no doubt frantic when he finally learned of the gravity of Dashiell's illness, visited her at home not long before her death and attended the January 25 funeral. Mary Dashiell later sent him a keepsake, a celluloid court-plaster case that her daughter had cherished. "I cannot tell you how deeply I appreciate your kindness in sending me something which belonged to Bessie and which she enjoyed," he wrote in reply. "This token will be a continual reminder of her beautiful life and death, and I shall never look at it without being thrilled with the thought of her patient endurance of such continued suffering, and tender regard for those about her, lest they should be made unhappy or sad by her pain."

His grief had two consequences, one short-term and one of truly visionary sprawl. The shock of Dashiell's death left Rockefeller in no shape to attend college; slated to begin his first term at Yale University during the period following Dashiell's death, he spent the time instead tending to the grounds of Forest Hill, the Rockefeller estate in Cleveland. The raw adolescent wound of losing a beloved soul mate, especially for a boy who probably didn't have many, could only leave a deep and lasting scar, and although many a mourner vows to channel her or his grief into practical good, few in history were as well placed as John D. Rockefeller Jr. to translate such resolve into something of enduring impact, and few were as good to their word. As a young adult, he dedicated much of his philanthropic effort to the conquest of cancer; those efforts began five years after Dashiell's death, in 1896, with dabbling support for William Coley's research (the two men remained friends throughout life), grew prodigiously with his family's creation of the Rockefeller Institute for Medical Research (now Rockefeller University), and led ultimately to a multimillion-dollar gift that allowed creation of Memorial Hospital (now Memorial Sloan-Kettering Cancer Center) at its present site in New York City. Asked many years later how he became interested in cancer research, Rockefeller replied, "I think it goes back to Bessie Dashiell. . . . Her death came to me as a great shock."

Dashiell's swift death left her physician no less shaken. Only ten years older than Bessie, William Coley was too young, too professionally unseasoned, to shrug off the case. In his very last scientific paper, written nearly half a century later, he reiterated that Dashiell's case made a "deep impression." He felt helpless in the face of Dashiell's illness, and was naive enough to think he could do something about it. The case sobered Coley not only because of the speed with which the cancer killed, but because of the crude, puny, and utterly ineffectual obstacles hurled by her doctors to impede its fatal course. Medicine, as it was practiced then at New York's finest hospitals, had nothing better to offer than morphine, brandy, and the bone saw, and nothing to show for the effort. William Coley appears to have been one of those stoic personalities who betray no emotion, not because he didn't feel any but probably because he felt too much; when he related Dashiell's case history to surgical colleagues several months later at the New York Academy of Medicine, cataloging each decrement of her mortal illness with the detached precision of the keen clinical observer he had already become, he hinted at his own frustration only in his concluding remarks. "A disease that, starting from an insignificant injury, can attack a person in perfect health, in the full vigor of early maturity, and in some insidious, mysterious way, within a few months, destroy life, is surely a subject important enough to demand our best thought and continued study."

Coley's "continued study" began almost immediately. Curious about such a starkly malignant disease, he decided to search the hospital records for past cases of sarcoma to learn more about this generally rare disease. In the course of this research, he stumbled upon an unusual case—an aberration, really—that stopped him in his tracks. And that is how, several months after the death of Elizabeth Dashiell, in the spring of 1891, Coley found himself in the midst of a most unusual epidemiological manhunt on Manhattan's teeming Lower East Side. For several weeks, during the hours after work, he searched the tenement neighborhoods, climbing interminable flights of stairs, inflicting his pidgin German on startled tenement dwellers who opened their doors to an earnest young doctor, certainly not the kind of man regularly seen paying house calls in those precincts. At each door he inquired after a German immigrant, a house painter, last seen at New York Hospital in February 1885. The man's name was Stein, and he probably bore a large, telltale scar behind his left ear.

"Nature often gives us hints to her profoundest secrets," Coley said on another occasion, "and it is possible that she has given us a hint which, if we will but follow, may lead us on to the solution of this difficult problem." Nature had given Coley a hint; now he determined to follow it down whatever avenue, up whatever steps, through whatever door, to find this man Stein; and through Stein to explore the biology of an apparent medical miracle.

Banging on tenement doors may seem like a profligate waste of a surgeon's valuable time, to say nothing of a professional hazard to his valuable hands, but medicine was different at the turn of the century in many respects, and in Coley's case his willingness to augment knife and gown with medical gumshoeing speaks well of his ability to practice a new kind of medicine. If there has been a reluctance to accord him recognition as the father of immunotherapy, it may in part relate to a tendency to hold this thoroughly nineteenth-century physician to twentieth-century scientific standards, and perhaps also to overly extravagant claims made on his behalf by champions of alternative approaches to cancer treatment. In order to appreciate what William Coley accomplished, it is essential to appreciate who he was—and just as important, who he was not. He was a wonderful clinician, a superb surgeon, an esteemed colleague of the major medical workers of his generation, including William Welch of Johns Hopkins, Harvey Cushing of Harvard, and the Mayo brothers. One thing he was not, however, was a scientist; he never trained in any form of laboratory work, and never truly understood the rules by which the increasingly scientific approach to medicine was being played. Yet in straddling two distinct medical eras, he managed to be a modernist and old-fashioned all at once; and in order to understand why Coley's anti-cancer vaccine received the largely indifferent reaction it did, it is essential to understand Coley as a man ahead of the times and yet behind them, too.

His roots were strictly rural. William Bradley Coley was born on January 12, 1862, in the small Connecticut community of Saugatuck, just north of Westport, in a district that formed part of a heavily wooded, gently rolling sixty-five-acre parcel of land granted to the Coley family by King George III in 1763 and informally known as "Coleytown." His father, Horace Bradley Coley, taught in a one-room schoolhouse and farmed corn, onions, and potatoes to make ends meet while his mother, Clarina Wakeman Coley, tended the house. The family traced its American roots back to Samuel Coole, who arrived in the Massachusetts Bay Colony in 1631. Theirs was a clan of solid (and stolid) citizens: militiamen, selectmen, church deputies, known for their farming, teaching, and upright lifestyles.

An empathetic physician throughout his career, Coley may have acquired the skill involuntarily, beginning at an early age. His immediate family suffered devastating losses to the kinds of common mortal illnesses that would, by the end of Coley's career in medicine, be regarded as novelties. Horace Coley's first wife, Polly Sophia Wakeman, died in 1854, probably of puerperal (childbirth) fever; his second wife, Clarina, mother of William Coley (and sister of the recently deceased Polly), succumbed to typhoid fever when her son was only nine months old. Of Coley's first two stepsisters, one later died of peritonitis, the other of an unknown fever; Coley's sister Carrie, probably the most beloved of all in his immediate family, died in 1892 after a life-threatening pregnancy and an emergency abortion late in the first trimester (which Coley reluctantly performed himself), and Coley also lost his second-born son Malcolm, who died in 1901 at age five of an acute gastrointestinal infection, after less than a week of illness. Following the death of William's Coley's natural mother, Horace Coley remarried a woman named Abbie Augusta Gray; a much-beloved stepmother to William Coley during the crucial years of his upbringing, she too died, in November 1879. By the time Coley turned eighteen, in other words, he had lost two mothers.

By that age, however hardened by the deaths around him, Will Coley had developed into a slender, handsome young man of five feet eight, brown-haired and brown-eyed, of notoriously shy demeanor. Intelligent and deeply religious, he attended a private academy in Westport, where his first exposure to the classics would initiate a lifelong love affair with great literature and account for two of his greatest extracurricular passions: reading and the collection of rare books.

Nothing about the upbringing or education of William Coley advertised the imagination with which he perceived the importance of his toxin therapy or the genial tenacity with which he defended it against many detractors over many decades. When he set off by train to New Haven for his first year at Yale College, having been accepted into the class of 1884, it was like crossing some intellectual Bosphorus and setting foot upon a newly discovered and exotic continent of books, ideas, society, culture—all the more so for someone who, if truth be told, probably reminded no one of a worldly, urbane dandy. As the journal he began to keep suggests, life before Yale was virginal in every respect ("Played cards! I never knew the name or value of a card before but learned so quick," he wrote, in a brief verbal frisson, familiar to all undergraduate initiates to the pleasures of sin, in February 1882; the wonder is that it took until his sophomore year to make the discovery). The language quoted in his diary is of someone just learning that he has a voice, much less testing its limits: painfully earnest and literal, rarely witty, uncomfortable with strong opinions, Coley's journal rarely dares to scratch beneath superficial observation. "Completely carried away with it," he wrote glowingly of Longfellow's poem *The Courtship of Miles Standish.*

Coley had never played cards before going to college, had never gone to the theater, had never smoked. He found time each week to teach Sunday school, read his Homer and Aeschylus for classes and his *Paradise Lost* and *Samson Agonistes* with greedy pleasure during breaks, and supplement his always threadbare finances by ghostwriting compositions for less gifted students while tutoring others in math and physics. He must have been an excellent tutor: he once commanded $32 for several sessions, which sum covered roughly 20 percent of the annual tuition at Yale. He was not "tapped" for membership in a secret society, did not compete in sports, did not join clubs (except the freshman debating society), and decided against attending the Junior Promenade one year because he could not afford the $4 ticket. As for recreation, he tended toward the more economical diversions: ice-skating, boating, most of all walking. As his classmates later recalled, Coley was "very quiet and reserved at Yale—no extracurricular activities and never athletic." Of his personality, all we are told is that he possessed a "merry laugh."

To anyone who has made the acquaintance of that humble and self-effacing creature, the surgeon, Coley's ultimate migration to that branch of medicine seems almost temperamentally precluded by his earnest indecisiveness. His initial exposure to the medical profession came by way of his extended family; when he was home on break from college, he would sometimes accompany his uncle, Dr. Joseph Henry Wakeman, on house calls in the Redding, Connecticut, area in the doctor's horse-drawn buggy. But he was far from decided upon a career in medicine. In January 1884, just a few months short of graduation, he described a fitful night in bed trying to solve the problem: "I lay awake a good part of the night listening to the fierce blast of the wind and rain upon the roof and meditated whether to study the

Chemistry Optional next term and take Medicine as a profession or to take Law. I could not come to a satisfactory decision, although I did not give up until nearly morning."

After much fretting and procrastination, he punted—he decided to accept a post as "assistant principal" (a glorified term for an over-worked instructor) at the Bishop Scott Grammar School in Portland, Oregon, where he could set aside some money and ponder the decision anew. The school was run by a Yale graduate named Joseph Wood Hill, who happened to be the husband of a distant cousin, Jessie, which fact undoubtedly complicated negotiations when Coley insisted on timely installments of his $600 a year salary, always seemingly in arrears. After taking the Northern Pacific railroad across the continent, he remained two years in Portland, teaching everything from Greek and Latin to mathematics, gazing out the window of his room at Mount Hood, saving close to $750 with which to further his education, and still pondering in which profession to invest it. Here his reading brought him into contact with two influential figures, one medical and the other spiritual.

The latter, interestingly enough, was Giordano Bruno, the radical sixteenth-century Italian humanist. This Renaissance scholar championed the Copernican, sun-centered view of the solar system; believed the universe was infinite and composed of many systems like our solar system; posited an atomic-based concept of matter; spoke of the Bible as a book of moral but not scientific authority; lectured on a peaceable utopia where all religions coexisted in an atmosphere of harmony and dialogue; dabbled, fatefully, in mysticism; and aroused the fury of ecclesiastical authorities by steadfastly sticking to these radical, unorthodox, and clearly erroneous beliefs. Bruno did not, like Galileo, mumble his defiance under his breath to save his skin, for which insolence Bruno was ultimately burned at the stake in Rome's Campo de' Fiori in 1600. Coley filled several pages of his notebook with admiring observations about Giordano Bruno. "Drank deeply of the spirit of the Renaissance," he wrote. "Accepted the discoveries of Copernicus, and used them as a lever to push aside antiquated systems of Philosophy." Coley never thought of himself as a martyr, but the lifelong vicissitudes of winning acceptance for his work might easily have reminded him of Bruno's travails.

Another book exerted a more immediate impact. Coley happened to read the autobiography of J. Marion Sims, the celebrated surgeon who had helped found the New York Cancer Hospital in 1884. "I have been asked many times why I studied medicine," Sims wrote in a passage Coley dutifully copied into his notebook. "There was no premonition of the traits of a doctor in my career as a youngster; but it was simply in this way: At that day and time, the only avenues open to a young man of university education were those of the learned professions. . . . A graduate of college had either to become a lawyer, go into the church, or to be a doctor. . . . I would not be a lawyer; I could not be a minister; and there was nothing left for me to do but to be a doctor." Not only did Coley reach the same conclusion, but twenty years later, this Oregon schoolteacher would play a major role in shaping the hospital founded by Sims into the city's—and ultimately the nation's—preeminent cancer research institution, Memorial Hospital.

In September 1886 Coley began his medical education at Harvard; it says something about the somewhat less rigorous standards of medical training in the days prior to Abraham Flexner's historic 1910 report, which urged massive reform in medical training and modernized American medicine, that Coley was allowed to enter the three-year program at Harvard Medical School as a second-year student largely on the strength of spending several months during the summer of 1886 accompanying his uncle on rounds in a horse and buggy. At that time many medical schools were not affiliated with a university, most required only two years of study, and hardly any used a standardized examination to test the competence of would-be practitioners before unleashing them on an increasingly wary public. Harvard was certainly among the best of the lot, but it too was struggling to absorb and incorporate the profound changes in medical science, of which there were many in the latter half of the nineteenth century: Joseph Lister had introduced the concept of treating surgical wounds with antibacterial substances (antisepsis) in 1866, followed soon after by the technique of using sterile conditions (asepsis) in the operating theater, and these concepts remained so new during the period of Coley's education that their merits were still heatedly debated in some American hospitals; Louis Pasteur had proposed the germ theory, thereby implicating infectious microorganisms as the cause of infectious disease; and Robert Koch had, just four years before Coley's matriculation, identified the mycobacterium that caused tuberculosis, a disease then claiming roughly 5 million lives worldwide each year. Medicine was in the throes of the most profound changes since the Renaissance, although not all the changes had sifted down to the schools.

Nonetheless, Coley managed to engineer for himself—and it was quite as much his doing as Harvard's—a remarkably liberal, unusually modern medical education. Whereas the conventional curriculum involved little more than an endless series of lectures on anatomy and physiology, chemistry and pathology, Coley had extraordinary good luck in obtaining on-the-job training. During the summer following his first year of medical school, in 1887, he happened to be visiting two friends from Yale who served on the staff of New York Hospital, and they invited him on the spot to fill in for a doctor who had taken sick leave, which is how—at age twenty-five, still without a medical degree, and at a moment's notice—Coley found himself patrolling the corridors of a major metropolitan hospital as "Acting Junior Surgeon" for six weeks. The title almost certainly exaggerated largely factotum responsibilities, but in an era that had barely begun to embrace the educational wisdom of exposing students to actual clinical situations in the wards of hospitals, Coley found himself not merely exposed but plunged into real doctoring: dressing wounds, handling the "etherizing" at operations (that is, serving as anesthesiologist), observing the modern technique of surgical asepsis, acquiring hands-on experience during a hands-off era of medical education. His very first medical paper discussed heat-related illnesses, based on the many sunstroke victims he treated during the summer of 1887. To his sweetheart he wrote with an apprentice's pride, "I sewed on a finger the other day and expect to amputate one today."

Even Coley could not have realized at first how fortuitous that summer's employment would turn out to be. He not only landed in one of the premier private hospitals in New York, but in fateful proximity to two of the best surgeons in the country. Robert Fulton Weir, described in Walter Graeme Eliot's *Portraits of Noted Physicians of New York, 1750–1900* as "one of the three greatest surgeons of the day in N. Y. City," was a wiry, gray-bearded fellow with a hooked nose and sharp, hooded, deep-set, and lively eyes, who in formal portraits appears like someone too restless to dress well, pose well, or let go of a raptorlike intensity that stiffened his face with impatience. In addition to a teaching appointment at the College of Physicians and Surgeons, Weir later served as president of the New York Academy of Medicine. The other was a disciple of Weir's named William Tillinghast Bull. The name conjures up an image of Taftlike bulk and Ashmolean sideburns: Bull was indeed a heavyset young man thirteen years older than Coley, his mustache slicked to pinpoints, hair parted smartly down the middle as befits the ladies' man and befriender of nurses he was said to be, a sharper dresser than Weir, with less delicate hands and chilling, distant eyes. He achieved early celebrity in

1884 when, pioneering a form of emergency surgery at the Chambers Street Hospital that has lamentably become ever more essential to twentieth-century medicine, he is said to have performed the first exploratory operation to treat an abdominal gunshot wound; Bull enjoyed such local celebrity that when he became ill with cancer in 1908, the diagnosis and his subsequent months-long struggle against the disease was front-page news in the *New York Times* and other papers. His influence on Coley is obvious: the young surgeon's second paper, "Treatment of Penetrating Shot-Wounds of the Abdomen," appeared the following year in the *Boston Medical and Surgical Journal*, forerunner of the *New England Journal of Medicine*.

The historian Paul Starr has written of the transformation of American hospitals "from places of dreaded impurity and exiled human wreckage into awesome citadels of science and bureaucratic order," and Coley had the good fortune to enter the field just as the foundation stones of that massive new edifice were set into place; indeed, he later claimed to have entered medicine "at the most opportune time in a thousand years." Weir and Bull, for example, aggressively championed the use of aseptic conditions in the operating theater, and Coley learned the technique at their elbows. With each passing year, medicine became more of a science-driven enterprise. Bacteriology's assault on infectious disease represented the leading salient, and collateral fields like microbiology and immunology moved swiftly through the breaches that breakthroughs against infectious diseases created; aseptic techniques allowed surgeons to treat otherwise minor but problematic conditions like hernias and fractures without fear of mortal infection. And significantly, Coley's career would enact in miniature one of the emerging and most enduring conflicts of twentieth-century medicine, when for the first time the duties of the conscientious physician sometimes worked at cross purposes to the needs of the rigorous researcher. With the rise of "scientific medicine" came a newly hyphenated medical creature, the physician-researcher, who straddled with great difficulty the chasm between bedside and laboratory bench.

With the summer experience under his belt and, a year later, his Harvard education behind him, Coley accepted an invitation to be a surgical "Interne" at New York Hospital, beginning in January 1889 (as a last grand adventure before New York, he signed on in August 1888 as ship's surgeon to the *Kennard*, a barkentine that plied the human trade bringing cheap immigrant labor from the Azores to the textile mills surrounding Boston). Coley's first year in New York was exhilarating, depressing, determining. He arrived in November 1888 and found an apartment at 18 West Sixteenth Street, just next door to New York Hospital, then located on Sixteenth Street between Fifth and Sixth Avenues.

This venerable institution, founded five years before the Declaration of Independence was signed, had relocated to Sixteenth Street in 1875 and received a handsome face-lift, with a redbrick facade and mansard roof. There were nine house physicians at the hospital, who hustled to attend some 375 patients daily; Coley served until July 1889 as a "junior walker," or intern, graduated to senior walker that summer, and was promoted to house surgeon early in the spring of 1890. His skill in surgery must have been considerable, for in addition to his internship at New York Hospital, William Bull secured appointments for Coley at the Post Graduate Medical School, where he had an instructorship in surgery, and at the Hospital for the Ruptured and Crippled, where he would serve for some forty years, ultimately as chief surgeon (this latter is now the Hospital for Special Surgery, where New York Mets and Jets and other professional athletes take their well-heeled joints and bones for repair). There he championed the use of the "Bassini procedure" to treat hernias, a technique he performed thousands of times during his career. At one point in the midst of this intense apprenticeship, Coley's beloved Alice Lancaster broke off their relationship, though they would ultimately reconcile and marry on June 4, 1891, after several intense months of activity that would forever shape the trajectory of Coley's future years.

With early success came lifelong paradox. The boy who milked cows and plowed the fields of his father's Connecticut farm later ministered as personal physician to a rich dowager on one of her European vacations; the boy who walked six miles to Sunday school at the Methodist church in Easton became the man about town who insisted on a chauffeur, in part because he never learned to drive a car. The teenager who shunned college social events for want of money (and, probably, want of social confidence) would later join fifteen medical societies and ten social clubs, reorganize the Sharon, Connecticut, golf club, and exchange chummy letters with robber barons and their kin. The boy raised in a culture of thrift and deprivation became the successful surgeon who perpetually lived beyond his means, who dabbled in real estate speculation, and whose financial adventurism was so ill-advised that near the end of his life he could barely pay for medical treatment for his dying wife.

Yet here is the most fascinating paradox of all. The man who made quite a good living by use of medicine's crudest utensil, the knife, stumbled upon its subtlest of healing tools: the blood, with its cells and its wondrously potent ensemble of molecules. Because he was never trained as a scientist (and never really became a great one during his lifetime), Coley failed to cast his work into a rigorous, molecular idiom; nor could he have, given the technical limitations of the day. Whatever credit he deserves—and he deserves a great deal—derives from the fact that he was an exceptionally conscientious physician who made careful observations, believed what he had seen, and clung to those beliefs with a tenacity that emulated what he regarded as Charles Darwin's greatest virtue as a scientist: doggedness.

Of all his many qualities, that was perhaps the one that served him best. And he would have a lifetime of controversy to put his doggedness to good use.

It would not be inappropriate to regard Coley's approach to medicine as "Darwinian," but in a temperamental rather than scientific sense. Coley openly acknowledged Charles Darwin as the inspiration for his own medical intuition. "The one quality of mind which was of the greatest help to him [Darwin] in making his discoveries," Coley once told a group of students, "was that of never letting exceptions pass unnoticed."

Coley's own career as a medical detective began with an exception. Shaken still by the death of Bessie Dashiell, hoping that "some little light might be thrown" upon the treatment of these particularly malignant forms of cancer, Coley began a systematic search of the medical records at New York Hospital over the previous fifteen years, looking up each and every case of sarcoma. He consulted his senior colleagues, Bull and Weir, for cases from their private practice; their casebooks, several of which are preserved in the rare books collection of the New York Academy of Medicine Library, are gorgeous, hand-lettered souvenirs from a distant, barely recognizable medical world. Weir's 1886–87 casebook, for example, is a block-sized, gilt-edged tablet of thick ledger paper, its title page, "Clinical Records," hand-lettered and tinted with the care and precision of a medieval manuscript, its epigraph *"Ne tentes, aut perfice!"*—loosely, "Nothing ventured, nothing gained"—brimming with the interventionist optimism of the new medicine. Sandwiched in among cases of hernia and leg amputation and harelips, of "spurious valgus" and even anthrax, all duly recorded in flowing pen-and-ink script, Coley found the occasional cases of sarcoma, and he filled a tiny leather-bound notebook of his own with the details of each case.

Bit by bit, over the course of several months, even as he performed his regular medical duties at New York Hospital, Coley "collected" (the quaint nineteenth-century verb for scholarship, as in gathering cases from disparate published sources) approximately ninety cases of sarcoma. Many came from an 1879 paper by the Philadelphia surgeon Samuel Gross, immortalized in one of Thomas Eakins's greatest paintings; many others were culled from Weir and Bull. Coley was able to determine follow-up histories for forty-four of them. By twentieth-century standards it was a roughshod bit of research, but it represented a minor triumph in an era when medical record-keeping shared much more with a medieval manuscript library than with modern, computerized case histories. Hospitals did not routinely maintain record-keeping departments; doctors kept their own records, usually in a single, perishable notebook.

Coley's analysis revealed two qualities persistently evident throughout his work in immunotherapy—one positive, the other not so. Following the example of Darwin, he specifically zeroed in on the exceptional cases. Unfortunately, as would happen on more than one occasion when it came to publishing his observations, the numbers in the papers didn't always add up. Of the 44 sarcoma cases with follow-up, for example, Coley reported that 26 suffered fatal recurrences (presumably following amputation), 9 were alive (the duration of survival ranging from three to ten years), and 4 recent cases were alive, twelve to eighteen months after surgery. That left 5 apparently unaccounted cases, about which Coley says nothing.

On the other hand, Coley's efforts—however imperfect—demonstrate that even at this early age, he realized the crucial significance of follow-up, of "the subsequent history," as he put it, "which is so very important and at the same time so difficult to obtain." Then as now, periosteal sarcoma—the type that ravaged Bessie Dashiell—was known to be, in the words of one physician, of "the most intensely malignant character"; even amputation usually failed to prevent recurrence of the cancer, which was likely to appear within a year. If Coley found any comfort in this dreary scholarship, it was that Bessie Dashiell's cancer was extraordinarily rare; of the two dozen cases of bone sarcoma that Coley managed to assemble, not a single one occurred in the bones of the hand. Still, the message was unrelievedly grim. Coley observed that "the large proportion of cases in which fatal and often speedy recurrence follows operation is sufficient to make the surgeon almost lose faith in his art in the treatment of this dread disease."

Almost but not quite: Coley refused to indulge his despair, and his curiosity led him to a cancer case that would forever change the course of his career. He found it in William Bull's files, in the extraordinary case of a German immigrant named Fred K. Stein. Coley must have read this chart with barely containable excitement.

Fred K. Stein's problems began in the fall of 1880, just about the time Coley was discovering cards at Yale, when a small red spot appeared on his left cheek, just in front of his ear. He had been generally in good health, save for two unusual episodes—a "carbuncle" that appeared on top of his head when he was ten, causing his hair to fall out, and a "chancre" when he was sixteen that required six months of mercury treatment (probably a case of syphilis). By June 1881 the "spot" on his cheek had grown to the size of a hen's egg, requiring surgical removal. The cancer recurred quickly and grew this time to the size of a pigeon's egg, requiring a second, seemingly successful operation the following September. Two years later, in the autumn of 1883, Stein noticed the growth had returned just below the lobe of his left ear, and by the time he sought help at New York Hospital the following June, the growth looked ugly and ominous. It measured four and a half inches in diameter, according to hospital records, and had the superficial appearance of a bunch of grapes.

Bull operated again on June 5, 1884, but by this time the tumor had grown so large that he couldn't close the wound; indeed, several skin grafts were attempted but failed to take. By September 15 Bull resorted to the knife a final time, when the recurrent tumor became complicated by a large ulceration. Microscopic analysis at this point indicated "round-celled sarcoma." In the course of that last extensive surgery, Bull discovered that part of the cancer had become attached to the carotid artery, which carries blood to the brain, and thus not all the "tainted tissue" could be removed. As it was, Stein hemorrhaged during the operation and emerged from the ordeal with a gaping wound five inches long and four inches wide, virtually a crater on the left side of his neck partially held together with silver sutures. Bull deemed the case "absolutely hopeless."

Fred Stein was well on his way to becoming one of those nineteenth-century horror stories that dissuaded all but the desperate poor from entrusting their lives to hospitals. He had undergone four operations; his cancer had recurred four times, and now could not be en-

tirely removed. He'd lain in a hospital bed for months. One final
indignity remained. On October 12, one week after another failed skin
graft, Stein developed a raging fever and broke out in an angry red
infection that galloped across his neck and face. The infection was
quickly diagnosed as erysipelas (air-eh-SIP-ehl-us). This disease,
caused by *Streptococcus pyogenes,* a bacterium related to the germ
that causes strep throats, was one of the most common postoperative
infections in nineteenth-century hospitals, with a not inconsiderable
mortality rate. Sherwin Nuland writes that erysipelas, known also as
"St. Anthony's fire," "spread a furious redness through its victim's
incised tissues with enormous rapidity, killing more often than not";
Shelby Foote, in his history of the Civil War, mentions how outbreaks
of the often fatal disease afflicted wounded soldiers abandoned on the
battlefield. Hospitals in England sometimes became so contaminated
with erysipelas and gangrene that their administrators voted to raze
the buildings and build anew rather than attempt to eradicate the
tenacious infections.

Stein's attack was quite severe, spreading rapidly over his face and
neck; no sooner did the first attack subside than a second took hold
two weeks later. The hospital staff placed him in isolation, and the
records note that "after each [attack of erysipelas], the cicatrization
[healing] of the ulcer has rapidly proceeded and during the attacks the
flabby and apparently sarcomatous granulations have been absorbed.
There is now left a healthy ulcer the size of a silver dollar." In other
words, to the utter astonishment and delight of Bull and his colleagues, the diseased tissue simply disappeared. Just as surprising, the
wound began to heal, forming a healthy scar. On February 26, 1885,
Stein was discharged into the teeming immigrant hordes of New York
and for all intents and purposes disappeared from medical scrutiny.
"That was the last note in the history; nothing was known about the
end-result," Coley recalled later. "I felt it was extremely important to
know this, so I started out to trace the patient. . . ."

It is a great pity that Coley never described this episode in greater
detail in any of his papers, for it may in retrospect be seen as one of the
great unsung epidemiological moments of nineteenth-century medicine, as inspired in its own way as Edward Jenner's observation about
milkmaids. At the time Coley rediscovered this case, in the late winter of 1891, no one knew what had become of Fred K. Stein. The man

might still be at his old address; he might have returned to Germany; he might have suffered another recurrence and died. Medical follow-up was not considered the key datum it is today, and perhaps it was precisely this episode that impressed upon Coley its essential importance.

It took Coley, according to his daughter's recollection, at least several weeks to track him down. The German immigrant community in 1891 centered around the Lower East Side, and since that was where Stein last lived, that was where Coley commenced his search. (Coley studied German at Yale and apparently read German medical journals on a regular basis, so he probably spoke the language at least passably.) He also probably climbed stairs, knocked on doors, and did the kind of shoe-leather epidemiology they wouldn't teach in medical schools for another generation or two. One day Coley found himself face to face with a long-faced, dark-haired man with a fulsome growth of black beard. As soon as Coley saw a prominent, jagged, craterlike scar running from just below his left ear down his neck, he knew he had his man.

The young surgeon fetched Fred Stein back to New York Hospital, where Bull examined—no doubt with considerable surprise—and photographed a patient deemed inoperable and near death seven years previous. The German explained that he had enjoyed excellent health since his discharge in 1885, and that the cancer had never come back. Like a speck of dust warping the clear lens of statistics, Fred K. Stein wandered back into the streets of Manhattan.

Nature had given Coley his hint. "If erysipelas, a disease produced by a specific organism, could cure a case of undoubted sarcoma when occurring accidentally," he reasoned, "it seemed fair to presume that the same benign action would be exerted in a similar case if erysipelas could be artificially produced." Coley was "so strongly impressed with the case that I determined to try inoculations in the first suitable case."

Fred Stein's remarkable recovery sent William Coley back to the library to examine a considerable, though idiosyncratic, literature about the role of infections on simultaneous disease. Medical interest in such cases grew in the cracks between other, more respectable avenues of inquiry—the solid and celebrated recent triumphs of bacteriology, the emerging science of immunology, and the long-standing medical interest in odd and very rare instances of healing known as spontaneous remission (cases where advanced tumors improbably disappeared, often in individuals who at the same time were battling a severe bacterial infection). Indeed, the notion that certain infections sometimes conferred health as well as disease attained the respectability of a title, if not the respectability of science per se, in the late eighteenth century; the physician Erasmus Darwin, grandfather of Charles, reflected this fascination when he (and many others) used the term "laudable pus" to indicate the desirable ooze of inflammation and infection around a surgical wound that somehow assisted a favorable outcome.

Similar observations had long attended the phenomenon of spontaneous remissions, which represent a fascinating, if controversial, chapter in medical history. Although many of the observations date from a period when medicine was prescientific, the fascination has lived on well into the twentieth century, and the enigma of spontaneous remission has perhaps best been captured in two beautifully written papers by a prominent Canadian pathologist, William Boyd. "A moment's thought," he wrote in 1957, "is sufficient to convince us that in biology, as in other fields of science, nothing is really spontaneous, for every event must have a cause."

These various threads suggest a long history of manipulating the immune system, knowingly or accidentally, to treat illness—an empirical history that clearly predates any sophisticated molecular understanding of the immune system itself. In *A History of Immunology*, Arthur M. Silverstein points out that the term *immunity* gained widespread acceptance in the nineteenth century after Edward Jenner's successful smallpox vaccination experiments, but the concept of acquired immunity—and indeed vaccination, the first form of immune manipulation—existed both in theory and practice long before. Observers in the Middle Ages developed enough of an idea of acquired immunity to understand that if an individual survived exposure to a lethal disease once, he or she "acquired" protection against a second exposure to the same disease; this protection was attributed to various factors, including (according to one seventeenth-century commentator) "a balsamic blood." The idea of vaccination, in practice if not name, may have originated in China, and certainly achieved the status of accepted folklore in Asia Minor and parts of Europe by the beginning of the eighteenth century.

In 1714, in a well-documented episode, two doctors communicated to the Royal Society of London details of a practice in Constantinople known as "variolation," in which crusts of pustules from mild cases of smallpox were used to establish a preventive infection in others. This led to the famous Royal Experiment of 1721–22. "This experiment involved nothing less than the first clinical trial in immunity," Silverstein writes, "in which the efficacy of inoculation was tested first upon condemned prisoners and then upon a group of orphans, in order that the prince and princess of Wales might be reassured and permit the inoculation of their children, which in fact took place in 1722 following the successful clinical trial." The royal inoculation was a success, despite the misgivings of Sir Hans Sloane, the royal physician; in a remark as unintentionally insightful as it was mistaken, Sloane voiced deep foreboding about the idea of "raising such a commotion in the blood." The remark survives as one of the best characterizations of what is now known as immunotherapy: deliberately causing a healthy biological commotion by injecting material that elicits or enhances the immune response.

A second, rather more tenuous thread of human experimentation involved desperate measures to treat malignant disease; in the field of cancer, the very first immunotherapists were eighteenth- and nineteenth-century doctors who deliberately infected patients with one deadly disease in the hopes of ridding them of a second, more imminently deadly affliction. If this practice appears nearly homicidally empirical to our eyes, it also highlights a medical distinction that pertains to this day: diseases like cancer tend to be complex and difficult to treat, and have historically inspired more than their share of dire interventions.

As in Coley's serendipitous case of Fred Stein, the connection between infection and spontaneous regression began with tantalizing observations. As early as the seventeenth century, F. Hoffmann in his 1675 treatise *Opera Omnia* made note that outbreaks of erysipelas seemed antagonistic to other, previously existing diseases. In 1783 there was a report in the literature by Wenceslaus Trnka de Krzowitz noting that a case of breast cancer disappeared following malaria. Arsène Hippolyte Vautier, in 1813, described several cases of cancer that appeared to be cured by gangrene, and a prominent nineteenth-century Parisian doctor named S. L. Tanchou "collected" no fewer than 300 cases of breast cancer in which success had been claimed following treatment with an infectious agent or deliberately soiled dressing, which led to the short-lived use of what was called "laudable pus, setons, or issues" in its treatment. The underlying suggestion was that prior exposure to severe infectious disease was seen by some to inadvertently vaccinate individuals against malignancy. As one observer colorfully summed up the theme, "He who escapes the Scylla of infectious diseases will most probably succumb to the Charybdis of carcinoma in later life."

All these rare, random, and marginally scientific observations led up to what may qualify as one of the most peculiar contretemps in nineteenth-century medicine, all the more so because of the splendidly respectable surroundings in which it took place. The discussion occurred at the Royal Academy of Medicine in Brussels around 1851, and its main protagonist was a Belgian physician named A. Didot, who championed a bizarre method of cancer treatment known as "syphilization"; as the term creepily suggests, doctors deliberately induced syphilis in patients in an attempt to vaccinate against future occurrences of cancer. At mid-century, several terminal cancer patients so infected had supposedly improved (with slender documentation, no surviving pathology evidence, and the general state of medical knowledge, these reports have the fiber of folklore—possibly meaningful, but highly suspect until inspected more methodically). In any event, the improvement—if any—was transient.

Didot himself treated an inoperable breast cancer patient with pus from the chancres of another syphilitic patient; according to a report published in 1852, syphilis developed over the course of a year, and the cancer coincidentally diminished in size, after which the patient's syphilis was cured following treatment with mercury (the long-term side effects of treatment with this known poison apparently did not merit comment). The rationale, such as it existed, for this radical remedy was Didot's claim (now considered dubious) that since antiquity, prostitutes had been conspicuously free of uterine cancer. Another well-known Parisian doctor named P. Ricord, who specialized in syphilis, voiced vigorous opposition to these experiments, and they not surprisingly soon fell out of favor. As late as 1883, however, another physician named Verneuil reported on numerous observations of concurrent cases of syphilis and cancer, asserting that definite cures had occurred in some cases. The underlying philosophy has survived into the twentieth century; in 1929 several cancer patients were treated at Johns Hopkins with tuberculosis, and Henry J. Heimlich, the physician credited with popularizing the lifesaving Heimlich maneuver, has noted that "malariotherapy"—deliberate infection with a curable form of malaria—was used on tens of thousands of neurosyphilis patients in the U.S. between 1918 and 1975.

Helen Coley Nauts, Coley's daughter, has produced a splendid monograph documenting many of these unusual medicinal interventions. The eighteenth- and nineteenth-century physicians who resorted to such flagrantly dangerous treatments on the basis of such flimsy evidence would today be branded quacks, bounced out of the nearest academy, and perhaps land on one hop in jail. But there must have been a kernel of truth in what they reported, for many respected observers made note of the same phenomenon. Just as many of these crude treatments were doomed to grotesque failure, a portion of the observed remissions were almost certainly credible and probably due to a fortuitous stimulation of the immune response. In a well-known case that would have been infamous for its iatrogenic ghastliness were it not even more celebrated because it was reported to have worked, a French physician named Dussosoy treated breast cancers by soaking gauze dressings in the pus of gangrene and applying them to ulcerating tumors, or made small incisions and deliberately added an inoculum of gangrene; according to nineteenth-century medical reports, whose splendidly acute clinical observations are unfortunately almost uniformly undermined by the absence of sophisticated pathological analysis, these tumors were said to have disappeared. "Here," wrote Tanchou, "gangrene seems to have replaced live cautery, caustics, or the scalpel."

Exceptional cancer case histories, especially spontaneous "cures," have always held special fascination for surgeons, probably because they of all physicians are acutely aware of the limitations of their craft in arresting a malignancy. The occurrences of spontaneous regressions, most agree, are so rare that they barely merit serious attention; but even in the nineteenth century, some surgeons noticed that favorable outcomes in cancer cases were often accompanied by infection. In fact, in marked contrast to the lessons of asepsis, some surgeons around the turn of the century actually recommended "suppuration"—creation of a pus-forming, festering infection—as part of the surgical treatment for sarcomas, and the suggestion remained in standard textbooks for many years. More recently, in 1966, after nearly a century of such hints, Tilden C. Everson of Chicago and Warren H. Cole, a surgeon at the University of Chicago, reviewed and documented some 176 reported cases of spontaneous regression, which they defined as "a partial or complete disappearance of a tumor in the absence of treatment which ordinarily is considered capable of producing a regression." Everson and Cole suggested that several forms of cancer—melanoma, lymphoma, neuroblastoma, and choriocarcinoma—are especially prone to spontaneous regression and might possibly be more susceptible to immune manipulation. Their report ultimately led to a symposium sponsored by the National Institutes of Health in 1974 to explore the phenomenon, although this meeting may have been more successful at producing metaphors than hard science. The former chief of the Breast Clinic at Johns Hopkins, Edward F. Lewison, came up with the most appealing phrase, referring to spontaneous remissions as "whispers of nature," and the immunologist Gustav J. V. Nossal was most emphatic about where the whispers were coming from. "I'm going to put my money on the immune system," the Australian told the meeting. "We've got a hobbyhorse—let's thrash it for what it's worth."

Back in the late nineteenth century, as Coley's library reading soon revealed, the loudest whispers seemed to come about in the presence of the bug that caused erysipelas, *Streptococcus pyogenes*, but these whispers echoed for the most part in the halls of bacteriology. This field, which concerned itself with the single-celled organisms that caused infectious diseases, provided the side door through which Louis Pasteur, Robert Koch, and Emil von Behring stumbled upon the kingdom of immunity. "Strep erysipelatis," as the bacterium was then sometimes known, was responsible for erysipelas, and in Europe too a correlation had been observed between accidental cases of erysipelas and stunning cancer regressions. Coley claims not to have been aware of work in Europe using erysipelas at the time he tracked down Fred Stein. But by the time he gave a lecture to the Surgical Section of the New York Academy of Medicine on April 27, 1891, not long after he found Stein and preceding by just one week his own first human experiment with live erysipelas bacteria, Coley had done his homework. "That erysipelas has an influence upon malignant disease has long been recognized in a general way," he noted, "but only recently has there been any scientific attempt to determine the nature and limits of that relation."

The first to capitalize on these observations was the German scientist W. Busch, who noticed in 1866 that both the swollen nodes of lupus and malignant tumors seemed to diminish in size during an erysipelas infection. In 1868, in the optimistic empiricism sweeping biology, Busch deliberately infected a woman suffering from inoperable sarcoma with the bacteria. The procedure was grisly in the extreme: after the patient's tumor was cauterized with a white-hot iron, she was moved to a hospital bed whose every previous occupant had contracted erysipelas, apparently from the bedding. The tumor shrunk by half within a week, and lymph nodes in the neck also diminished in size, but nine days after the infection began, the patient collapsed and died.

In 1882 Friedrich Fehleisen, a surgeon attached to the renowned university clinic in Würzburg, Germany, reported *Streptococcus pyogenes* as the cause of erysipelas and also described animal experiments in which erysipelas infections seemed to cause transplanted tumors to melt away. Having observed these striking results in dogs, Fehleisen extended this novel approach to humans; he inoculated seven patients suffering from inoperable cancers with pure cultures of erysipelas. The lone sarcoma patient among the seven responded, and the size of the tumors decreased. Among six cases of carcinoma, Fehleisen claimed three responses, including a recurrent breast tumor that completely disappeared and another of the breast where the tumor, reportedly "the size of two fists," decreased by half. The duration of these responses was not recorded.

The idea of using bacteria to treat cancer, wild as it may seem, actually enjoyed an enthusiastic—though short-lived—rebirth on the fringes of the bacteriological triumphs sweeping through biology. Following the successful example of smallpox vaccination, late-nineteenth-century physicians had become smitten with the idea of using attenuated (or weakened) microorganisms for vaccination against infectious disease; it was not unreasonable to entertain the hope that such protection might work against malignancies, too. What is interesting about this rather feverish avenue of inquiry is that the leading bacteriologists of the day, because they worked primarily with bacterial pathogens, were perforce led to the branch of the immune system that is generally most effective in dealing with these particular infectious agents: antibodies. Pasteur's success in inducing acquired immunity to fowl cholera in 1880 and to rabies (a viral disease) later on, as well as von Behring's discovery of a component in blood serum that neutralized diphtheria in the 1890s, were effects "mediated," as scientists like to say, by factors in the blood that came to be known as "antitoxins" and are now recognized to be antibodies. Unbeknownst to the physicians who attempted to treat cancer, antitoxins (that is, antibodies) were for the most part consigned to a secondary role. Antibodies are especially adept at attacking free agents in the bloodstream, as it were—viruses, bacteria, their toxic by-products, and other unwanted rubbish that have not entered (i.e., infected) cells.

The other, less-traveled path of immunology led to so-called cellular immunology, where cells known as lymphocytes have the ability to identify cells harboring pathogens (and cancers) within and then destroy them.

Science is not different from any other field of endeavor when it comes to taking the path of least resistance, so for nearly half a century after Pasteur, Koch, and von Behring, the attention of the field that would ultimately be called immunology concentrated almost exclusively on these so-called serum factors (or antibodies). Indeed, the very name of the field at that time, serology, suggests just how fixed the focus was on the serum, or liquid and noncellular fraction of the blood. Cells by definition were not part of the serum, and by extension were not part of the science; in fact, the conscientious immunologist at the turn of the century would take great pains to eliminate all cells before getting down to the hard work of studying the serum. With the advantage of hindsight, the delayed investigation of the cellular side of immunology might be seen as an early hint that this century's emphasis on reductionism can have a downside, namely a narrowness of intellectual vision that creates blind spots that can persist, as in this case, for decades.

In 1888, another German named P. Bruns conducted a survey of therapies using erysipelas; this paper made a deep impression on Coley. "He has collected 14 cases of undoubted malignant disease," Coley wrote several years later, "in which erysipelas occurred, either accidentally in the course of the disease, or was produced by inoculation. Of these 5 were sarcoma (diagnosis confirmed by the microscope) [and] 3 were epithelioma, which 6 cases were either carcinoma or sarcoma. Of the 5 cases of sarcoma, 3 cases were *fully* and *permanently* cured" (Coley's emphasis).

We can tell by the time interval that young William Coley's standards for what constituted a cure were less than golden: these patients were barely two or three years out from treatment when he pronounced their cures full and permanent. But who could blame him for his excitement? By 1890, Coley later claimed, no less than forty cases had been recorded of the "entire disappearance of malignancies during an accidental erysipelas attack." To witness such beautifully aberrant and healing biochemistry in action, to see apple-sized tumors shrivel and disappear, was so rare, so pleasantly confounding, that it was like reversing time and running the film of malignancy backward toward a miraculous, disease-free beginning.

Against the background of Coley's utter despair over Elizabeth Dashiell's swift death and his utter naïveté as a young physician, one reads in those two italicized words volumes about William Coley's enthusiasm. "Having satisfied myself that the mortality from erysipelas uncomplicated was very small," he later wrote, "I determined to inoculate the first case of inoperable sarcoma that should present itself. I had but a short time to wait. . . ."

2

"THE MAN WHO DOES THE MOST WORK DOES THE BEST WORK"

Case reports of this sort are often referred to with derision by investigators who characterize them as anecdotal. Ironically, the same investigators give credence to anecdotes when they are presented in the Bible, in the Decameron, or by Homer. I submit that "story telling," particularly "true story telling," is one of the best and most applicable teaching elements in our armamentarium and wonder if it is not counterproductive that stories have gone out of style in "modern" scientific endeavors. One could ask what would have happened if Withering had not heard anecdotal reports regarding the effect of the digitalis leaf?
—Bernard Waisbren Sr., 1987

🖎 *His name was Zola,* and he was a drug addict, probably from the painkillers he took for the disease that was only weeks away from finishing him off. In April 1891 this Italian immigrant—age thirty-five or forty (depending on which Coley paper you read), profession unknown—sought treatment at New York Hospital in desperately poor condition. Like Fred K. Stein, Mr. Zola—his first name has not survived—suffered from a sarcoma of the neck. One year earlier, in 1890, a "Professor Durante" of Rome had removed the growth; within a year it roared back, and William Bull had attempted surgery again. In March 1891 he removed a large piece of tumor "about the size of an orange," according to the pathologist's report, but the remaining cancer was so deep-seated that Bull declared the disease inoperable. This, in time, became the standard criterion for a referral to William Coley: inoperable cancer, nothing else to be done, patient expected to die.

Here was as sorry a case as could be imagined. The cancer in Zola's neck had already spread, probably to the right lung (he had a hacking cough that wouldn't go away) and certainly to the right tonsil; there, another ill-placed tumor ("the size of a small hen's egg") sat like a boulder, all but blocking the man's throat. He could not talk, could not eat solid food, and could barely swallow liquids, which he promptly regurgitated through his nose, leaving him emaciated and cachectic. Meanwhile, the neck wound where Bull had operated the previous month had not healed; there, too, the cancer was judged to be rebounding. Barring a miracle, Mr. Zola was not expected to survive the summer. "Apparently," Coley told colleagues later, "he had only a few weeks to live."

At this point, feeling there was nothing to lose, Coley proposed his first foray into what we would now call nonspecific immunotherapy. No documents survive, but he discussed the idea with William Bull, and Bull apparently consented to this unusual experiment. Coley then explained to the patient that he wanted to infect him intentionally with erysipelas. If a disproportionate number of immigrants seem to have suffered the role of medical guinea pig in these early treatments, it is at least in part because the nineteenth-century hospital was shunned by all but the poor, for whom the luxury of a private physician was not financially feasible; one of the seldom-noted contributions of the immigrant class, as invaluable as it was involuntary, was to serve as human subjects in experimental medical treatments such as this. Zola consented, and in the beginning of May Coley and his colleagues began what turned out to be an unexpectedly arduous and protracted effort to infect this poor fellow with the lethal erysipelas.

If Zola took a risk, so too did Coley, and this first excursion into unconventional "biological modulation" deserves to be considered in several contexts. The first is Hippocratic: penicillin would not enter clinical use for another fifty years, and in 1891, erysipelas was one of the leading causes of hospital-acquired postoperative infection and death; it was not at all a benign procedure, and in the era before antibiotics, it might reasonably be considered to trespass Hippocrates' most sacred medical boundary: First, do no harm. The second, related context is ethical: it is certain that such a therapy, grounded upon virtually no scientific knowledge of immune function and suggestive of no known mechanism of action, would never win institutional blessing in the 1990s; and yet at the same time Coley's experiment has much in common with modern testing of radical new cancer treatments, which hew disturbingly close to the line that separates reasonable clinical testing from the use of humans for outright experimentation. That inevitably leads to the third context, which concerns the proper design of a meaningful clinical experiment and involves patient selection: in 1991 as well as 1891, patients selected for such study are almost always terminal, inoperable, and beyond hope, which may introduce an ethically defensible but problematic wrinkle into whatever results might be obtained. Doctors today, as did Coley one hundred years ago, routinely bemoan the fact that patients so gravely ill may be less likely to muster as favorable a response as patients at an earlier stage of disease, especially in therapeutic approaches that depend crucially upon the biological resources of the patient's own immune system; and yet physicians usually are not permitted to test new treatments against the early stages of a disease like cancer until it has been shown to be effective against the most dire cases. Again, an ethically defensible catch-22.

The fourth context has to do with the history of science: though medical science rightly and necessarily proceeds on the basis of rational knowledge, some of the highest peaks in the landscape of medical progress have arisen out of serendipity, from the accidental discovery of penicillin in 1928 to, citing a recent example in the field of immunotherapy, the fortuitous discovery that alpha interferon is highly effective against hairy cell leukemia. Luck plays a much larger role in the process of discovery than many researchers would care to admit, and William Coley was about to get very lucky.

The fifth context is personal: although Coley would throughout his career distinguish himself for his surgical skill and innovation, his youthful interest in a cure for cancer would come to color the rest of his life with both exhilaration and despair. That he stubbornly clung to his beliefs in the face of relentless criticism was the ultimate expression of passion and daring in this quiet, button-down man, and it has been true of every true believer since, every researcher who has seen a large and established tumor, in mouse or man, shrink and even disappear following an experimental treatment.

Finally, it is impossible to view this work and ignore current knowledge about immune function, which adds a layer of paradox to what was to transpire in a tenement room in lower Manhattan. Recent laboratory investigations using Coley's toxins have consistently suggested that heat-killed bacteria of the sort used in the Zola experiment, which Coley called strep erysipelas and is now known as *Streptococcus pyogenes*, by themselves produce a barely detectable effect against tumors in animals. Which makes the results of that first experiment even more remarkable, serendipitous, and enigmatic than has been generally recognized.

On May 3, 1891, Coley initiated the American era of cancer immunotherapy by deliberately infecting Signor Zola with cultures of erysipelas. Erysipelas being so contagious and potentially dangerous, hospitals were reluctant to host the experiment; the historic first inoculations took place not at New York Hospital but in Zola's home in the Lower East Side (probably the "Mulberry Bend" area near Canal Street, where most of the city's Italian population resided). In attendance were two other doctors, and also Zola's niece, on hand to serve as a nurse and who justified all the concern about contagion when she later contracted erysipelas from her uncle.

Coley's very first case advertised both the dramatic successes that nourished his optimism for four decades and the serious shortcomings that kept his critics well fed for just as long. In this first experiment, the indecisive Coley of school days took over. He dabbled to a fault. He used two different preparations of erysipelas, and he administered the treatment in two different ways. According to one technique, the bacteria were grown on a gelatin-like culture; Coley then applied the treatment by creating small lacerations in the skin ("scarifying") and rubbing live bacteria into the open wounds. In the other, the bacteria were grown in "beef tea"—similar to what we would call beef bouillon—and injected by hypodermic needle. The amounts must have been miniscule, because it took twenty-three hours for Zola's body to register the merest sign of perturbation. Even then, the alarums were modest: a slight temperature (100.5°F), a slightly elevated pulse, minor chills, nothing dramatic. These middling symptoms subsided after several days. Three times Coley repeated the inoculations over the next two weeks. Three times Mr. Zola's immune system murmured in response—slight temperatures, slight chills. The cultures, it turned out, were inert, and Coley hadn't provoked much worse than the equivalent of a mild case of the flu.

On the fifth attempt, Coley switched to a fresh batch of bacteria prepared by two colleagues at the College of Physicians and Surgeons, now part of Columbia University. Not only did he inject a larger amount of bacteria this time, but he varied the location, giving most of it directly into the unhealed neck wound and injecting the remainder at several points under the skin. The injections prompted "intense local redness." Within eight hours, Coley noted severe chills, vomiting, and an intense headache; but again, Zola's temperature barely topped 101°F. Coley nonetheless persuaded himself that he detected changes; "the tumor of the neck diminished in size," he later wrote, "and the general condition improved." On June 2, one month into treatment, the tonsil tumor had "appreciably diminished in size," he wrote in a communication to the *Annals of Surgery,* and Mr. Zola's cough had abated. He could swallow food again. His general condition appeared to his doctors to be "excellent." In the absence of unambiguous changes, such as the complete disappearance of the tumors, these must be considered highly subjective observations. Although he discontinued the treatments in August after sixteen inoculations, Coley felt encouraged. Was such optimism justified? He wouldn't even begin to know if the effect was enduring until autumn, but the fact that Zola managed to stay alive through the summer, given his desperate wasting condition in the spring, probably struck all his doctors as a minor victory.

During that same spring and summer, Coley treated one other patient, a sixteen-year-old girl. At the time the inoculations began, her leg just above the knee bulged out to the breadth of a baby's head, with a sarcoma the size of two fists projecting from a previous incision wound; Coley and his colleagues noted "marked" improvement in the lower portion of the tumor, but they ultimately concluded that this patient's disease was too far advanced to justify continued treatment.

Coley proclaimed himself greatly heartened by these first two attempts at treatment. In fact, he allowed himself to utter the single most dangerous word in the vocabulary of the oncologist. "In both cases the antagonistic effect was well marked," he wrote not long after, "and improvement, even if temporary, was sufficient to make me believe that, had the cases been less far advanced when the treatment was begun, it would not have been too much to have looked for a permanent cure." Here, too, he betrays his youth; certainly nothing resembling talk of a cure on the basis of such preliminary and incomplete evidence would slip past modern peer reviewers.

But in this premolecular world of medicine, seeing was believing, and Coley had seen several advanced tumors shrink in size. He had become in one short summer of youthful excitement America's first true believer in immunotherapy. He would not by a long shot be the last.

❧

As it turned out, Coley had even more reason to be optimistic than the results described in his 1891 paper implied; interestingly, in a talk he gave many years later to the Royal Society of Medicine in London, Coley provided a somewhat different perspective on the Zola case. "I worked continuously from May to October, 1891, to produce an attack of erysipelas, without success," he admitted in a 1909 talk; in reality, he hadn't even managed to properly infect Zola and the teenage girl. Frustrated by his inability to trigger a rip-roaring infection after months of trying, Coley then asked his friend and New York Hospital pathologist Farquhar Ferguson, who had plans to spend the summer of 1891 in Europe, to visit the Berlin laboratory of Robert Koch; perhaps Koch, the foremost bacteriologist in the world, could provide a fresh culture of erysipelas—preferably a fresh, *lethal* culture isolated from a patient that had succumbed to the disease. By the beginning of October, when Coley received this satisfying virulent gift from Koch, they had to start from scratch. Zola's tumors had rebounded to their previous size, and on October 8 Coley returned to Zola's bedside to try all over again. This time he not only succeeded in causing the disease but nearly killed the poor man and his niece in the process. Both came down with raging infections.

Within an hour of receiving an injection of Koch's germs directly into the neck tumor, Zola's body shook violently with severe chills for forty minutes, and he was racked by nausea, vomiting, and severe pain; his temperature rocketed to 105°F. Within twelve hours, a typical red swatch of erysipelas appeared on his neck and spread across his face and head in both directions, meeting on the far side. Then began a series of physiological transformations so stunning that, reported by someone with less integrity and honesty than Coley, they might well have inspired sneering disbelief.

"The tumor of the neck began to break down on the second day," the surgeon noted, "and a discharge of broken-down tumor tissue continued until the end of the attack. At the end of two weeks the neck tumor had disappeared and the tonsil tumor had decreased in size." Mr. Zola, desperately ill in May 1891, gained weight, arose like Lazarus from his tenement bed, and resumed as normal and hopeful a life as could be attained by an immigrant in nineteenth-century New York. The neck wound healed completely, the tonsil tumor stopped growing—it never went away but neither did it become larger—and in

1893 Coley could report, "The patient's general condition at present (nearly two years) is very good, although he is suffering from a confirmed morphine habit which he had contracted previous to the inoculations." Coley last saw Zola in October 1895; he was still in excellent condition, with the tonsilar tumor still present but seemingly no longer malignant. Soon after, Zola returned to his native Italy and later died in his homeland, eight and a half years after the first injections, of what was described as "a local recurrence" (the irony here is that if indeed the same malignancy recurred and caused death, the initial diagnosis was probably correct, although we will never know with certainty).

Initiating a curious phenomenon that has bedeviled immunotherapy on into the modern era, Coley had had the good and bad fortune of achieving a remarkable success in his very first case. And that was all it was: a single case. Medical researchers refer to these, with entirely appropriate disdain, as anecdotes (unless it happens to be *their* anecdote, in which case they tend to be a little more magnanimous about its significance and implications); Zola's recovery was interesting, of course, but possibly unrelated to any general and predictable reaction, and certainly of no statistical significance whatsoever. Its success remained inexplicable. The tyranny of the anecdote, the boon and bane of immunotherapeutic interventions, had formally begun.

The history of medicine has long been marred by claims similar to Coley's, made by some well-intentioned and more often by less-than-well-intentioned men willing to merchandise any hope (no matter how slender or preposterous) to the hopeless, and willing to conflate possibility (no matter how slim) with probability among those driven by illness to bet their very lives against odds that would shame even the greediest bookmaker. Despite some sporadic early successes, it is quite possible that Coley's work would have won him a place in that dubious pantheon of discredited cancer messiahs, purveyors of surefire cures that horrify modern sensibilities. Historians have referred to the 1890s and early 1900s as the "Golden Age of Quackeries," and folk remedies for cancer included herbal plasters, liquid paraffin, and even live frogs. Among the treatments attempted by reputable physicians of the day (including, in some cases, Coley himself) were a dubious drug called chelidonium; "electric puncture," which involved running an electrical current directly through breast tumors, for example;

"katophoresis," which involved using electrical currents to carry mercury into cancerous tissue; direct injections of alcohol into tumors;
injections of lymph gland extracts from animals; and even injections
into the womb of carbide of calcium, or quicklime, which among
other atrocious side effects led to a dangerous buildup of acetylene gas
in the uterine chamber. What separated Coley from these other practitioners was that he attempted, although not very successfully, to
frame his medical research in the idiom of science and indeed befriended (and enlisted the financial support of) wealthy patrons who
could support his research.

His relationship with the Rockefellers began at Bessie Dashiell's
deathbed with John D. Rockefeller Jr. In 1896, Coley began to receive
some modest research funding from the Rockefellers (the very first
dollars the family spent on cancer research), and over the years he
consulted with representatives of the family on directions in cancer
research, fitted John Sr. with a truss (and received notes from him
urging a visit to their Florida seaside mansion, with the coy addendum
that "the latch-string is always out" at the many-roomed mansion),
and responded to John Jr.'s occasional requests for "those little devices" (the euphemism by which America's wealthiest man sought to
replenish his supply of condoms). For more than forty years, Coley
enjoyed the friendship of both Rockefellers.

But Coley established an even closer relationship during his early
career with the Huntington family. Archer M. Huntington was the
stepson of Collis P. Huntington, of "Big Four" robber baron fame, and
he provided funds for the construction of a special pavilion at the New
York Cancer Hospital (later Memorial Hospital) when Coley needed
an isolated ward in which to continue his dangerous erysipelas experiments. The hospital at that time occupied a magnificent four-story,
turreted structure designed in the style of a French Renaissance châ-
teau, which stood (and still does, though barely) at the corner of Central Park West and West 106th Street. It was in this building, known
informally as "the Bastille," that many of Coley's pioneering experiments with his cancer vaccine took place. Later, in 1902, Coley arranged a gift from the Huntington family that established the first
foundation for cancer research in the United States.

The initial erysipelas experiments stalled, due to a most unexpected reason. For a deadly pathogen, the erysipelas bug only rarely
lived up to its villainous reputation. Between 1891 and 1893 Coley
attempted to infect twelve patients, all diagnosed with inoperable and
incurable malignancies. Despite rubbing, injecting, and scarifying

with live germs in "many attempts extending over many weeks," Coley failed to produce any outbreak whatsoever in four carcinoma patients; as before, he nevertheless claimed to have observed "slight temporary improvement." In four cases only, all sarcomas, was he able to rouse a roaring infection; these included Zola, a thirty-eight-year-old woman with a twice-recurrent sarcoma of the breast (she had several partial responses before dying), and a forty-year-old man with an inoperable tumor in his back, diagnosed as sarcoma, which had spread to the groin. Coley first treated this gentleman on April 21, 1892, almost a year after Zola, and it took near-daily injections over a four-week period to trigger a reaction. Finally, the man spiked to a 105.5°F fever and developed the telltale rash. It is worth recording Coley's reaction in detail, for it must have been moments like these that sustained him for many years to come:

"From the beginning of the attack the change that took place in the tumor was nothing short of marvelous," he noted. "It lost its luster and color and had shrunk visibly in size within twenty-four hours. Several sinuses formed the second day and discharged necrosed tumor tissue. A few days later the tumor of the groin, which was about the size of a goose egg and very hard when the inoculations were begun, broke down and discharged a large amount of tumor tissue. Three weeks from the date of the attack of erysipelas both tumors had entirely disappeared." As before, the effect ebbed and flowed; on July 1, two months after the initial injections, a small local recurrence appeared in the man's back, joined two weeks later by six small nodules. Coley resumed inoculations, but couldn't produce a bona fide attack of erysipelas until mid-November, by which time the main tumor had grown to five inches by three and a half inches, with a marked recurrence in the groin. "On November 14," he wrote, "after an injection of 22 mm. [a "minim" equals 1/60th of a dram, or about one drop] of a culture previously used without effect, a moderately severe attack of erysipelas developed, during the course of which the tumors both in the back and groin disappeared."

Back and forth the battle went. Following yet another recurrence, Coley surgically removed the growth in the man's back; once more the cancer returned. Finally, following another three-week course of the "toxic products" of erysipelas, all nodules disappeared and the man regained his normal weight and general health. He remained well for three and a half years, ultimately succumbing to an abdominal tumor that Coley termed "probably metastatic."

After the first twelve patients, the ledger read as follows: four carcinoma patients neither contracted a robust infection nor showed a major response, but all eight sarcoma cases responded, including two complete responses where even metastatic tumors had disappeared. Beguiled by these apparent successes, and bedeviled by the failure to achieve more consistent infections (and therefore responses), Coley reached a sensible conclusion: "The uncertainty of being able to produce erysipelas when desired (failure having occurred in two-thirds of my cases) coupled with the fact that frequently repeated injections of the liquid cultures of the erysipelas germs had produced a marked beneficial effect upon the tumors, led me to attempt some plan of isolating and using the active principle of the germ." Coley had become a little more cautious in the interpretation of his results, but not much, as is evident in a carefully worded summation in an 1893 paper that appeared in the *Post-Graduate*. "The clinical and experimental data above referred to," Coley wrote, *"conclusively* prove that the germ of erysipelas contains a powerful principle, which has a marked antagonistic effect upon tumors, more especially sarcoma, and which, in a certain proportion of cases, is curative of the same." Ignoring for a moment the unwarranted optimism of the word "conclusively," Coley's summation could be restated as follows: There is something in erysipelas that has antitumor activity, more often against sarcomas than carcinomas; in a few cases it seems to induce long-lasting remissions.

Against as formidable a disease as inoperable cancer, that was saying something, and in truth it had all been said before, mostly by researchers in Europe. But the worst of all possible clinical outcomes forced Coley into temporary retreat, and then a fatefully fruitful modification. Two of those first twelve patients died—not of cancer, but of erysipelas infections that had raged out of control. The medicine turned out to be worse, or at least swifter, than the disease. Although he never reflected at length on this unhappy development—"Shortly afterward," he admitted in passing, "I had two deaths, both in far-advanced cases, and the patients had been duly warned of the risks of inoculation"—these deaths forced him to change the formulation of the vaccine.

The two deaths plunged Coley into a medical and ethical quandary. He felt the experiments so far were "absolutely confirming" of previous random observations of what he somewhat naively described as a "curative influence"; he saw what he considered unquestionable evidence that fast-growing tumors could at least be held at bay for a while (not an unreasonable standard, since a century later, the principal rallying point for any new chemotherapeutic treatment is its ability to cause, however temporarily, shrinkage in an established, bulky tumor). But the deaths underscored the dangers of using live germs as a form of medicine. Coley knew, too, that the pioneer in the field, Friedrich Fehleisen in Germany, had given up on erysipelas injections as a treatment because of precisely those "practical difficulties"; what he probably didn't know is that Fehleisen was forced to resign his position in the Würzburg clinic because of patient fatalities (he ultimately ended up in California). Taking stock toward the end of 1892 of what had been learned so far, Coley realized that it was no easy trick to cause an infection with erysipelas, and that on rare occasions success could be fatal. "And, most important, I had been impressed with the fact that repeated injections of the living bouillon cultures of the streptococcus of the erysipelas had an inhibitory action upon the growth of the tumors which, while only temporary, was nevertheless distinct."

This line of reasoning led Coley to an underappreciated but crucial insight about the mode of action of erysipelas that flows in a meandering but unmistakably unbroken line through this century's medical research and leads to the molecular identity of key components in the immune system known as cytokines, the use of molecular biology to characterize those molecules, and the hastened development of new technologies (specifically biotechnology) to generate those rare molecules in sufficient abundance to test their mettle against cancer and other diseases in the modern clinic. What William Coley began to realize was that "a portion at least of the curative action of the erysipelas lay in the toxic products of the erysipelas, which might possibly be utilized without producing an actual attack of erysipelas." The patient deaths turned out to be a blessing in disguise, because they forced Coley to abandon live germs. He would try something different, and in so doing he stumbled upon one of the most powerful stimulants to the immune system ever discovered.

Coley began to suspect that it was a *toxin*—a substance, a "factor," what we would now suspect to be a molecule of some sort—produced by the bacteria that had such dramatic tumor-melting properties; but it is also a subtle and open-minded insight because of that seemingly innocuous phrase, "a portion." It left the door open to the possibility that there might be more to the responses than just the toxins.

This line of reasoning freed Coley from the use of live bacteria, and he immediately began tinkering with a less dangerous approach. Although the word "toxins" began to appear increasingly in his papers and talks, it may be more useful to modern ears to think of the evolving preparation as a kind of vaccine—not a preventive vaccine, like the crude but highly effective treatment pioneered by Edward Jenner in the eighteenth century, but a different kind of vaccine, one that used a killed microbe to stimulate an antagonistic response against certain forms of cancer that were already present in the body: a therapeutic vaccine. In the summer of 1892, working with lethal strains of erysipelas for maximum strength, Coley's friend Alexander Lambert (later president of the American Medical Association and personal physician to Teddy Roosevelt) grew the bacteria in the laboratory and then inactivated the microbes in one of two ways—either by killing them outright with heat or by removing them by use of a porcelain filter whose pores were so small that it separated the single-celled bacteria from the fluid in which they had grown. Captured in this leftover fluid, Coley believed, was what he called the "active principle."

Coley tried these preparations in four cases of inoperable sarcoma. The preparations triggered high temperatures (in the range of 103.5°F) and modest, though temporary, effects on the tumors. Something even more powerful was necessary, he felt, and it was at this point that Coley decided to spice up the recipe. He added a second ingredient to the vaccine—a bacterium known at the time as *Bacillus prodigiosus*.

The idea did not drop out of the sky. Once embarked on the idea of a bacterial vaccine, Coley closely followed developments in the field, particularly in the European journals, and thus became aware toward the end of 1892 of experiments by a French doctor named G. H. Roger of the Pasteur Institute, who had reported that the virulence of the erysipelas germ increased markedly when it was grown in the presence of *Bacillus prodigiosus*. "Roger had never used the *Bacillus prodigiosus* alone or with the streptococcus of erysipelas on the human being, and had never, as far as I know, suggested it as a therapeutic agent," Coley said later, at least in part to secure the priority of the radical step he was about to take.

It was a fortuitous choice—a "shot in the dark," as one molecular biologist later called it—but that should not detract from an appreciation of this lucky development; by adding *B. prodigiosus* to his vaccine, Coley managed to select what is known as a "gram-negative" bacterium (negative because it fails to take up a stain used in diagnosis). These gram-negative microbes manufacture and shed, almost like dead skin, a substance in the cell wall known as an endotoxin, one of the most powerful inducers of an immune response known to biological science. And this substance became very well known to biologists (though not in time to do Coley's reputation much good) because, fully eighty years after first employed by Coley to treat human disease, scientists discovered that the endotoxin produced by *B. prodigiosus* stimulated the immune system to manufacture a host of powerful molecules, such as tumor necrosis factor, interferon, and interleukins. So potent was tumor necrosis factor, for example, that, as the name implies, it had the remarkable ability to necrose, or kill, tumors in laboratory animals; so potent, too, that in sufficient quantity and in the right circumstances, it could plunge animals and people alike into shock, causing a very swift death. And humans, as the essayist Lewis Thomas once pointed out, are "perhaps the most sensitive to endotoxin of all animals. Less than one millionth of a milligram will produce shaking chills and high fever in a normal adult."

Coley was a generation or two ahead of the complicated molecular biology of endotoxins, but his mixed bacterial formula was very much of its time. Even as Coley prepared to treat his first patient with this new mixture, researchers elsewhere began to report similar observations. A Dutch physician named C. H. H. Spronck performed, if anything, a more systematic investigation of a preparation very much like Coley's at roughly the same time at the University of Utrecht. Preliminary experiments with dogs, published in October 1892, showed that in at least some cases the mysterious factor in erysipelas seemed capable of dissolving advanced tumors. Spronck went on to treat twenty-five inoperable tumors in man. Of equal importance, Spronck argued—and Coley was aware of this work—that on the basis of his animal experiments, the effect of the mysterious factor in erysipelas was "systematic," as he put it (*systemic*, in modern immunological parlance). In other words, even when Spronck injected his toxins into the dogs at a location distant from the tumor, the canine tumors shuddered to a halt and liquefied into dead tissue. Whatever acted upon the cancer, it appeared to circulate freely throughout the body. A clue to this systemic effect appeared in Spronck's autopsies of the dogs. Under the microscope, some of the tumors looked besieged by a large number of "polymorphonuclear leukocytes," indicating that an inflammatory or immune response had occurred. There, in Spronck's microscope, lay a faint scientific hint that these occasionally surprising responses might in some way be related to the immune system—

and that the toxins might, in Hans Sloane's apt phrase, have stirred their own form of "commotion in the blood." But it really wasn't much of a clue, and since it wasn't the type of clue to flatter the main currents of immunological thinking in the last decade of the nineteenth century, it went largely unexplored. Nor, to be blunt, was Coley the right person to explore it.

At the beginning of 1893, as he prepared to use mixed toxins for the first time, William Coley straddled two medical worlds. On the one hand, he fancied himself a medical modernist, adventurous and forward-looking, willing to try new and experimental treatments for old, intractable diseases; on the other hand, he retained an old-fashioned conviction that there was a proper way to do things, which included his stubborn belief in the *art* of medicine. Every patient and each disease was a unique landscape that demanded and deserved exploration with those most rudimentary of a doctor's surveying tools: touch, sight, smell, the wisdom of the senses. This was especially true of cancer, the disease that would dominate his working life.

"You should embrace every opportunity to study the character of every tumor you meet with," he told a group of medical students a few years later. "Look at it thoroughly, feel of it carefully, examine its relation to the overlying skin, whether adherent or not, find out whether it is attached to the deeper structures, ascertain its mobility, its consistence, its relation to bone or soft parts, and after you have done all these things and have also in a certain number of cases been able to find your diagnosis confirmed or disproved by the microscope, you will probably be surprised by your own knowledge and ability to make a diagnosis. . . ." Here was the voice of the palpating and probing clinician, at precisely the historical moment when biomedical science began to pride itself on diagnostic instruments more sophisticated than a pair of hands.

By January 1893 Coley had just turned thirty-one and had, one month earlier, become a father for the first time with the arrival of his son Bradley. Soon to occupy a handsome brownstone on West Thirty-fifth Street just off Fifth Avenue, he ranked among the most promising young surgeons in New York, with appointments to the staff of the Hospital for the Ruptured and Crippled in November 1891 and to the recently opened New York Cancer Hospital in 1892. By 1900, when most doctors earned roughly $1,500 a year, he was well on his way to earning the very considerable sum of $30,000.

Blessed as he seemed to be with professional and personal success, Coley was about to learn one of medicine's most humbling lessons: physicians whom the gods wish to torment are allowed to have a smashing success in their very first attempt at a controversial new treatment. Zola was a mere overture. Coley prepared to try his new mixed bacterial vaccine for the very first time. There was unfortunately no shortage of good candidates, children usually, with inoperable sarcomas and little hope. One of them was a sixteen-year-old boy, of German background, "with good family and personal history." And a massive sarcoma, about the size of an eggplant, poking out from his lower abdomen.

The case came to Coley in January of 1893 as a referral from a New York physician named Lemuel Bolton Bangs, who felt the young man might be a candidate for artificial erysipelas infection, which was tantamount to saying that nothing else could be done. The boy's name was John Ficken, and not terribly much is known about him. He had been bedridden with a large, inoperable tumor attached to the abdominal wall that created an enormous bulge in his abdomen; the growth visibly protruded from just beneath the skin, from his navel down to the pelvis. A biopsy revealed the presence of spindle cells, a sign suggestive of malignant sarcoma (no slides exist to corroborate the diagnosis). Coley decided this patient represented the first opportunity to test his "toxic products."

On January 24, 1893, Coley administered the first injections of his toxins directly into the boy's tumor. Sobered by the deaths of patients infected by erysipelas, Coley had by now abandoned the use of live bacteria; his colleague Alexander Lambert, working in a laboratory at Columbia's College of Physicians and Surgeons, grew cultures of both *Streptococcus pyogenes* and *Bacillus prodigiosus* bacteria and, filtering them through the microscopic pores of porcelain, strained out the microbes. What remained was a deep purplish red fluid that contained, presumably, the "powerful principle." Thus decanted, the fluid—known ever after as "Coley's toxins"—passed on to Coley for treatment.

As he would do with all his patients, Coley began with small doses—0.5 cc, a tenth of a teaspoon—and gradually increased the dosage over a ten-week period, at an interval of two or three days, until he provoked a strong reaction. This approach, known as titration, essentially meant that Coley determined each patient's optimal dose by trial and error. Ficken's side effects mimicked those of the earlier erysipelas injections: violent chills within twelve hours, nausea, vomiting, headache, fever, and local redness and swelling at the injection site. "The chill and tremblings were extreme," medical records noted. But so too was the patient's response. Frequent examination and measurement revealed that the tumor had shrunk about 80 percent, and the boy's general condition seemed to improve. Satisfied, Coley halted the treatment on May 13, 1893; a month later the tumor no longer visibly protruded. To Coley's old-fashioned but seasoned hand, the tumor's size had apparently diminished to less than two inches in diameter. The boy in the meantime had gained ten pounds.

So Coley had obtained a response. But as he knew, a response was only half of the equation in calculating the worth of any cancer treatment; the other issue was duration.

If Coley's belief in his toxins depended in part on the continued welfare of John Ficken, his conviction only grew stronger and firmer over a great number of years. He sent the boy home in 1893 and continued to stay in touch, updating his progress year after year, and although colleagues wearied of hearing about the same old case, this medical anecdote stuck like a thistle in everyone's skepticism. John Ficken grew up, left home, found a job as a bartender and an apartment in Greenwich Village, and happily disappeared into the fabric of life in New York City. The Ficken who was inoperable and at death's door at the time Coley first treated him was alive and well, in perfect health, when Coley described the initial results of the mixed bacterial vaccine at Johns Hopkins University in 1896; was alive and well in both 1900 and 1907, when Coley hauled him along while giving talks at the New York Surgical Society; in 1909, when Coley claimed roughly 150 successes (in some 500 inoperable patients treated) in a talk to colleagues at the Royal Society of Medicine in London; in 1913, twenty years after first being treated, when Coley updated those results in a presentation to the International Cancer Congress in Brussels. Indeed, cancer was not foremost among his medical worries when the orthographically challenged Ficken wrote Coley in 1910, apparently about a truss. "I want to see you this summer at the Cliniq Ruptiures & Kreples," Ficken wrote, referring to the Hospital for the Ruptured and Crippled. "I was up there 4 times but failt to meet you I want a new Belt."

He lived until 1919 when, apparently unbeknownst to Coley, he suffered a heart attack on a subway car and died in Grand Central Station. Of his forty-seven years on the planet, twenty-six came after Coley treated his otherwise untreatable disease.

Buoyed by a success the dimensions of which would not become clear for decades, Coley treated another five patients during the first few months of 1893 with the toxins. None had as promising a response as John Ficken. Coley, perhaps gilding the results with founder's pride, pronounced their responses "sufficiently encouraging to make further experiments exceedingly desirable." In his first report on the new vaccine approach, which appeared in the *Post-Graduate* in 1893, he wrote: "While the treatment of malignant tumors by erysipelas inoculations or by injections of the *Toxic principles* of erysipelas is still in the experimental stage, we cannot but recognize that a long step in advance has been made and there is good ground for confidence that further investigation along these lines will be followed by even more brilliant results than those already obtained." Reaction to such immodest optimism was swift, and unforgiving.

The December 15, 1894, issue of the *Journal of the American Medical Association* featured an editorial applauding recent "gratifying" news from the superintendent of immigration's office reporting a decrease in the number of immigrants arriving in the country and a seven-page, state-of-the-art review article on the natural history of yellow fever, a communication that managed to mention the word "mosquito" not once. Sandwiched between these quaint assertions of scientific and social certitude, in a journal emerging as a dominant voice of medical authority, appeared an editorial about Coley's toxins. The commentary attacked not only the toxins but also Coley, and in remarkably personal terms:

> There is no longer much question of the entire failure of the toxin injections, as a cure for sarcomata and malignant growths. During the last six months the alleged remedy has been faithfully tried by many surgeons, but so far not a single well authenticated case of recovery has been reported, so far as our reading has extended, and the personal experience of surgeons of our acquaintance with whom we have conversed, demonstrates that in all cases in which they tried the erysipelas toxin the result was

no improvement. . . . It now seems as if any assertion, no matter how absurd, needs only some strong voice or lucid pen, to make the profession swallow it greedily. It is the age of the sensation monger, and the seeker after notoriety may enjoy a temporary celebrity by a very easy process. He has only to announce the sure cure of some hitherto incurable disease by some foreign chemic product, or microbic mystery and the thing is done. The celebrity may be short-lived and suffering humanity deluded by false hopes, but the story was a pleasant one while it lasted.

Coley, proprietor of an admirably restrained indignation, fired off a quick reply to the 1894 *JAMA* editorial. "That a few physicians, in a very limited number of cases, with indifferent preparations of the toxins, have failed to obtain good results will not, I am sure, have great weight in the minds of the scientific portion of the profession, in determining the failure or success of this method of treatment of sarcoma," he wrote. But that was a misplaced sentiment. As of 1894, the toxins would forever bear the stigmatic adjective "controversial," and by 1896 the *Medical Record* issued its "final" verdict: "To hold out the hope of cure by this or any other means in conditions which are non-curable cannot be too strongly condemned." Coley would forever after battle two persistent misconceptions about the vaccine bearing his name: that no one else could get them to work, and that they did not work at all.

The problem, such as it existed for the medical profession, was that Coley was not one of the fly-by-night hucksters. Quite the contrary: he was well on his way to becoming one of the most accomplished surgeons in New York (Joseph Lister, after reading an 1897 paper by Coley describing 360 consecutive hernia operations without a single infection-related death, applauded such pristine aseptic technique as "enough to cause gladness in the heart of any man who loves his fellow man") and already possessed a reputation for integrity such that colleagues would have invited more discredit on themselves than Coley by dismissing him as dishonest. Dismiss him they did, as many felt they must do, but for more impersonal, scientific reasons. The cures he continued to claim were not due to his so-called mixed toxins, they maintained, but unfortunately reflected simple, straightforward, garden-variety medical self-delusion. The cancer cases for which he claimed cures were not cancer at all.

Of all the criticisms, the charge of misdiagnosis was the most galling to Coley. His exasperation and cynicism, rarely on display, simmered through in a remark destined to evoke little sympathy among his peers. He insisted that his diagnoses were accurate, writing with rare sarcasm, "These results are indisputable facts, the diagnosis having been established, not only clinically by the leading surgeons of the East, but confirmed microscopically by the best pathologists; and if further evidence were needed, the fact that a number of them had rapidly recurred after operation would be enough to establish their malignancy. If the diagnosis is still doubted, after having been subjected to all these tests, it seems to me that we had better at once abandon the discussion as to the treatment of sarcoma, and employ our time in learning how to make a diagnosis." On another occasion, with a logic sure to infuriate his fellow surgeons, Coley made an unpleasantly legitimate point: "If we claim that the cases that are cured by the toxins or the x-rays are mistaken diagnoses, we are logically driven to the same explanation in cases that are cured by operation."

But the *JAMA* editorial, and the sentiment behind it, was not without merit. Even in Coley's hands, the toxins only worked in a minority of patients, and editorials like the *JAMA* piece formed one rhetorical tile in a mosaic of doubt that over the years added up to one negative message: no one besides Coley could get the toxins to work. As we shall see, that was never true. But there was another, more scientific problem that may help explain the doubts.

W here did Coley's work fit into the burgeoning science of "serology"? How did Coley's work fit into the immunology of his day? The answer is: not at all. Coley was not an immunologist, and never became one (they did not offer a Ph.D. in immunology at Harvard Medical School until 1974). Although the action of the mixed vaccine is now believed to have been immunological, Coley did not know it. Immunologists, for their part, had no reason to pay much attention to Coley's work; whatever practical wisdom had been acquired about immunity grew out of Edward Jenner's success with vaccination and the emerging antitoxin treatments. Immunologists typically tried to remove all blood cells before beginning their experiments on serum. Coley and immunologists in effect were walled off from each other by their complementary ignorance.

It is always dangerous to analyze retrospectively historical work through the highly corrective lens of modern knowledge, but it is worth a quick detour to consider Coley's toxins in the context of

Jenner's vaccine. There is a similarity of approach in the use of first live and then attenuated microbes, but also great differences in physiological effects and especially in the manner in which the new therapies were received. Late in the eighteenth century, according to the well-known story, Edward Jenner noticed that milkmaids infected with the mild viral illness known as cowpox, or vaccinia, seemed immune to fatal infections of smallpox. "What renders the cowpox virus so extremely singular," Jenner wrote, "is that the person who has been thus affected is forever after secure from the infection of smallpox." Having made that singular observation, Jenner performed an extreme experiment: he decided to inoculate several neighborhood children, of pauper or peasant background, with the pus of cowpox scabs, and then challenge them with live, virulent smallpox. ("What he did would be forbidden today," writes Robert Desowitz, "but in eighteenth-century England there were no committees on human experimentation and few rules.") It became clear that children thus vaccinated were spared death from the far more lethal smallpox. Though properly hailed as the father of vaccination and a major medical pioneer, Jenner's experiments hardly distinguish him from the practitioners who used gangrene, syphilis, or malaria as a medication—except, of course, that vaccinia induced a much milder disease and a demonstrable, reproducible immune effect that uniquely prevented smallpox. In other words, it worked.

"Vaccination" had the additional and indisputable advantage that it did so much more often than Coley's cancer vaccine, even though Jenner was as much at a loss to explain why vaccination worked as Coley was to explain why his toxins didn't more often or more completely. And perhaps equally important, Coley would have as much trouble today observing the beneficial effects of erysipelas infection against cancer in a world aswim in antibiotics as Jenner would have of finding a milkmaid in today's dairy industry. Theirs were situational observations unique to their time, their cultures, their medical milieux.

Vaccination, arguably the most successful clinical (and public health) strategy in the history of medicine, grew out of folklore, empiricism, and medical naturalism: shrewd observation accompanied by almost total scientific ignorance. Indeed, it took more than a century for science to explain the successes of Jenner's vaccine, and given its long-term and irrefutable success, it gives pause to consider all the things that Jenner *didn't* know. He didn't know that smallpox was caused by a virus (viruses would not be identified as distinct, disease-causing entities until more than a century later). He didn't know that it was the immune system that conferred protection against the deadly virus (that would not be demonstrated until well into the twentieth century), nor that it was immune system proteins called antibodies that played a major role in neutralizing the virus (antibodies were not discovered until 1901). He did not know that T cells also played a role in eliminating the virus (T cells were not generally recognized until the 1960s). And he did not even care to know that tumors, unlike viruses, did not flash the same kind of signal to the immune system and thus would never be so easily identified or eradicated by the immune system (indeed, whether cancer cells even display these surface signals, known as antigens, was until recently a hotly debated scientific point). So, public health implications notwithstanding, Coley unknowingly tackled a much tougher scientific problem than Jenner.

The bottom line is that vaccination against infectious diseases worked consistently and predictably, and that in turn had a profound impact on how immunological research evolved. By the late nineteenth century, the central controversy in the emerging science of immunology involved the relative importance of "humoral" (or serum-based) immunity, of the sort that accounted for smallpox and diphtheria antitoxins, and "cellular" (or cell-based) immunity, of the sort where cells also played a role in the protection of the body against infectious disease. This fierce professional debate has been ably captured in Arthur Silverstein's history of immunology, and it is germane to Coley's work for this reason. Submerged in this intellectual dispute is a crucial difference in immunological responses to infectious diseases and cancers. Most tumor immunologists now believe that longlasting protection against cancers, when it occurs at all, is achieved primarily by the cellular arm of immunity, and yet Coley's work landed on the losing side of a historic intellectual disagreement that resolved decades later, with both arguments in effect turning out to be true. But in Coley's day, components in the blood—"serum factors," they were usually called—seemed to offer a protective effect against disease, and immunology became the province of chemists, not biologists or doctors. Largely because of those successes, humoral immunotherapy as practiced at the beginning of this century was viewed to be, in the parlance of yearbooks, most likely to succeed in the battle against disease. As the French historian Ilana Löwy has shown, there was an immense amount of experimentation in the period between 1890 and 1914 with "immunological" medicine, but it was an immunology synonymous with antibody—an immunology, in short, to which Coley's toxins appeared unrelated. Only around 1910, when the German biochemist Paul Ehrlich proposed his "magic bullet" concept of chemotherapy against disease, did immunotherapy begin to subside into the generally flat landscape of promising but unproven therapies. And so Coley initiated his most important series of experiments in an intellectual environment utterly incapable of making scientific sense of them.

By the spring of 1896, when Coley traveled to Baltimore to give a talk to the Johns Hopkins Medical Society, the lines had been fairly well drawn for a battle that would last at least until Coley's death forty years hence, and beyond. At this talk, Coley reported on 160 cases and claimed that "nearly one-half" of ninety-three cases of sarcoma treated with the toxins "showed more or less improvement." Far less success, he admitted, was noted in cases of melanoma, carcinoma, and osteosarcomas. Doctors at St. Luke's and Methodist Episcopal Hospitals in New York, and from Massachusetts General Hospital in Boston, referred patients to him; some of the cures were nothing short of remarkable.

John Finney, who would later serve as head of Johns Hopkins Hospital, noted that even in cases that failed to result in a cure, the toxins in his hands had "great value" in reducing pain and improving the comfort and condition of patients. And William Welch, destined to head the medical board of the Rockefeller Institute, said, "I have been very much impressed by this personal statement from Dr. Coley, and I see no way of gainsaying the evidence which he has brought forward, that there is something specifically and genuinely curative in his method of treatment. A single undoubted cure of a demonstrated cancer or sarcoma by this treatment would be enough to establish the fact that the treatment exerts some specific curative effect, for the spontaneous disappearance of undoubted malignant growths of this character is almost unknown."

But not everyone enjoyed similar success, and in all but a few quarters (or so it has been generally assumed) the toxins were dropped almost as quickly as they were tried. Here too is a lesson as meaningful today as it was painful to Coley in his: once a novel therapy acquires the odor of inefficacy or controversy, it becomes extremely difficult to rehabilitate; physicians by and large do not like to toil at

improving somebody else's idea (and burnishing someone else's reputation), and even with the most optimistic and generous interpretation of his results, Coley never helped more than about 40 percent of
the patients who received the toxins. Despite initial reports of occasional successes, and despite the fact that the pharmaceutical firm
Parke, Davis & Co. began to manufacture a commercial version of the
vaccine in 1899, mainstream medical opinion first adopted a guard-
edly optimistic and cordial position on the treatment, then decidedly
cooled, the entire transition taking little more than a year or two.

By 1896 many prominent physicians had reached the conclusion
that the toxins were worthless. Typical was the sentiment expressed
by Nicholas Senn of Rush Medical College in Chicago, at the May
1895 annual meeting of the American Medical Association in Baltimore. "The treatment of inoperable sarcoma and carcinoma with the
mixed toxins, as advised and practiced by Coley, has been given a fair
trial in the surgical clinic of Rush Medical College, and so far it has
resulted uniformly in failure," Senn told colleagues. "Although I shall
continue to resort to it in otherwise hopeless cases in the future, I
have become satisfied that it will be abandoned in the near future and
assigned to a place in the long list of obsolete remedies employed at
different times in the treatment of malignant tumors beyond the reach
of a radical operation."

Two other physicians of impeccable credentials immediately tried
to replicate the treatment and immediately reported failure: the Philadelphia surgeon William W. Keen, author of the most authoritative
surgical textbook of his era, failed to see any response in seven patients, and Roswell Park of the University of Buffalo reported similar
results, though more gently. "This simply shows how different men
with the same method get different results," Park acknowledged, "and
that the man who does the most work does the best work. If others
can obtain anything like the results obtained by Dr. Coley the millennium is at hand, and there is opened up for patients with these terrible
conditions an era of hope to which they have heretofore been strangers."

The millennium was a long time coming. It would be another forty
years before a follow-up editorial would appear in the *Journal of the
American Medical Association*. This little-noted article stated, in the
less hyperbolic language of a mature journal, that Coley's toxins in
fact appeared to be effective against certain types of cancer. By then,
however, other more promising technological approaches had captured the imagination of physicians for cancer treatment, especially

radiotherapy. Contrary to general impressions, many other physicians reported success with the toxins, including Henry Meyerding, an orthopedic surgeon at the Mayo Clinic (which routinely recommended use of the toxins at least through the 1930s) and Howard Lilienthal of New York's Mt. Sinai Hospital; the Lister Institute in London continued to manufacture the toxins at least through 1943. Coley struggled to remain open-minded; but after hearing this familiar litany of criticism for nearly twenty years, he told the Surgical Section of the Royal Society of Medicine in London in 1909 (quoting at length from, as he put it, a "recent and unpublished" paper):

> It is natural that any new method of treatment of disease should stand a certain definite test before it can hope to secure recognition. When it comes to the consideration of a new method of treatment for malignant tumors, we must not wonder that a profession with memories overburdened with a thousand and one much-vaunted remedies that have been tried and failed takes little interest in any new method and shows less inclination to examine into its merits. Cold indifference is all it can expect, and rightly too, until it has something beside novelty to offer in its favor. Sixteen years ago, when I began to use the toxins for inoperable sarcoma, I did not expect the profession to adopt the method. I was perfectly willing to wait until its great objection of novelty had given way to time, and my own results had been duplicated and confirmed by other observers. No one could see the results I saw and lose faith in the method. To see poor hopeless sufferers in the last stages of inoperable sarcoma show signs of improvement, to watch their tumors steadily disappear, and finally see them restored to life and health, was sufficient to keep up my enthusiasm. That only a few instead of the majority showed such brilliant results did not cause me to abandon the method, but only stimulated me to more earnest search for further improvements in the method.

More revealing still were the remarks Coley made at the conclusion of a talk he gave at Guy's Hospital in London two years later. In response to prolonged, enthusiastic applause, the stoic's curtain momentarily parted. In a comment engorged with both gratitude and remorse, he remarked, "I have never in my own country received such a hearty reception as you have given me today."

The truth of the matter is that he had never in his own home institution received so hearty a reception, and therein lies a revealing clue as to why the toxins may not have enjoyed wider acceptance. Perhaps the most eloquent American exponent of a rival cancer therapy, radiation, was James Ewing, and Ewing would soon become medical director of Memorial Hospital—would become, in effect, William Coley's boss.

3

BLEAK HOUSE

*If we are going to waste our weary years in bothering
about the conversion of all the skeptics we meet, we are going
to have a busy time of it. Your position in these matters is
so assured that nothing that these new invaders of the subject
can do or say will dislodge you. Let them go their own way
and you will find that in the end it meets your own.*
—Harvey Cushing, letter to W. B. Coley, May 1921

At the beginning of 1896, just as Coley prepared to publish his
first report on a series of patients treated with the mixed toxins, word
arrived from Germany of a wondrous new discovery that blazed so
brilliantly with power and promise that, to all but a few stalwart be-
lievers, Coley's work faded into the washed-out background, like a
photograph taken with too much light. Wilhelm Roentgen—working
in the same Bavarian university town where Friedrich Fehleisen had
identified the microbe that causes erysipelas—discovered a previously
unknown form of radiation, which he promptly called x rays. Few new
technologies have made such a broad impact on so large a portion of
the population in so short a period of time.

Radiation—in the form of x rays and later radium—assumed a scientific and social popularity that, in some cases, evolved into a truly
suicidal cultural craze. As medical historian Stanley Joel Reiser describes it, the public infatuation with x rays inspired cartoons, poetry,
and a new accessory for the fashionable woman about town: x rays of
one's hands, bedecked with jewelry. With the subsequent discovery of
radium in 1898, frivolous, dangerous, and largely unregulated forms
of radiotherapy became commonplaces of the early-twentieth-century
apothecary, including radium-containing candies, liniments, and
creams. One popular patent medicine known as Radithor, a concoc-

tion of purified radium bottled in distilled water, was touted as a cure for dyspepsia, high blood pressure, impotence, and many other illnesses, and reached hundreds of thousands of consumers. Only in 1932, when it became apparent that unregulated consumption of radium led to slow poisoning and decomposition of the body, did the Food and Drug Administration begin to restrict the use of radium in patent medicines.

The naïveté of the public toward radiation, and its unappreciated dangers, reflected to a certain degree the medical profession's own naïveté. The use of radiation as a diagnostic tool won quick, and deserved, acceptance; its role as a therapeutic agent, however, remained very much more ambiguous, although doctors tended at first to tout its virtues more than its potential dangers and helped to foster an overly generous trust in radiation as a harmless and benevolent agent of healing. A prominent Boston surgeon named Ernest A. Codman reported shortly after the turn of the century that 8,000 patients had been x-rayed at Massachusetts General Hospital in Boston without any showing signs of the skin irritation known as dermatitis. To concerns that x rays could harm living tissues and damage the skin, Codman was perhaps the most esteemed voice in a chorus of physicians insisting that x rays were not intrinsically dangerous, and that injuries, when they did occur, were almost always due to operator error.

Whatever other criticisms might be lodged against William Coley, having a closed mind in matters medical was not among them. He swiftly embraced the new technology when the ungainly first-generation instruments became available around the turn of the century; in fact, when the board of Memorial Hospital turned down his request to acquire the first x-ray machines for the institution, he went outside administrative channels to procure two of them for the hospital in 1901 with a gift from his friend Archer Huntington.

But just as he was one of the first doctors in New York to attempt radiation treatment of cancer, he was one of the first to insist on conservative interpretation of the emerging data. Precisely because of his experience with the toxins, Coley more than most physicians weighed the value of any therapy not simply on immediate effects but on the *duration* of the healthy response. As early as 1903, in the journal *Medical Record,* he admonished colleagues to temper the enthusiasm that even he still shared. "We find abundant evidence that the x-rays have an inhibitory action on all forms of malignant tumor . . . ," he wrote. "Yet all these cases have been too recent to be classified as cured. In fact, *sufficient time has not yet elapsed in a single case of cancer treated with the x-ray to justify us in regarding it as cured.*" The caution no doubt won him few admirers for several reasons: first, he openly questioned a therapy that inspired almost faddish devotion among his medical peers, out of all proportion to any demonstrated long-term efficacy and in almost total ignorance of its long-term dangers; second, he seemed not to hold similarly high standards for his own vaccine approach, which would for the next thirty years always be compared against this new technology, and always be found wanting. Such a comparison, unfortunately, was biologically unjustified, because there existed a fundamental difference between the two forms of treatment. Radiation by definition was local; the toxins, for reasons that would remain a mystery for decades, exercised both local and systemic effects. Unlike radiation, the vaccine attacked both the devil that was known, in the form of visible tumors, and the devil that remained unknown, in the form of invisible metastases.

As his view of radiation's virtues became more measured, however, Coley found himself further toward the edge of the medical mainstream of his day, and it is impossible to view the contemporary medical perception of Coley's cancer vaccine without placing it in this larger context. The advent of radiation as a rival technology had an impact on the medicine he practiced, the reception to the toxins, the institutional battles he fought, and the personal antagonisms that ultimately brought him so much psychological and physiological grief. The debate over the relative merits of radiation therapy versus Coley's cancer vaccine offers a fascinating view of the battle between two rival, emergent medical technologies. This battle was never more vigorously joined than at Memorial Hospital in New York, and inevitably became enmeshed in Coley's complex relationship with the foremost pathologist in New York City (indeed, in the entire country), the estimable James Ewing.

Ewing is a pivotal figure in the history of Coley's vaccine. To his admirers and disciples, he exemplified the unyielding, dispassionate, even harsh allegiance to scientific principles that helped make American biomedical research so fabulously productive in the twentieth century. To his detractors, he became seduced by a new technology and fell victim to the very scientific biases he so vocally despised in others. He has sometimes been demonized by Coley's daughter Helen, who has publicly alleged that Ewing virtually hounded her father into ill health if not to death, but the relationship between these two distinguished and occasionally dogmatic men was much more complicated than that.

There was much history between the two, much affection and respect in the early days, and ultimately much rancor and stubbornness from both quarters as the two men grew older. To review their correspondence over the years is to be reminded of a troubled but nonetheless resilient marriage, of partners of like mind who embark on their shared journey on extremely intimate terms, who grow irreconcilably estranged and lacerate each other with a frank and bitter hostility that knows the other's vulnerabilities only too well, who then feel remorse and affect wobbly rapprochements until some trivial incident sets off all over again the cycle of accusation and recrimination. The tragedy is that, under other circumstances, one can imagine a happier marriage of Coley's clinical skills and imagination with Ewing's critical conservatism and insight that might have produced great advances in cancer treatment, might even have given a substantial push to immunology, particularly to the study of cellular immunology. It is therefore worth detouring for a moment to explore their relationship, because it sheds light—thanks to a good deal of surviving and previously unpublished documentation—on just how institutional politics and personality clashes can color the background of a medical debate seemingly waged purely on the basis of scientific merit. In the somewhat narrow field of inoperable bone cancers, two rival technologies—both unproven—came into conflict, and the story of Coley and Ewing is also the story of how one of the courses of treatment emerged intact while the other was essentially abandoned.

Ewing is justifiably revered as one of the towering figures of American medicine in the first half of this century, a man referred to in the newspapers as the "dean of American cancer authorities" and "Mr. Cancer," and known affectionately to his staff simply as "The Chief." Born in Pittsburgh on Christmas Day of 1866, he attended Amherst College and trained as a medical doctor at the College of Physicians and Surgeons, but gave up his private practice after several years—"He had few patients and concerned himself more largely with the laboratory," wrote one colleague in reminiscence—and thereafter dedicated his career to that precinct of medicine that is often said to be best suited to scientific (and dispassionate) intellects: pathology. In the best sense, he was an opportunistic pathologist, studying malaria when troops returning from the Spanish-American War were bivouacked on Long Island, studying pernicious anemia when a close friend was afflicted, and most sadly studying toxemia in pregnancy after that illness claimed the life of his wife Catherine while she was pregnant for the second time. That death, friends and colleagues later maintained, turned Ewing into a tragic, reclusive, possibly embittered figure. "After the death of his wife Dr. Ewing withdrew from society," observed his colleague Fred W. Stewart, who succeeded Ewing as head of pathology at Memorial Hospital. "His social contacts became more limited, and there began a period of almost exclusive devotion to the laboratory and to his teaching. He practically lived in his laboratory, rarely leaving until late at night, reading and writing almost constantly."

A man of modest height, with a long, mournful face and cold eyes, he had sustained a baseball injury that had left him with a permanent limp and then suffered the Jobian indignity of becoming afflicted with an illness that tormented him all his life, an extremely painful neurological disorder of a facial nerve known as "trigeminal tic douloureux," which, judging from pictures taken over the years, seems literally to have pinched and squeezed all the joy out of his face. At one point, in the 1920s, he endured injections of pure alcohol into the nerve tissue in an attempt to alleviate the pain, and he finally allowed Harvey Cushing, the eminent Boston neurosurgeon, to sever the nerve surgically in 1926. Quite apart from the nerve disorder, Ewing had more than his share of personality tics as well, dressing in baggy, old-fashioned clothing, eating the same meal night after night, and popping his set of false teeth out at moments designed most to discomfit whomever he happened to be with, including at least one U.S. president, Herbert Hoover. His complete and utter devotion to work may explain why he claimed never to have realized, until a newspaper article made the obvious even more obvious, that the downtown apartment building in which he lived had become a whorehouse, or at least close enough to provoke a police crackdown on prostitution.

Aside from the fact that both men suffered from chronic and progressive conditions, and the irony that Coley's first scientific paper was about a form of radiation (ultraviolet) while Ewing's was about a cell of the immune system (leukocytes), Coley and Ewing were utterly unalike. Coley cultivated a genial, civil, highly social disposition, eloquent bordering on prolix, an Anglophile adherent to the "moral virtues"; Ewing exhibited a crisp, sharp-tongued, incisive intelligence and, the record suggests on more than one occasion, a bit of mean-spiritedness. Ewing made no secret of his distaste for surgeons; Coley, like many clinicians, never entirely trusted the judgment of men who only looked at slides and never at patients. To the medical community at large, Ewing was undoubtedly the more respected scientist, but Coley was probably the more beloved.

They formed the most important mutual admiration society in cancer research at the turn of the century, at least in New York. Ewing became interested in cancer pathology and diagnosed a number of Coley's cases while working as a pathologist at Loomis Laboratory, an independent lab located in lower Manhattan near Bellevue Hospital. When the lab later became affiliated with the Cornell University Medical College, Ewing came with it—indeed, was the prize plum in the package—and quickly established a reputation for brilliance. Although there exists little documentary evidence to support an oft-repeated suggestion that Coley repeatedly pushed, and apparently finally succeeded, in getting Ewing a "foot-hold" in the pathology department of Memorial Hospital, there can be no doubt that the two men shared a vision of the hospital's mission to which they devoted years of public and clandestine effort—wooing benefactors and scheming to line up the support of members of the board of directors for their ambitious plan. There, by dint of medical skill and political savvy, Ewing began to leave an indelible imprint on the institution. He was, Coley said in 1908, "by far the best man in America to direct cancer research work," and few would quibble with that assessment over the next thirty years.

For a short time, Ewing even assisted on a bit of research tangential to Coley's toxins, and this work merits mention. It was not until 1907, fourteen years after John Ficken got his bracing jolt of the toxins, that scientists in the United States published anything resembling a systematic investigation of the properties of Coley's vaccine. These animal studies were done by Silas P. Beebe and Martha Tracy, both MDs working under the auspices of the Huntington Fund for Cancer Research, which had facilities at the Loomis Laboratory. By then the Loomis lab was headed by Ewing, who in a related bit of research studied a transplantable tumor, known as a lymphosarcoma, to show that this particular cancer was not caused by an infectious agent. The tumor, implanted into dogs, formed the basis of animal experiments testing the antitumor activity of bacterial toxins, and the aim of the work, according to papers read by Beebe and Tracy at the June 1907 meeting of the American Medical Association in Atlantic City, had "the hope of determining the rationale of this method of treatment, and if possible placing it on a more scientific basis."

Over a period of several years, they surgically implanted these tumors into dogs and then treated the animals with a number of bacterial preparations—not just Coley's bacterial mixture, but each component separately, either *Streptococcus pyogenes* or *Bacillus prodigiosus*. They also tested two other common microbes, *Staphylococcus aureus* and the familiar intestinal bacterium *E. coli*. What they learned (and what would be substantiated in tests half a century later) is that *B. prodigiosus* did not merely increase the potency of the erysipelas germ but possessed distinct antitumor activity of its own—indeed, seemed by far the most potent component of the two microbes in Coley's vaccine. The staph germs did nothing to perturb or inhibit the growth of tumors, while the mixed toxins led to "a rapid and complete recovery [regression] of all tumors under treatment."

Tracy and Beebe were outstanding scientists (Tracy went on to serve as dean of the Women's Medical College of Philadelphia in 1920, an unusually grand achievement for a woman in American medicine at that time), and two aspects of the study merit particular mention. As early as 1907, Tracy and Beebe suspected that the tumor-killing effect could be traced to what we now know as bacterial endotoxins—compounds unique to gram-negative bacteria and as specific to each bacterial species as fingerprints; if introduced into dogs (or humans), these unique substances could prove profoundly toxic. With this insight, they jumped several decades ahead of their time. Their other important observation was that, even when injected at a distance from the tumors, Coley's toxins caused tumors to shrink. "Such action, while chiefly local, is at the same time something more than this," they noted, "for it is repeatedly observed that tumors at a distance from the site of injection undergo regression simultaneously with those inoculated, while in one instance the entire treatment was by inoculations at a distance from the tumors."

Chiefly local, but something more than this—this stellar observation, although probably flawed for reasons unbeknownst to Tracy and Beebe, nonetheless separated and identified two crucial phases of the immune response, local and systemic. In their experiments, bacterial toxins stimulated both.

It was a lovely paper, one of the first attempts to give some scientific backbone to what later would be called the "Coley phenomenon." But its findings were undermined by two problems, one scientific and the other financial. The strategy of transplanting lymphosarcoma tumors, while reasonable at the time, is now considered to be a questionable method for studying cancer, and illustrates just how tricky it can be to use animals to study cancer therapies. According to modern biologists familiar with the use of these transplanted tumors in mouse studies, this type of cancer is notoriously unpredictable, and so shrinkage of the tumor cannot always be attributable to the treatment. In addition the tumors, as we now know, were ultimately rejected not as cancers but as foreign tissue transplants; at best the bacterial endotoxin may have accelerated this process of rejection. The other problem was more mundane: these experiments cost more than the Huntington Fund's annual budget and nearly had to be abandoned in midstream for lack of funds. Beebe, a respected but cantankerous biologist, threatened to resign, and this forced Coley into the embarrassing position of requesting a last-minute gift from the Rockefellers, who bailed him out.

But even the Rockefellers began to distance themselves from Coley's scientific endeavors—they absolutely insisted, as a condition of the grant, that the source never be revealed. By this point the Rockefellers were deeply involved in setting up the institute, and they opted to concentrate their resources for cancer research there. Ultimately the lack of funds cut off a promising avenue of research, for as Beebe and Tracy speculated, "the absorption of such dead tumor cells may give rise to some sort of antibody in the body fluids, thus raising the resistance of the animal against tumor cells not yet destroyed by the toxins." This suggestion hints at nothing less than acquired immunity against cancer—a suggestion that remained unpursued.

Whatever Ewing thought of the toxins in 1907, his opinion seems to have turned decidedly negative by 1912, when he published "The Treatment of Cancer on Biological Principles." The full range of Ewing's intellectual gifts are on display in this paper; his great scholarship and incisive critical mind remain a model to any medical age. He seemed favorably disposed to the investigation of biological approaches, but remained rigorous and skeptical. His knowledge of the literature appears prodigious, his ability to succinctly summarize work and identify weak points leaps from almost every sentence, and the wit, on occasional but unmistakable display, is biting and sardonic. What makes this paper particularly revealing, however, is its great sin of omission: Ewing managed neither to mention nor to cite a single instance of Coley's work, which by that time claimed close to 200 durable cures in inoperable and supposedly hopeless cancer cases, using a decidedly "biological" approach. Given that they worked at the same institution and debated these same issues, it is difficult to attribute this to oversight or space limitations. It seems at best mischievously ungenerous. But it is a time-honored practice in science to emarginate an unpopular thesis simply by not citing it.

Ewing was perfectly capable of such intemperate behavior. Although a privately published history of Memorial Sloan-Kettering speaks of him as "quiet, introspective, and an exceptionally able man," a review of press clippings, scientific papers, and correspondence reveals another side of his personality: prickly, opinionated, quick to pull the trigger, whether the issue was scientific, medical, political, or personal. Some of his public remarks, with an odor verging on malice, seem especially inappropriate for someone who occupied so influential a position in American medicine. At the 1939 dedication of the newly constructed $5 million Memorial Hospital at its current site at York Avenue and Sixty-eighth Street, Ewing took the opportunity of this happy occasion to blame labor unions for holding up progress in the war against cancer, going so far as to suggest that working-class people, especially members of labor unions, were less "deserving" of medical care at the new facility than other citizens. "I see in this morning's newspapers a statement regarding the attitude of the labor unions at the World's Fair which describes them as 'tigers tasting blood,' " he told the gathering, according to the account in the next day's *New York Times.* "I greatly regret to think that the progress of the care of cancer should meet at this time a substantial obstacle from a portion of the population who will come here no doubt asking for help and expecting to get just as good treatment as more deserving persons." A self-proclaimed general before the war on cancer was officially declared, he assured a Harvard Club gathering in 1938, "If I had a million dollars, I could find a solution for gastric cancer. . . . If I had another million dollars, I could find a solution for breast cancer. Then there is the form of cancer known as lymphosarcoma, which annually claims many deaths. At present we know nothing about this form of the disease. It is only a name. If I had another million, I could find a solution for lymphosarcoma." Ewing, alas, was among the first to embrace cancer research's fatal equation: money equals a cure.

Nonetheless he was "the pathologist's pathologist," and under Ewing's leadership, according to one admirer, the study of cancer became a "sound, thorough, and virile science." Around 1910 Ewing began to use the department of pathology at Memorial Hospital as a bully pulpit for the rigorous, tough-minded science he preached; almost immediately, his outspoken pronouncements and vinegary demeanor brought him into conflict with Coley. There were at least two scientific reasons that Ewing found himself at odds with Coley. First, Coley believed cancer was parasitic in origin—that is, caused by microbial organisms yet undiscovered (indeed, Coley clung to this belief like a terrier to a favorite bone for the whole of his life, although his reasoning in support of it was highly unpersuasive). Ewing disagreed emphatically, indeed said he personally regretted "extremely" that any member of his cancer staff might hold such a view. Second, Ewing simply did not believe the toxins were any good; he thought they represented badly done science, and told Coley so in no uncertain terms. And these two areas of disagreement became ignited by the sparks thrown off by a third, almost daily source of friction: the use of radiation in the treatment of cancer.

Ewing's administrative star at Memorial Hospital was so firmly hitched to the future of radiation that he could be said to fairly glow with enthusiasm for this form of therapy. He had befriended a Canadian entrepreneur named James Douglas, who had studied for the ministry, then medicine, then mineralogy and geology before rising to the leadership of Phelps-Dodge Corp., the Arizona-based mining company. In 1907 Douglas's daughter Naomi developed breast cancer, and Ewing accompanied Douglas and his daughter to Europe on a heartbreaking pilgrimage in search of the exquisitely rare radium; despite receiving radium treatment in London, Douglas's daughter died in 1910. That same year, Ewing proposed a meeting that would include Douglas, Coley, and John Parsons, head of the Memorial Hospital board, to discuss a plan to reshape the institution and in effect relaunch it as a hospital devoted exclusively to cancer research. Douglas ultimately made a gift of $100,000 in 1912, sweetened by a $390,000 trust that became effective two years later, with the condition that Ewing be made chief pathologist at Memorial.

Around the same time, in 1913, Douglas established the National Radium Institute, which mined 8.5 grams of radium in Colorado, and several years later Douglas donated nearly half of this radium to Memorial Hospital (this at a time when there was said to be barely a gram of the precious mineral extant in the entire United States). Meanwhile, Ewing became president of the Medical Board in 1913, when Memorial became officially affiliated with Cornell Medical College. Reflecting the naive enthusiasms of the day, Douglas himself used radium recreationally, rubbing his wife's feet with a radium-based ointment and drinking so-called radium water—possibly Radithor—as an elixir. He died in 1918—"of pernicious anemia, and perhaps of radium poisoning," according to one account—and Ewing dedicated his massive textbook on cancer, *Neoplastic Diseases,* to Douglas, "a man of ideas, ideals, and affairs."

In 1913, when Ewing became de facto medical director of Memorial Hospital, he exercised considerable power over the course of research at the institution and the publications that emanated from it. Some of the surviving correspondence between Ewing and Coley confirms that the two could be considered coconspirators in the campaign to persuade a divided and reluctant board to support their long-range vision for the hospital as a center committed exclusively to cancer research. But almost from the instant that the desired change took place, Ewing began to undermine Coley's reputation with the board; by 1917 the hospital's former medical director, Richard Weil, privately confided to Coley that Ewing was allowing Coley's detractors to "go out of their way . . . almost maliciously . . . to depreciate the toxins." In addition to this behind-the-scenes campaign, Ewing never passed up a public opportunity to express reservations about the bacterial vaccines and, as the years went by, more and more came to articulate the major criticism of Coley's toxins: that all of Coley's cures were in fact misdiagnosed cases of diseases other than sarcoma. It took considerable gumption for Ewing to press this brief, for a number of those original diagnoses had been microscopically confirmed by Ewing himself when he served as chief pathologist at the Loomis Laboratory. But as Ewing added near the end of one letter, "I often wish you would stop writing so much about the pathology and treatment of bone sarcoma. Everybody knows your views, and your very frequent publications must appear to many like advertising."

Ewing's rise to the post of medical director more or less coincided with Douglas's gift, and in this new position of power he increasingly criticized Coley's vaccine work. Even a brief taste of the correspondence between Coley and Ewing goes a long way toward explaining the deep-seated and probably irreparable divisions between these two headstrong and talented individuals, divisions that began as a scientific difference of opinion but soon spilled over into bureaucratic and personal skirmishes as well. It is worth quoting from these letters at some length because they have never been published before and came at a time when people not only wrote frank letters but seemed heedless of the impression, much less the paper trail, they would leave for posterity. The correspondence gives an unusually vivid picture of the way a scientific dispute unfolds (Ewing, who almost always fired off his letters on half-sheets of the pathology department's letterhead stationery, typically began his broadsides with "My dear Coley"; Coley, perhaps formal or possibly just chilly, invariably began with "Dear Dr. Ewing . . ."). Moreover, the letters illustrate that medical research often becomes inextricably entangled with the politics and bureaucracies of institutions, and modern clinicians will quickly recognize some areas of conflict as pertinent to medical research at the turn of the century as they are today: we read about surgeons quarreling with other doctors, and we see how the neutral statistics of medical research can be influenced by fierce behind-the-scenes battles over such bureaucratic details as the assignment of cases.

Coley, for example, long suspected that Ewing was assigning some of his bone cancer cases to other physicians who were more eager to use radiation, as indeed proved to be the case once in 1924 when he discovered that a boy with bone sarcoma had been sent to one of Ewing's disciples, Frank Adair, then an assistant in the Breast Clinic (and later executive director of the hospital). Outraged that a patient with bone cancer had been assigned to an expert in breast cancer, Coley threatened to resign, and Ewing—uncharacteristically contrite—backed down. "You are quite right about the Feldman case," he admitted. "We tried to put one over on you . . . and it did not work." Ewing, for his part, became increasingly impatient with what he considered Coley's antiquated diagnostic skills, scolding him at one point for confusing benign giant cell sarcoma, malignant periosteal giant cell sarcoma, and myeloma, adding that "these are three well recognized and different diseases, with which your article shows you to be unfamiliar in spite of years of experience." On one occasion, even Coley was forced to agree. "I was never more sure of a clinical fact than I was that the tumor in this case was primary in the abdominal wall," he wrote of a misdiagnosis on April 24, 1917. "However, I now see that I was mistaken and am perfectly free to acknowledge it. . . . This only proves the fallibility of the clinician, which I have never hesitated to accept."

Fallibility was not a concept Ewing welcomed into his vocabulary or from his staff, certainly not when it concerned a diagnosis. One of his most persistent complaints was that Coley misrepresented Ewing's pathology reports. "Your letters are hardly courteous," he wrote on January 15, 1917, complaining about the amount of time Coley's requests to the pathology department were taking. "Moreover, you have not made the best use of my reports, but have used them in the past to discredit me and confuse the whole subject of microscopical diagnosis. I must therefore refuse to cooperate with you in any further attempt to confuse this subject by misuse of pathological reports made on small pieces of tissue. If, however, you are willing to make a proper study of these tumors and to employ microscopical data in the proper relation, my services are at your disposal."

The following day, Coley thanked Ewing for his candor ("It is the first time I have ever had an intimation that I have written anyone a letter lacking in courtesy") and expressed surprise that Ewing's lab felt overworked and underappreciated. "Inasmuch as you have brought up the question of courtesy, and cooperation, and the proper use of microscopic reports," Coley continues, "I should like to mention the dogmatic and arbitrary manner in which you have tried to discredit the accuracy of the diagnosis in cases in which I have from time to time shown, successfully treated with the toxins, in some cases going so far as to say that your own original diagnosis of malignancy was in error. Your hostility has been so open and unfair that it has been remarked on by most of the men who attend the conference [that is, the hospital meetings of physicians to discuss cases]." So filled are these letters with petulance and bickering that it is easy to forget that there lies at their heart a serious and important medical dispute: the accurate diagnosis of cancer cases, which must precede any accurate assessment of treatment. Without access to the contested pathology slides, it is difficult to reach any definitive conclusions. As in many disputes of the sort, there was probably a bit of truth on both sides. There is little question, however, that Ewing equated the reputation of radiation therapy with the reputation of the hospital itself; indeed, around 1915, he instituted what would prove to be a disastrous policy of using radiation as the primary treatment for all bone cancer patients on the ward at Memorial.

This led to one of the bitterest exchanges over radiation, which occurred in 1922. In reviewing a paper prepared by Ewing, Coley hurled the ultimate insult at Ewing, accusing him of behaving in an unscientific manner. "You do not hesitate to lay down very definite rules," Coley wrote, "telling the clinician just how to treat the several varieties of the disease [bone cancer]. If your conclusions were based upon a very careful study of the clinical data and an analysis of the end results of the various forms of treatment (particularly the results obtained at the Memorial Hospital during the past seven years), there would be some justification for *rules and conclusions*; but as it is they are entirely at variance with the end results obtained by the use of radium or x-ray alone, you do not hesitate to entirely ignore them and to advise radium or x-ray as a method of choice in all cases. If it were merely a laboratory problem in which the main object was to determine the effect of radium or x-ray in these several varieties of long-bone sarcoma, there could be no criticism, but, *while saving the life of the patient still remains to the clinician the chief object in the choice of method of treatment,* I think my criticism is well-founded." Even Ewing seemed to agree ("I feel we are not very far apart in the whole matter," he wrote in reply), but he insisted on his right to tell others how to treat patients. "I am not a mere microscopist," he snapped. There in a nutshell is a classic tension of twentieth-century medicine: clinician versus scientist, patient care versus research.

Nowhere is the ill will surrounding the radiation issue more explicit, however, than in a July 1917 communication. In the course of passing along his thoughts on the text of a paper Coley submitted to the hospital's publications committee, Ewing sharply criticizes Coley's understanding of the treatment and prognosis of bone sarcoma as "quite incompetent and out of date," and then adds a damning postscript: "Janeway [Henry Janeway, head of the Department of Cancer Surgery and Radiation Therapy] and I feel that this report will do harm both to radium and to the Memorial Hospital, but we see no way of preventing it except by curtailing your use of radium and this we do not want to do except as a last resort. Gradually we shall get enough cases with better results, and other institutions will report other good results, so that you will be discredited in the end."

In private, at least, there was no pretense. Displaying the same breezy confidence with which they dispensed their medical opinions in public, Coley's superiors at Memorial predicted, and awaited, his downfall. Soon a prestigious scientific project presented a forum for this much-anticipated demise.

In 1920 three of the nation's leading cancer experts organized a research initiative that, by accident or design, had as a secondary, unstated agenda the aim of laying Coley's toxins to final, discredited rest. Ernest A. Codman—the same Boston surgeon who had spoken so reassuringly about the safety of radiation two decades earlier—began the process of organizing a project known as the Bone Sarcoma Registry, with the help and blessings of Ewing and Joseph Colt Bloodgood, a respected surgeon at Johns Hopkins Medical School. Codman had taught at Harvard Medical School, practiced at Massachusetts General Hospital, and become a strong advocate for follow-up studies to measure the true value of medical treatments. He was, in the words of medical writer Lawrence Altman, "decades ahead of his time as a surgeon and reformer of the medical care system."

The purpose of the registry, as Codman outlined it in a July 1920 letter to members of the American College of Surgeons, was to record each and every case of bone sarcoma in a central clearinghouse of information, to insist on the highest standards of diagnosis (including, when possible, microscopically confirmed tissue samples examined by qualified pathologists), to record the form of treatment, and to follow patients diligently over time to determine the success of those treatments, all the while sharing information among participating physicians. Insofar as it attempted to place the treatment of these rare, difficult-to-diagnose, and often swiftly fatal cancers on a firm statistical basis, the Bone Sarcoma Registry represented a step toward modern, statistically meaningful medical research. As a practical matter, it also offered the possibility of vetting William Coley's diagnoses and keeping dubious cases out of the registry.

Codman, however, wrestled with a delicate problem. He found himself in the awkward position of soliciting rather fervently the participation of Coley, who (as Codman himself acknowledged in a follow-up letter) had treated "probably more living cases than any man in the world." A registry without Coley's bone sarcoma cases would from the outset suffer an egregious gap, an almost preemptive blow to its credibility. At the same time, Codman harbored undisguised animus toward the toxins, and made sure Coley knew it. "My personal experience with your toxins has been bad," Codman admitted in his letter, "and you must pardon me if I confess skepticism in regard to your results, if at the same time I admit that I know men whose observation I trust who speak well of your methods. I have the most earnest desire to do you full justice, for I have always felt that your work was most admirable in its scientific spirit and in its earnest desire to relieve these hopeless cases. That your treatment has a profound systemic effect I have no question but I am inclined to attribute the successful cases to errors in pathologic diagnoses." And later: "I feel that you are an admirable and courageous enthusiast."

The vocabulary damns with kindness—"earnest," "admirable," worst of all "enthusiast." With such praise, Codman painted Coley as a much-esteemed colleague whose enthusiasms had gotten the best of him, a true believer in an era that belonged to the tough-minded scientist. Joseph Bloodgood conveyed much the same message in a later letter; he reminds Coley that "when we come to make comparative statistics, we must eliminate all elements of error, especially that due to mistaken diagnosis." It was the same criticism that had dogged Coley since the days of the 1894 *JAMA* editorial. While Coley's critics acknowledged that he had achieved many successful treatments—"I must admit you still have more to your credit than anyone else," Codman almost grudgingly conceded—they promptly reeled their praise back in by intimating that the cases had been misdiagnosed.

If Codman found himself in an awkward spot, so too did Coley. To decline to contribute cases to the registry would be to appear to lack confidence in both his diagnostic skills and his toxins. Yet here were three people with a highly public history of disparaging his diagnoses, and now Codman, Bloodgood, and Ewing formed the three-member committee that would accept or reject cases for the registry. As they continued to parry over ground rules, Coley wrote in a 1921 letter to Bloodgood, "There is only one way in which the evidence which I have presented in this letter as well as in former papers, can be weakened or nullified, and that is by assuming that most of the successful cases were examples of errors of diagnosis. However, one might think that the experience of more than thirty years, enriched by the fortunate opportunity of having personally observed nearly three hundred cases of sarcoma of the long bones, might entitle that person's opinion to some weight in the question of diagnosis." He concluded this same letter saying, "I do object very strongly however, to either you or Codman, throwing out as 'errors in diagnosis' all cases that you have not personally passed upon."

If it was any consolation to Coley, he wasn't alone with these misgivings. The Mayo Clinic declined to participate in the registry, and as Charles Mayo confided to Coley in a 1923 letter, "I think Ewing and the whole group are hopeless from being long set in their ways." At one point in this behind-the-scenes maneuvering, Coley appealed to William H. Welch, director of the School of Hygiene and Public Health at Johns Hopkins and one of the most respected physicians of the era, asking him to review Coley's sarcoma slides to give an independent judgment. "*Confidentially,*" Coley told Welch, "Ewing, for years, has done everything he possibly could to obscure or discredit the toxins, even at the expense of going back on his own diagnosis. His enthusiasm for radium has practically destroyed his ability to see clearly or to think straight." Smelling controversy from afar, Welch would have nothing of it. "I think these men [Codman, Bloodgood, and Ewing] are entirely competent, and that their methods of reaching conclusions will in the course of time yield valuable results," he wrote in a long and collegial letter, adding at the end, "I hate like everything to disappoint you, for I have the highest regard for you personally and for the great work that you have done."

Codman continued to cajole Coley, at turns flattering and antagonizing, and Coley continued to balk at depositing his notes and his slides with the registry. The question of accurate diagnosis was absolutely legitimate, but Coley saw it as one more campaign to discredit the toxins—and one more campaign where the pathologists were calling the shots, not the clinicians. "It is true we may call the disease by different terms, every decade," Coley told Codman at one point, "but the clinical history and the clinical signs of sarcoma of the long bones are the same as they were in 1879 and sufficiently characteristic to enable any intelligent surgeon to make a diagnosis in the great majority of cases." Reluctantly, Coley ultimately agreed to participate. According to the ground rules of the registry, Codman and company would sit in judgment on every new diagnosis Coley submitted. Thus, the stage had in effect been set for a scientific referendum on the efficacy of Coley's toxins. To that public and professional drama, Coley added another, more private concern. As early as 1920, Coley had confessed to Codman the fear that he might not live long enough even to see that day of reckoning.

Against the background of these professional challenges, and unknown to all his colleagues, Coley simultaneously grappled with an unfolding personal drama that caused great distress; oddly, it can also be said to reflect a misdiagnosis on his part of elephantine proportions. Its origins dated back to March 1913, when Coley's longtime patient and patron Archer Huntington received an anonymous parcel in the mail—intended, apparently, to discredit Coley. It contained an anonymous letter suggesting that Coley was a sick man, accompanied by a slender volume called *Acromegaly: a Personal Experience.* The 160-page volume, bound in red leather and embossed in gold print, had been written by an English physician named Leonard Portal Mark, ominously identified on the title page as "late president of the West London Medico Chirurgical Society." Telling his wife that he wanted to check the family farm, Coley traveled up to Connecticut, took a room in the Sharon Inn, and read the volume over the weekend.

It must have been unsettling for Coley to read Mark's dreary roll call of symptoms: "queer feelings" and "queer fevers," headaches that turned his "cranium into a temple of discords," and incredible weariness ("the acromegalic may be overcome by the most intense feeling of fatigue"). Mark described how his feet grew too big for his boots, and how his glove size went from seven and a half to nine and a half; how his eyesight deteriorated, his bones thickened, his tongue became enlarged; how his lower jaw grew so beyond its normal bite that he could no longer chew food properly and suffered constant indigestion. In a passage Coley could hardly have failed to notice, Mark observed that his surgical instruments "are now too small, and prevent my using them as deftly as I might."

Leonard Mark remained in ignorance of his condition (or at least in denial, for he admits there were many clues) until he was forty-nine, at which time Pierre Marie, the French physician credited with discovering the condition, happened to be leading a delegation of doctors during a visit to London and spotted Mark in a crowd. He immediately pointed him out to colleagues as a "typical acromegalic." Now Coley, fifty-one years old at the time and trailing a similar history of clues behind him, may well have nodded in agreement as he read Mark, who noted, "I was exceedingly annoyed with myself for not having made the discovery sooner, and felt I must be branded as an ignoramus for not knowing more about acromegaly."

To see pictures of Coley as a handsome young man, resembling nothing so much as a dashing cavalry officer in a Civil War daguerreotype, is to understand instantly how this revelation must have crushed his spirits, not least because he had failed, skilled diagnostician though he was, to recognize the evidence of disease staring him in the face each day. But the evidence now was unmistakable: a benign tumor, slowly but insistently pressing on the bulb of the pituitary, was in the process of transforming his once handsome face into a mask of drooping skin. Already the bones of his face had become noticeably enlarged. He wrote about it in a journal that was discovered after his death but has since been lost.

"In the journal," his daughter Helen later recalled, "he said that he had never feared death, but he had hoped that it wouldn't be by inches and with great pain, which is just what this terrible book suggested. He said that he'd always hoped to see me grow up to womanhood and see Brad established in his profession, but he was very grateful for all that life had brought him so far and hoped that he could have accomplished more than he had." Nonetheless, the realization that he was an acromegalic came as "a cruel shock," she said on another occasion, and he "determined not to discuss it with his wife, or friends, and so faced it alone, determined to work harder than ever." Once he recognized the disease, Coley also recognized its implications, and he quietly began to tone down his lifestyle and moved his family to a less expensive apartment, anticipating a day when he could no longer maintain his surgical skills.

All in all, by the time he turned sixty, Coley had good reason to feel isolated, alienated, and dismissed by his peers. Those pressures and more—mismanaged finances, the separation and divorce of his son Bradley—built up and came to a head toward the beginning of 1922. When both Coley and Ewing appeared to be slow in sending pathology slides to the Bone Sarcoma Registry, Ernest Codman fired off a scolding letter on February 6, complaining of the lack of cooperation from the two men. "I have not registered a single case from the Memorial Hospital as yet," he fumed. "Ewing has been either too sick or too busy and you have been perhaps too distrustful of me or perhaps too fond of golf. At any rate, you have done more interesting experiments at your institution than have been done elsewhere, and yet your institution has not registered a single case in the Registry!"

Given Coley's habitual overwork, there may have been more than a grain of truth to the accusation; casting it in such insulting terms, however, stung the surgeon, and his reply made clear that it was the pathology department at Memorial Hospital (Ewing's department, that is) that had dropped the ball. Still, here was Coley, the youngest fellow ever elected in his day to the American College of Surgeons, replying to Codman with almost pathetic defensiveness. "As a matter of fact," he wrote, "I have played only nine holes of golf since September and while I believe you had time to go on a duck-shoot this fall, I had to give up several attractive shooting invitations to stay in town and grind, largely on sarcoma of the long bones. I have had nothing but weekend vacations for the last four years, so you see, my shortcomings in regard to the Registry are not due to too much fondness for sports but to too little time."

According to Coley's daughter Helen, Codman's needling letters caused great "pain and worry"; she would later claim in public talks that "Codman's letter was the final straw that precipitated an ulcer with an almost fatal duodenal hemorrhage about February 12, 1922. He never regained his former vigor in the remaining fourteen years of his life, and after further traumatic episodes with Ewing or Codman, he had other hemorrhages."

Coming as it does from a blood relative, that interpretation may understandably overstate cause and effect. Nonetheless, during that February 1922 medical crisis, Coley lost three pints of blood and came very close to dying. At about this same time, he was surprised to learn that the "self-appointed triumvirate" of the Bone Sarcoma Registry (as he referred to Bloodgood, Codman, and Ewing) had elected to throw out all his past successes; "Do not trouble to send us," Codman had written toward the end of 1921, "any slides of material from cases *now dead.*" Only living cases would be considered, and only after the "steel barriers" of "their Registry" consented to include them. Unhappy with the tilt of the playing field but out of the game altogether if he did not compromise, Coley felt he had no choice but to participate under these more restrictive conditions; he must have known that the results would probably decide, as scientifically as medicine in that day could, whether there was any value to the toxins or if he was destined to appear the diagnostic buffoon his critics implied he was. "I am 60 years old," he confided to William Welch, "and according to the present outlook, I may have but a few years, or none at all, in which to furnish this additional evidence required to convince the remaining skeptics."

The hemorrhage confined him to bed for several weeks. To while away the hours, the teenage Helen Coley read aloud to her father. The book he most wanted to hear, she recalls, was *Bleak House,* the Dickens novel about a legal controversy that outlives all its protagonists.

4

THE METHOD OF CHOICE

*Still one may remember that in medical science as
well as in politics, minorities have a surprising way of
becoming majorities in a very short time.
What we need most is more facts rather than opinions.*
—William B. Coley, "Some Thoughts on the Problem
of Cancer Control," *American Journal of Surgery, 1931*

William Bradley Coley arose early, as usual, on the morning of
May 25, 1934, a cool and overcast day in Manhattan, and no doubt
glanced at the papers, as was his habit. For someone born during the
Civil War, it must have been strangely disheartening to read in the
spring of 1934 about yet another bellicose crescendo building in
the world, the papers this time filled with the rumblings of Mussolini
and Hitler, and to realize that the same old human predilection for
aggression had survived into a new century so much more proficient
at expressing it.

But not all the thunder that day traveled from across the sea. Ohio
National Guard troops had opened fire on a crowd of strikers at an
automotive plant in Toledo, and a sheriff in Blue Earth, Minnesota,
threatened to fire tear gas into a crowd of angry farmers protesting a
farm foreclosure sale. The widow of architect Robert McKim committed suicide, and Clyde Barrow, slain two days earlier in Louisiana, was
about to be buried, while crowds lined up around a Dallas funeral
home to view the body of his partner in crime, Bonnie Parker. In news
from the world of science, two amateur zoologists were headed back
to the Bronx Zoo from Indonesia with two specimens of Komodo
dragon, which, according to an account in the *New York Times*, "are
credited with being descendants of the famous prehistoric monster,
the Tyrannosaurus Rex." And in that curious way that newspapers

96

can often miss the news whose impact may affect readers long after gangsters are buried and scandals of the rich fade, medical experts from Minnesota, Baltimore, Boston, and New York would gather at Memorial Hospital later that same day to discuss the best treatment for bone cancer.

By the 1930s the public viewed cancer, along with heart disease and pneumonia, as one of the three most feared diseases. "And most baffling of these," *Time* magazine opined in 1931, "is Cancer, which is rapidly overtaking the other two for the rank of World's Worst Disease." The gathering at Memorial took place more than forty years after Coley's first use of his cancer vaccine, and as he dressed for his trip to the "Bastille" on the Upper West Side that Friday morning, he might well have mused that the meeting represented one of the last occasions at which he would defend the toxins. In many ways, he was the same doctor he had been in 1890—a medical naturalist, a meticulous observer, most of all a conscientious clinician. He planned to argue, perhaps for the last time in so formal a scientific setting, with all his principal critics on hand to disagree, that despite the many doubts that persisted, he could still in certain cases make a cancerous tumor implode upon itself, self-destruct in a slurry of dead tissue, and permanently disappear.

These final years had been particularly bittersweet for Coley. In 1933, when he retired from Memorial Hospital after forty years, there was a gala dinner at the Waldorf-Astoria; the master of ceremonies was John Finney of Johns Hopkins, and on hand to provide fulsome testimonials were the dean of Yale University, Frederick S. Jones, Charles Mayo, and even James Ewing, who graciously attributed to Coley "miraculous cures in substantial numbers all over the world." Stepping gingerly from this bath of adulation, Coley resorted to a self-deprecating form of thanks, noting that he had for several years been fitted with a hearing aid and had "been able to hear but one-half of the flattering tributes the speakers have paid me; had I heard more, I am sure I should have been too overcome to respond at all."

But there had been many distressing moments, too. Not only had he invested all his savings in stocks in the early 1920s, but his broker had persuaded him to buy on margin, and when the market crashed in 1929, he hastily had to borrow $10,000, an enormous sum in those chaotic days, to stay afloat. Thus, at age seventy-two, he was still forced to grind away as a surgeon to pay off debts and make ends meet. He had for many years suffered recurrences of the serious, hemorrhagic duodenal ulcers that had first afflicted him in 1922, and scar tissue from these ulcers caused an intestinal blockage in 1931 that required surgery, which Coley endured under local anesthetic. Finally, there was the acromegaly.

If he looked in the mirror prior to leaving home that day, he could not possibly have escaped noticing the ravages of disease. He looked his age, and more. As his daughter later admitted, his "increasingly grotesque appearance caused people in the street to turn and stare." As a young man, Coley's face was hard and handsome, a silvered cameo of dash and vigor, but had now grown soft and jowly, pathologically so; his lower lip looked swollen and bloated, and his face sagged with folds of flesh. The slow-growing tumor in his pituitary had caused an inappropriately high secretion of growth hormone, and the results were plain to see in the mirror. His emeritus status, the impaired hearing, the bloated features—even his appearance conspired against him, invited his peers to see him as an old and addled bystander shuffling along the shoulder of the highway as younger colleagues sped by in the headlong rush of medical progress. "The work of convincing the world at large or the smaller medical world of the truth of a theory," he'd told some medical students, "requires both time and patience." Patience he still had, in copious amounts. But time—he was running out of that.

He took a cab from his home at 655 Park Avenue—he still always traveled by car, never walked—to Memorial Hospital, where he had first joined the staff in 1893, when it was still called the New York Cancer Hospital. The hospital was located at the corner of Central Park West and 106th Street, and he no doubt walked past the pavilion area, where he had conducted the first erysipelas experiments in 1891, to get to the auditorium. There he would find Codman, who years earlier had confided in a letter that one particular cure claimed by Coley had occurred "in spite of" the toxin therapy. There would be Bloodgood of Johns Hopkins Medical School, proprietor of 45,000 diagnostic slides of cancer and proprietor also of oft-stated doubts about Coley's diagnoses. And of course there he would find the medical director of Memorial Hospital himself, the imperious James Ewing, his onetime friend and coconspirator in shaping the hospital into the research powerhouse it had become.

By now Ewing had become the consummate insider. Back in the 1920s, when John D. Rockefeller Jr. sought advice about the best directions to follow in cancer research, James B. Murphy, head of the Rockefeller Institute, had replied without hesitation, "Back Ewing at Memorial," and Ewing acted with the bravura of a man who knew he had the full confidence and support of the Rockefellers. By 1934 he certainly knew that John D. Rockefeller Jr. had agreed to underwrite the multimillion-dollar relocation of Memorial Hospital to the Upper East Side, and he also knew that Rockefeller was particularly delighted that Ewing had agreed to give up his position at Cornell University Medical College to head this new Memorial. According to *The Legacy of Bessie Dashiell*, the limited-edition history of Memorial Hospital produced by the Laurence Rockefeller–associated Woodstock Foundation, Ewing and Rockefeller "were of similar nature: quiet, introspective, and wholly dedicated to the task at hand. Apparently no such bond existed between John and Dr. Coley, without whom none of this might have happened. Coley was more than a competent scientist, and in some matters was well ahead of his time. But he also was intuitive, and an entrepreneur and promoter of sorts—vitally essential talents for his purpose, but somewhat outside the usual Rockefeller pattern." Coley was the outsider now: old, "intuitive" (i.e., unscientific), belittled for his stubborn belief that cancer was caused by microorganisms and his "advertising," as it was seen, for the toxins.

After years of sniping, medical colleagues usually didn't even bother to invite Ewing and Coley to the same meeting anymore, but they didn't have much choice on this occasion. Indeed, there was a delicious irony at the core of the 1934 symposium: the topic of the day was a rare and extremely lethal form of bone cancer known, after its discoverer, as "Ewing's sarcoma," and the physician who claimed by far the greatest success in its treatment, among virtually all the bone tumor experts in the country, was William Coley.

Of all the events in those last few years, none assumes the importance of the modest and little-known symposium that took place at Memorial Hospital in May 1934—not that it brought greater fortune to Coley or greater acceptance of the toxins, but because it addressed the issue of Coley's vaccine in as scientific a manner as could be accomplished in that day. Indeed, the gathered physicians, among the most skilled and celebrated of their generation, would debate the merits of all bone cancer treatments.

In 1934 the general medical consensus was that amputation represented the best of a bad lot of treatments—"the method of choice," as physicians put it—to use against bone cancer, although there was considerable sentiment for the use of radiation therapy. Ewing's sarcoma was a rare form of bone cancer that develops in the lining of the marrow; because the blood itself is made in the marrow, this particular sarcoma led to an especially insidious complication because it spread quickly and easily via the blood, and amputation was recommended as soon as possible after diagnosis, although many doctors believed radiation to be of value. Coley planned to deliver two messages on the subject that day, both of which contradicted prevailing sentiment. First, he would dampen the general enthusiasm for radiotherapy by suggesting that radiation simply didn't work against bone sarcomas in general, citing a series of more than 100 patients at Memorial Hospital in which not a single cure had been effected. (Coley's son Bradley, who succeeded his father as head of the Bone Tumor Service at Memorial Hospital and could hardly be called a proponent of the toxins, reinforced this message; at the symposium, he would report that in 70 cases of Ewing's sarcoma treated at Memorial by radiation alone between 1923 and 1933, 65 were dead, and of the 5 who remained alive and disease-free, only 1 had passed the five-year mark). Second, not surprisingly, Coley would take the opportunity to point out that the toxins performed quite differently.

Coley opened his talk by tossing an olive branch—but a barbed one—in the direction of Ewing, acknowledging his seminal early work on this form of cancer. "While many of the earlier writers discussed *endothelioma* or *endothelial myeloma*," Coley began, "Ewing was the first, in 1921, to give an accurate histological and clinical description of these tumors, and to point out that they merit a distinct classification and possibly a different form of treatment," he stated. Then Coley used Ewing's own words—that "an infectious agent may be connected with it"—as an excuse to climb once again on his wobbly hobbyhorse: that cancer was caused by microbes.

Having disposed of that obligatory bit of business, Coley went on to summarize the long-term results of the toxin treatments in cases of bone cancer. To physicians treating these cases, to patients suffering from them, and to scientists (or anyone else) curious about the role of the immune system in treating disease, the results inspired two paradoxical reactions: great interest and great caution. To begin, Coley discussed 44 cases of Ewing's sarcoma that had been accepted into the Bone Sarcoma Registry. Twelve had been treated by other doctors with radiation; none had survived five years. Coley himself had treated 32 cases with the bacterial vaccine, and 12 remained free of disease from five to twenty-one years after treatment. To punctuate the point, Coley introduced two of these long-term survivors to his colleagues at the symposium.

The first was a young man of twenty-two known as "H. S." In March 1920, as an eight-year-old boy, he had been brought to the Hospital for the Ruptured and Crippled with a rapidly growing tumor in the fibula, one of the long bones below the knee, which was diagnosed as round-cell sarcoma. By the time Coley first saw the patient in May, the cancer appeared to have spread to lymph nodes in the groin; despite this hint of metastases, Coley recommended amputation and then treated the boy postoperatively with toxins. By October, a very large tumor ("the size of a child's head," Coley said) developed in the hip bone, and chest x rays confirmed fears that the cancer had already spread to the lungs. A radium pack was applied to the hip, but the nodes in the lung were left untreated. Within several weeks, however, the mass in the groin diminished and finally disappeared, and the boy bounced back to normal health. Coley preserved the microscopic slides from this case; James Ewing later pronounced it Ewing's sarcoma, and it was accepted into the Bone Sarcoma Registry.

An even more remarkable recovery occurred in Captain George Brodhage, a small mustachioed and bespectacled man, hair slicked and parted in the middle, who appeared on crutches in the Memorial auditorium. The improbable saga of this man, known to the medical world as "G. B.," was accepted into the Bone Sarcoma Registry in 1926 and became widely known as the Christian and Palmer case, after the two Public Health Service surgeons in New York who first treated him. G. B., a thirty-two-year-old mariner, underwent an amputation of his left leg at mid-thigh in September 1925 after a bone biopsy suggestive of cancer; the disease was later diagnosed as Ewing's sarcoma by, among several doctors, Ernest Codman and James Ewing himself.

Within three months of amputation, metastatic tumors began to crop up, the first just above the belly button, a second near the groin, and a third, fist-sized growth in the stump. Beginning on January 5, 1926, Drs. S. L. Christian and L. A. Palmer administered Coley's toxins for six weeks, mostly shots into the buttocks, with a few directly into the tumor growing in the stump. All three growths diminished in size. Over the next several months, when the tumors began to rebound, the doctors recommenced toxin therapy. By May and June of that year, after the shots had again been suspended, the tumor in the stump rebounded dramatically until the circumference of the leg reached an astounding thirty-one inches, nearly the size of a man's waist, with metastatic growths cropping up in the clavicle, scalp, cranium, and neck vertebrae. In the face of this overwhelmingly grim situation, Christian and Palmer began a third concentrated course of toxins on August 5, 1926, and all these tumors again regressed.

This time, the patient seemed to have turned a corner. By late November he had gained thirty pounds, all the tumors had disappeared, and he left the hospital on December 5. The following year, at Coley's suggestion, the patient received two booster courses of injections. He received no treatment other than the toxins, and he remained alive and disease-free at the time of the 1934 symposium (indeed, he was alive when last traced in January 1953). Not only did the case demonstrate the effectiveness of Coley's toxins against advanced metastatic cancer (the man received no other form of treatment) but it suggested that repeated administration of the vaccine at selected intervals over an extended period of time could nudge an utterly hopeless case back toward health.

Anecdotes do not make for good science, and Coley knew this, too. So he presented the results of 115 cases of Ewing's sarcoma that had been treated at Memorial Hospital and the Hospital for the Ruptured and Crippled to illustrate the point that, at the very least, the best statistical evidence available did not support the use of radiation alone. In 56 cases deemed "operable" by doctors, where there was an attempt to avoid amputation at first, 26 patients were treated by radiation alone; none survived five years. Another 26 were treated with radiation and Coley's toxins; 2 survived five years. Four other patients were treated with toxins alone, he continued, and 2 had survived five years. Another 13 operable cases underwent amputation followed by a prolonged course of toxins; 7 of them, or more than half, survived five years. In other words, there were 11 five-year cures in this series of 69 operable patients; Coley's toxins were a component (usually with amputation) in all 11 cures, radiation in only 2. Perhaps not coincidentally, it had been suspected as early as 1915 that radiation suppressed immune function.

The results in "inoperable" cases—that is, cases where the cancer was inaccessible to surgical intervention (and thus with grimmer prognoses) or where the patient refused amputation—were, if anything, even more remarkable. Twenty-four inoperable cases were treated by radiation alone at Memorial Hospital; 21 were dead, and the other 3 had undergone treatment too recently to evaluate. By contrast, 22 patients had been treated with the toxins, alone or in combination with radiation. There were 12 five-year cures.

It cannot be emphasized strongly enough that these results did not by any stretch of the imagination constitute statistically significant proof of anything. For the 1930s, they represent an admirable attempt at meaningful biomedical investigation, but the modern clinical trial—and the methodological constraints imposed upon such a trial to provide a meaningful statistical conclusion—did not even begin to emerge until the 1950s. Coley might have been forgiven for portraying these as definitive results, but in fact he did not. His critics by the same token might reasonably have argued that the results were not statistically significant, and they too would have been correct. But even an untrained observer, attempting to discern an encouraging pattern in such admittedly inconclusive results and a clue about which direction to pursue next, might reasonably have inferred two very distinct and obvious trends in the data: that radiation alone seemed to offer no beneficial effect against these bone cancers and that Coley's cancer vaccine, alone or in combination with another form of treatment such as amputation or radiation, appeared to offer benefit—in some cases, significant, long-lasting benefit.

"The results obtained by early amputation followed by prolonged treatment with Coley's toxins," Coley told his colleagues, "would seem to justify us in regarding this as the method of choice." But knowing full well that skepticism about the toxins existed almost independent of the data, Coley proposed a way to address this impasse. He told the symposium that he was "willing to have the question settled by a comparative study of the five-year cures listed in the Bone Sarcoma Registry. If more than half of the five-year cures of endothelial myeloma and a large proportion of the osteogenic sarcomas are my cases, and all have been treated with toxins, this fact would seem to establish the value of the toxins far more conclusively than could any argument one might advance."

In the proceedings of the 1934 meeting, which were later published in the *American Journal of Surgery*, there is no record of a response to Coley's proposal. But reading the papers at a distance of half a century, one is struck by the failure of the top cancer researchers in the country to come to grips with the bald contrast in the data. It does not exaggerate to suggest that the conclusions reached in that room on that day could have had an impact on the treatment of patients with not only Ewing's sarcoma, an admittedly rare cancer, but by extension other bone cancers, other sarcomas, and perhaps other more common cancers, to say nothing of the impetus even occasional successes might have given to basic researchers curious about the immunological mechanisms responsible for these unusual regressions. Which is precisely why James Ewing's comments are so illuminating to the clash of medical cultures on display at the symposium.

Ewing spoke after Coley. He seems to have grown fond of the sound of his own authority, to judge from his crisply autocratic proclamations in public forums, and his peremptory remarks at the Memorial symposium toe that same hard line. In a characteristically brief, blunt, and aggressively opinionated talk, Ewing confined his remarks not to treatment but to a tangential theme: the use of biopsy in diagnosing sarcoma (this must have been of more than passing interest to Coley, whose fateful biopsy of Elizabeth Dashiell began his lifelong fascination with bone cancers). Ewing was against biopsies, in no uncertain terms. He believed they were unnecessary and dangerous (because they might promote the spread of the disease) and should be an intervention of last resort, not least because they interfered with the optimum administration of radiation. But in the midst of his very short talk, Ewing made the following extraordinary statement: "I venture to urge that every case of persistent unexplained pain in a bone should be regarded provisionally as sarcoma and treated by radiation." To the suggestion in Coley's data that over a hundred patients had been treated with radiation at Memorial Hospital without a single five-year survivor, there is no record of Ewing's response.

Diagnosis—even microscopic pathology, even by the expert pathologists of the Bone Sarcoma Registry—is not an infallible art, and it is not possible to state with any certainty that the data presented at the 1934 symposium would stand up to modern scrutiny. Indeed, there is good reason to treat it with caution. But assuming that the organizers of the Bone Sarcoma Registry had sufficient medical confidence in their own guidelines, the emerging data would at the very least seem to invite a reappraisal of clinical thinking. Radiation alone resulted in not a single definitive remission; Coley's vaccine was involved in roughly 50 percent of the five-year responders, including 12 of 22 inoperable sarcoma patients—many of whom, although not all, had satisfied the criteria for inclusion in the Bone Sarcoma Registry. Predictably, many physicians chose to remain unpersuaded by the evidence. Bloodgood reiterated his belief that the evidence "justified" radiation treatment, and Channing Simmons of Boston insisted that the value of Coley's toxins "remains to be proved." "Few surgeons have been able to obtain such good results as Coley," Simmons said. "On the other hand, Coley has probably had the largest experience in the treatment of bone tumors of anyone in this country."

But Ernest Codman, for so long Ewing's straw man in questioning

Coley's methodology, seems to have undergone a religious conversion after Coley's presentation. In his postsymposium summation, he retreated almost wholly from two decades of adversarial sniping, and even Ewing must have been surprised by the totality of Codman's conversion, which he signaled by saying that with the use of the toxins "occasional miracles have occurred, and in Coley's own undiscouraged hand these miracles have not been infrequent." But Codman was only warming up.

"It seems to me to be beyond a doubt," he continued, "that he [Coley] has convincing evidence to show that the mixed toxins are of value in the treatment of sarcoma, and especially in the type now called Ewing's sarcoma, with which his paper deals." Perhaps of even greater significance were the following remarks by Codman, recorded for posterity in the *American Journal of Surgery:* "Just as it has seemed quite justifiable for the Memorial Hospital during the last decade to test out the value of the radiation alone in inoperable cases or in patients opposed to operation," Codman wrote, "so it seems even more indicated that some great clinic should try out Coley's toxins during the next decade. Unquestionably they produce a profound constitutional effect, and their administration is followed by a marked increase in the production of lymphocytes. It may be that the activity of the lymphocytes accounts for the occasional miracle which follows this treatment. It is time for some great hospital to apply its laboratory resources to the wholly justifiable and distinctly hopeful purpose of giving this treatment a fair trial under favorable conditions."

A fair trial under favorable conditions. That was all Coley had desired.

Codman's remarks came as a remarkable endorsement of Coley's toxins. *The occasional miracle.* But even endorsements are just expert opinions: the important thing is that Codman recommended exactly the kind of trial that spectators of medicine assume lies at the heart of the scientific enterprise: something cannot *appear* to be true, but must be proven through well-designed experimentation. Codman's recommendation begs an obvious question: was such a trial attempted?

The answer is no, and the reasons are complicated. One possible explanation came from Henry W. Meyerding of the Mayo Clinic. According to Meyerding, the treatment of choice at Mayo for bone cancers was amputation followed by postoperative irradiation and use of Coley's toxins; of 28 patients with osteogenic sarcoma in one study, he reported that 14 remained alive, and 7 had survived for more than five years after surgery. "In every case, according to the custom started by Dr. Charles Mayo, when the patient goes home, I write the local doctor that toxins are expensive but should be given," Meyerding told the gathering. "Invariably, when I write to a doctor, he says, 'what is the use of giving toxins? they are not good.' With this prejudice against the toxins very few will take them." Prejudice and expense undoubtedly contributed to reluctance—nonscientific reasons, to be sure, but pervasive and pragmatic ones that could not be ignored.

Even so, why didn't "some great hospital," in Codman's words, accept the challenge to undertake a scientific trial of the toxins? Why not at the Mayo Clinic, where the toxins had been used successfully for more than two decades? The cost of such a study during the Depression might be one reason (Meyerding: "I too would like very much, if the economic situation would permit, to try, under real scientific control, in a carefully selected group, radium, x-ray and toxins alone.") Or why not at Memorial Hospital, where William Coley had spent much of his distinguished career? The history of sarcoma treatment is a silent answer. Following the 1934 symposium, the preferred treatment remained amputation for many years. It is safe to surmise that with a depression, a world war, and medical questions of more urgent resolution on the horizon, the proposed study was never attempted. By the end of World War II, the rise of chemotherapeutic agents offered the tantalizing possibility of preferentially targeting cancer cells for elimination, and these drugs ultimately became the next new promising breakthrough in treatment. Indeed, the belief in chemotherapy has been especially rewarded in the treatment of bone cancers, because the drug methotrexate has been recently shown to increase the five-year survival rate in osteosarcoma to approximately 85 percent.

Part of the answer may lie in the direction that basic immunological research had taken since the beginning of the century. As Codman implied in his remarks, lymphocytes may have been involved in the action of the toxins, but immunology had still not cleared its head from what would be a fifty-year hangover dating from the debate between the cellularists and the humoralists; lymphocytes were white blood cells, but as late as the 1960s, there were papers in the literature that referred to lymphocytes as "auxiliary cells," certainly not as important as antibodies in the immune response. Oddly enough, James Murphy of the Rockefeller Institute had done some of the most ambitious and imaginative explorations of the role of lymphocytes in infectious disease and cancer, but he was a great admirer of Ewing, not Coley.

Otherwise the seed of Coley's insight fell on sterile intellectual ground; it would not be until 1942 that Karl Landsteiner and Merrill Chase would first suggest a prominent role for white blood cells (also known as "mononuclear" cells) in immunity, and not until the 1960s that the immunological role of these cells would begin to receive more than passing attention. Arthur Silverstein argues in his history of immunology that this delayed appreciation for cellular immunity cannot be ascribed to technical or intellectual limitations, but rather to the fashions of scientific research at that time—and, one might add, to scientists and doctors like James Ewing who set those fashions. "The pioneering cell transfer experiments of Landsteiner and Chase," Silverstein writes, "establishing the critical role of mononuclear cells in cellular immunity, were well within the technical competence of investigators early in this century. But the notion of cellular immunity was out of favor, and few investigators in that environment were stimulated to pose the questions that might have led to such studies." More even than Coley, Codman explicitly posed the question, but no one in American medicine felt an urgent need to address it.

Part of the answer undoubtedly resides with Coley himself. First, he professed no particular taste for the harsh and sometimes vicious debate; "I will leave the answering of such arguments," he told an audience in 1909, "to others with greater love for disputation than myself." Second, in his passion to find treatments that would help his many bone cancer patients, he left himself open to criticism from the increasingly scientific wing of medicine; Almroth Wright, a leading British immunologist, dismissed Coley's approach as "unscientific." Finally, to the claim that the vaccine worked inconsistently or unpredictably (and often not at all), he offered no answer, and had none to offer. For the hundreds of inoperable cancer patients treated with Coley's toxins over the first fifteen years of their use, there was never a standardized formulation for the medicine (at least fifteen different preps were used), never a standard dosage, never a standardized form of administration (sometimes injections into the tumor, sometimes remote from the tumor), never a recommended amount of time for duration of treatment, never an accompanying and ongoing series of animal experiments to test dosages and variations in the treatment protocol. There was not even a standardized form upon which to record the results. Although Coley's very name (and, inextricably, his prestige) was attached to the treatment, especially to the weak and ineffectual "Coley's Fluid" marketed by Parke-Davis for so many years, he made puzzlingly little effort at what we would today call quality control. When many other physicians complained that they could not replicate his results, it never occurred to Coley until very late in the game that inconsistent preparations could have led to so much damaging confusion, though the confusion he always acknowledged.

"What is the explanation for these variable results?" he himself asked at one point. "Why should the toxins behave so differently in these cases, causing some of the very worst and most hopeless ones to become permanently cured, and others showing little or no effect?" Coley's answer to that question, unfortunately, undermined even further his cause. He believed—and stated, repeatedly, from a paper in 1896 to the last one he wrote, published forty years later—that the cause of cancer was microbial. His insistence on what appeared to be a wholly inaccurate and certainly unfashionable theory did not enhance his credibility, nor did his fascination with a cancer researcher named Glover who was later revealed to be a quack. By the time Coley attended the 1926 international symposium on cancer at Lake Mohonk, New York, his daughter, who accompanied him, realized that he was treated as someone on the fringe of cutting-edge research.

Another part of the answer rests with the medical establishment of that era. Paul Starr has written that in the "crucial decades" after the Flexner report (roughly between 1910 and 1930), the medical philosophy—and funding patterns—of the Rockefellers' General Education Board and other foundations favored only certain medical schools and institutions, consolidating medical authority in a few hands. "Though the board represented itself as a purely neutral force responding to the dictates of science and the wishes of the medical schools," he writes, "its staff actively sought to impose a model of medical education more closely wedded to research than to medical practice. These policies determined not so much which institutions would survive as which would dominate, how they would be run, and what ideals would prevail." Extensive correspondence between John D. Rockefeller Jr. and his medical advisers makes clear that the family's longstanding support of Coley's research, begun in 1896, had wavered by 1912, when the Rockefeller Institute's Board of Scientific Advisors concluded that research proposed by Coley was "not of such promise as to be worthy of support." Simon Flexner, head of the institute, reportedly felt that "that kind of study would be largely a waste of money." The institute's energies were instead directed toward chemotherapy and synthetic drugs. The Rockefellers, meanwhile, clearly endorsed the Ewing approach to scientific medicine at Memorial Hospital, and Ewing endorsed radiation therapy; the fact that Coley was a "mere" medical practitioner placed him at odds with the prevailing emphasis on laboratory research.

Influential figures like Ewing seem in retrospect so seduced by the clean, scientific nature of a new technology like radiotherapy that they suspended scientific judgment in assessing its efficacy and persisted in testing radiotherapy against certain cancers long after it was clear that it exerted no positive effect; later on, even Ewing's closest colleagues in effect apologized for The Chief's almost slavish devotion to radiation. As Fred W. Stewart admitted in a memorial tribute to Ewing in the 1940s, "Whatever the future of radium or of radiotherapy may be, whether these methods grow in usefulness or become discarded, it must be admitted that Ewing insisted on their trial to the utmost. . . . He was most antagonistic to surgery for surgery's sake, to extensive surgical work in the face of incurable disease, and to surgical intervention by surgeons who lacked understanding of the natural history of cancer or the total inefficacy of their ill planned procedures for interrupting this natural course. He was less vocal in the face of ill planned or ill conducted radiotherapy but justified it in the name of clinical experiment."

Finally, one cannot exclude the role played by the politics and personalities of medicine—not a definitive role perhaps, but a factor impossible to dismiss. Directions in science are often charted and pursued before the data become unambiguous, and there is always the possibility that the defining personalities and defining institutions of any given era can set the compass heading by dint of forceful leadership when definitive scientific evidence is still wanting. Anyone who believes that scientific discourse is dispassionate, impersonal, and by definition objective is invited to read Silverstein's engrossing account of the highly personal feud between those fabled microbe hunters Robert Koch and Louis Pasteur, their respective institutes, and their proxies. Recent scholarship suggesting that Pasteur may have dissembled about methods and data and stolen ideas from competitors reminds us that even when dealing with the most towering and celebrated intellects, the quality of the data—and not the reputation of the scientist generating it—must always assume paramount importance.

Given the immense promise of certain immunotherapeutic approaches currently under investigation, and especially given the possibility that a few lives might have been much prolonged if Coley's approach had held up upon wider testing, about the best that can be said about the 1934 symposium is that medicine was presented with excellent clinical motivation to explore a promising complementary treatment to therapies now widely conceded to have fairly limited success in the treatment of certain cancers—namely radiation and chemotherapy—and an excellent research opportunity as well to learn more about the immunological cells and molecules involved in this reaction. It was an opportunity lost, probably forever.

The question left hanging over the meeting—a question of the sort that medicine normally rushes to address whenever there is an important hypothesis to test and a reasonable way of testing it—remained unanswered. Why were physicians so resistant to Coley's approach, and why was it ultimately never tested? Those questions were no less pertinent in William Coley's day as they are to us now, and in a hauntingly puzzled manner Coley himself tackled them head-on at the 1934 symposium:

"I have had no difficulty in getting the family doctor to carry out the treatment under my direction," he said in reply to a question from the audience,

> but I freely admit that there are many difficulties in the way of persuading surgeons in general to undertake the treatment as a routine measure. First, there is the lack of beds in most hospitals to permit the treatment to be carried out there. Every surgeon is anxious to have his beds occupied by acute cases, with a rapid turnover, and is unwilling to have them filled with sarcoma cases that require a number of weeks or months for treatment. In the second place, the surgeon, probably, has had no experience with the toxins, and he is unwilling to give the time to looking up the subject and learning something about the method and end-results. It is a lot easier for him to send the patient to a radiologist who is always near at hand, and thus be relieved at once of all responsibility. It takes a great deal more time, thought and perseverance, to treat a case with toxins than by a simple amputation or by irradiation; but if a careful study of the end-results of the Bone Sarcoma Registry shows not a single case of osteogenic sarcoma or endothelial myeloma cured by irradiation alone, and a very considerable number cured by toxins alone or by toxins combined with irradiation or surgery, then I believe the profession should feel that their duty to the patient requires them to learn more about the method.

If Codman's remarks proposing a large clinical trial hung over the meeting like a challenge, Coley's remarks about "duty to the patient," viewed over half a century of but limited successes in the treatment of cancer, now read like an elegant, carefully worded Hippocratic curse. Certainly by accident, perhaps even haphazardly, he had stumbled upon a treatment that seemed to help otherwise hopeless patients, and although he could never explain when the vaccine would work, or why, or for how long, that sense of "duty to the patient" obliged this unusually conscientious clinician to continue its use, at considerable cost to his reputation, his health, and ultimately his legacy. Among the singularly lethal cancers he treated, probably no other surgeon in the world could claim as many miracles. Even if they were only occasional, Coley may be forgiven for having thought that surely *any* miracle against such a grim and often hopeless disease deserved more active and systematic investigation than a few laudatory and long-forgotten remarks buried in a dusty journal.

William Bradford Coley never proved that the mixed toxins that bore his name truly worked. But against a disease where more conventional therapies in his day placed a dismal second to his approach, Coley's toxins were never formally tested as a possible treatment for bone sarcoma. As we shall see, in 1977 the American Cancer Society committed an unprecedented $2 million for clinical tests of a drug called interferon against cancer on the basis of preliminary tests on only 14 human patients. In 1985, amid a firestorm of publicity, the National Cancer Institute spent perhaps $10 million to test another anticancer therapy, the combination of interleukin-2 and LAK cells, after 11 of 25 cancer patients responded to the combination therapy; to date, only one of those patients has survived as long as the dozen or so Coley patients in the Memorial Hospital series.

We are not privy to all the medical decisions that were made in the wake of the Memorial Hospital symposium—decisions at the level of the medical community, where consensus must sometimes be forged out of paradoxical or incomplete information; decisions at the level of institutions, where the study urged by Codman and others might have been pursued; or even at the level of the bedside, where each and every day, doctors presume to honor the trust placed in them when they make profoundly fateful decisions about which treatment offers the best hope to seemingly hopeless patients with bone sarcoma and other deadly cancers. We do know, however, that no lengthy, in-depth, prospective study has ever been conducted to prove or disprove the efficacy of Coley's toxins. We do know that even James Ewing, in the fourth edition of his classic textbook on cancer, *Neoplastic Disease,* published in 1940 and the first to appear after the 1934 symposium, finally admitted that "radiation alone is generally unsuccessful with typical osteogenic sarcoma"; that "in some recoveries from endothelioma of bone, there is substantial evidence that the toxins played an essential role"; and that "Coley's toxins have been used with other methods in certain cases of osteogenic sarcoma which recovered." Despite all that, he added, "I have been unable to form any definite estimate of the part played by this agent in this disease."

We know there appeared a rehabilitating account by the Council on Pharmacy and Chemistry in the *Journal of the American Medical Association* that, some forty years after the same journal had written off the toxins as snake oil and its inventor as a shameless hustler, cautiously informed physicians that recent reports from the lab seemed to justify a role for vaccine therapy in cancer cases. "It appears," the council report reads, "that (1) undoubtedly the combined toxins of erysipelas and prodigiosus may sometimes play a significant role in preventing or retarding malignant recurrence or metastases; (2) occasionally they may be curative in hopelessly inoperable neoplasms; (3) probably their value is rather strictly limited to tumors of entodermal [sic] or mesodermal derivation and, more particularly, in the case of bone tumors, to those exhibiting little or no osteoplasia. The Council has, for these reasons, retained Erysipelas and Prodigiosus Toxins-Coley in New and Nonofficial Remedies, with a view to facilitating further studies with the product, especially in connection with its use as a prophylactic in conjunction with conservative or radical surgery. Its use in definitely inoperable cases may be quite justified, in many instances, as a desperate attempt to combat the inevitable."

And finally we know that, although the message of Coley's toxins had by now been transformed from a whisper of nature into a murmur of promising data, fewer and fewer physicians seemed inclined to listen.

According to one retrospective analysis of Coley's cases, the vaccine obtained cures in at least 20 percent of the advanced sarcoma patients, numbering in the hundreds, treated by Coley and other physicians. If, as his critics have implied, Coley failed science, perhaps science failed Coley a little too, by not accepting the scientific—and, one is tempted to add, *ethical*—imperative of giving the vaccine the "fair trial" Ernest Codman proposed.

෪

Coley left no record of his reaction to the Memorial symposium and its aftermath. His daughter says he was immensely pleased by the discussion, but he didn't have much time to savor it: he had less than two years of life left, and they were marked by illness, financial pressures, personal despair, and but a few triumphs. Among the latter were his election as an honorary fellow to Britain's Royal College of Surgeons, an honor conferred previously upon the likes of Harvey Cushing, George Crile, and the Mayo brothers, Charles and William. John D. Rockefeller Sr. contributed $250 so that Coley could make the trip to England, and William Mayo wrote a congratulatory letter after the *JAMA* editorial that read in part, "I was pleased to see that the American Medical Association has finally recognized the character of your work with regard to the serum you recommended so many years ago and which the profession was slow to accept as having actual curative value—a fact which, as you know, we have demonstrated in the Clinic."

But in the winter of 1935 Alice Lancaster Coley became ill and was diagnosed with colon cancer. Coley's daughter secretly sold part of his rare book collection to raise the money to pay her medical bills. Nothing speaks louder of Coley's own conviction that his vaccine had distinct and limited indications, specifically to sarcoma, than the fact that he apparently never attempted its use on the carcinoma ravaging his beloved wife of forty-five years.

The following year, while the still-ailing Alice visited Atlantic City with her two sisters in April 1936, Coley himself became seriously ill. Still operating to pay the bills, the fleshy effusions of acromegaly now superimposed on the sag of aging flesh, he may have suspected the end was near, but it came more swiftly than he could have imagined. On April 14 he'd had lunch with his daughter Helen and returned to his office, but fell suddenly ill with sharp intestinal pains. Despite his protests, he was admitted to the Hospital for the Ruptured and Crippled by Eugene Pool, his physician; while in the hospital, Coley dictated the final changes in what would be his last scientific paper. It went off to the *Glasgow Medical Journal* on the SS *Aquitania*.

The following day, when the pain became very intense, Pool insisted on emergency surgery. That afternoon Coley returned to the operating room where he had performed surgery for so many years, this time as a patient. Under local anesthetic, with his daughter Helen holding one hand and his son Bradley the other, he underwent an exploratory operation. From behind the sheet that blocked his view of the proceedings, Helen recalls, her father demanded to know, "Well, what are you finding?" Pool was finding irreparable woe: extensive gangrene had attacked the visceral tissue, due in part to adhesions from a previous operation that cut off the blood supply to the gut. No one needed to explain the implications to Coley. Failing rapidly, he died at the hospital around 2 A.M. on the morning of April 16, 1936, at age seventy-four. John D. Rockefeller Jr. sent flowers to Coley's widow and paid a visit to the home the day her husband died, and James Ewing was an honorary pallbearer at the funeral. Misunderstood to the end, Coley was eulogized by the *New York Post*, damningly, as the "leading American exponent of the germ theory of cancer."

5

THE COLEY PHENOMENON

The day will yet come when posterity will be amazed that
we remain ignorant of things that will seem to them so plain.
—Seneca, quoted by William Coley in
"The Idea of Progress," 1920

For *someone known to be* an unreconstructed optimist, William
Coley proved to be an extremely unlucky fellow, even in death. His
old nemesis James Ewing died in 1943; a "pathologist's pathologist" to
the very end, Ewing diagnosed his own fatal bladder cancer, and with
his passing one of Memorial Hospital's most persuasive voices in favor
of radiation therapy was silenced. But the influential post of medical
director went to a physician named Cornelius P. Rhoads, who during
World War II headed the medical division of the army's Chemical
Warfare Service. The provenance is telling: the next great development in cancer treatment derived from the chemical agents that made
up the poisonous mustard gases first used in World War I, and Rhoads
became a leading "evangelist," in the words of one historian, for their
use in what we now know as chemotherapy.

Rhoads appeared on the cover of *Time* magazine in 1949, hailed as
"an outstanding symbol of medicine's determined campaign" against
the disease. At one point in 1942, Rhoads seriously entertained the
idea of conducting a trial of Coley's vaccine, as initially recommended
by the chairman of the 1934 symposium at Memorial. He made plans,
according to surviving correspondence with Coley's heirs, to "proceed
immediately" with the production of experimental batches of the vaccine at Memorial, outlined a series of animal studies, and proposed
treating a human case of lymphoma or breast cancer (in addition,
Gregory Shwartzman, a respected scientist at Mt. Sinai Hospital and
the researcher who discovered the "Shwartzman reaction," apparently

agreed to participate in the experiments, volunteering to perform laboratory tests for free).

For some reason, however, Rhoads changed his mind, and by 1949, when *Time* paid its visit to Memorial Sloan-Kettering, Coley's toxins did not rank among the experimental strategies considered promising enough to merit mention in the story, which in addition to chemotherapy noted the use of penicillin-like molds and viruses as potential cancer treatments. Moreover, Rhoads had apparently contacted Parke-Davis in 1950 and suggested that they needn't continue production of the toxins, as Memorial would undertake the task. Parke-Davis, after half a century, dropped "Coley's Mixed Fluids" from its product line. When Memorial later ceased to make them, no source in America existed to provide material to doctors still using them or for a clinical trial, at Memorial or anywhere else.

Then, in 1963, came a final blow to the vaccine's credibility. In the wake of the scandal over thalidomide, a sedative that caused birth defects in European women who took it during early pregnancy, the Food and Drug Administration tightened its procedures for testing "new drugs," a category to which Coley's toxins now became assigned. The American Cancer Society's Committee on Unproven Methods of Cancer Management, which was established in 1954 and began to publish an annual brochure known as *Unproven Methods of Cancer Management,* added Coley's toxins to this ignominious list in 1965. Though it was removed ten years later, the damage had been done. To the postwar generation of oncologists that knew nothing of William Coley—nothing of his sterling record as a surgeon, nothing of his role in establishing the first cancer research fund in America, nothing of his efforts with James Ewing to define the mission of what is now Memorial Sloan-Kettering Cancer Center as one of the leading cancer research institutions in the world—the vaccine known as "Coley's toxins" had been bureaucratically stigmatized and lumped together with such controversial therapies as Krebiozen, laetrile, mistletoe, and orgone boxes.

As several groups of researchers have learned since 1965, physicians desiring to test the toxins now need to file an Investigational New Drug (IND) application with the FDA, a procedure that not only incurs considerable cost but subjects the clinicians to review and monitoring by the FDA throughout a lengthy testing process. These restrictions are designed to protect citizens from unsafe or unscrupulous remedies, but they can sometimes have the collateral effect of discouraging clinical research by all but the wealthiest of institutions and pharmaceutical companies; a further discouragement in terms of the toxins is that, without patent protection on a hundred-year-old treatment, there is little incentive for a drug company to invest upward of $200 million to test the vaccine and find that it works, only to watch competitors stream into the market.

But in fact, there has been little interest in doing so. Quite independent of whatever odor of quackery attaches to the toxins, the sad truth—sad for Coley and his heirs, but sad also for what it says about the ways of science—is that the world of molecular medicine has passed Coley by, with barely a tip of the hat.

Coley's work might well have disappeared into obscurity—into even greater obscurity, rather, for it is accorded only fleeting mention in the standard histories—were it not for the heroic efforts of his daughter, Helen Coley Nauts. In 1938, on a visit to the family's country home, Okeden, in Sharon, Connecticut, where William Coley is buried, she rediscovered a trove of her father's papers stored in a barn on the property—all his medical records and case reports, his correspondence with the Mayo brothers and Joseph Lister and Peyton Rous, his professional and personal diaries, the latter dating back to Yale and his musings about Giordano Bruno and Oliver Wendell Holmes ("He'd never thrown away a thing!" she said later). Nauts hauled boxes of the material back to New York, where she piled them, from floor to ceiling, in one of her daughters' bedrooms and patiently worked her way through the entire archive with the intent of writing her father's biography. "I would take one box at a time, and I went through 15,000 letters and papers in two years," she said.

As she worked through these voluminous records, Helen Nauts arrived at several key insights. "These ideas hit me hard—more than they hit Father, I guess," she told me when I visited her office, located in a room of her apartment on Park Avenue. By then she was eighty-six years old, but so feisty and blunt-spoken that she seemed less to have inherited her father's gentle demeanor than that of his sharp-tongued, imperious antagonist, James Ewing. "One was the need to get a stable preparation, the variation of potency. The need to teach how to give it, and the importance of technique in administration. And the fact that so many of the patients were *terribly* far advanced when they were first treated." Displaying a flair for epidemiological legwork that mirrored her father's initial foray into the field of cancer vaccines, Nauts traced patients, contacted coroners, and collected death certificates to determine the outcome of many cases. Several patients who received Coley's toxins are still alive as of this writing; one of them, retired radiologist William Curtis of Seattle, was treated in 1921 as a twelve-year-old bone cancer patient. "It's always been my feeling," he said in an interview, "that although the toxins didn't give me any relief as far as the size of the tumor went, there's a very good chance that it increased my resistance to the tumor cells, so I never got any pulmonary metastases."

Nauts cataloged the cases according to the types of cancer treated and then correlated the results with each of the fifteen different preparations of the toxins (each with its varying potency) that had been used on patients. She began publishing her conclusions and, reviving a Coley tradition dating back to 1896, received a small grant from the Rockefellers—Nelson, this time—to form the Cancer Research Institute, which has grown into a highly respected organization with the stated goal of advancing the field of cancer immunotherapy. From her first appearance in the scientific literature, dating from a talk she gave in July 1945 at the annual meeting of the American Association for the Advancement of Science and printed one year later in the journal *Cancer Research,* she has gone on to publish some eighteen monographs detailing the use of Coley's toxins and related topics, such as the role of fever in cancer remission. In 1953 Nauts and two colleagues summarized thirty dramatic cures of inoperable cancer and claimed that the toxins could achieve an 80 percent rate of five-year remissions as an adjuvant in inoperable cancers if given for a period up to six months following surgery. It is a remarkable accomplishment, not least because in publishing her conclusions, Nauts demonstrated how a dedicated layperson can make significant contributions to scientific discourse. There can be no ambiguity as to her ultimate mission, of course—nothing less than to rehabilitate her father's reputation and revive use of the toxins in modern medicine.

These retrospective studies suggest the crucial impact that production technologies can have on the success of a new drug or treatment. Helen Nauts has asserted, for example, that the commercial version of Coley's toxins, which Parke, Davis, & Company began to manufacture in 1899, was prepared incorrectly for many years and performed inconsistently in the hands of many practitioners; it almost certainly was inferior in potency to the preparations Coley himself routinely used for many years, which were prepared at the Loomis Laboratory by first Bertram Buxton and later Martha Tracy. "Father didn't have the sense to have it tested comparatively," Mrs. Nauts says. "And it was the letters I read that brought this out. One man in Pawtucket, Rhode Island, had *both* kinds of toxins, and he went home and he injected himself. And he wrote Father and said, 'You know, I have to use eight times the dose of the Parke-Davis product to get the same reaction that I get with a small dose of the Tracy.' So Father then got Parke-Davis to get in touch with Tracy, and they changed it a little bit, and got it a little bit better. But not much." It is a minor point, but Parke-Davis's inconsistent product may be an instance where a pharmaceutical company lacked the technical expertise to mass-produce a sophisticated biological product. Nor is this surprising. Parke-Davis had the unenviable task at the turn of the century of manufacturing a stable, consistent, heat-inactivated bacterial vaccine, which is not dissimilar to challenges that proved daunting a half-century later to an industry that, with far more sophisticated technologies of fermentation, tissue culture, and immunochemical assays, struggled to produce the first killed-virus polio vaccines of the 1950s.

Nauts also drew attention to the fact that the longest remissions, including many permanent cures, were associated with repeated long-term use of the toxins; some treatments lasted three or four years. Another key observation, and one of the most interesting, involved fever; high fever often correlated with a favorable response. This observation has been independently reported by many observers, with the result that hyperthermia has become an area of keen interest in medicine today. A large, multicenter European epidemiological study is currently underway in an attempt to see if a personal history of high fever may be correlated with a lower risk of melanoma, as was suggested in a smaller 1992 German study.

Since the 1950s there have been about half a dozen attempts to test Coley's (or Coley-like) toxins in a modern setting—animal studies led by H. Francis Havas at Temple University in the 1950s, and human trials by Barbara Johnston at New York University in the 1960s; by Herbert Oettgen and colleagues at Memorial Sloan-Kettering in the 1970s; by Rita Axelrod and colleagues again at Temple in the 1980s; and by Klaus Kölmel of the University of Göttingen in 1991. Although these studies are not without methodological blemishes and have been faulted for failing to heed important constraints that even Coley appreciated, especially in the selection of suitable patients and the duration of treatment, it is noteworthy that most have consistently reported promising, though admittedly ambiguous, responses (they are discussed in detail in the notes). Nonetheless, it would be fair to say that a definitive, well-designed, controlled study of Coley's toxins against sarcoma, for example, remains to be done; and given the rise of new, more specific immunotherapeutic interventions, probably never will be.

In all likelihood, it is neither Helen Nauts's unstinting efforts nor the several inconclusive trials that have led to the increased respect that Coley's experiments have more recently been accorded. Rather, nearly half a century of basic research in immunology, quite unrelated to (and quite unconcerned with) the vaccine, has shed enormous light on complex mechanisms of the immune response that were utterly unfathomable in Coley's era. And it is this research that provides an avenue to answer two key questions about Coley's toxins: Did they really work? And if so, how? At least some researchers do not hesitate to answer the first question in the affirmative. The molecular era of immunology, to which the rest of this book is devoted, has begun to answer the second.

Proof of the toxins' scientific worth remains mired in the miasma of conflicting detail about diagnosis, effectiveness, and follow-up, and there is surely no consensus on what Lloyd Old has come to call the "Coley phenomenon." It is not uncommon still to hear prominent immunologists dismiss the therapy as "rubbish," and National Cancer Institute researcher Steven A. Rosenberg—like Coley, trained as a surgeon, and like Coley, criticized by colleagues for his aggressive attempts at immunotherapy—probably reflects the feelings of many cancer researchers. In his book *The Transformed Cell*, Rosenberg mentions the toxins in passing, but dismisses the vaccine—in a few sentences, and somewhat incorrectly—by saying that the treatments "were abandoned when others could not reproduce his findings." However, Old and Herbert F. Oettgen of the Sloan-Kettering Cancer Research Institute contributed a chapter to the 1991 edition of *Biologic Therapy of Cancer*—a cancer textbook coedited, incidentally, by Rosenberg—in which they write, "Those who have scrutinized Coley's results have little doubt that these bacterial toxins were highly effective in some cases." Radiation therapy and later chemotherapy, they continue, held greater appeal for doctors because they were "both more predictable and comprehensible. Reluctance to accept claims for the effectiveness of Coley's toxins stems from the fact that virtually no studies of this approach have been carried out according to modern standards of clinical evaluation."

To this day Coley's work remains forbidden fruit to many researchers—the results look tantalizing, but the methodology remains too thorny to swallow whole. Still, modern biologists seem more prepared to make a place for Coley at the table today than during the gloomy days when the toxins appeared on the blacklist of the American Cancer Society. Frances Balkwill, a researcher at the Imperial Cancer Research Fund in London, devoted considerable discussion to Coley's work in her 1989 book, *Cytokines in Cancer Therapy*. While stating that "one cannot fail to be impressed by the case-histories described by Helen Coley Nauts," she correctly noted that "there was no standardization of techniques or dosing, and the existing evidence for clinical efficacy was scattered and never presented properly." Nonetheless, in an interview in London, she suggested that the toxins may have triggered a "cocktail" of cytokines that had greater therapeutic activity than any single cytokine, and also implied that modern oncologists would do well to learn from Coley the physician.

"I can't explain why Coley got the results he did," she said, "but I'm convinced he got them. I think also that over the years Coley's work was in some sort of disrepute. The reasons for that are multiple. One is that the commercial preparations were not quantifiable, and they were very variable. You couldn't quantify his toxins; even he couldn't quantify his toxins. But if you read his papers very carefully, you will find that he would titrate them in each patient so that he got what we call the 'cytokine flu,' or the cytokine toxic shock response. And if a patient failed to respond, he would give the patient a few days' rest and then start again. Which is quite extraordinary. I mean, he could teach people doing cytokine therapy an awful lot if they went back and read those papers."

At least one immunologist suggests that in a hundred years of cancer research, no one has done better. "Coley's toxin was the *ultimate* in heterogeneous responses," says Michael Osband, chief of pediatric hematology-oncology at the Boston University School of Medicine. By heterogeneous, Osband intended to suggest a particular phase of the immune response that is general, or nonspecific, and occurs during the initial moments of immunological distress; it is the general alarm of a system that knows something is wrong but hasn't yet identified the problem. "And when you look at the history of cancer immunotherapy, I think, to be blunt, we are not much farther along in comparison to Coley. . . . That 20 percent to 40 percent of patients will respond to Coley's toxins is, in my mind, incontrovertible, and the same percentage of patients—and probably the same actual patients—who would respond to any effective immunotherapy." One immunologist, requesting anonymity, goes even further. "The reason why nobody does Coley's toxins today is because in the era of recombinant DNA and sexy gene therapy, you can't raise money to fund that company," he says, "and the FDA wouldn't know what to do with it. The FDA is into, 'Okay, show me on a gel that you got one band.' But that's very nonphysiological. That's where everybody's wrong."

Perhaps the strongest recent endorsement of Coley's work appeared in the May 7, 1992, issue of the British journal *Nature,* where Charlie Starnes, a molecular immunologist trained at Stanford and working now at the biotechnology company Amgen in Thousand Oaks, California, called for a return to the use of Coley's toxins in the treatment of certain cancers. In making the case for renewed experimentation and clinical trials, Starnes analyzed a series of inoperable cancers that had been treated with the vaccine alone and no other form of treatment (an analysis, incidentally, considered far too restrictive by Coley's daughter); he concluded that approximately 10 percent of inoperable sarcoma patients achieved remissions lasting twenty years or more with the use of Coley's toxins. "Although certainly considerably less than perfection," Starnes wrote, "such odds would provide hope for an otherwise inoperable patient."

A close reading of the data in Starnes's *Nature* article suggests that 30 of 154 evaluable patients with sarcomas or lymphomas—just under 20 percent—were indeed free of disease twenty years after treatment. Starnes chose to interpret these results with especial rigor, he said in an interview, because sarcomas tend to be less aggressive than other cancers, and the natural course of the disease is highly unpredictable. But if Coley's overall results in inoperable tumors, as interpreted by Starnes, were measured against what passes as the gold standard of current cancer therapy—that is, no evidence of disease for a period of five years after treatment—Coley's toxins performed remarkably well when they were the sole form of treatment: they "cured" 73 of 154 sarcoma and lymphosarcoma patients in Starnes's grouping, better than 47 percent. One could argue that this arbitrary grouping is statistically meaningless because of the nonrandom selection of the cases; but in the absence of a controlled, randomized trial, those are the frustrating choices that remain.

One of the virtues of Starnes's research has been his reluctance to make extravagant claims for the toxins. In a longer review article published in 1994, Starnes and colleague Bernadette Wiemann made the interesting point that, rather than providing a global immunological treatment for all cancers, Coley's results may inadvertently have left us clues to which cancers are immunogenic—that is, which cancers are vulnerable to immune attack. They have also theorized that the toxins seemed to be particularly effective against cancers that originate in the embryonic tissue known as the mesoderm. This would suggest, they say, that the vaccines might be effective against not only sarcomas and lymphomas but also a portion of other more common cancers that have mesodermal roots, such as ovarian and renal cancer.

"Given these striking results," Lloyd Old asked colleagues at a 1993 symposium, "why hasn't Coley's approach been forged into a widely available therapy with predictable benefit for cancer patients? This is a question we will leave for future historians, but there are clearly a number of reasons why progress was less than optimal and why the approach even became controversial for a period." Old reiterated the usual reasons: preparations weren't standardized, chemotherapy and radiation therapy looked more promising. "The best reason, however, is that science had to catch up with the Coley phenomenon and that the cellular and molecular language of inflammation and immunity had to be understood before the forces that Coley unleashed could be predictably translated into tumor cell destruction."

What "cellular and molecular language" did the immune system speak in reply to Coley's rude injection of bacterial by-products that made tumors disappear? Or, to phrase the question another way, what didn't scientists know in Coley's time that could possibly account for the effects he saw, and when did they begin to know it?

In the sixty years since Coley's death, immunology and cancer research have seen a series of pennies drop; one discovery after another has moved us closer to a reasonable hypothesis about what kind of molecular commotion this crude but pioneering cancer vaccine might have stirred up. In the 1930s and 1940s, for example, a biochemist at the National Cancer Institute named Murray J. Shear, inspired in part by Coley's work and fascinated by the power of bacterial toxins, conducted a lengthy series of experiments in mice using *Serratia marcescens,* one of the two bacteria in Coley's vaccine. Having confirmed the antitumor effects of *S. marcescens* in mice, Shear painstakingly searched for the active ingredient of this anticancer effect and managed to isolate a single factor, a single entity, that appeared to trigger the dramatic effects observed by Coley and a host of others. In 1943, in one in a long and heroic series of biochemical papers, Shear reported that the endotoxin from *Serratia* turned out to be a component of the bacterial wall; since the molecule was composed of fatty groups (lipids) and sugars (saccharides), Shear began to refer to it as lipo-polysaccharide—LPS, for short. Ten years before the discovery of the double helix, which gave molecular biology its namesake molecule, Shear gave immunology not so much a molecule as an *inducer* of molecules. The appearance of LPS, as would occur in certain bacterial infections, would unleash a torrent of unknown and strikingly powerful immune substances that could make experimental tumors shrink and disappear in a fashion that virtually mirrored Coley's clinical descriptions half a century earlier.

Shear's work in turn inspired further research. The notion that bacteria and their constituent parts could provoke a generalized, or "nonspecific," immune response triggered research in several directions. In 1959 Lloyd Old and Baruf Benacerraf showed that mice infected with another bacterium, the attenuated tuberculosis microbe bacillus Calmette-Guérin (or BCG), seemed to enjoy enhanced immune resistance to transplanted tumors. Ten years later, a flamboyant French physician named Georges Mathé, in a report that set off what is now widely viewed as an unfortunate frenzy of ill-advised clinical work, claimed that BCG could extend the life of leukemia patients. Hundreds of patients received BCG, most without benefit. Lost amid the gloom that accompanied BCG's meteoric rise and abrupt fall as an anticancer agent is the fact that one durable and indisputable use has survived the test of time. In 1976 Alvaro Morales in Canada first tested BCG in the treatment of superficial bladder cancer, against which it has proven to be highly effective.

In 1975 Elizabeth Carswell and colleagues, working in a group headed by Old at Sloan-Kettering Cancer Research Institute in New York, would finally identify one of the molecules induced by LPS. It was a protein that made its appearance extremely early in the immune response, sending signals between cells of the immune system (a cytokine), and since it had the power to cause tumors to shrivel and die in mice, they called it tumor necrosis factor (TNF). With this discovery, Coley's scientific heirs truly brought Coley's toxins into the world of molecular immunology, but TNF was only part of the story.

In 1978 another penny dropped in upstate New York. Michael Berendt, Robert North, and colleagues at the Trudeau Institute in Saranac Lake published a series of articles in the *Journal of Experimental*

Medicine, again citing the early work of Coley, showing that while endotoxin (and later tumor necrosis factor) might cause tumors to shrivel up and die, the TNF molecule alone did not and could not cure animals of cancer. Something more was needed, an immunological comrade in arms, to prevent the disease from recurring. In further experiments, the Trudeau researchers went on to implicate a white blood cell known as the T lymphocyte (or T cell) in tumor regression, and speculated that TNF disrupted the blood vessels feeding a tumor, causing so much tumor cell destruction that T cells could recognize and effectively destroy all remnants of the tumor.

But what connected the commotion of the early, nonspecific immune response (where TNF raised a big ruckus) to the later, more precise cellular immunity, where T cells seemed to play the dominant role in achieving long-lasting remissions? In 1989 workers in two laboratories—one at the Wistar Institute in Philadelphia and the other at Hoffmann–La Roche in New Jersey—provided a clue when they discovered a powerful molecule that helps orchestrate the earliest moments of this immune response. The molecule, now called interleukin-12, makes its appearance in response to bacterial endotoxins, among many other stimulants; not only does IL-12 provoke the release of other immune molecules, but more importantly, it seems to act as a link that oversees the transition from early to late in the immune response, from general to specific, from nonspecific hubbub to the methodical elimination of unwanted visitors by killer T cells. It thus may suggest a possible answer to the question raised by the Trudeau researchers; it serves as a molecular bridge from the early, nonspecific immune response to the later, specific immunity conferred by T cells.

Finally (although in immunology, the word "finally" has a shelf life of a month or two at best), in 1990 researchers in Denver showed that bacterial exotoxins, of the sort produced by gram-positive bacteria such as erysipelas, function as "superantigens." For reasons still unclear, these bits of protein provoke an unusually broad, multipronged T cell response. Charlie Starnes of Amgen has argued that superantigens, in addition to their ability to expand certain populations of T cells in a nonspecific manner, can also compromise liver function and cause a buildup of endotoxin in the general circulation. He believes that this may be at least partially responsible for the fact that under certain conditions, superantigens have been shown to enhance the already potent biological effects of endotoxin by as much as 50,000-fold or more.

In other words, a half century of research almost willfully unconcerned with Coley has bit by bit provided a molecular framework upon which to hang a reasonable hypothesis to explain the "Coley phenomenon." It is still not known—and may never be—exactly what the toxins did, but Lloyd Old has suggested that they triggered the release of a broad array of molecules, ranging from painkilling endorphins to fever-inducing substances like interleukin-1 to hormones that stimulate the production of immunologically active blood cells (colony stimulating factors) to cytokines like TNF and interferon. The research has shown, directly or by implication, that the vaccine involves multiple cytokines and probably also lymphocytes—involves, in short, a complex and multidimensional immune response. Neither Coley nor his critics had the foggiest notion of the cells, the molecules, and the order of interaction involved in the immune response. Which is why Coley lived the nightmare of every conscientious clinician: believing what he had seen, but far too ahead of his time to explain it.

A definitive molecular explanation of the toxins' overall activity remains elusive, and seems destined more for historical rather than clinical revisitation, but it has progressed sufficiently that Drew Pardoll of Johns Hopkins School of Medicine, a talented young researcher whose group is working on an antitumor vaccine using gene therapy, made the following remark at a recent meeting. "One of the things I've come to appreciate," he said, "is that many of the things that we are doing are recapitulating things that Dr. Coley did a century ago."

Charlie Starnes goes even further. As he and his colleague Bernadette Wiemann wrote not long ago, "Today, over a hundred years from the time Coley began his career, we have managed to piece together a reasonably distinct understanding of at least some of the mechanisms responsible for the phenomena which Coley and his contemporaries observed. An absolute lack of such understanding contributed much to the original controversy surrounding the treatment known as Coley's toxins. Today, with far less justification, the treatment has still yet to receive even the academic, let alone clinical attention that it deserves. . . . It is entirely possible that what Coley accomplished for his patients with the use of his vaccine was the best that anyone could ever do, even today. A more complete and thorough understanding of the problem of cancer does not necessarily guarantee us a more effective method of treating it."

If Coley's toxins were effective against Ewing's sarcoma, as suggested by the results presented at the 1934 symposium, a potential pool of 3,000 patients in the United States alone might qualify for treatment each year. If the toxins were effective against soft-tissue sarcomas and lymphomas, as Coley's results suggested and Starnes has argued, they might bring relief to some of the roughly 60,000 Americans who come down with those cancers each year. And if they were effective, as Starnes proposes, against all cancers of mesodermal origin, like kidney cancer (22,000 new cases each year) and ovarian cancer (26,000 new cases each year), the universe of potential beneficiaries grows.

Those are all huge ifs, almost ridiculous leaps of argument and faith. Unfortunately, such threadbare speculation is the ultimate legacy of William Coley, along with the proper clinical test that was never done in the past and almost certainly will never be done now.

II

THE
PATRON SAINT
OF CYTOKINES

Leucocytes may be regarded as mobile unicellular glandular bodies which set free their secretions in the humors of the organism. But little is known of the nature and functions of the substances they secrete.
—Alexis Carrel and Albert Ebeling, 1922

6

IN SEARCH OF AN INTERFERON

*Scientific reasoning is an exploratory dialogue that can
always be resolved into two voices or two episodes of thought,
imaginative and critical, which alternate and interact.*
—Peter Medawar, *Pluto's Republic*

⚮ At the end of June in 1956, a young Swiss biologist boarded a
train in Zurich clutching his valise, his typewriter, and most important, a letter inviting him to spend a year as a guest researcher at the
prestigious National Institute for Medical Research in England, a laboratory renowned for its excellence in biological research and known
informally as Mill Hill for the modest knoll upon which it sat on the
northern outskirts of London. Relatively new to the game of science,
Jean Lindenmann set off on this adventure with the freshest and
healthiest of illusions. "I myself was extremely naive at that time," he
would say many years later. "I thought, well, when you do research,
you *find* things."

Lindenmann's first few months in London, however, schooled him
more on the nature of unpleasant discoveries. On July 1, his very first
day of work at Mill Hill, Lindenmann signed a piece of paper relin-
quishing in perpetuity all patent rights to whatever he might "find"
during his research in England to Britain's Medical Research Council,
the government agency that funded and oversaw biomedical research;
he was more than happy to do so at the time, although he has had
occasion to wonder since then. His research went very badly at first,
and several months after his arrival, burglars ransacked his modest
North London rental flat the day after he was married, and walked off
with his typewriter and his wedding ring. (It bespeaks Lindenmann's
reticence that his colleagues at Mill Hill knew nothing not only of the
robbery but of the marriage.) All he was left with was his new bride,

131

his naïveté, and his enthusiasm for a new collaboration that had been hatched over an afternoon cup of tea at the lab and was roaring along toward a discovery most pleasant.

When he left Zurich for London, Lindenmann was thirty-one years of age, single, and an anointed star of the Swiss scientific establishment. Lindenmann's father, a civil engineer, was Swiss by birth (but traveled widely for his work), and his mother was French; Lindenmann was born in Zagreb, in the former Yugoslavia, and as a youth spent time growing up in Paris, Belgrade, and other European cities—"typical Swiss, as you can see." In 1937 his parents sent their son to neutral Switzerland for schooling; once war broke out two years later, the teenage Lindenmann lived in a Zurich boardinghouse for the duration of the conflict, attending first high school and then university while living on his own. By the time he entered the University of Zurich, intending to study physics, he had become a handsome, quietly astute, and somewhat taciturn young scientist. He soon switched to medicine, receiving his medical degree in 1951 and then embarking on a research career in bacteriology. Under the tutelage of a famously imperious professor named Hermann Mooser, considered the preeminent biologist in postwar Switzerland (and with a temper as mighty and awesome to behold as the Alps), Lindenmann showed sufficient promise to receive a prestigious stipend from the Swiss Academy of Medicine to study a year in England.

Each research era has its meccas, its hallowed destinations for certain lines of inquiry, and at the time Mill Hill towered like a great cathedral in the landscape of biology, during a period when the mores, accent, and signal achievements of biology were decidedly Anglican. Specifically, Mill Hill was a citadel of research in the field of virology, and viruses were exactly what Jean Lindenmann planned to study. Lindenmann did not know a great deal about viruses on the eve of his departure for London, as he would be the first to admit. But no self-respecting biologist could afford to ignore viruses, because virology opened the door to two of the most exciting areas of biology of that era, molecular biology and immunology. "Viruses," as one scientist would puckishly remark, "know more immunology than immunologists."

Like a great many virologists in the 1950s, Lindenmann had become fascinated with a mysterious biological phenomenon known as interference, which had first been described in the scientific literature by two British scientists in 1937. It was a phenomenological wonderment, a parlor trick in a lab dish. Simply put, interference was an invisible shield against viral infection, a protective effect imparted to cells (or animals) that had previously been infected by another, usually weaker kind of virus. In the original study, for example, monkeys infected by Rift Valley fever virus surprisingly resisted subsequent infection by the normally fatal yellow fever virus. Somehow, the monkeys biologically braced themselves against the second, more serious infection. An explanation of this kind of immunity—this *interference*—had obvious public health ramifications, especially during an era of surging optimism engendered by the success of the first-generation polio vaccines. Indeed, American researchers in the 1940s had even attempted a short-lived clinical trial of interference, infecting volunteers with the flu virus in the hope that it would blunt a second, more severe infection (in an irony probably lost on the biotechnology companies that later stumbled down this same path, this early experiment was halted because of intolerable side effects).

Interference was like a religious experience—biologists who witnessed the phenomenon became entranced by its power and took it on faith that the biological mechanism underlying it could ultimately be explained. For Lindenmann, the religious conversion occurred in 1955, in the year prior to his departure for England. His boss Hermann Mooser, an expert in rickettsial diseases, had learned that interference occurred with the bacteria-like rickettsiaes as well as with viruses, and proposed an experiment to Lindenmann. Would interference occur, the senior scientist wondered, even when the first wave of infecting virus was physically blocked from entering vulnerable target cells? In order to answer this question, Lindenmann performed an elaborate—and, in hindsight, amusingly complicated—experiment involving eggs, viruses, and red blood cells; the details are not so important here as two aspects of the work. First, Lindenmann happened upon a useful technique in a paper that had appeared in a relatively obscure Australian journal, reported by two presumably Australian virologists named Margaret Edney and Alick Isaacs. These scientists described a technique of inactivating enzymes in influenza viruses so that they would stick to red blood cells, like cockleburs to cloth, and not be able to penetrate the cell wall or enter into target cells, yet still cause interference. Second, Lindenmann—who performed these experiments mostly when Mooser was away on an extended trip abroad, doing routine diagnostic work in the lab by day and his interference research after hours—satisfied himself that interference could indeed occur even when the viruses were prevented from physically entering target cells.

Lindenmann had a paper already written upon Mooser's return, but the senior scientist, as rigorous as he was cantankerous, reviewed the data with an acid eye and decided that the experiments failed to conclusively prove the point. Mooser decided to withhold the paper from submission for publication. And so, on the train ride to London, interference was both an intellectual itch and a sore point to Lindenmann. "It was sort of an unfinished business, still lingering on in my mind," he said. *"Not* an obsession. But I had some investment in it, yes." As everyone would later learn to their dismay, Mooser felt he had an investment in it, too.

When he arrived in London, Lindenmann reported to the head of the Division of Bacteriology and Virus Research at Mill Hill, a towering figure in the field named C. H. Andrewes. "CHA," as he was informally known in the lab, had in the 1930s discovered the virus that causes influenza; building upon that work, he had by 1948 almost single-handedly organized and implemented a global "early warning" network of laboratories to monitor outbreaks of influenza under the auspices of the United Nations' World Health Organization. The network's headquarters—Room 215 at Mill Hill, in the World Influenza Centre laboratory—became the focal point for identifying new flu strains in order to prepare seasonal flu vaccines. It was in a tiny room next door, between the World Influenza Centre laboratory and a water closet, that Lindenmann set up shop. He did not work on influenza, however. Andrewes asked him to attempt to grow the virus that causes polio in the kidney cells of rabbits; there had been a report that such a technique was possible, and it would have immensely expedited the commercial preparation of polio vaccines. Lindenmann struggled the entire time he was in England to solve the problem, got nowhere, and even viewed his failure as something of a merit badge. "I spent quite a lot of time trying to do this, and it just didn't work," he said. But the negative results "probably indicated to Andrewes that I was a rather serious chap who would not simply for the hell of it get positive results just to show off."

If Lindenmann husbanded even a faint ember of resentment over the Mooser episode, a chance encounter in the cafeteria at Mill Hill in early August, about a month after his arrival in London, served to fan the coal—and his interest in interference—back to life. One of the great attractions of a small, clubby place like Mill Hill was the way it

flung virologists, biochemists, immunologists, and other scientists together informally, not least over a seemingly endless succession of
daily breaks for tea, and it was over a cup of tea one afternoon that
Lindenmann found himself in an increasingly animated conversation
with a short, intense, and to his mind serious, almost sad-looking
biologist with dark good looks. The name flew by him during the
introduction, but clearly the scientist enjoyed considerable esteem
among Mill Hill's staff. During their polite chitchat, Lindenmann
mentioned that he had just finished some experiments on viral interference, and the man's dark dancing eyes lit up; indeed, the fellow
began to grill the Swiss scientist almost aggressively, with needling
curiosity for details, on the exact techniques he had used in these as-
yet-unpublished experiments. The man seemed positively obsessed
with the method Lindenmann had used to inactivate the viruses prior
to attaching them to the red blood cells. Lindenmann, unfailingly polite and a good scholar with excellent recall, patiently answered each
question. When the issue of the "obscure" Australian paper and inac-
tivation of the viruses came up, the grilling became infinitely more
picayune. The man was *very* keen on this technique, so keen that at
one point Lindenmann found himself thinking that only the authors
of the paper themselves could possibly ask more detailed questions.

And with that, the scales fell from his eyes.

"This was the same man!" Lindenmann suddenly realized. He had
just become the victim of a leg-pull by Alick Isaacs. He had just met
the man to whom his reputation, in the shorthand ampersands of scientific citation, would forever be linked, for better and for worse:
Isaacs & Lindenmann.

Their initial encounter is telling. Colleagues found Lindenmann
quiet, serious and sober, even plodding, sometimes inflexible (although to be fair, these less than generous impressions may have had
more to do with his junior status, his phlegmatic nature, and his halting English than his sense of humor, which comes across in his writing as wry, subtle, and sophisticated). In any event, he made the
perfect foil for Isaacs's unique brand of humor, and perhaps also for
Isaacs's unique style of science. One is tempted to see Lindenmann as
the scientific straight man to Isaacs the improvisational scientist in
what would prove to be one of those very brief, intense, and immensely fruitful collaborations, almost like a brief love affair that is
never forgotten and after which life is never quite the same. The analogy is not entirely far-fetched, because Isaacs was legendary for seduc-
ing other scientists with his infectious enthusiasm into assisting him

on pet biological projects, and the minute Lindenmann mentioned interference, Isaacs turned on his very considerable charm. And the leg-pull was vintage Isaacs.

"Alick was very amusing, and could keep a very straight face sometimes," said Susanna Isaacs Elmhirst, Isaacs's widow. "He'd startle people by saying something absolutely outrageous with a very straight face. And he was an *extremely* nice person. It was surprising if somebody didn't like him." Lindenmann liked him immensely, and upon further discussion, the two men agreed on the spot to commence a new series of experiments on viral interference in September, as soon as Isaacs returned from the summer holidays. In so doing they formed a collaboration that, in its twists and missteps, its blunders and intuitions, could serve as a primer of scientific discovery: they set out in the wrong direction, for the wrong reasons, and ended up stumbling upon a prize that would win both of them enduring scientific celebrity. Less well known is the fact that the hurly-burly of precisely that fame left one of them almost wishing they hadn't found interferon and the other one not entirely sure that they had.

In the fall of 1956 Alick Isaacs was everything that Jean Lindenmann was not: zestfully verbal, every bit the family man (by then with three children), Jewish by birth (but a confirmed and contented atheist by the time he left home), a superb writer, the heart and soul of a party, almost compulsively open and compassionate, and an accomplished scientist of boundless imagination, somewhat given to risk and chance taking. He was also so humane that colleagues tend to describe his science and personality in the same breath. "His sense of humor was quite wry and teasing," recalled David Tyrrell, who first encountered Isaacs when they worked in the same hospital in Sheffield in the 1940s, "but he did always seem, particularly in those days, to find life fun and exciting and interesting, and certainly his style of doing work was to do experiments—he particularly liked to do them himself—do them carefully, and really then let his imagination play around with the results and try and think what it was that *might* be going on to explain the observations he'd just made." Asked later about Alick Isaacs's strengths and weaknesses as a scientist when they worked together, Lindenmann replied without hesitation, "During the time I was there, I believe he had no weaknesses. I couldn't point out a single definite weakness." "Most scientists are pedestrian," said another

Mill Hill colleague, Anthony Allison, "and Alick was not. He had original views about *everything.*"

Isaacs was a brilliant scientist, one of the brightest lights of his generation, but it later turned out that he also suffered from manic-depressive disorder, and it is impossible to recount the story of interferon—its discovery, its reception by the scientific community, and its many ups and downs—without at least considering a direct or indirect role of the illness on the way the story unfolded. There is no question that some prominent scientists were skeptical about the discovery, expressing doubts that were no doubt amplified by Isaacs's private demons and making the unfortunate commingling of these scientific and personal moments of doubt an especially poignant and long unappreciated aspect of the story. Because of the battles he waged against his disability, at least from 1958 on, Isaacs's contribution can be seen as even more heroic.

Isaacs's upbringing was as rooted to place as Lindenmann's was peripatetic. Born in Glasgow on July 17, 1921, he was the eldest of four children of a striving first-generation Jewish immigrant of Lithuanian peasant background named Louis Isaacs. The family surname dates from a bureaucratic unpleasantry visited upon Barnet Galinsky, a Lithuanian peasant fleeing anti-Semitism around 1880; Galinsky explained his name to a British immigration official as "Barnet, son of Isaac," and it ended up as Isaacs. Louis Isaacs, Barnet's son and Alick's father, began as an itinerant peddler but worked himself up into becoming a prosperous shopkeeper who saw in his children the chance for social advancement that he himself had never enjoyed. Young Alick Isaacs thus straddled two singular cultures, raised in an Orthodox household with a kosher kitchen, yet possessed of a charmingly broad Scots accent and pictured occasionally in kilts.

"He enjoyed school, was a popular pupil, made friends easily and did well but not brilliantly," his brother Bernard later put it, and "if any sign of the future scientist was to be detected, it was in the concentrated thought which he would devote to the solution of a chess problem." He never studied biology until medical school, but as soon as he settled on medicine as a career, Isaacs began to collect prestigious prizes and fellowships the way other college students collect beer cans. He began medical studies at the University of Glasgow in 1939 and then received a fellowship in 1947 to study at the University of Sheffield under Charles Stuart-Harris, one of the leading virologists of the day. In addition to encountering the phenomenon of viral interference for the first time in Stuart-Harris's lab, a subject to which he would return again and again during a very productive scientific life, he met his future wife Susanna Gordon, who was charmed by his Scots accent (and somewhat less impressed by his ability to draw blood, which occasioned their first meeting; she had a fever of unknown origin, and Isaacs had been asked to draw a sample of her blood for tests). They shared a love of music, an appreciation of keen wit, and a sense of estrangement from their respective families. At Sheffield, Isaacs opted to pursue medical research rather than clinical practice, a decision that created a rift with his father, which widened into a deep and enduring estrangement several years later when he announced his intention to marry Gordon. A formidable intellect as well as a wonderfully accomplished physician, Sue Gordon would go on to become a prominent child psychiatrist in Britain, but her many virtues paled, in the eyes of the Isaacs clan, against one overriding flaw: she was not Jewish. Following their marriage in Australia in 1949, Alick Isaacs had virtually no contact with his parents or siblings and was in fact disinherited by his father. Making light of even painful domestic frictions, he ever after used a pet phrase to characterize relations with his family: "All assistance short of actual help . . ."

By that time, however, Isaacs had found worthy surrogates in the world of science. He received a prestigious Rockefeller Foundation Travelling Fellowship to study in Melbourne, Australia, with the eminent virologist and immunologist F. Macfarlane Burnet; during a burst of expatriate creativity, Isaacs and Australian colleague Margaret Edney produced half a dozen papers on viral interference in a two-year period. Although a virologist by training, Burnet was much more: he stood at one of the busiest intersections of the life sciences during the immediate postwar period, where virology met immunology and molecular biology, and he would win enduring fame and adoration for what is known as the clonal selection theory of antibody formation. This formidable hypothesis, later ratified by experimentation, holds that humans are endowed at birth with an enormously large repertoire of distinct antibody-producing cells warehoused, as it were, in the lymph nodes and spleen; it was the chance encounter with an invading microbe that would turn on, or select, the appropriate antibody, so that in a sense each bug picked its own poison off the shelf. The sheer sprawl of Burnet's intellectual curiosity made a lasting impression on Isaacs, who cultivated an insightful familiarity with all three rapidly advancing fields when he was invited by C. H. Andrewes to return to Mill Hill and work on influenza.

Influenza monopolized the bulk of Isaacs's time and energies during the early 1950s, but he continued to dabble in the problem of viral interference. Inspired perhaps by Burnet, he made sure to keep abreast of new developments in immunology and molecular biology, wrote witty unsigned vignettes for the *Lancet*'s "Peripatetic Correspondent" column, and penned many stylish, shrewd, sharp-tongued editorials on the subject of virology, immunology, and science in general. A scientific expert, he wrote in one of countless exuberant bon mots, "is one who has spent considerably more than the customary one-year period studying his subject." Entertaining, provocative, and bristling with adventuresome ideas, Isaacs quickly assumed a prominence in British science. Jean Lindenmann's appearance at Mill Hill gave him the excuse to rescratch the old itch of interference, this time with some new technology, some new ideas, and an eager new collaborator.

During his fellowship in London, Jean Lindenmann scraped along on a weekly allowance of five pounds; he washed his own clothes, ate all his meals in his flat, and rarely joined the crowd for lunch or a pint of ale at the Three Hammers because money was so tight. He therefore recalled with special vividness the gambling that took place in Room 215 in Mill Hill during the fall of 1956. He and Isaacs wagered six pence—"a lot of money back then," Lindenmann added—on the results of the experiments they performed.

The betting was partly an exercise to force them to think through all potential outcomes, all the reasons why an experiment would or wouldn't work. "In the course of a true collaboration between two workers," Lindenmann later remarked, "it is often so that one 'plays' the role of the enthusiast and the other the role of the sceptic. This does not mean that the 'sceptic' mistrusts the experiments, nor that the 'enthusiast' is ready to cut corners. . . . Since Alick, at that time in permanent high spirits, always put his bet on the hoped-for outcome, what else could I do but put my bet on a different outcome? Nevertheless, I was happy to lose most of the time."

No gambler would ever have bet on the improbable and wholly unplanned series of experimental detours that ultimately led them to a stunning observation, and it is essential to view experiments of this sort, before the development of the powerful and precise tools of molecular biology, as exercises in great imagination. No one in the 1950s was going to have the technology to lift, as with tweezers, a molecule out of a lab dish; the best that could be done was to create a little theater of biological activity in a lab dish, limit the number of characters, and manipulate them in such a way that the momentous exclamation, the denouement of the drama, could be decisively attributed to one or another of the actors. That in a fashion is exactly what Isaacs and Lindenmann did. Moreover, the papers they published on their experiments told only half the story: the experiments that worked, as is so often the case, grew out of unpublished experiments that failed in an interesting way.

They began in September, with a reasonably interesting but flawed premise. Isaacs had closely followed developments in the field of molecular biology, where the study of viruses that prey only upon bacteria (known as bacteriophage, or just phage) had presented researchers with a genetic window upon the biochemical nature of heredity—specifically the role of deoxyribonucleic acid, or DNA, in conveying that genetic information. Viruses lent themselves well to such studies because they were extremely simple: a naked thread of genetic material, either DNA or RNA, sheathed in a ball or tube of protein, incapable of replicating—or, in a sense, of being "alive"—unless inside a cell. James D. Watson and Francis Crick had done their epochal work showing the structure of DNA to be in the form of a double helix at another Medical Research Council laboratory, just up the road in Cambridge. Just as any explosion will cause all within earshot to stop and look in the direction of the blast, everyone in biology oriented themselves—and the problems they hoped to investigate—to the great noise coming from Cambridge in April 1953.

Like everyone else who had been stumped by the problem of interference, Isaacs believed that the viruses did something to the infected cell—blocked the entry of other viruses, used up crucial resources, made some ill-formed, half-finished viral protein inside the cell that somehow interfered with a second wave of viruses. In the vernacular, these first visitors were uninvited guests who cleaned out the refrigerator before the next wave of guests arrived. Isaacs's teatime conversation with Lindenmann inspired an idea for a photogenic experiment that might prove the point, an experiment that derived wholly from molecular biology.

Influenced by the ascendant role of the nucleic acids DNA and RNA in biology, impressed by the new technology of electron microscopy, and especially impressed by a famous set of electron micrographs in the 1950s showing the empty shells of bacteriophage clinging to the outside of cells like shells of cicadas on the trunk of a tree (an experiment that proved viruses replicated by way of their nucleic acids), Isaacs proposed an imaginative variation on this theme. "He thought that in order to induce interference," Lindenmann recalled, "a nucleic acid has to be injected into a cell. But how do you show that the outer membrane of the virus particle remains on the outside?" Isaacs proposed—and Lindenmann is quite clear that Isaacs conceived of the experiment—attaching inactivated flu virus to red blood cells to prevent them from entering the cell, as Lindenmann had done the previous year. Then perhaps they could see if these two elements, exposed to target cells, could interfere with a subsequent infection. If Isaacs's hypothesis was correct and nucleic acids were the agents provocateurs of interference, they might be able to glimpse evidence of this by looking at the viruses stuck to specially prepared red blood cells. In other words, would viruses "handcuffed" to blood cells still be able to induce interference, and if so, what would the scene look like in an electron microscope?

The target cells in these experiments customarily took the form of a thin carpet of embryonic cells known as the "chorioallantoic membrane," a diaphanous layer of living matter that upholstered the inside of the shell of a ten-day-old chicken egg. Isaacs's aim was to emulate the famous bacteriophage snapshots of classic phage genetics, which showed intact viruses on the outside of cells before infection and only the empty outer shells of the viruses following infection. A good set of snapshots capturing the "before" and "after" of interference would, in Isaacs's view, imply that the virus's injected genetic material alone, and not the entire viral particle, was sufficient to cause interference.

That this "beautiful idea" of Isaacs's, as Lindenmann would later characterize it, was doomed to fail, as we shall see, is superfluous now. First, it got them into the lab doing experiments, and second, it forced them to deviate in two significant ways from business as usual in an interference experiment. One of the deviations obliged them to perform the messy task of eviscerating red blood cells as a preliminary step to the main experiments; the other led, by sheer convenience and chance, to a different and ultimately liberating way to perform the experiment, which allowed them to interpret the biology of interference from an entirely different perspective: a change in stage direction, as it were, that changed their view of the drama.

First, the drudgery. Red blood cells (or erythrocytes) possess the iron-containing molecule hemoglobin, and those tiny but dense atoms of iron were enough to scatter the relatively weak beams of electron microscopes in those days. So, to get the clear photographs they were sure would be so revealing, Isaacs and Lindenmann painstakingly removed hemoglobin from each and every red blood cell, creating what Lindenmann called "erythrocyte ghosts." Hideous though this step was, it promised an unobstructed view of the virus particles as they attached to the target cells; indeed, this step repeated an experiment performed by Isaacs and another colleague, Heather Donald, in 1954. Robin Valentine, Mill Hill's gifted young electron microscopist, quickly obtained very nice pictures of flu viruses perched upon the so-called ghost cells prior to infection. They had their "before" photograph.

The second wrinkle illustrates how an infinitely subtle refinement in technique can change a modest, almost derivative experiment into an epochal observation. In most laboratories researchers studied viral interference by injecting viruses directly into ten-day-old eggs, specifically into that same membrane lining the allantoic cavity; they would later poke a hole in the egg and remove a bit of fluid, which could be tested to show the extent of viral replication. This is the approach Lindenmann had used in Zurich the previous year. But Isaacs and Lindenmann obviously couldn't stick a whole egg in the path of an electron microscope. Mindful primarily of obtaining the best pose for their picture, they adopted a technical shortcut that was shunned by most purists in the field.

It seems so trivial that it merited exactly four sentences in the "Methods" section of the paper they ultimately wrote about the 1956 experiments, yet it changed the experimental conditions in a dramatic way—"a crucial, *crucial* manipulation," their colleague Derek Burke would later point out. Isaacs and Lindenmann poured the embryo out one end of their fertilized eggs and then, using forceps and a considerable amount of ovoid dexterity, gently tugged out the entire remaining membrane. This left them with a small patch of living cells, roughly the size of a cookie and as thin as cellophane, which could be cut into smaller, two-centimeter-square segments and kept alive at least two or three more days when placed in a test tube and bathed in nutrient fluid. And that is exactly what they did in the next round of experiments. Instead of using eggs like almost every other scientist struggling to explain interference, Isaacs and Lindenmann put tiny pieces of membranes into test tubes, added the so-called erythrocyte ghosts with heat-inactivated flu viruses stuck to them, and then passed on the materials to Valentine, who would take the actual pictures. Unwittingly, they had changed the dramatis personae of their experiment—their three-character play now featured viruses, ghosts, and, in a last-minute substitution, pieces of membrane instead of whole eggs.

The heat-inactivated viruses, attached to the ghosts, induced interference as predicted, which delighted Lindenmann; this essentially confirmed his earlier experiments in Zurich, which Mooser had refused to submit for publication. Interference, he said later, "could be demonstrated as readily as I had done, but in a more elegant, less messy, more economical and more flexible manner." But, unbeknownst at first to Isaacs and Lindenmann, the mixture of cells, nutrient fluid, viruses, and red cell ghosts in the test tube created a proteinaceous muck that was the equivalent of smearing grease on the lens of an ordinary camera. "It was just impossible to say whether they were empty particles or full particles," Lindenmann said. "So really, this beautiful idea of Isaacs's just didn't lead anywhere."

And yet it did.

Even before they learned that the "after" pictures would come out hopelessly murky, Isaacs and Lindenmann had noticed something unusual, which provoked an additional, unplanned experiment almost to kill time while waiting for Valentine to produce the second set of electron micrographs. They wanted to see how many times the "damn ghosts," as Lindenmann began to refer to them, could induce interference. They tested this by removing the original piece of membrane from the test tube after twenty-four hours or so and dropping in a fresh one to mix with the remaining ghosts. When they transferred in this new square of membrane—something they could not have done with whole eggs, of course—they observed something rather unexpected. Normally, inactivated viruses of the sort stuck to the ghosts lost the ability to cause interference over time (Isaacs had coauthored the paper on this very fact during his days in Melbourne, so he knew it very well); in this case the ghosts continued to block the second wave of viruses with undiminished vigor. This looked "very suspicious," Lindenmann recalled: either Isaacs had been wrong before, or he and Lindenmann had changed something subtle about the experimental conditions.

"I believe that at that moment two mental movements were simultaneously present in our minds," Lindenmann later noted. "One was to go ahead with Isaacs's idea and pursue the electron microscopy work. This, however, brought interruptions in our part of the work, since delays were inevitable—waiting for our turn on the microscope, for pictures to be developed, occasional breakdowns, and so on. So the other mental movement was gravitating around the phenomenon of interference itself: The ease with which manipulations could be performed must have impressed Alick, and it certainly impressed me." When they began to concentrate on the "suspicious" results, they realized that the only truly novel thing about this new experiment was that the phenomenon of interference had stayed with the fluid; when they put a fresh membrane in the old fluid, this new membrane resisted infection longer than anticipated. In following that bouncing ball, as it were, Isaacs and Lindenmann shifted their focus, slowly but decisively, from the viruses to the membrane cells themselves. They began to realize that maybe it wasn't the first viruses that interfered with the second viruses. Maybe it wasn't the viruses at all. Maybe it was something the cells produced—"a sort of intermediate product of interference," as Lindenmann put it—that played the lead role. At that moment of preliminary epiphany, the official experiments on interference truly commenced. "Then it went really fast," said Lindenmann of the rush of experiments that followed. "Then it must have been a matter of three weeks."

For all his prolific writings, Isaacs apparently never recorded the thrill of reaching this juncture, but he surely savored the moment. Lindenmann assumed that scientists made discoveries like this all the time; Isaacs knew better. He also knew exactly what he was looking for, even before he found it, and in a gesture brimming with bravado and not a little arrogance, on November 6, 1956, he wrote in his laboratory notebook, "In Search of an Interferon." Lindenmann coined the name, incidentally, as "laboratory slang"; he had studied physics prior to biology and facetiously proposed "interferon" as an envious biologist's revenge for the muons, pions, and other atomic particles newly christened by his colleagues in physics—a bit of wordplay that appealed to Isaacs's impish, tongue-in-cheek sensibilities. As it dawned on them that the property of interference resided in the cell and was a substance of some sort, Isaacs one week later headed a page in his notebook, "Getting Nearer to Interferon." And then, toward the end of November, they performed what Lindenmann regards as the "crucial" experiment—a way of restaging their drama with only two actors, as it were, and waiting to see which one spoke the line "Interferon." They dropped the red cell ghosts from their script, guessing either the viruses or the cells caused interference. "We did not proceed systematically," Lindenmann later admitted, "but according to intuition or to preconceived ideas." It now became simplicity itself—or, as one interferon researcher later put it, not without admiration, "an almost idiotically simple experiment."

❧

The gist of the groundbreaking experiments was indeed quite simple: they dropped a piece of chick membrane into a nutrient-supplied test tube and then added flu virus. As expected, these inactivated viruses induced interference in the cells. Within hours of adding the first virus to the test tube, however, they removed the speck of membrane, washed it thoroughly to remove any clinging viral particles, and dropped this laundered confetto of cells into a clean, fresh test tube with fresh nutrient solution. Then they did nothing. They just let this "conditioned" piece of membrane steep for twenty-four hours in the fluid of the clean tube before removing it. Then—and this was the "idiotically" beautiful and simple step—they dropped a fresh, virginal piece of membrane into this second test tube and added fresh, live flu virus. The live virus failed to infect the virgin cells, even though these cells had never seen a first, conditioning, interfering dose of virus. Isaacs and Lindenmann then realized that the cells had been protected not by something in viruses but by something *in the fluid of the test tube.* That something, that "factor," had been "liberated," as they later put it, by cells in the first piece of membrane in the fluid, almost like a tea bag releasing its essence into a cup of hot water. They had found their interferon.

What they didn't fully appreciate is that it was a message wholly dictated by and released from the cells, a message spread via the fluid. When they later published the results, their first papers speculated that interferon might be an abortive viral protein that blocked infection. They had fair warning to consider an alternative explanation. Isaacs had dinner with his old mentor Macfarlane Burnet in London on June 25, 1957, and, as he wrote to Lindenmann the very next day, Burnet "seemed very impressed about interferon but had no suggestions, except that he thought it was a cellular rather than a virus product."

Burnet's intuition proved sound. Interferon ultimately turned out to be a molecule made by cells in response to infection by viruses; under duress, these cells oozed the molecule into the broth of biochemical liquid outside the cells in order to alert other cells in the vicinity to steel themselves against impending viral attack. It was, in short, a cell-to-cell signal, a neighborhood alert, a civil defense siren at the level of cells. Many years later, this family of molecule—there are now known to be many—would become known as cytokines.

As is often the case, this kind of molecule had left footprints in the test tubes of earlier researchers, and Isaacs and Lindenmann were not the first to have glimpsed a molecular beast of this sort. As early as 1928, two tuberculosis researchers named Margaret Reed Lewis and Arnold Rice Rich had reported that in certain circumstances white blood cells could be made to migrate, an early hint of signaling molecules, and in 1944 Valy Menkin of Duke University claimed to have purified a substance from pus that sent a signal to the brain to induce fever, which he called "pyrexin." In 1953 Ivan L. Bennett Jr. and Paul B. Beeson at Yale pinned down the identification of a factor they claimed was different from Menkin's fever-inducing molecule, calling it "endogenous pyrogen" and showing that it could be extracted from white blood cells (it later was renamed interleukin-1). That same year, Rita Levi-Montalcini and Viktor Hamburger of Washington University in Saint Louis isolated a substance that sent orientation signals to growing nerve cells, which they called nerve growth factor. As Lindenmann now admits, he was totally unaware of that previous work, and believes Isaacs was, too.

These were such isolated and tenuous observations, as research ahead of its time often is, that they often failed to make much of an impression on mainstream biologists and never appeared to be part of a larger, more general biological phenomenon—not, at least, until the 1960s. Lewis and Rich's work came too early, Menkin is said to have been too secretive, Hamburger and Levi-Montalcini influenced neurobiologists but not immunologists.

Lindenmann admits he was too young, and possibly too phlegmatic, to appreciate the significance of what they had stumbled upon and "took this more or less in stride." Not so Isaacs. "Alick immediately realized that it was something of great significance." And he immediately attempted to capitalize on their very good fortune.

First, he importuned a Mill Hill biochemist to perform some crude initial tests on the factor they had identified. Enter "Calamity Joe." If Lindemann was Isaacs's straight man, Derek Burke was Isaacs's fall guy, the man asked—without anyone knowing it—to do the impossible: purify interferon. At the time Burke was a rangy, high-strung, rather serious twenty-seven-year-old chemist learning the biochemistry of viruses while working on the first floor at Mill Hill in the chemistry department; Isaacs privately referred to him as "Calamity Joe" because of his penchant for laboratory mishaps. Nonetheless, they had successfully collaborated previously, and Burke was perfectly capable of performing the quick-and-dirty laboratory tests needed in order to publish the interferon work, so Isaacs approached him in February 1957.

"Alick came along," Burke recalled, "and said—I remember his words precisely—'And we're doing something rather interesting on interference. Would you like to help?'" Burke took the bait. He helped them pin down what they had a bit better, but not much. When he added ammonium sulfate, which will precipitate out a protein from a solution, he produced a slightly more purified form of the interferon. When he added trypsin to a test tube containing the interfering factor, interference no longer occurred; since trypsin digests protein, here was a clue that interferon was a protein. But these clues were not exactly definitive, and it wasn't until August 1957 that Burke established with reasonable certainty that they indeed were dealing with a protein. By then it was too late for publication—Isaacs and Lindenmann had prepared and submitted two papers on the experiments the previous March to the *Proceedings of the Royal Society* (Valentine was coauthor on one of them), and publication was imminent.

Isaacs, according to Lindenmann, was supremely confident about the result, but predicted—correctly—that many scientists would doubt the work. "Publish and be damned," he ruefully confided to Lindenmann just before the papers appeared in September 1957. The defiant remark resonates with both public bravado and private fear; neither sentiment was misplaced. C. H. Andrewes later said that the discovery of interferon "must be accounted one of the outstanding landmarks in biological science in its decade." But if Isaacs was right about his science, as indeed he would prove to be, he was also right about the skepticism of his peers.

The first in-house presentation of the results occurred at Mill Hill before Lindenmann left. "Alick was very enthusiastic," Lindenmann recalled. "In retrospect, one might suspect that he was in the upswing phase of his manic-depressive mood. This could have impressed those who knew him longer, and had also seen him in lower spirits, much more than it did me, who had never seen him differently. My feeling was that, having once convinced himself, he could become impatient with others who presented the very same objections he had already overcome in his own mind. I don't recall any particularly vicious comments from anyone, just the ordinary type of skepticism—whether the controls were carefully planned, whether the differences seen were statistically significant, whether any unsuspected bias might have skewed the results."

Like skaters testing the limits of a partially frozen pond early in the season, Isaacs and Lindenmann were the first to glide toward the middle of the cytokine pond and were on very thin ice. Not long after the first papers appeared, the whispers began, both within and beyond Mill Hill, about "misinterpreton" and "hypotheticon" and "imaginon."

Every discovery in science has at least a second act, even if it is only a disappearing act. The afterlife of interferon is one of the most fascinating and instructive in modern biology, in part because the public perception of the molecule as a therapeutic bust is completely at odds with the scientific perception of an exceptionally rich "model system" that has been used to understand the inner life of cells. The factor discovered by Isaacs and Lindenmann has turned out to form part of an extraordinarily interesting family of molecules (more than twenty different forms of interferon exist in humans alone). In addition, the interferon story has established a pattern (a paradigm, if you will) of pharmaceutical expectation and disappointment that has held depressingly to form for many of the cytokine molecules discovered in its wake. Moreover the story, as it subsequently unfolded at Mill Hill and elsewhere, previewed all the melodrama and tensions one associates with a biology of more rambunctious recent vintage: press conferences, patent battles, hype and expectation, disputes over credit and priority, a rush to enter clinical trials, horrific technical obstacles to purification, crushing disappointments for its two discoverers, all thistles bunched around the flowerlike beauty of the molecule and its potent biological activity, which has turned out to be fascinating in ways not even Isaacs and Lindenmann could have imagined. Finally, most sadly, the discovery took an exceptional toll—as discoveries often do—on the two decent men who made it.

Once a "factor" is discovered, it is the biochemists who fill out the birth certificate with the nitty-gritty details. Derek Burke's enduring contribution to the interferon field—entirely unintended and thoroughly cautionary for every biochemist who followed in his halting, staggering footsteps—was a profound, almost Beckett-like form of scientific frustration, a drama without denouement that might have been called *Waiting for Protein*. Burke was the first scientist to attempt to produce and purify interferon, a herculean task because, as it would turn out, very few molecules of interferon account for its potent antiviral activity. In so doing, he became patriarch to a long line of biochemists who would bang their heads against walls, build towering purification columns that would be the envy of Mies van der Rohe, and lie awake in bed at night trying to figure out a way to obtain quantities of pure interferon. "Immense, just immense," he said curtly when asked to describe the problem. "It was a mountain, yes."

Little realizing that the impasse was more in the nature of technology rather than of intellect or will, and that such a feat would not be technically feasible for twenty years or so, Burke and Isaacs optimistically published follow-up papers in 1958 that, if nothing else, illustrate the gullibility of peer reviewers; one claimed to reveal "Some Factors Affecting the Production of Interferon" (in the *British Journal of Experimental Pathology*), the other, in *Nature*, "Mode of Action of Interferon." In truth, they couldn't reliably produce any interferon, and they certainly had no idea how it worked. No less an authority than Burke himself now pronounces the 1958 paper on purification as badly flawed. "What we purified," he admitted, "was chick albumen!"—in other words, a contaminant. There was no dishonor in failing to purify a cytokine in the 1950s and 1960s. Indeed, the very difficulty testified to something new and important—no one had ever before glimpsed so much biological wallop in such tiny amounts of protein.

If Burke's sleep was roiled by nightmares about purification, the second act for Isaacs and Lindenmann became a living nightmare of an entirely different, and far more serious, sort. Within a year of their triumphant publication, damnation of a sort found both of them, just as Isaacs had grimly predicted. In June 1957, after an almost magical year of collaboration with Isaacs, Jean Lindenmann returned to Zurich and made the first public disclosure of interferon outside of Mill Hill, at a symposium in Bern. The mildly skeptical squall in the audience in Bern was nothing compared to a storm brewing back in Zurich. In a fit of scientific jealousy, Hermann Mooser accused Lindenmann and Isaacs of stealing his ideas about interference as reflected in the unpublished experiments by Lindenmann in 1956 and not crediting him with having inspired their work. Isaacs was perfectly capable of dismissing the allegations, and did so with a strong and persuasive letter to Mooser in November 1958.

Mooser was a man capable of prodigious and unforgiving indignation, however, and his wrath found a nearby and convenient target in Lindenmann, with disastrous implications for Lindenmann's career. Their relationship chilled considerably after the interferon work was published, and in 1959 Mooser finally forced Lindenmann to resign his university position on the eve of completing work on his advanced degree. The codiscoverer of interferon, now with two young children, was in effect excommunicated from Swiss science, hard-pressed even to land a job in a government laboratory in Bern. "This was a difficult moment for me," Lindenmann said, with considerable understatement. "Mooser died in 1971, aged 80," Lindenmann later wrote, "still believing that his not being recognized as the true discoverer of interferon was a grievous injustice."

Lindenmann all but dropped out of the interferon field, although not before publishing several more strong papers during a brief three-year stay at the University of Florida in the early 1960s. Shrewder perhaps than Burke, he realized that "progress in interferon research would rely mainly on biochemistry, and as long as the best biochemists were not able to come up with the formula for interferon, then it was useless to do too many biological experiments." He also realized that as the visiting junior scientist with no track record, he would forever remain in Isaacs's shadow when it came to the discovery of interferon. But at least one interferon expert believes the sober, taciturn Swiss scientist was essential to the collaboration. "It is my opinion that the collaboration . . . was fundamental to their discovery of interferon," the Finnish virologist Kari Cantell has written, "and would not have been made by either working on his own."

Lindenmann's travails paled compared to the purgatory that Isaacs entered late in 1958. While still wrestling with Mooser for his scientific honor, Lindenmann had sent a letter to Isaacs and received in reply a grim dispatch from C. H. Andrewes. "Alick is ill," Andrewes wrote from London in the fall of 1958, "suffering from a very severe depression," and had been forced to take an extended leave from work.

Isaacs's problems were both intensely private and, paradoxically, highly public. Several family events had placed him under unusual strain in the spring and summer of 1958, but the stress of the interferon controversy may well have played a contributing role. Long before "hype" and "science-by-press-conference" entered the standard vocabulary of scientific derogation, interferon blazed into public consciousness in newspapers, in magazines, in the scientific press, and on television, setting a standard by which all subsequent wonder drugs, miracle molecules, and magic bullets would be measured—and similarly, found wanting. Science journalism was different in those days: a reporter called Isaacs after the work was published in the fall of 1957, but Isaacs said he didn't want to discuss it, and the matter ended there. Almost a year after publication, however, at the annual meeting of the British Association for the Advancement of Science in August 1958, interferon was "disclosed" to the world at large by way of a press conference. Isaacs and Derek Burke also donned white tie and tails to describe it at the Royal Society's annual "Conversazione," a kind of science fair in formal dress where scientists discussed the year's most interesting work; they appeared on television, when such was a rarity for scientists, to discuss the factor. Fatefully, someone—probably Isaacs, although the origin is unclear—described interferon as the "antiviral penicillin." "Alick didn't shrink from publicity," Burke recalled. "He was quite a good publicist. And he talked about it very engagingly and persuasively, of course. So, yes, he raised the stakes. And then had trouble delivering, and that could have been part of—*part* of—the depression."

The initial hopes raised about interferon as an antiviral agent played upon a sensitive nerve almost impossible for someone outside Britain to appreciate. Alexander Fleming had discovered penicillin in 1928 at St. Mary's Hospital, a few miles down the road from Mill Hill, but through a complicated series of scientific and commercial misadventures related to World War II, the process used to manufacture this truly miraculous antibiotic had been developed and patented in the United States. As a result, the British found themselves forced to pay onerous royalties on a discovery made in London. During the 1950s bitter debates broke out in Parliament about the effect the penicillin royalties had on the British balance of payments, and the mere mention of the "antiviral penicillin" not only fired feelings of pharmaceutical patriotism in the United Kingdom but evoked the widespread sentiment—in political as well as medical circles—that such intellectual piracy should never be allowed to occur again. There existed therefore a great deal of pressure to obtain a patent for interferon, but this was not simply a matter of submitting the proper forms. In order to patent the "antiviral penicillin," British researchers had first to establish its efficacy as a drug; in order to prove its efficacy, they hastily had to organize clinical trials; and in order to conduct such trials, they needed substantial quantities of a drug that, as Burke's sleepless nights attested, was devilishly hard to quarry. Thus the Scientific Committee on Interferon, formed in 1959, with—just back from medical leave—Alick Isaacs as chairman.

High-profile (and high-stakes) committee work of this sort had the unhappy side effect of heaping a great deal of pressure—scientific, political, and organizational—on Alick Isaacs at a time when his shoulders were perhaps not quite broad enough to bear such a heavy load.

According to at least some of the people with whom he worked closely, he was never quite the same man and the same scientist after the initial bout of depression; he suffered recurrent manic depression, according to his wife and colleagues, and it is an open secret that he seems to have manifested manic-depressive behavior after his return to Mill Hill in the spring of 1959. On first thought, it may seem almost an act of cruel institutional indifference to place a fragile and still recovering psyche like Isaacs's at the center of interferon's controversial and uncertain development. On the other hand, Isaacs insisted on sitting at the head of the table. More than one former colleague has remarked that during his manic phases, Isaacs was unusually productive, and he certainly wanted to be in the lab to push the interferon story further along toward the first clinical trials. These trials, conducted in 1961 and published the following year in the *Lancet*, were by contemporary standards rather well conceived, but of a scope and impact far short of the promise implicit in an "antiviral penicillin." The committee pooled a group of 38 adult volunteers, none of whom had been vaccinated against smallpox and many of whom were employees at the various labs participating in the study. In double-blind fashion, each volunteer received injections at two sites on their arms, one with monkey interferon and the other with a neutral fluid; twenty-four hours later, these same two sites were cut and rubbed with vaccinia virus, the kind used in smallpox vaccine. The thought was that interferon would disrupt normal vaccination, an interference that could be measured by the extent of inflammation around the site of vaccination. Virtually every vaccine injected into the control site "took" (37 of 38), while only 14 of 38 vaccinations took after prior treatment with interferon; the participating scientists interpreted this to mean that "a highly significant degree of protection by interferon was found." Isaacs proclaimed the experiment a success—"Work on interferon has just passed a critical stage," he wrote in *New Scientist*—but he alone seemed convinced of interferon's future in medicine. The results hardly reminded the drug companies of the Lazarus-like cures effected by penicillin in its very first uses.

Perhaps because of his pedigree by way of Burnet, Isaacs always viewed interferon in an immunological context, and the irony here is that immunology as a field was still so immature, even within the lofty confines of Mill Hill, that hardly anyone saw a role for interferon as anything other than a curiosity. Upon the arrival of Peter Medawar as director in 1962, Mill Hill became a hotbed of immunology research; Medawar had just received a Nobel Prize in 1960 for earlier experiments on immunological tolerance, specifically the way fetal tissues could be "trained" not to reject transplants of foreign tissue, little realizing that interferon and other molecules like it played essential roles in the process of rejection and tolerance. Interestingly, Isaacs proposed a collaboration with his friend Leslie Brent, who had worked with Medawar on the Nobel Prize–winning experiments.

"When I was there, Alick was already in a sort of slightly manic-depressive phase," Brent recalled in an interview. "And he could be tremendously bubbly and enthusiastic, full of ideas, some of them quite wild. And as a result, I think, one tended to be a little bit on one's guard. . . . For example, I think he wanted me to do a joint experiment with him on tolerance and interferon, to see whether one could actually induce tolerance. And I was—well, I was embroiled with lots of other things at the time, and I didn't think it had a great chance of success, so I rather backed off that, and we never actually did those experiments." Brent is probably correct in suggesting that the proposed experiments might not have yielded much. But Isaacs's instincts were sound; Brent noted that interferon, in its gamma and beta forms, has turned out to be "clearly important" in the very processes that he and Medawar studied. "I think interferon's universality was certainly not appreciated, and I don't think it could have been, really."

Nonetheless, Isaacs viewed that first clinical trial through an immunological, not a virological, lens. "Interferon can be thought of as the body's natural means of defence against virus infections," he wrote in 1962. "It can be likened, therefore, to antibody which is produced by the body as a defence mainly against foreign proteins. In medical treatment, antibody is not very often given directly, except for gamma globulin in infections like measles, and antitoxin in diphtheria. Instead, people are stimulated actively to produce their own antibody, by the injection of vaccines. It may be that one day interferon injected or otherwise given to patients will prove to be less useful than a means for stimulating people to make their own interferon, if a harmless way of doing this can be found." This was Isaacs at his best—he was one of the earliest to place interferon, and by extension all the cytokines to follow, in the same arena with antibody, a daring assertion in the 1960s and one that invited ridicule from the immunology community, but one that has been borne out by three decades of subsequent research.

But even at the best of times, Isaacs's enthusiasms were not universally shared, and when he added, "The committee on interferon expects to have enough work to keep it busy for some time to come," he was essentially writing as a committee of one.

The reality was quite different. British pharmaceutical companies viewed the first clinical trials as discouraging, and their interest in producing interferon cooled. According to Norman Finter, a member of the committee, Isaacs "complained bitterly that the companies were grossly underestimating the potential for interferons, and that these would have been developed much faster if American rather than British companies had been involved." Hailed as the viral penicillin in 1958, interferon's future by the mid-1960s had already fallen into eclipse, and there was no reason to expect the prospects to change quickly. No one could purify it; no one could produce it in quantity; and by the mid-1960s, even at Mill Hill, annual reports of scientific work by the virology division hardly mentioned it. The molecule, while scientifically interesting, no longer seemed to promise huge therapeutic benefits—or huge commercial rewards. In the early 1970s Macfarlane Burnet remarked to a colleague in Australia that it was a pity that interferon had been discovered as an antiviral agent, because it possessed so many other activities.

Perhaps the least understood aspect of the interferon story, however, is just how huge the scientific rewards were, how quickly they became apparent, and how little they mattered to a public that, as always, is more appreciative of practical applications than of basic knowledge. Once they started looking for it, researchers found that interferon had left a calling card almost everywhere they turned. In 1961 Ion Gresser, then at Harvard, reported in a largely overlooked experiment that leukocytes, or white blood cells, made interferon; nobody could appreciate it at the time, but the observation nudged the molecule closer and closer to the realm of immunology. In 1962 Kurt Paucker and Kari Cantell, two Europeans working in the Philadelphia laboratory of Werner Henle, became the first to show that interferon had unusually powerful "cytostatic" effects—it could prevent cells from growing. Shrewd experiments by Joyce Taylor and Joseph Sonnabend in Isaacs's lab at Mill Hill in 1963 established that interferon acted on the nucleus of the cells it affected; it seemed to turn on genes inside a cell to do something under conditions of threat that they didn't otherwise do, and it therefore developed that interferon was only the most visible link in a biochemical relay stretching from the surface of the cell to its headquarters. In 1965 E. F. Wheelock of Western Reserve University in Ohio identified a new form of interferon in leukocytes that clearly established an immunological role for the molecule (this "immune interferon," now known as gamma interferon, was one of the molecules triggered by Coley's toxins). And by 1974 Ian Kerr, also at Mill Hill, had teased part of the biochemical signaling mechanism apart, reporting an early example of what has become, two decades later, an enormously fruitful area of biological investigation—and commercial investment—known as signal transduction. Only in retrospect can the early interferon work be seen as lighting slow-burning fuses that led, inexorably, to bursts of research that would ultimately illuminate virtually all cytokine research a decade later and be central to cell biology in the last two decades of the twentieth century. The problem wasn't that interferon didn't "do" anything, as some critics had implied; to the contrary, it did too much.

But the line of experiments that transformed the field—indeed, created a huge vacuum of need that helped tug the nascent technology of genetic engineering out of its academic nest and into full flight—began in Paris in 1966. Ion Gresser, born in New York and named after one of the dialogues of Plato, left Harvard to become an expatriate American researcher at L'Institut de Recherches Scientifique sur le Cancer in Villejuif, France, just outside Paris. A man of Thoreauvian curiosity and intellect, with a prickly sense of independence that went along with very astute scientific intuitions, Gresser has worked on interferon for more than thirty years and has consistently opened up new areas of inquiry, often years before they were fully appreciated by others. But even he did not appreciate an unexpected observation in the laboratory that, like the original discovery of interferon, at first inspired grave doubts.

At the time biology enjoyed a resurgence in the theory that viruses caused cancers, as had been demonstrated in several animal tumor systems. One of these was the Friend leukemia virus, which caused cancer in mice, and on one of his periodic trips to visit his family in New York, Gresser obtained some of these viruses from Charlotte Friend, the discoverer of the virus, to perform some experiments with interferon in mice. His thought was that as an antiviral agent, interferon might block or delay the development of virally induced leukemias in these mice, as indeed proved to be the case. In January 1968, in a set of follow-up experiments ably recounted in Sandra Panem's *Interferon Crusade*, Gresser tested interferon against a different kind of laboratory cancer—a nonviral, transplanted tumor. He assumed, he admitted later, that interferon would have no effect on these cancers, and when all the animals treated with interferon survived the experiment unharmed, Gresser accused his technician, Chantal Maury, of having forgotten to inject these mice with tumors. In fact, she had. When they repeated the experiments in the summer of 1968, their work in part delayed by the student riots in Paris, they obtained the same results. Gresser repeated these experiments many times before publishing in 1969, by which time he had created a monster. Interferon prevented cancers in mice.

Cancer changed the whole ball game for interferon once again. It went from the "antiviral penicillin" to the "magic bullet." For those who cared to believe the results (and there were many who did not at first), interferon cured mice of cancer. Gresser recalls sitting in his lab with his colleague Pernilla Lindahl and saying, "Why don't they believe us?" And she replied, "Don't worry, one day not only will they believe us, they will do the same experiments and nobody will remember that we did them!" She was right.

The year 1967 was a watershed in several ways for the field. The idea that interferon could blunt the headlong progress of cancer changed the way the field unfolded. Like Isaacs before him, Gresser at first encountered overwhelming skepticism. And as Isaacs had found, there was no easy way out in explaining what interferon was and how it did what Gresser claimed it did. At the same time, the field was virtually frozen in takeoff. Still, technology had not caught up with the biology of these potent molecules. Biochemists still could not isolate it in a test tube or purify it; they still could not produce it in bulk; they knew it possessed all sorts of biological activities and yet still could not pin down its very existence. The snide whispers about "misinterpreton" had by no means died down. A decade after its discovery, interferon tantalized like a kind of fata morgana, a mirage wavering in the distance, just off the shores of achievable biology, just out of reach.

It is against this backdrop that the field lost its most respected figure. In January 1964, after several years of uneven work, Alick Isaacs suffered a crippling brain hemorrhage that left him with reduced visual skills and reawakened more pronounced episodes of psychological distress. "After this," C. H. Andrewes wrote later, "he underwent a series of episodes of profound mental disturbance. . . . In between his bouts of illness Isaacs was still bubbling over with fresh ideas about interferon, but the quality of his work was not what it had been."

Isaacs's battle for intellectual normalcy, for a return to the wit and brilliance of his earlier career, is without a doubt the saddest chapter of the interferon story. A long string of manic-depressive episodes required periodic hospitalizations and occasional medication. Relegated to a small two-person Laboratory of Interferon Research shared with Joseph Sonnabend, he was a shadow of the scientist he had once been. Photographs from the 1960s no longer hint at the impish humors within; he had aged tremendously under the stress of an illness that, in his many lucid moments, he understood only too well. Unbeknownst to his colleagues, he attempted to take his own life on at least two occasions.

"He was just beginning to regain his form just before he died, which was very sad," Sue Isaacs recalled. She sat in the cold blue afternoon light of a London winter that filled the kitchen of the house in Hampstead where her husband suffered the final hemorrhage, reliving the phone call to the doctor, the trip to University Hospital, the anxious last hours. At her suggestion, he had commenced psychoanalysis the previous autumn. "He went, and he was ever so much better. Gradually he got a lot better and came home for Christmas in 1966. And he was as though maybe things were going to be okay. We were going to be able to settle into life. And we could talk about anything, and however long we'd got, we got. And Christmas was always good in our house, just because it was a nice occasion." Even thirty years later, a tiny bit of hopefulness began to creep into her voice as she continued. "He'd been home for Christmas, and I think he stayed home or had perhaps just gone back for his last stay in the hospital. And one had the feeling that *really* would be the last stay. . . . I don't know if he had started work that January. I should think not. Not quite. But whatever he was going to do, really we felt we'd got him back again, then. And then"—she paused—"then he had another hemorrhage." This second, even more catastrophic bleed struck on the evening of January 24, 1967. Two days later, Isaacs died in University Hospital, London, at age forty-five.

"The chief loss was that the field kind of lost its patron saint, its rallying point," said Robert Friedman, a former colleague. "I mean, he was the only one in the field who had any degree of scientific acceptability and respect. All the rest of us were too young, and by then it was considered a marginal area of science. And the field, in losing him, had lost some of its credibility, and it had lost its rallying point, spiritually and geographically, because Mill Hill was the place where all the work was being done. If you wanted to learn about interferon, you went and worked at Mill Hill. And once he was lost, there was no one to pick that up." So by 1967, ten years after the discovery of interferon, its discoverers had departed the stage. Lindenmann was back in Switzerland, but out of interferon research entirely (or so he thought), and Isaacs was dead. The final and ultimate sadness is that at least one close colleague believes that this brilliant and imaginative scientist, among the first to wander fearlessly through the fogs that marked the dawn of cytokine research, may himself no longer have believed that interferon existed.

Not only did the molecule exist, but it can now be seen as protagonist in a kind of archetypal biological narrative, one that has been emulated but never surpassed by all the potent molecules to follow—a "round unvarnish'd tale," Isaacs once called it, quoting Shakespeare. "The glimpses we have of the road ahead," he added, "make us eager to press on with the journey." His would turn out to be a little cross on the side of a very long and winding road.

7

LYMPHODREK

. . . blood is knotted to life—in fantastical wise—
to its passions, to its commotions.
—Piero Camporesi, *Juice of Life: The Symbolic*
and Magic Significance of Blood, 1995

☙ *The spectacular rise of interferon* as a purported anticancer drug, and even the haste with which venture capital by the millions poured into the biotechnology industry in the late 1970s, can in some ways be traced directly back to a horrible tragedy on a Finnish highway in the mid-1960s. At the time, Hans Strander had been working in Finland as a graduate student for about a year. He was a large-boned, barrel-chested Swede with a mop of dark hair and well-worn laugh lines crowding the corners of his eyes; having trained in medicine and immunology at the Karolinska Institute in Stockholm under the estimable tumor biologist Georg Klein, Strander had arranged to pursue his doctoral thesis in Helsinki with an enterprising young scientist named Tapani Vainio.

Vainio headed the blood group department at the State Serum Institute and, undaunted by the problems that stymied interferon research elsewhere, astutely recognized that interferon potentially held the key to understanding how cells receive messages from other cells—could, in short, serve as a model system for cell biology. Strander eagerly assisted Vainio on this ambitious line of inquiry, but the work had barely gotten under way when, on August 16, 1965, the thirty-five-year-old Vainio drove off from his home in a suburb of Helsinki without fastening his seatbelt and was broadsided by another vehicle at the first intersection. He never regained consciousness and died that same day.

Grief was the crucible for the scientific alliance that ensued. Suddenly bereft of his charismatic mentor and also stripped of a topic for his graduate thesis, Strander wondered whether to return to Stockholm or to stay in Finland and work with another scientist he'd been collaborating with on a somewhat more prosaic project. Fatefully, he openly discussed this dilemma with his colleague, a quiet and self-deprecating virologist named Kari (pronounced "Cory") Cantell. The more the two men talked, the more they realized that continuing their collaboration made sense. Planning to spend only a year in Finland, Strander ended up staying five more years working at Cantell's side, on a rather less ambitious but nonetheless daunting project. He could not guess at the time, but it would turn out to be one of the most thankless and extraordinary biological enterprises in the world, one that would forever change the field of cytokine research, the medical applications of recombinant DNA technology, and the commercial emergence of biotechnology.

Cantell would not impress anyone at first blush as having the candlepower of his friend Vainio. Strander was "typically nordic, robust and calm, a viking"; Cantell, in Cantell's own view, was quiet, stubborn, and a bit of a loner. He looked more like a government functionary, which in a sense he was; his laboratory screened blood and other medical specimens sent from throughout the country for evidence of viral infections, and Cantell had also established a small name for himself by developing a mumps vaccine for the Finnish military (mumps was an especially dangerous disease in Finland, where many young men grew up in sparsely populated areas and were never exposed to the virus until they joined the armed forces, by which age the mild childhood illness turned into an adult disease with dangerous complications). Cantell was a short and compact man with horn-rimmed glasses, short brown hair, and a tight, lean, almost ascetic face. He possessed a low, sonorous voice that was an especially appropriate instrument for the fatalism, the philosophical ambiguities, and the intellectual modesty he so often expressed. Early in his career, when he was invited to write a review article for a prestigious American journal, he wrote back suggesting that there must have been some sort of misunderstanding. He viewed science as beset by fashion, little realizing that the work he and Strander initiated would lead to one of the noisiest fads of biological research in the 1970s.

At the time of Vainio's death, Cantell occupied a modest office at the State Serum Institute in Helsinki, and it was in a small, wood-paneled laboratory immediately adjacent to this office that Cantell and Strander performed a kind of cellular alchemy that became, for a while, a wonder of the world; they figured out a method for producing what soon would be the entire world's supply of human interferon suitable for use in patients. Their intent from the beginning was to produce human interferon from white blood cells, or leukocytes. "We saw these cells as potentially a rich source of human interferon that might be used in therapy," Cantell noted many years later, "and so our main aim was to optimize the production of this sort of human interferon in a way that could be taken up to a very large scale." Cantell's production method, as heroic as it was destined to become obsolete, almost single-handedly kept the interferon field alive for fifteen years, and it all took place in that little room, looking out upon the Mannerheimintie, a busy thoroughfare leading north out of the city center. At a time when other cytokines were merely whispering their existence to biologists in other labs, Cantell and Strander had established a cottage industry; they harvested cells by the billions, goosed as much interferon out of the cells as they could, filled ampoules with this crude preparation, and shipped it to the few researchers in the world lucky enough to be on their very selective mailing list. "Many visitors were surprised," Cantell said later, "when they came here, expecting to see a factory and instead seeing that it was so small and modest."

Perhaps because he is so humble about his own talents, Cantell does not mind an empirical, trial-and-error type of science; another way of putting it is that he possesses enough respect for nature not to try and outthink it. And so when he decided at the beginning of 1963 to study the medical potential of human interferon, he played no hunches: he tried dozens of different types of human cells in cell culture, and attempted to induce the production of interferon by infecting these cells with many different viruses under many different culture conditions. It is the kind of science fraught with failure.

Following up on an observation by Ion Gresser in 1961, Cantell tried white blood cells from humans in May 1963; rummaging around in a laboratory freezer one day, he discovered by chance a rare strain of Sendai virus, which for reasons that remain a mystery to this day tickled "sensational" amounts of interferon out of white blood cells. The term "sensational," however, demands qualification. By Cantell's estimate, the average human blood donation contained roughly a billion white blood cells, and he calculated that it required white blood cells culled from no less than 70,000 human donors to yield 250 milligrams—a quarter of a gram—of pure human interferon. That was the mountain that lay before them. Strander stayed alongside Cantell every step of the way for five years, and the two scientists developed a close and enduring friendship that persists to this day—a friendship that had a very direct bearing on the early clinical testing of interferon.

Interferon's thousand-mile journey to the cancer clinic began with a single, mistaken assumption. By the time Strander returned to Stockholm in 1969 and began to practice oncology at the Radiumhemmet, or cancer ward, of the Karolinska Hospital, he and Cantell had already hatched plans to test their interferon against cancer. But which kind of cancer? In the late 1960s viruses had come to dominate the thinking of researchers looking for the cause of cancer, and Hans Strander began to play with the notion that several forms of cancer—including osteogenic sarcoma, a nasty malignancy that typically attacks children—might have a viral origin. Bone cancer therefore became a natural target for a known antiviral substance like interferon. After Gresser's experiments in Paris in 1966 and 1967, animal studies seemed to lend even greater support to that view. So just when a small but aggressive band of researchers began to put two and two together, to clamor for pure interferon to try against cancers of presumed viral origin, Hans Strander alone among virtually all the oncologists in the world had already established a Scandinavian pipeline to the only laboratory making it, had indeed distilled the precious protein with his own hands alongside Cantell, and therefore enjoyed privileged access to a substance so scarce, and potentially so powerful, that within a few years it would be a household synonym for medicinal, or at least pharmaceutical, gold.

The problem is, the work got under way well before anyone realized that viruses did not cause bone sarcomas. Nonetheless, after initially testing leukocyte interferon in a melanoma patient in 1969, Strander went on to treat at least twenty-eight patients in Sweden over the next four years, including eleven youngsters with osteosarcoma, with the Cantell preparation. The preliminary results from these trials remain controversial, although the Australian immunologist Gustav Nossal once referred to Strander's initial efforts as "action-oriented empiricism of a heroic order."

Strander and Cantell never claimed more than that humans could tolerate high doses of interferon, and that a few isolated cases looked "promising." The real problem, in retrospect, was how other people interpreted these results. In the early 1970s, for example, Strander compared fourteen bone cancer cases he had treated against "historical controls"—that is, against general statistics on the survival of osteosarcoma patients—and concluded that interferon treatment might confer prolonged survival. Cantell later admitted that the use of historical controls was a "bad mistake." The key point here is that a politically influential researcher at Sloan-Kettering Cancer Research Institute named Mathilde Krim organized a famous meeting in New York in 1975 to promote the production of interferon for further clinical testing, and the genie was out of the bottle. Two years later, the American Cancer Society authorized the purchase of $2 million worth of leukocyte (or alpha) interferon, virtually all of it from Cantell's lab, and the National Institutes of Health also purchased large amounts for testing. The perceived medical need for interferon—a need, incidentally, that was never proved beyond supposition—had the effect of creating a demand in search of a technology. If anything, this new demand only highlighted an old dilemma: there was barely enough stuff to test. As Cantell himself put it, "Our shortage of interferon seems to be a chronic and incurable disease."

Immunologists, if they took note at all of Cantell's efforts, were destined to remain unimpressed—the "viral penicillin" had precious little to do with the central issues of immunology in the 1960s, and Cantell's little distillery seemed more like plumbing than serious scientific inquiry. And yet there was an important message, at least inferentially immunological, buried under the trillions upon trillions of white blood cells that Cantell culled and milked in his Helsinki laboratory—in a sense, a fundamental shift in focus that interferon research anticipated by a good decade. Because of Cantell's method, the Finnish Red Cross made a point of separating the white blood cells from the rest of its donated blood. They kept them, used them, manipulated them, studied them. And this observation, made purely on the basis of practical considerations and virtually insulated from the concerns of immunology, invites a detour into the immunology of the late 1950s and 1960s, and its almost incidental intersection with the important observation about leukocytes collected and processed in Cantell's laboratory. The clinical experiments with interferon paralleled, and in some respects anticipated, a significant broadening of immunological thinking. If molecules like interferon served as signals, it begged several obvious questions: where did the messages come from, and where did they go? The answer, in both cases, was cells.

Immunologists are fond of citing Mephistopheles' wondrous exclamation in *Faust:* "Das Blut ist ein ganz besonderer Saft" ("The blood is an absolutely remarkable fluid.") But in fact it took a very long time to discover just how wondrous a juice it is, and you could convey a rather irreverent but nonetheless revealing history of immunology in the twentieth century simply by describing the parts of the blood that were considered expendable by researchers busy pursuing "more important" scientific questions.

From the very first application of the centrifuge to human blood, researchers acquired the ability to separate human blood into several component fractions. If you spin blood in a centrifuge, you are left with three layers. Normally the bottom 40 to 45 percent of the test tube will be red, packed with the heavier red blood cells. These are "dumb" cells, without a nucleus, essentially single-celled hod carriers ferrying oxygen to the tissues of the body and wastes back to the lungs and kidneys; and yet even this proletarian stratum in the society of blood possesses some sophisticated internal divisions. In 1901 in Vienna, Karl Landsteiner (later of the Rockefeller Institute for Medical Research) showed that these dronelike cells fall into distinct and mutually exclusive groups, which he classified as A, B, and O, on the basis of small molecular markings on the surface of the cells known as blood group antigens (the fourth blood group, AB, was identified the following year). The distinction became crucial when blood transfusions became a routine part of modern medicine.

The top 45 percent or so of the tube comprises a thick, straw-colored liquid, almost the color of honey mustard or an apricot jam; this is the plasma, also sometimes known as the "serum." It was from this layer that Louis Pasteur and Emil von Behring fished out the antitoxins for rabies and diphtheria, respectively, and where for many decades thereafter biologists prospected for other similarly powerful "serum-based factors" in what appeared to be the mother lode of the blood's component parts. Not without reason: the key elements of immunological activity against toxins reside in this layer, namely the antibodies, and before the turn of the century, a number of laboratories—including New York City's Public Health Department—produced and sold antitoxins against several diseases. But this effort did not in the long run yield much success; as historian Arthur Silverstein has noted, the heyday of serum-based medicine was relatively limited and short-lived.

Separating these two larger layers is a thin, watery, barely perceptible yellow-pink buffer of biological material, as blurry and indistinct as the seemingly insignificant boundary separating two bold color fields in a Mark Rothko painting. Known as the "buffy coat," this thin layer was so imperceptible that for many years it was routinely lumped together with the red cells by hospitals and blood banks, which fractionated blood in order to get red blood cells for transfusions. Unfortunately, what turned out to be half the immune system lay in that thin layer, and often caused inexplicable adverse reactions in patients. Packed in this barely discernible layer are all the nucleated cells, which by convention have come to be called "white blood cells"—lymphocytes, macrophages, neutrophils, all specialized cells with specific and essential functions in inflammation and immunity. As our hard-won knowledge of molecular markers upon their surface now tells us, these cells can be classified into subsets upon subsets, each with distinct and specialized functions. To cite one example of nomenclature that has unfortunately impinged upon public consciousness, certain T lymphocytes possess a surface marker known as CD4; these perform the role of "helper" cells in an immune response, and it is this subpopulation of white blood cells that declines inexorably in AIDS patients. In similar fashion, so-called killer T cells possess the CD8 marker, natural killer cells the CD56 marker, and so on.

Michael Lotze, an immunologist and surgeon at the University of Pittsburgh, has likened serology in the days prior to molecular biology to "paleo-immunology." All that began to change with dramatic swiftness in the late 1970s, in part when immunology met the enabling technology of genetic engineering. But like most great rivers of change in science, this one meandered through a number of different jurisdictions and gathered momentum with the confluence of several smaller streams of thought, some of them reflecting basic biology and others reflecting sheer technical innovation. One significant tributary flowed out of the research on interferon, with its notion that cells talk to each other by secreting molecules known as cytokines, of which interferon was one of the earliest and certainly the most publicized of avatars; in the general chatter of molecular conversation, as biologists were in the process of discovering, perhaps the chattiest of the lot were the immune cells. Just which cells were doing the talking, what they were saying, and when they felt the urge to speak—all the specifics of the dialogue—remained a muffled conversation behind a closed door, a separate room to which no one yet held the keys. But after many decades during which immunologists focused almost exclusively on serum as the enchanted realm of immunology and on antibodies as the principal keepers of the peace (a bias that filtered down to the public in the sense that many children who lined up for their polio shots in the 1950s, and their parents, still think of antibodies and immunity as interchangeable terms), biologists began to appreciate the fact that acquired immunity was a two-character play, with another lead actor in the immune response: lymphoid cells.

There are any number of places to begin the story of the cellular part of immunity. The traditional nativity scene is set in Messina, Sicily, where the Russian émigré biologist Elie (or Ilya) Metchnikoff demonstrated a role for cells in the immune response. Metchnikoff—likened in Paul de Kruif's classic, though hyperbolic, *Microbe Hunters* to "some hysterical character out of one of Dostoevski's novels"—showed that by inserting a thorn into the transparent larva of a primitive marine creature like a starfish, he could observe a remarkable immunological drama transpire under the microscope. "Wandering cells," as he called them, migrated to and congregated at the scene of injury. He marveled as they slowly engulfed bits of the intruding thornlike amoebas and eventually digested the debris. Metchnikoff called those cells phagocytes (from the Greek for "to eat"), because they seemed to surround, attack, and ultimately devour the foreign material; they are now more commonly known as macrophages and monocytes and neutrophils.

But in science, as in life, timing is all: Metchnikoff had the bad fortune to announce his discovery of the phagocyte in 1883, at the height of public acclaim for Pasteur and Koch, at precisely the historical moment when "antitoxins" to typhus and diphtheria had astonished the world with the practical value of antibodies. "The discovery of antibodies," Debra Jan Bibel notes in her review of this and other crucial experiments in her book *Milestones in Immunology*, "pushed phagocytes out of the limelight." In one of those unfortunate "either/or" scientific debates that, ostensibly in the interests of clarity, can muddy a field for a generation, immunologists at the turn of the century felt compelled to decide whether cells or antibodies played *the* critical role in immunity. As time would ultimately show, the question itself had been framed incorrectly. They both played critically important, complementary roles in defending an organism against injury and infection.

The fact that immunological cells, especially lymphocytes, had received short shrift became sufficiently obvious by the 1930s that Arnold Rich decried "the complete ignorance of the function of this cell" as "one of the most humiliating and disgraceful gaps in all medical knowledge." But the rehabilitation of the cellular arm of immunity, it is generally agreed, took a decisive turn due to experiments at the Rockefeller Institute in the 1940s, when Karl Landsteiner and Merrill

W. Chase "deconstructed" an immunological response in animal experiments, demonstrating a surprising role for cells.

The Rockefeller researchers looked at two types of immune response, the almost immediate welts typical of an allergic response and the delayed welts that occur, for example, in response to the injection of tuberculin antigen under the skin, which take one or two days to form. They established these allergic or hypersensitive reactions in guinea pigs—a recall test similar to current skin tests for allergies or tuberculosis—and showed that delayed reaction to tuberculin involved immune agents other than antibody. When they removed the "exudate" (basically, the fluid produced by inflammation) that oozed from animals challenged by antigen and then centrifuged the material, they found they could cause the same intense immunological reaction in other animals by transferring the "sediment" in their test tubes, but not the clarified fluid. The sediment, of course, contained cells—white blood cells, to be precise. Hence, leukocytes—a loose term denoting a mixed population of cells—played a role in the immune response. Only in the 1950s, however, did cells truly begin to emerge, as it were, from the shadow of antibodies, and they did so not always because of careful reasoning and bold scientific hypotheses. Sometimes it was an amusing combination of blind luck and the grudging acceptance of unexpected and otherwise inexplicable results.

Perhaps the most startling development, not least because it trumpeted such unique news in so obscure a venue, occurred in 1956. *Poultry Science* is not normally considered a must-read for immunologists, but Bruce Glick, Timothy S. Chang, and R. George Jaap published a two-page article in the journal that began to connect healthy immune function to the activity of small but significant glands. These three scientists identified a key organ without which chickens are unable to mount an immune response. Like many accounts in the literature, this article barely hinted at a very serendipitous scientific discovery with immense immunological ramifications.

Glick, then a graduate student at Ohio State, had been pursuing one of those seemingly modest dissertation topics that add one more anonymous brick of ballast to the great ship of graduate studies. It had to do with a tiny and, if history was any judge, by no means compelling mystery of bird anatomy. Birds possess a small, blind sac called the bursa of Fabricius, which sits near the excretory tract known as the cloaca; that the function of this bulb of tissue had remained a mystery since its discovery by Hieronymus Fabricius in the sixteenth century suggests how little urgency attended its explication. Glick, in short, had set out to explain something that had remained inexplicable, to no one's great distress, for four hundred years.

The bursa was believed to function something like a lymph gland in young chickens and might also act like an endocrine gland. Nobody knew. But it is a time-honored strategy in biology to remove a component—an organ or tissue in the old days of gross anatomy, a single gene in the "knockout mice" of today's more precise molecular erasures—and simply see what happens. So Glick removed the bursa of Fabricius from a number of birds: adult chickens, young chickens, newly hatched chicks. Nothing happened. All the birds remained in perfect health. Glick sheepishly returned all his experimental subjects to the university's pool of surplus animals.

About this same time, Glick's friend and colleague Timothy Chang served as a teaching assistant in a class at Ohio State on immunology, and as part of a classroom demonstration on antibody formation, he fetched a few chickens from the same university animal pool to perform a standard experiment for his students. He first injected heat-killed salmonella bacteria into the chickens as a kind of vaccination and planned to wait a couple of weeks until, predictably, antibodies would have formed. The prior exposure, of course, had been intended to induce protective immunity, as it had in countless previous classroom demonstrations, but on this occasion, many of the chickens died and none developed antibody. Chang was dumbfounded. This flew in the face of a century of immunology—none of the chickens had developed protective immunity.

As Chang and Glick puzzled out what had happened, they soon tumbled onto a delicious, highly improbable coincidence. Chang had not only selected the very chickens from the animal pool upon which Glick had previously experimented; he had managed to select only chickens whose bursas had been removed shortly after birth. That accident of timing gave them a crucial clue to the biology of immunology: among all the chickens whose bursas had been removed, only hatchlings later became immunologically impaired, incapable of making antibodies! By complete serendipity, Glick and Chang had implicated Fabricius's obscure gland in the early and obviously determining control of antibody formation—at least in birds.

Realizing the implications, Glick, Chang, and their colleague George Jaap hurriedly repeated their accidental experiment in more systematic fashion. They surgically removed the bursas of both white leghorns and Rhode Island red chickens when they were two weeks of age, then exposed these same birds to bacteria three to four months after birth. As expected, the birds without bursa generally failed to develop antibody. Somehow the bursa controlled antibody production, and it exercised this control fairly early in an animal's life, because the gland typically shriveled and atrophied in chickens in two or three months. Glick and colleagues fired off their paper to the journal *Science*, whose editors replied in a form letter that the research was "of no general interest." So the report appeared in *Poultry Science* in 1956.

There is no bursa of Fabricius in humans, but as the Ohio State scientists noted in their paper, there might be some similarity of function between the bursa in chickens and the thymus, another gland of obscure function that humans *do* possess. The gland, so named because its pyramidal shape resembles a thyme leaf, is located just behind the breastbone; it is largest at birth, about the size of a petite coin purse, and then tends to shrivel into insignificance by age ten, when it can be removed without consequence. Of no dispute, however, was the fact that this little purse fairly bulged with lymphocytes. The problem was that as late as the 1950s and 1960s, immunologists did not quite realize that lymphocytes were in fact the most valuable coin of their realm. The great British immunologist Peter Medawar pronounced in 1963 that "we shall come to regard the presence of lymphocytes in the thymus as an evolutionary accident of no very great significance."

One of the earliest blows struck for the significance of lymphocytes came from England. James Gowans, at the University of Oxford, described in 1959 a tracking experiment in which he followed the movement of cells in rats and showed that "effete small lymphocytes," as he called them, constantly trafficked back and forth between the general circulation and the lymph system—the first strong suggestion that lymphocytes were major players in the immunological commerce between the circulation and lymph glands. As researchers began increasingly to focus on the role of glands in immunology in the late 1950s, several outstanding immunologists trained their sights on the thymus—Robert Good, then at the University of Minnesota; Byron H. Waksman, then at Harvard; Jacques F. A. P. Miller, then at the Chester Beatty Research Institute in London; and Macfarlane Burnet in Melbourne. Three years after the *Poultry Science* paper, Robert Good (he would later become director of Sloan-Kettering) learned of Glick and Chang's work, realized the crucial significance of these glands at an early age, and immediately began to explore more thoroughly how the thymus might play a role in defining the immune response.

In the early 1960s Jacques Miller in London connected two crucial dots. Miller's interest in the thymus came from a tangential direction; he originally thought the gland might play a role in the development of leukemia but soon found it worthy of exploration in its own right. In a beautifully precise and thoughtful set of experiments, Miller removed the thymuses from one set of mice within sixteen hours of their birth. As controls, he removed the thymus from five-day-old mice and as an added control performed "sham-thymectomies"—a faux form of surgery that involved all the steps of thymus removal, including incision and opening the chest, short of actual removal of the thymus itself, to make sure that whatever results he observed could be ascribed to the thymus, and not to any immune or inflammatory response related to this neonatal surgery.

Miller's data spoke volumes about the early importance of the thymus to proper immune function. The animals whose thymuses had been removed within a day of birth succumbed to laboratory infections at a much higher rate than expected. Miller also noticed that while the number of white blood cells rose to normal adult levels in the sham-operated animals, they remained abnormally low in the animals without thymuses. The thymus, he concluded, somehow controlled the population and function of white blood cells, and white blood cells clearly played a major role in protecting animals against infection. Like the Ohio State experiments, the thrust of Miller's work suggested that removal of a once seemingly inconsequential organ very early in an animal's life permanently crippled proper development of the immune system.

In looking back at these experiments, it is amusing to see the strain and discomfort of talented researchers like Miller as they tried to tailor their correct but unfashionable results to fit the scientific fashions of the day. One of the most notable things about Miller's paper on the mouse work, for example, which appeared in the *Lancet* in 1961, is that in trying to explain the unexpected, Miller struggled mightily against the prevailing bias in immunology. Even in the 1960s, the concept of antibody still remained magnetic north to all immunologists, the orientation against which everyone set their research compasses. "It has been suggested that the thymus does not participate in immune reactions," Miller wrote somewhat defensively, ". . . because antibody formation has not been demonstrated in the normal thymus." And this: "Thymectomy in the adult animal has had little or no significant effect on antibody production." In other words, if the thymus couldn't be linked to antibody production, it was presumed not to have anything to do with immunity. Antibodies not only remained the star soloists of the immune system; they were still perceived as the *only* soloists.

But change was on the way. If Miller didn't force that change in thinking at first, he certainly did through a great run of experiments in the 1960s after he returned to Australia, culminating in 1967 and 1968 with papers coauthored with Graham Mitchell that established unambiguously that the lymphocytes that passed through the thymus were, in the hierarchy of the blood, the most important immune cells of all. Glick and Chang and Miller, each in their own way, had identified the source of the two main rivers of what is now known as the acquired (or specific) immune response—the response that protects us from return engagements by familiar pathogens, so that we only contract measles or mumps once. It was not a single experiment, but an accumulation of work circling around the twin poles of bursa and thymus that brought immunological cells front and center, and it is reflected in the origin of the names that are used now to distinguish these key cells. B for bursa, T for thymus—therefore, B cells and T cells. These once obscure glands lent their initials to the most important cells of immune memory, because these lymphocytes, created in the bone marrow (itself a prodigious birthing center of the immune system), went to school in the glands and received their education in the earliest days of an organism's existence. What was true of Rhode Island red hens and Jackson Laboratory mice was also, roughly speaking, true of humans. One family of lymphocytes reported to the bursa of Fabricius in birds (and to the bone marrow, it would turn out, in humans) to get their immunological marching orders on which antibody—which of a million potential proteins—to manufacture, and they became known as B cells. A completely different family of lymphocytes reported to the thymus for its education, and these so-called T cells would then roam the body via the bloodstream, return to the lymph nodes and the spleen (as Gowans had observed), and thus perform immune surveillance.

Once activated—and in immunology, the qualifier "activated" suggests a biological phase change every bit as profound as that from liquid to gas in chemistry—these two classes of cells complement each other. A single activated B cell (or "plasma" cell) can in one hour churn out 10 million antibodies, which are particularly adept at neutralizing toxins, viruses, and bacteria floating freely in the bloodstream. Activated T cells, on the other hand, churn out more of themselves. Humans possess at minimum a trillion T cells, and when a few of them encounter a biological trigger in the form of antigen, they incite themselves to form millions of identical clones of themselves. Some of these special T cells possess a biological version of x-ray vision that even B cells lack; they can identify normal body cells in which viruses, for example, have taken refuge. The most potent of the lot, the killer (or cytolytic) T cells, indeed have the capacity to kill those infected cells to control a disease or cancer. By the mid-1960s, it had become clear that these two cellular arms did business with each other: Jacques Miller and Graham Mitchell proved that certain T cells, now known as "helper" T cells, "were essential to allow B cells to respond to antigen by producing antibody." In other words, helper T cells sat atop the blood's organizational chart—they told B cells when to make antibody.

B cells and T cells form a kind of liquid, extracerebral memory. Forged in the dark damp of bone marrow, educated in the wet pulp of spleen or lymph node, these sophisticated cells remember foreign proteins that have passed by before—telltale molecules on the surface of the measles virus, for example—and this molecular memory allows the cells to respond more quickly and decisively upon a second encounter. When these "educated" lymphocytes, either B cells or T cells, encounter an inciting and very specific molecular flag, known as an antigen, on a second-time invader, something remarkable happens: they convulse into reproduction ever more rapidly, copying themselves millions of times in a couple of days in what is known as a clonal expansion. Each activated B cell, each activated T cell, has molecular antennae acutely attuned to the immunological threat underway; their specificity might be likened to frequency bands on a radio, each cell tuned to a single frequency, each quiescent when there is no signal to receive. But the moment they detect the signal, they undergo explosive proliferation, deploy, and ferret out each virus or other pathogen broadcasting antigen at that single, narrow frequency. This is what gives immunity its devastating specificity and lifesaving speed. By 1993, half a century after Arnold Rich lamented biology's neglect of the lymphocyte, the Australian immunologist Gustav Nossal could write that "the lymphocyte is perhaps the best understood of all living cells."

Each good answer in science usually begets another good question. B cells and T cells do not act in a biological vacuum. In pondering the activity of immune cells, the syncopated arrival of first neutrophils, then macrophages, then natural killer cells, then B cells and T cells on the scene of some immunological insult—the scraped knee, the respiratory virus passed by a sneeze—one is tempted to liken their orchestrated comings and goings to the social organization of ants. Different immune cells have different responsibilities—surveillance, emergency service, cleanup, security, and dronelike roadside maintenance of the living tissues. Such sophisticated activity requires exquisite timing and coordination, and hence all these cells have evolved ways to communicate, to coordinate activities, to launch into action, to call off the hunt.

As Alick Isaacs and Jean Lindenmann were the first to discover, it is hard to eavesdrop on cells, harder still to convince colleagues of the legitimacy of such faint molecular whispers. But answers would come when there was a way to study the behavior and activity of white blood cells in a test tube, and that hurdle began to be surmounted in 1960 when Peter C. Nowell at the University of Pennsylvania noticed serendipitously that white blood cells in a lab dish divided wildly when exposed to a common plant extract known as phytohemagglutinin, or PHA—an extract that, in this case, came from ordinary red kidney beans. Out of these experiments developed a technique of tissue culture to keep lymphocytes alive and multiplying in a laboratory dish long enough to see what kind of substances would activate these specialized cells. This allowed researchers to hold a powerful microphone to a test tube, and what the immunologists heard amazed them.

As immunologists came to appreciate that cells, especially lymphocytes, gave Faust's "wondrous juice" a special immunological kick, a small group of visionary cell biologists began, like both disciples and penitents, to repeat the travails of Isaacs and Lindenmann. They reported the existence of a staggeringly large number of "factors"—a vague term meant to denote a molecule that demonstrated remarkable powers in a laboratory dish but resisted all attempts at isolation and identification. These mysterious entities were so elusive, so beyond the customary yardsticks of laboratory science to measure, that, as Byron Waksman and Joost Oppenheim once acknowledged, "for a considerable period the study of these nonspecific mediators was not accepted by immunologists as worthwhile." In an immune context, these factors activated, suppressed, stimulated, inhibited—they were, in short, powerful immunological verbs in search of precise molecular nouns. The nouns were a long time coming.

There were, as always, early hints about the importance of cells (and of the signals between them), and the pioneers here may well stretch back to the 1920s and earlier. Alexis Carrel, the Nobel laureate at the Rockefeller Institute who is remembered mostly for his work in transplantation surgery, published a theory in the 1920s suggesting that leukocytes functioned as circulating dispensaries, nursing and nourishing the growth of tissues with substances he called "trephones"; even that novel hypothesis followed upon work by an endocrinologist named Marcel Eugene Emile Gley, who suggested that secreted factors he called "harmozones" acted like hormones upon other cells, promoting their growth.

But it was the discovery of interferon that really lit the fuse for an explosion of discoveries, and although Isaacs and Lindenmann had been criticized for coining a breezy name like interferon for their factor, biology—and especially those of us outside it—might wish that their flair for naming had been more infectious, given the terminology that followed. Acronyms began to pile up in the literature: MIF and MAF, LIF and LAF and BAF, EP and LP, TNF, LBF, NAP, and LMF, to name a few. Once immune cells could be studied in a dish, something like 300 separate biological activities were ascribed to activated immune cells. The noise was deafening, the biochemical equivalent of crowd sounds. It began, as Joost Oppenheim of the National Cancer Institute has put it, as an era of "phenomenological detection," destined to remain confusing and, to be frank, not terribly respectable until biochemistry and, ultimately, recombinant DNA technology brought quantification and order. "Instead of doing biochemistry, everyone was giving acronyms to these things," said Steven Mizel, a researcher at Bowman-Gray Medical School in North Carolina. "It became an immunological Tower of Babel."

If the casual reader is numbed by the following recitation of molecules and their activities (we will keep the anesthetic brief), so too were immunologists who struggled to make sense of them. In 1965 two researchers at Royal Victoria Hospital in Montreal, Shinpei Kasakura and Louis Lowenstein, detected the presence of a factor that caused white blood cells to multiply; it was called, variously, "leukocyte mitogenic factor" (LMF) or "leukocyte blastogenic factor" (LBF). Like the first interferon researchers nearly a decade earlier, Kasakura and Lowenstein could neither produce nor purify the factor; they could only describe an activity that happened in glassware, and they were among the first to employ an all-purpose phrase that would appear, like a fig leaf, in the literature of factors for more than a decade: "At present, the nature of this factor is unknown."

In the next two years, two independent groups of researchers—Lewis Thomas and John David at New York University, and Barry R. Bloom and Boyce Bennett at Albert Einstein College of Medicine in New York—detected a factor that inhibited the movement of macrophages, normally among the earliest immune cells to appear on the scene; it became known as "macrophage migration inhibitory factor" (or MIF), not to be confused with the macrophage activating factor (or MAF) identified five years later by Carl Nathan and colleagues at Cornell University Medical School, which induced macrophages to engulf and destroy cells that had been infected by bacteria. And MAF was not to be confused with LAF, a "lymphocyte activating factor" secreted by macrophages that triggered the proliferation of the white blood cells known as T cells, which was reported by Igal Gery and his colleagues at Yale in 1971. Less confusing at first, because it had a slightly less elliptical name, was a molecule isolated by Nancy Ruddle and Byron Waksman at Yale in 1967 called lymphotoxin, a factor produced by activated white blood cells in vitro that killed cells; however, this was later renamed tumor necrosis factor beta (TNF beta), not to be confused with tumor necrosis factor alpha (TNF alpha), the factor reported in 1975 by Elizabeth Carswell and colleagues in Lloyd Old's group at Memorial Sloan-Kettering Cancer Center. This latter molecule, however, *was* confused with "cachectin," a factor reported about ten years later by Anthony Cerami's group at Rockefeller University, which reportedly caused severe wasting (cachexia) in infected animals. A brief debate erupted over this last molecule, but it quickly subsided when molecular analysis showed that cachectin possessed the exact same amino acid sequence as tumor necrosis factor. Molecular "arbitration" of the same sort would ultimately decide a great number of disputes, where multiple teams of researchers had all grasped different parts of the same elephant.

Immunologists wearied of all the acronyms. As Igal Gery and Joost Oppenheim noted in a spirited 1994 review article, "The uncharacterized lymphocyte-derived activities were termed 'lymphokines' in 1969 by Dumonde *et al.* and 'lymphodrek' by more irreverent investigators." Their point was that if you grew lymphocytes in a lab dish, the fluid surrounding the cells—known as the "supernatant"—contained an invisible stew of factors, and unless you could separate out and identify the molecules, it was all molecular borscht or "lymphodrek." Or, as Oppenheim likes to say, "Til you *got* it, you ain't got it."

Despite the inherent confusion of this alphabet soup, the emerging message had a profound impact on immunology in the 1970s, as it would begin to have on medicine and commerce a decade later. Cells of the immune system not only communicated with each other; they produced, especially when aroused, a roar of white noise in the form of molecules passing back and forth. These signals acted like hormones, in the sense that they either stimulated white blood cells to multiply or emphatically suppressed their activity; attracted other cells to the scene of immunological insult or inhibited the usual suspects from crashing that scene; induced profound changes locally or issued a generalized alarm that could affect the entire organism, such as molecules that traveled to the brain and induced fever. In 1974 Stanley Cohen proposed the word "cytokine" as an umbrella term under which all these homeless factors could huddle; he defined them as the family of molecules manufactured and secreted by a variety of cells engaged in immunological or inflammatory responses. (In the two decades since that first provisional definition, cytokines have subsequently also been recognized as a key to identifying several subspecies of immunological cells; certain white blood cells, depending on their activity, are said to produce a particular "cytokine profile," which can be used to identify them as surely as the colors of a football uniform.)

In the fifteen years from 1964 to 1979, reports of new factors numbered in the hundreds, and MIF and BAF might also have stood for the comic-book-style sound effects of the debate among immunologists trying to make scientific sense of this incoherent soup. The debate reached a climax of sorts in May 1979, at the second International Lymphokine Workshop, held in Ermatingen, Switzerland, when immunologists tried to impose a little nomenclatorial discipline on their proliferating acronyms. In an admirable but ultimately flawed attempt at simplification, a group of young turks at the meeting believed that all the activities that had been observed over the previous fifteen years could be attributed to just a few molecules, and in a decision that falls somewhere between group consensus and smoke-filled backroom edict, these biologists decided to reclassify these factors and designate them as either interleukin-1 (if they originated in macrophages during the early, nonspecific phase of an immune response) or interleukin-2 (if they originated in lymphocytes during the late, specific phase). The moral of this ill-fated exercise in etymology is that attempts to simplify an inherently complex system may lead only to greater confusion. At last count, immunology had discovered sixteen different interleukins, some of which by consensus don't belong in the grouping, and there is no guarantee that the end is in sight. The naming ceremony in Switzerland remains controversial to this day.

More important, if truth be told, the discoverers of MIF and MAF and BAF and EP all found themselves crowding into the same boat occupied most famously, if not first, by Isaacs, Lindenmann, Burke, and all the other pioneering interferonologists of twenty years earlier; in a sense, they all rowed through that scientific fog first encountered by the interferon crowd. They were tantalizingly close to some incredibly potent molecules, and yet couldn't get their hands on them in a scientific sense—couldn't purify the factors or produce them in quantity for adequate study. But about a year before the immunologists met in Ermatingen, Kari Cantell received a visit in his Helsinki laboratory from a charming, well-to-do Swiss gentleman whose name, he later confessed, "meant nothing" to him. The visitor proposed an unusual collaboration, and by the time they finished talking, the two strangers had agreed to embark on a project that would in effect solve everyone's problems.

8

"THE CLONING OF INTERFERON AND OTHER MISTAKES"

It is not unreasonable to imagine a time when several kinds
of pure interferon, in different dosage forms—over-the-counter
or by prescription—would be purchasable at neighborhood
pharmacies. We might use it to ward off colds all year round
and flu in flu season. We might not have to worry about rabies,
about shingles or hepatitis or warts or cold sores or eye
infections or other familiar ailments. We might be able to
regulate our immune systems at will, maintaining our
resistance to a variety of diseases in all seasons and at all
ages. We could keep our transplanted parts and organs from
being rejected. And perhaps we can count on interferon
to cure and/or prevent cancer! Was there ever a
panacea-hawking, snake-oil salesman at any
carnival who promised more of his
product? Yet it could all come true some day.
—Life magazine, July 1979

In March 1978 Kari Cantell received a call at his laboratory in Helsinki from Switzerland. Calls to this lab from strangers were not unusual; as the lone producer of clinical-grade interferon in the world, Cantell's files brimmed with letters from cancer sufferers seeking a miracle cure. Patients showed up unannounced at the laboratory and phoned his home at all hours. He had fended off requests from wealthy industrialists, colleagues in the scientific community, desperate patients offering "*very* high sums of money," he recalled, and even one curious visitor from Italy who, on behalf of a terminally ill child, seemed to be making an offer Cantell couldn't refuse, which he none-

178

theless did. The gentleman calling from Switzerland made a rather more agreeable proposition.

His name was Charles Weissmann—"a name then unknown to me," Cantell admitted later. Weissmann identified himself as a professor of molecular biology at the University of Zurich, professed great interest in interferon, and invited Cantell to Zurich to give a talk on interferon "as soon as possible." Cantell, who has never particularly relished giving talks, politely declined the request. Weissmann asked if Cantell was collaborating with any other molecular biologists trying to isolate the gene for human interferon. Cantell not only said he was not; he seemed, to Weissmann's ears, not even to grasp the question or understand the magnitude of change going on in molecular biology that would prompt it.

Stymied in his attempt to lure Cantell to Zurich, Weissmann asked if he could visit Cantell in his Helsinki laboratory. Ever gracious, Cantell extended an open invitation. He did not, however, expect Weissmann to ask if the "day after tomorrow" would be convenient. Only then did Cantell begin to sense the urgency with which he was being courted to participate in a highly speculative and high-risk endeavor. "Utopian" was the word that crossed his mind as he pondered the project—not only in the visionary sense, but also in terms of how impractical such a vision appeared to be.

Weissmann would have made an impression on any audience, especially an audience of one. In a world where it is increasingly rare to find a personality droll in even one language, Weissmann manages to be wry in five. A medium-sized, well-dressed, vigorous man, with large-rimmed glasses and thinning dark hair that turns curly and gray at the neck, he has the air of a bon vivant and the track record of a shrewd, meticulous scientist who always gravitates toward interesting, fundamental problems. Born in Switzerland and raised during the war in Brazil, he is exceedingly cosmopolitan even by Swiss standards; his father was a well-known film producer and distributor in Switzerland. Weissmann learned English from his governess, with whom he spent more time, he said, than with his parents during his early years (he speaks several languages fluently); he took scientific vows of poverty when he decided to pursue biological research, but this is tempered by the fact that he and his wife live in great splendor on a vast and beautiful estate in the hills above Zurich, which he inherited from his parents and has filled with the art of David Hockney, Niki de Saint Phalle, and other modernist luminaries collected over the years. He received training steeped in the European tradition, even while on

American soil, for his great mentor during a seven-year stint at New York University in the early 1960s was the Spanish-born Nobel Prize–winning biologist Severo Ochoa. Weissmann exudes both great charm and the sense that he doesn't suffer fools gladly.

Weissmann arrived as advertised two days after that first phone call and once again outlined his plan to "clone" the gene for leukocyte interferon—that is, use the latest in recombinant DNA technology to find the gene. With the gene in hand, the biologists theoretically could copy it, study it, and use it to make limitless quantities of interferon. Even a lone wolf like Cantell found it irresistible to join the pack chasing this gene. He had something Weissmann wanted—a particular set of cells, those buffy-coat leukocytes he'd been skimming like cream off blood donations to the Finnish Red Cross. Those cells theoretically contained an abundance of the indispensable starting point for their experiments: messenger RNA for the interferon gene. And Weissmann had something Cantell had sought for decades—a new technology that promised to magically transform that message, through the new miracles of recombinant DNA, into the dream that had eluded visionaries like Cantell and even Alick Isaacs for more than twenty years: unlimited amounts of pure interferon. But as Weissmann explained the plan, it was as if he were speaking in tongues, a foreign language full of "clones" and "vectors" and "plasmids" and "shotgunning."

"He just looked blank when I outlined what we were proposing to do," Weissmann recalled. "He was totally surprised that there even was the *possibility* of—I mean, he'd never heard of cloning. He'd never heard of recombinant DNA. He was totally disbelieving that this could be done. It was cultural shock for him." Cantell may have seemed overwhelmed by the technical aspects, but he understood only too well the big picture. "It looked so fantastic," he recalled, "that I was willing to do whatever I could for that collaboration." In retrospect, neither party seems bothered by what Weissmann neglected to mention in those very first conversations (indeed, seems not to have brought up for years), which is that he was proposing the collaboration not on behalf of his university lab group back in Zurich but for a newly formed biotechnology company called Biogen, one of among a handful of companies in the world daring to translate the new technologies of biology into commerce.

"During his visit to my laboratory," Cantell noted later, "Weissmann said nothing about Biogen or the commercial implications behind the collaboration, and these dawned on me little by little much later. I was interested in the collaboration only from the scientific point of view, in line with my own previous studies on the production of interferon, and again did not think about making money from it. Nevertheless, no doubt, I would have collaborated with Weissmann in just the same way, even if I had known from the very beginning of his connection with Biogen. I considered his chance of success to be very remote, but the importance of the project was obvious. There is a saying in Finland: 'Salmon is such a tasty fish that it is worth trying to catch one even if in the end you have no success.' " Like all good fishermen, Cantell steeled himself for failure while secretly hoping they would hook the elusive gene.

His pessimism may have rubbed off on his guest. Weissmann vividly recalls Kari Cantell's lack of faith in the new science. "I gained the impression that he did not believe that the undertaking would be successful," Weissmann wrote several years later. "But then again, neither did I."

Weissmann's interest in interferon actually dated from the previous summer, from June 1977 to be precise, when he took his "annual walk" with an old friend, the esteemed Yale researcher Peter Lengyel at the Gordon Conference meeting at the Tilton School in New Hampshire. Lengyel and Weissmann had worked together in the early 1960s in Ochoa's laboratory at New York University and had remained good friends ever since. Hungarian by birth and a biochemist of excellent reputation, Lengyel had studied the action of interferon in mice for a number of years and attended the famous 1975 interferon meeting in New York, where the possibility of using recombinant DNA as a way to produce the protein had been mentioned as an almost laughably distant prospect. In retrospect, this leisurely stroll through the New England woods comes down to us as a bucolic postcard from an Old World way of biology destined to perish almost by the end of that very summer. Science has always had its competitive, rude, and self-interested interludes, but interferon in particular seems to have kicked off a period of rambunctious, hell-bent, and occasionally cutthroat races, at turns brilliant in its science and mischievous in its manners, an era marked by rivalries between academic laboratories and biotechnology companies, and sometimes even within laboratories and within companies. As both Weissmann and Lengyel would learn, gentility became a professional liability in this new biology.

Weissmann had never given a moment's thought to interferon before. Lengyel spoke glowingly about the surprisingly broad range of effects interferon exerted upon cells, and how newcomers interested in the protein had been thwarted by the never-ending nightmare that had tormented the old-timers; in 1977, some twenty years after Alick Isaacs and Derek Burke had so confidently published their ill-fated papers on the purification of the molecule and its mechanism of action, interferon remained unpurified, its mode of action if anything a deeper and more provocative mystery. If this walk had taken place in almost any other year, the discussion might quickly have shifted to another subject, but 1977 represented a watershed. It was the year in which recombinant DNA was first used to clone a rare human protein. Lengyel knew about the new technology, but he also knew of tight restrictions on its use in the United States. "I was not interested in fighting these restrictions," he recalled, "and I didn't want to start any research with recombinant DNA because of this." The fact that Weissmann was based in Europe "made a difference," however, and they began to discuss the possibilities of collaborating to clone the gene.

Even as Lengyel and Weissmann strolled through the New Hampshire woods, a motley crew of thirty-something biologists for hire at a small California start-up company moved at a far less genteel pace a continent away, rushing day and night to accomplish a truly remarkable event in the history of biology. Using over-the-counter chemicals and some extremely arduous biochemistry, they would by August 1977 have synthesized a human gene, a snippet of DNA encoding the information for the human brain hormone somatostatin. Moreover, they would show that ordinary laboratory bacteria could be induced to read this man-made DNA and build the human protein in their primitive single-celled biochemical foundries. By the time this achievement had been announced and published, toward the end of 1977, every venture capitalist knew the name Genentech and every molecular biologist had more reason than ever to hope that any human protein, no matter how rare or powerful or elusive, could theoretically yield to this technology (one of the reasons everyone knew, of course, is that Genentech made sure they knew; the results of the experiments were leaked to the public prior to publication, and the company later called a press conference to announce their achievement, a ritual of public back-patting that was immediately added to the syllabus of every aspiring biotechnology company henceforth). And everyone knew that the people at Genentech weren't resting on their laurels; by the fall of

1977 they had already moved on to a more ambitious project, human insulin, and they had several more projects on the drawing board, including molecules like interferon and tumor necrosis factor—two medically promising examples of "lymphodrek" that had emerged from the explosion of work on immunological factors.

So the basic principles of genetic engineering, of "cloning" a rare and interesting gene, were much in the air. Find a gene; read its sequence of DNA letters (or nucleotides); pop this sequence, like a cassette, into bacteria; and turn these single-celled creatures into microscopic beasts of burden. As literate as a Rhodes scholar in the universal language of DNA, the bacteria would read the inserted human genetic text, understand it as instructions for the assembly of a particular protein, and perform the requested task. In theory.

With somatostatin, the proof of principle had been established. Indeed, the only thing holding up the field, as Lengyel had noted, was the uncertain regulatory climate. In fact, biologists the world over had in 1974 voluntarily subscribed to a moratorium on tinkering with recombinant DNA until federation regulations took effect in 1976. But the rules governing such research were quite strict (rules that Genentech, in an artful scientific slalom through technical loopholes, managed to avoid). The two-year delay, however, had the effect of bunching competitors at the starting line for a new era of biological experimentation, much as long-distance runners, thoroughbreds and pretenders alike, crowded the starting line of a celebrated marathon. Cloning, however, was no sport for long-distance runners; the next few years in molecular biology resembled nothing so much as a track meet in which all the events were dashes and sprints.

As Weissmann and Lengyel continued their walk through the woods of New Hampshire, Weissmann recalled, "it became evident that cloning of the chromosomal interferon gene would lead to a wealth of information unattainable by any other means." That was the high-minded, scientific motive. As Weissmann and others would soon learn, cloning that gene could lead to wealth, full stop. Practical concerns also made interferon a particularly promising and possibly remunerative target. Despite the heroic protein distillery set up by Kari Cantell in Helsinki, the scant amount of interferon he managed to decant for clinical trials was impure, and costly. One advantage of cloning is that if you could find the gene you were looking for and splice it into bacteria, the microbes actually did two jobs for the price of one: they manufactured the protein in copious amounts, and that protein could be purified with far greater ease (not least because new techniques of purification, like high-performance liquid chromatography, allowed exquisitely pure preparations). "I actually considered it, first of all, a technical challenge," Weissmann explained later. "I thought it an interesting challenge to try and clone a gene for which there were no data at the protein level or elsewhere. It seemed like an intriguing challenge to bypass that whole thing and really clone genes of which we knew nothing at all, except a final biological activity." A job description, by the by, that fit all the other cytokines awaiting molecular definition.

And yet, as Weissmann knew, all those clever technological tricks would be for naught if they couldn't prove they had interferon—if, in short, they didn't have a reliable assay. And here interferon's incredible potency provided extra incentive to the cloners. So powerful was the protein that if biologists injected a mere 50 nanograms—50 billionths of a gram—of the messenger RNA from interferon-producing cells into the egg cells of the African horned frog *(Xenopus laevus,* a lab favorite for this sort of thing), the frog cells would read the message and begin to make detectable levels of human interferon. Peter Lengyel explained all this to Weissmann, and Weissmann then proceeded off the top of his head to outline the experiment. They would attempt to identify the mouse interferon gene (mouse rather than human because it was thought to be easier). By the fall of 1977, they were underway. Lengyel had sent to Weissmann's lab a little tube of RNA believed to be rich in the message for interferon, and now it fell to the molecular biologists to work their newfound alchemy.

"I think every time a new method is created," Weissmann said years later, "you get a spurt of activity where people again exercise their creativity. After a while, all the easy ideas have become exhausted, and then it's tough going again. You know, it's like with an apple tree. In the beginning you shake it and 90 percent of the apples come down. Then you have to shake it a lot more to get out another 8 percent, and for the last two or three apples you have to climb the tree and crawl along the branches, and that's what happens with every technical breakthrough. In the beginning anything you do is new and exciting. You know, when recombinant DNA was invented, anybody and his brother could sit down and do an original experiment. And did, until all the apples had been shaken off."

Interferon appeared to be the shiniest apple, for all the wrong reasons. The belief that interferon might play a role in cancer treatment created a pharmaceutical demand out of all proportion to any real activity suggested by the first early trials; yet into that vacuum of credible research rushed biologists with their gene-splicing enzymes, venture capitalists with their cash, journalists with their most breathless adjectives, while doctors cheered and patients prayed from the sidelines. At one point in the late 1970s the medical philanthropist Mary Lasker made the rounds of biotechnology companies to drum up interest in interferon. An old hand at pulling political and social strings to create medical consensus, Lasker was known to encourage stories in the media to generate public interest in her pet projects; by design or coincidence, the magazines *Life* and later *Time* turned interferon into a miracle drug with cover stories. The thrust of these stories played right into the hands of the biotech companies: if only doctors had enough of this remarkable molecule to test in people. Thus, as Weissmann suggests, speed—not originality—was at a premium in those early days.

That became especially true in the spring of 1978, after Weissmann had begun collaborating with Lengyel. Two venture capitalists employed by the Canadian-based International Nickel Company (INCO) organized a series of meetings in which they asked prominent molecular biologists in Europe to consider forming their own biotechnology company. The headliners included Weissmann and two Americans, Phillip Sharp of the Massachusetts Institute of Technology and Walter Gilbert of Harvard (both of whom would later win Nobel Prizes). To the obvious question as to why North American venture capitalists would venture so far from home, the obvious answer seemed to be government regulation; the moratorium against gene-splicing experiments had ended in the States when two INCO representatives, Raymond Schaefer and Daniel Adams, first approached Weissmann in 1977, but the regulations—and public apprehension—were expected to slow down the work. "Maybe at the back of someone's mind," Weissmann mused later, "was the fact that working with recombinant DNA in Europe at that time was still easier than in the United States. That was not explicitly stated. It might have been considered unbecoming or unflattering, but anyway, the way it was put is, they were trying to recruit the best people in Europe to make a counterpart to Genentech."

Impeccably organized as always, Weissmann arrived at an organizational meeting in Geneva in February 1978 with a two-page typewritten list enumerating potential projects for the fledgling drug company.

They were for the most part all easily shaken apples: fibroblast growth factor, epithelial growth factor, transferrin, colony-stimulating factors, human growth hormone, bovine growth hormone, somatomedins, luteinizing hormone, follicle-stimulating hormone, erythropoietin, hepatitis antigens, and of course interferon. ("A pretty good list, I must say myself," Weissmann remarked, having fetched and perused it in his office fifteen years later; many of the genes were ultimately cloned, though not all by Biogen.) "To the delight of the organizers," he recalled, "I revealed that our laboratory was already working on the cloning of mouse interferon and that I would be prepared to extend our efforts to the human system." Nothing was decided; no one was asked to commit, but all agreed to attend a meeting three weeks later in Paris.

On March 25, 1978, the same group of scientists convened in an airport hotel outside Paris. Once again, Weissmann had done his homework. He arrived with a remarkably detailed timeline for the development and production of interferon as a product, almost a mini business plan. He identified osteosarcoma and hepatitis as the first clinical targets, although anticipating a broader market for other malignancies and infectious diseases. He estimated 5,000 patients per year initially for clinical trials. He projected a treatment protocol of twenty doses of 3 million units each (remarkably close to current regimens). And he estimated that natural interferon of the sort Kari Cantell produced, if he could produce that much at all, would cost something like 10,000 Swiss francs per milligram, or roughly $10 million to $20 million a gram, while recombinant DNA interferon produced by genetically engineered bacteria might cost as little as 100 Swiss francs per milligram, or 100 times less. As for the timeline, "I estimated twelve months for the cloning," he said. "A further eighteen months for expression. And I guess I estimated about three years for the production." In other words, five and a half years to large, pure amounts of interferon. Weissmann went even further. He shared with the others his personal calculations of "confidence in achievement": he was 100 percent confident that they could clone the gene, 75 percent that they could do it within time, 75 percent that they could do it within cost.

It all sounded very organized, very upbeat, very *rational*—100 percent confidence is as good as it ever gets in science, an endeavor where one ostensibly confronts the unknown. In many respects Weissmann's projections were remarkably accurate. Unfortunately, there were events in Biogen's immediate future that neither Weissmann nor any of his astute colleagues could have foreseen. Within a year, the new company would be hurtling toward bankruptcy, rocked by internal scandal, and teetering on the brink of failure. But on that happy spring day in 1978, no corporate birthplace augured a company's future more brightly than the City of Light, and Weissmann returned to Zurich knowing he would play a central role in that future, for among Biogen's first projects, along with cloning human insulin and developing a vaccine against hepatitis B, was the cloning of the interferon gene.

It was also at this meeting, unfortunately, that business priorities introduced Weissmann to the rude ethics of commerce. He would be doing this project without Peter Lengyel. As Weissmann's colleagues made emphatically clear to him, there was not much pharmaceutical interest in *mouse* interferon. Weissmann was forced to terminate the collaboration, upon which they had already begun work.

"I had to find a partner who would collaborate with us on the human interferon project," Weissmann noted. "A quick survey of the literature had made it clear that the ideal partner would be Kari Cantell in Helsinki, who had transformed interferon from a laboratory curiosity into a potential clinical tool." Weissmann picked up the phone and called Cantell, "fully expecting to hear that he was already collaborating with one of the many potent cloning groups I knew to be interested in interferon. To my surprise he was not, and in fact I even gained the impression that he was not fully aware of the potential of the emerging recombinant DNA field." Within a day or two of returning from Paris, Weissmann was on an airplane bound for Helsinki.

Even at the relatively short distance of fifteen years or so, the molecular contortions and technical manipulations required to clone a gene in the late 1970s have about as much relevance to contemporary biology as the Wright brothers' sketch of a propeller does to the engines that power a 747; writ large, the basic principles remain the same, but the power, speed, and ease of more recent technology renders the similarity, if not unrecognizable, at least irrelevant. In biology, what used to take years of blood, sweat, and most of all scientific imagination can now be ordered out of a catalog, delivered overnight, and paid for by credit card.

For nostalgic as well as documentary reasons, we should therefore be particularly grateful for a witty and gorily detailed account of the cloning of the gene for leukocyte interferon provided by Charles

Weissmann, who wrote his essay "The Cloning of Interferon and Other Mistakes" while the saga, and its aftershocks, was still painfully fresh in his mind. Published in 1981 in the journal *Interferon*, the essay represents the kind of semipopular scientific memoir that, if attempted more often, would considerably enrich and enliven the literature of both the sciences and the humanities, for it not only chronicles the human thrills and frustration inherent in pursuing a significant (and highly visible) scientific problem but describes with admirable candor scientific work as it truly unfolds, a journey that involves many more dead ends, blind alleys, one-way streets, and stop signs than retrospective accounts normally suggest, especially those in the popular press, where readers are often left with the impression that science as a process is like taking a freeway to the exit marked "Breakthrough."

With Cantell on board in the spring of 1978, Weissmann and a postdoctoral fellow in his lab, Peter Curtis, began searching in earnest for the interferon gene; by the time they finished, nearly a dozen scientists in Weissmann's lab would take a crack at it. Genetic information is encoded in the long molecule known as deoxyribonucleic acid (or DNA); the actual instructions for making interferon (or any other protein) reside in the exact sequence of chemical letters running the length of a gene, which for convenience are abbreviated A, T, C, and G for the biochemical bases they represent (adenine, thymine, cytosine, and guanine, respectively). In nominating DNA as the molecule for heritable information, nature chose wisely and well; the two strands of the double helix are held together in the middle by chemical bonds between bases that are said to be "complementary." This simply means that A's biochemically interlock only with T's, and C's fit only with G's, so the double helix in reality is two letter-for-letter and word-for-word mirror-image transcripts lying side by side, one running head to foot, the other running foot to head: the ultimate palindrome, 3 billion letters long.

When a gene is "expressed," the genetic word in effect becomes flesh; those double strands of DNA temporarily separate so that the encoded information on how to make a particular protein can be copied and communicated to the cell (when a gene is "copied," which happens every time a cell in the body divides, the entire two strands— and thus the entire genetic transcript for thousands upon thousands of genes—de-helicate, separate, duplicate their texts, and then migrate into daughter cells). In either case, a single or separated strand of DNA can always reconstitute its opposite strand, or repair any gaps, because of the process called "hybridization," which merely refers to the ability of complementary segments or sequences of DNA—TATA and ATAT, for example—to gravitate toward each other, almost like magnets, and reattach. If a molecule had a long monotonous tail, a chemical appendage that read AAAAAAAAA, for example, it would stick like glue to a microscopic fly trap made of strands that read TTTTTTTTT, because A's and T's are said to "hybridize."

And in fact there is a family of molecules in each and every one of our cells that indeed swims around with a long AAAAAAAAA tail. This is not DNA, but the kindred nucleic acid known as messenger RNA. Messenger RNA—colloquially, "the message"—is the cell's go-between molecule. Figuratively speaking, it can pass through walls, specifically the wall that separates the nucleus from the rest of the cell. In bacteria and similar organisms without a nucleus, known collectively as prokaryotes, this is a wasted talent, but in all creatures with nucleated cells, known collectively as eukaryotes and including everything from simple yeast cells to humans, messenger RNA enjoys free transit to otherwise inaccessible cellular districts. RNA copies genetic information from a discrete gene in the nucleus (in eukaryotes, DNA is sequestered in the nucleus) and travels out to the cytoplasm, the cell's working-class district. Here, where ribosomes are the protein-manufacturing plants and enzymes the heavy machinery, messenger RNA delivers the work order from headquarters. The text of RNA, essentially a mirror of the DNA text sequestered in the nucleus, tells the cells in which order amino acids must be assembled in order to create the desired protein.

Different cells in the body perform different functions, and therefore different cells will dispatch distinctly different messages from the nucleus to the cell proper. Hence, Weissmann knew—as did every other molecular biologist—that finding the message for interferon meant identifying the cells that made it most often. That is why the road to the interferon gene went through Helsinki. Those white blood cells skimmed, culled, and caressed by Kari Cantell in Finland were arguably the best source in the world for the interferon message. Not only had Cantell learned how to coax these white blood cells into making unusually large amounts of interferon, but because white blood cells do not function like normal cells, they are not so busy making all the other humdrum housecleaning proteins that normal cells do. They were said to be an "enriched" source for the interferon message. And so it was to Helsinki that Weissmann's young guns—Peter Curtis first, a number of others to follow—flew, shortly after

Weissmann's initial pilgrimage, with flasks of reagents like phenol and buffers, the tools to trap the message.

The project began with a miscalculation. Weissmann's group made a preliminary estimate about how frequently the interferon message turned up in Cantell's cells, and their calculations more accurately reflected the pessimisms of the Finn than the arrogance of the cloners. "We had made very worst-scenario assumptions regarding the frequency of the message," Weissmann admitted, "and sort of created an obstacle course for ourselves." That would not become apparent until years later.

In principle, the experimental approach was quite straightforward. Cantell had discovered fifteen years earlier that white blood cells make interferon when they are infected by virus; if a cell suddenly begins to make a protein, there should be a noticeable jump in the messenger RNA molecules specific to that particular protein. In anticipation of Curtis's arrival, Cantell had infected the white blood cells with virus ("induced" them, in the parlance) and then cracked open some of the infected cells at three hours, others at six, still others at twelve and twenty-four hours after infection, in order to determine when the interferon messenger RNA was most abundant. Since leukocytes typically produced only a tenth the amount of overall message as normal cells, Weissmann and colleagues hoped that the proportion of interferon message in that tiny fraction would be much higher.

After some preliminary experiments, the Biogen group had good reason to feel encouraged. They calculated that the interferon message represented anywhere from a hundredth to a thousandth of all the messenger RNA in their preps—a remarkably high proportion, given that there are perhaps 100,000 genes in each cell, many of them more frequently expressed than interferon. Peter Curtis separated the leukocytes with a smelly, skin-searing chemical called phenol and precipitated out the messenger RNA with ethanol, then flew back to Zurich with the "message" sloshing around in five-liter buckets.

In October 1978 Curtis left the lab, and two superb Japanese researchers stepped into the front line of the interferon crusade: Shigekazu Nagata and Tadatsugu Taniguchi. A biologist of Weissmann's reputation of course attracted excellent young scientists, but Taniguchi in particular was a special talent. An Italian-speaking, music-loving biologist whose radical politics as a university student had in part forced his expatriation from Japan, Taniguchi embodied the stresses of modern biology in a peculiarly Japanese way: he struggled to balance a serene, almost Zen-like deference for authority with relentless intellectual energy and great ambition. With Weissmann's guidance, he became a stellar molecular biologist who would, in three short years and a cloud of dust, establish a reputation as one of the foremost cloners in the world—an ascent that for a time prompted mixed emotions in his otherwise proud mentor.

Indeed, the interferon gene attracted a superb cast of recombinant gunslingers; it was the *High Noon* of cytokines. Another of those young turks had just finished a celebrated project in California and was about to turn his formidable talents to interferon. His name was David Goeddel, and his talents were such that when he later resigned his post as scientist at Genentech, the event merited a story in the *Wall Street Journal.* In August 1978 Goeddel successfully cloned and expressed the gene for human insulin, and Genentech announced the result at a September press conference (Weissmann's notes indicate that the Biogen scientific board dismissed the work as an approach of "dubious value"). After tying up a few loose ends, Goeddel trained his sights on interferon, and unlike Biogen, Genentech's informal cadre of cloners didn't require roundtable discussions in airport hotels. "The decision to do it," Goeddel said, "was almost automatic."

The Genentech team was almost a year behind Biogen; on the other hand, Goeddel and his colleagues had cut their baby teeth on two tough human genes, somatostatin (in 1977) and now insulin. More important, once they had a gene, they knew exactly how to stick the DNA into bacteria and get the bacteria to manufacture (or "express") human proteins; in May 1979 they coaxed bacteria to express human growth hormone. Once that hurdle had been cleared, biotechnology became an exercise in fermentation and mass production at the scale of single-celled organisms, rather like brewing beer, where instead of yeast converting sugars to alcohol in huge fermenters (the original biotechnology, according to some), genetically engineered microbes converted the inserted gene into the desired protein in huge fermenters.

The Genentech group—young, tireless, and as single-minded as the "Clone or Die!" T-shirts they wore with pride—probably represented the most formidable competitors to Biogen, but interferon's sap was sweet enough to draw lots of flies. In this new age of recombinant DNA, interferon dusted off its traditional role as the prototype molecule for biological melodrama. Weissmann had positioned himself, and Biogen, exceedingly well with Cantell, but he was aware that many others were gunning for the same prize. Leroy Hood at Caltech was interested, and he had a huge lab at his disposal. David Mark, a postdoc in the Stanford laboratory of Nobel Prize winner Paul Berg, had left academia for Cetus Corp., based in Emeryville, California, with the specific aim of cloning interferon. Even Derek Burke, who had worked alongside Alick Isaacs at Mill Hill in the 1950s, mounted an effort at the University of Warwick in England and was rumored to be on the verge of cloning the gene, although he recalls the sickening feeling of having to compete with scientists who seemed ballistically shot out of the barrel of recent molecular biology. Indeed, one poor soul who followed David Goeddel to the podium at a cytokine meeting remarked, "I felt like a passenger on a provincial train pushed onto a sidetrack, while the luxury express thunders by at 200 miles an hour."

But of all the potential competitors, Weissmann probably most feared a researcher at Hoffmann–La Roche named Sidney Pestka. "I knew that Pestka was working on it," he said later. "Pestka had been involved in the purification of interferon. His group had really done pioneering work on the purification, and they were really well ahead of the game at that time." Pestka knew the molecule, could assay for it, and had all the wealth and resources of a giant pharmaceutical company to back him up; the only thing he didn't have, Weissmann realized with relief, was expertise in cloning. But then Weissmann didn't know that Roche had quietly managed to acquire just such expertise: they had agreed to collaborate with Genentech to clone the gene. "That," Weissmann said, "I learned later on."

As it turned out, Weissmann did not have to look beyond his own laboratory to find that recombinant fever had spawned competitors. Tada Taniguchi—in Weissmann's estimation, "one of the most gifted students to graduate from our department"—had become deeply involved in the project. He had prepared messenger RNA, tirelessly injected RNA into thousands of frog eggs, prepared the hybridization experiments, and traveled to Helsinki, where he astonished Cantell with his feral energy. ("They were like animals," Cantell later said of Taniguchi and Nagata. "They slept only enough to survive.")

So you can imagine Weissmann's reaction when one day in the fall of 1978 Taniguchi announced that he had been offered a job by the

Cancer Institute of the Japanese Foundation for Cancer Research and planned to return to Japan—to clone the interferon gene! "Although I was advised to continue working on interferon at the Cancer Institute in Tokyo by many people," Taniguchi later wrote, "I had some hesitation about the possible overlap of my work with that of my teacher." He had more than hesitation when he saw Weissmann's reaction. "I was very shocked at that," Weissmann recalled, "because this is not really the usual way to go about things. I told him that I couldn't prevent him from doing that because, you know, everybody is scientifically free, but I certainly disapproved of it heartily! I thought it was unfair for him to take a project which I had conceived really from beginning to end and which, as it turned out, could be executed exactly the way it was conceived, and then walk off with the ideas and continue in his lab. He was, I think, under considerable pressure from the Cancer Institute to do particularly this project, so he was caught between hammer and anvil, so to speak."

But so, too, was Weissmann; he knew how good Taniguchi was. After several days of discussion—rather intense days, to hear Weissmann's version—Weissmann and Taniguchi struck an agreement, and in so doing may have been the first molecular biologists to initiate what has since become a common biological practice: carve up the human genome like railroad barons, apportioning different territories to different groups. Nature provided them with a felicitous solution to their awkward dilemma. There were several different types of interferon, so Taniguchi agreed to leave leukocyte interferon to Weissmann. He would instead attempt to clone a different form of interferon, made by connective cells known as fibroblasts. "He was very embarrassed about the whole situation," Weissmann said in retrospect, "because he was specifically *told* to do this. And if you know the Japanese way of running science, it was an order." Taniguchi is *still* embarrassed about it, judging by the discomfort he displayed during an interview fifteen years later when asked about the episode. "We never discussed very much in details," he said. "I don't know whether shocking was a proper word—but yes, I remember there was some discussion."

November 24, 1978, found Taniguchi in Vienna, enjoying a last concert at the Staatsoper before returning to Japan the following day after years of exile. Among the keepsakes that traveled with him was a farewell album assembled by his former colleagues in Zurich, including an inscription from Weissmann: "Best of luck even with interferon!"

∞

The minions who remained behind after Taniguchi's departure were destined to gain, in Weissmann's words, "exposure to the vicissitudes of research." Cloning a human gene in 1978 was as much a test of physical fortitude and dexterity as of intellect. Weissmann was keenly aware of the competition, and competition sometimes dictated strategy. In this case he decided to streamline the project midway by, paradoxically, making it biologically dirtier. Instead of trying to purify the messenger RNA for interferon, as Peter Curtis had attempted, Weissmann now opted for a messier "shotgun" approach.

At one point, Weissmann himself flew to Helsinki and spent a very long night with Kari Cantell cracking open nearly a trillion white blood cells and collecting the messenger RNA lurking inside them. There were a lot of messages in that rather dirty RNA preparation. Some of those messages—not many, but perhaps one in a thousand—carried the genetic instructions for interferon. Back in Zurich, Nagata did the shotgun cloning. Every single message—for housekeeping enzymes, say, as well as interferon—was copied into DNA. Rather than try to preselect the interferon gene at this point, they "shotgunned" *all* their DNA, thousands upon thousands of miscellaneous messages—from A to Zed, from interferon to genetic static—into bacteria. Then they manually would toothpick thousands of bacteria onto culture plates. In this kind of brute force experiment, elegance takes a backseat to statistics, probability, and, if you have the manpower for it, speed. If the interferon message was a one-in-a-thousand message in their RNA, as they had predicted, they calculated that in order to play it safe, they needed to plate out about 10,000 colonies; about ten of those transformed bacterial colonies, they calculated, would contain the human gene they wanted. If they only plated a thousand, they would cut statistical corners too closely.

Almost every tool used in a cloning project of that kind plagiarizes from nature. In nature, certain viruses known as retroviruses (the human immunodeficiency virus, or HIV, is one) produce an enzyme that allows them to copy the letters of their genetic material, RNA, into DNA in infected cells. Known as reverse transcriptase, this enzyme allows biologists to emulate the trick of viruses in a test tube, to travel upstream against the usual flow of genetic information and convert RNA back into DNA. When Nagata added reverse transcriptase to the messenger RNA brought back by Weissmann, for example, all the single-stranded RNA messages were copied into single strands of DNA.

In nature, when cells divide, another enzyme known as DNA polymerase copies the single, separated strands of DNA into a complementary second twisting strand of DNA (in essence, a second helix, making it double and whole again) before repackaging into daughter cells. In the lab, these enzymes could be used to transform the single strand of DNA into a double helix of DNA (*complementary* DNA, or cDNA, it is called). For all practical purposes, this corresponds to a gene.

Bacteria naturally use a ringlet of auxiliary DNA known as a plasmid, which multiplies many times within the cell, to develop drug resistance; in the lab, cloners would splice the gene they wanted (interferon in this case) into a plasmid and, using the chemical equivalent of crowbars, pry open holes in the wall of bacteria long enough to allow the plasmids to squeeze inside, where—like the rest of the bacterial DNA—it would be copied multiple times with each cycle of cell division. The only unnatural thing about the entire process was that in 1978, because of safety concerns, laboratories in the United States—and voluntarily elsewhere—had to use a deliberately disabled, slowpoke version of the common lab workhorse *E. coli* known as *Chi 1776*, and hope that these lethargic bacteria could be goaded into copying the inserted DNA along with their own as they divided.

Finding the cells that contained an interferon gene, of course, was another matter entirely. Here, however, the same remarkable potency that had mesmerized Isaacs and Lindenmann rescued the cloners. Interferon's power allowed them to pool their clones—and search more efficiently. Weissmann had the very shrewd idea of collecting plasmids from up to 512 different toothpicked colonies, cracking them open, and then using RNA from leukocytes to search for matches—to "hybridize" to promising snippets of DNA. The RNA that found matches could then be injected into frog eggs. If a single colony among those 512 different bacterial strains contained the interferon message, the frog eggs would read the message and, remarkably, make human interferon. And the researchers would know it, because back during their famous walk in 1977, Weissmann had received assurances from Peter Lengyel that this exquisitely sensitive, virtually foolproof test could detect even minute amounts of interferon. Indeed, the very potency in small amounts that had been the bane of biochemists for two decades now worked in their favor. If the pooled colonies tested positive, they knew that at least one of the 512 clones contained the gene, and they would divide and conquer to find it; they would go back and divide the pools into 64 and then 8 and so on, narrowing the search each time to the pool that was positive; if the pool was negative, it probably meant none of the clones in that pool possessed the gene.

Probably, but not definitely. The assay wasn't quite *that* good, and the uncertainty drove Weissmann crazy. "The assay was plagued both by occasional false negative and, worse yet, occasional false positive results," Weissmann wrote in his account. "Nagata bravely toiled away injecting thousands of oocytes whose extracts were assayed in Helsinki. Time and again, one particular group of 512 clones gave a positive result, followed by several negative ones. Each experiment, from hybridization to the phone call from Finland, took about three weeks. We decided (and what choice did we have?) that in the case of discrepancies, we would consider a positive result as the correct answer, and a negative one as a failure of the assay." Four pools, each of 512 clones, appeared to be positive; each was broken into subgroups of 64, injected into the frog eggs, and screened for interferon again. Weissmann has never said so in so many words, but he must have grown wary of Cantell's assay technique, or tired of waiting, because after many months of shipping frog eggs to Helsinki, he began to perform those tests in Zurich. In retrospect, both Weissmann and Cantell believe that their bacteria not only contained the interferon gene all along but were making human interferon fully a year earlier than they realized. The false positives and negatives threw them off.

False positives were the least of Weissmann's problems, however. In February 1979, he fired off a telegram to Dan Adams, the CEO of Biogen, which read in its entirety:

UNABLE TO PAY BIOGEN POSTDOCS DUE TO LACK OF FUNDS AND HEAVY INDEBTEDNESS. WILL HAVE TO DISCONTINUE BIOGEN PROJECT BY END OF MONTH UNLESS FUNDS IN HAND BY FEBRUARY 25 AT LATEST. REGARDS, CHARLES.

Shortly after Weissmann's desperate telegram, Biogen had exhausted its very finite resources and CEO Dan Adams left the company. Founded with such lofty ideals in March 1978, Biogen was "technically bankrupt" less than a year later, its management in "total disarray." Walter Gilbert, the Harvard biologist, stepped in as acting director and devoted much of his time to a mission to save the company, appealing on the one hand to the scientists to continue working on their projects and courting on the other any and all possible investors. "Wally was just talking to everybody," Weissmann recalled. "Wally was really the soul of Biogen, because he was the guy who really held things together when the going was getting tough. He called me up and he talked for a long time, explaining how they would get money and that it was very important to keep the project running to bridge over this time."

Perhaps even more critical than Gilbert's was the role played by Moshe Alafi, a West Coast venture capitalist who had helped to found Biogen. Just as the company was about to disappear into that entrepreneurial vortex of too many good ideas and not enough capital, Alafi approached Schering-Plough, Inc., the New Jersey–based pharmaceutical giant, about a last-minute infusion of money; the idea appealed to Robert P. Luciano, who had joined Schering just a few months earlier, and he engineered a quiet bailout. Weissmann, Gilbert, and the others at Biogen may be forgiven for mistaking this intervention for a rescue, although it surely must have seemed like one at the time; at a distance of fifteen years, it represents one of the great steals in the pharmaceutical industry. In exchange for providing $8 million in cash to Biogen, Schering received 16 percent of Biogen and the rights to three drugs under development by the biotech company: alpha (or leukocyte) interferon, beta (or fibroblast) interferon, and the blood growth factor erythropoietin. Annual sales of Schering's alpha interferon alone are said to hover around $500 million in recent years. Perhaps not coincidentally, Luciano rose to become CEO of Schering by 1982.

The race to clone interferon was unique in recent biology—not because it was the first human gene to be cloned (it wasn't), or because it proved a principle (it didn't), or even because it was the first medically *significant* human gene to be cloned, which clearly it was not. Insulin in 1978 and growth hormone in 1979 take precedence, especially the latter since it replaced a protein otherwise unavailable to medicine. Rather, it was the first *economically* significant human gene to be cloned. It had the potential to be the first billion-dollar molecule, and as such the race to clone the gene became a spectator sport among not only biologists but also venture capitalists, Wall Street aficionados, the business press, and even the lay public, which for years had been bombarded with optimistic dispatches about the "antiviral penicillin"

and "magic bullets." In 1979, in an article in *Life* magazine, interferon was touted as a miracle just over the next hill.

As various teams got closer to the prize, Weissmann began to detect the problem of disinformation. "There is this squidlike action," he complained in an interview, "where someone squirts a lot of rumors which are supposed to prevent other people from doing an experiment." No sooner had Schering's bailout saved Biogen's scientific projects than word began to circulate that another team had cloned the gene. "I mean, rumors were flying thick about this guy's clone and that guy's clone, and they've almost cloned it," Weissmann recalled. He most feared Genentech, for good reason, but Genentech's progress had become undermined, if not derailed, by another standard faux pas of postrecombinant etiquette: a festering intracollaborative rivalry between the Roche biochemists and the Genentech cloners. David Goeddel would send clones to New Jersey to be assayed; weeks passed before he would hear the results.

"To be honest, it wasn't a good collaboration," Goeddel admitted. "I thought we had an arrangement on how we were going to collaborate. We would do the molecular biology, they'd do the protein biochemistry and the assays. And Pestka was always telling us how difficult everything was and that we could never do assays, we'd have to send them to him to do. So we believed him. He'd been in the interferon area. We found out later they were actually very easy to do. And we'd send them there, and we'd never get the results back. They would always be on the bottom of his priority list. And it was only when his postdocs, whom I got to know, started calling me and saying, 'Don't let Sid know that we're calling you, but your samples don't get high priority. The RNA he sends you from our preps are always the worst preps.' He held up our progress quite substantially here, is my view. If he was a good collaborator, we could have succeeded a lot sooner." Weissmann, who later examined internal Roche and Genentech documents as part of a patent lawsuit, substantiated Goeddel's account of the collaboration.

"No, it wasn't smooth at all," Pestka agreed in a later interview. But from his perspective it was another case of biology's rude new etiquette overtaking science of a more dignified sort. An excellent protein chemist, Pestka had been working with interferons since the late 1960s, had managed to purify both alpha and beta interferons, and in late 1978, had managed to isolate a partially purified clone of alpha interferon in his small lab. Genentech's molecular biologists kept demanding "more and more material," Pestka recalled, but he was reluctant to share this material, in part because there was so little material, he says, and because, like any scientist, he wanted to do it himself. "I think there was some mistrust," he conceded. "In fact, they started offering jobs to people in my lab. They hired one, and got inside information that way."

As so often happens, the real competition came out of left field. In September 1979 Weissmann received shocking news from Japan—quaintly, by mail. Tada Taniguchi wrote Weissmann to say his team had beaten them all to the punch. Working largely by himself and needing only three preparations of messenger RNA, Taniguchi cloned the gene in the preposterously brief time of four months. Taniguchi rushed into print, in a rather obscure Japanese journal, the first public account to describe the successful cloning of an interferon gene (the gene for beta interferon, per the previous agreement), and soon after arranged to work at Harvard to express the gene—that is, get bacteria to make the human protein. "On the way to Boston, I visited Zurich to meet with my teacher again," he later wrote, in a sentence rich with double meaning, "and I was so happy that I could 'come back' so soon."

Weissmann chose to be unconvinced by Taniguchi's results at first, and for an ironic reason. Taniguchi had used the same approach as Weissmann, the so-called hybridization-translation analysis, and since Weissmann's team was having so much trouble with it, they convinced themselves that Taniguchi must have encountered the same reliability problems. "I was disinclined to believe it because I considered the vagaries of the assay too great to warrant a firm identification," Weissmann said. He was not alone. "No one was clear if he really had it," Goeddel said. "He published it and announced it and, I mean, it turned out he was right, but he took a risk in rushing out that publication in the Japanese journal—I mean, he had it by plus-minus screening, and he's very good and he was right. But some people probably wouldn't have published that without a little more confirmation." With the serenity of one whom history has shown to be both first and right, Taniguchi demurred when asked about this. "That was the right one, and I was 100 percent confident," he said. "I don't gamble!"

Back in Zurich, it fell to Taniguchi's friend and former colleague, Shigi Nagata, to slog through all the candidate clones in search of the gene for leukocyte interferon. But how could they know for sure? The work was plagued by many false positives—too many signals suggesting success. That was the question that tormented them all as 1979 drew to a close, and so all warm bodies were marshaled to get over the hump. Especially after learning Taniguchi's news, Weissmann felt that only two forms of proof could be considered satisfactory: either they could determine the sequence of nucleotide letters of the DNA in their clone (an extremely arduous task, and fruitless at that time because the protein sequence was not available for comparison) or they could look to see if any of the clones were actually *making* interferon. That, as Weissmann admitted, was an "unlikely" possibility. But rumors continued to swirl about rival groups; an inky darkness, as it were, descended on the Zurich laboratory. Any shortcut would suffice.

A newcomer to the laboratory, Alan Hall, got this assignment. "My thought was, 'Well, maybe it'll work,' " Weissmann remembered. "So then I told Alan, just as a throwaway experiment, 'Why don't you take these colonies and grow them up, make an extract, and assay them?' So Alan sat down and made a calculation, and then went ahead and diluted the samples." This dilution of the extracts was important; calculated incorrectly, it could weaken the signal in the assay to the point that it became undetectable, like searching for a radio station with the volume knob set too low. That, of course, is exactly what happened, and it was a minor miracle that Weissmann had enough wits about him to smell an experiment in need of repetition.

The turning point occurred just before Christmas—indeed, just hours after the Institute for Molecular Biology's annual Christmas party. Hall's initial findings did nothing to contribute to the holiday cheer: he told Weissmann that all his attempts to detect interferon came out negative. "After that party," Weissmann continued, "he showed me the plates and they were essentially negative *except* the one at the lowest dilution. It was just a *shadow* of some protection. And then I asked him how he did the experiment exactly, and then I learned that he'd done these dilutions." That Weissmann could spot a "barely perceptible" positive signal following hours of merrymaking is impressive enough; but when he learned that Hall had diluted the extracts—had, in effect, accidentally weakened the signal so much that it disappeared in all but the last dilution—he knew that they would have to repeat this so-called throwaway experiment again before giving up on it for good. Professing all the while great doubts about this almost hallucinatory positive result, Weissmann requested that Michel Streuli repeat the experiments with undiluted extracts. Hall left for the holidays, and so too did Weissmann. "I went off skiing," Weissmann explained later, "because I had more or less written off the experiment. You know, we'll do it, but I dare not *hope* that this is really something. You know, that sort of attitude."

Thus it was that around noon on December 24, 1979, availing himself of a phone booth at the top of Parsenn, a mountain just outside Davos in the Swiss Alps, "just before taking the run down," Weissmann decided to call the lab back in Zurich to see if anything had come of the experiment. "And Nagata told me it was distinctly positive," Weissmann recalled. The news left him "thunderstruck." Weissmann flew down the mountain, hopped in his car, and sped back to Zurich immediately, stopping several times to call the lab again, begging for more details: back in Zurich, he went directly to the laboratory, where he gazed at one of the same interferon assays showing antiviral activities that had convinced workers back to the time of Isaacs and Lindenmann that this rare and elusive molecule had remarkably potent effects. "Indeed, the assay plate was beautiful," Weissmann wrote rapturously several years later, "and as I looked at it I experienced the feeling of utter bliss which comes only rarely in a scientific lifetime. I knew with certainty that we had achieved what I had hardly dared hope for, even though we still had to formally prove that the protective agent in the bacterial extracts was indeed an interferon-like substance."

Weissmann's excitement is genuine, but he may have expressed his scientific modesty with more than a little poetic license. Nearly a year earlier, he had drafted a patent application based on the technique he had "hardly dared hope" would be successful, and on that nail-biting drive back to Zurich he stopped to make one other phone call. He called Moshe Alafi, the California-based venture capitalist who had helped to arrange the bailout of Biogen several months earlier, told him the news, and requested that the company's patent attorney be dispatched from New York right away. The following day, which happened to be Christmas, Biogen attorney James Haley stepped off a plane in Zurich.

Confirmatory experiments, preliminary purifications, writing out the patent—"Everything was done between Christmas and the New Year!" Weissmann said. The patent especially was easy. "All the technical aspects had been written out," Weissmann said. "We knew what we were doing. If it was going to be positive, we knew how it would come out. The only thing that was missing was the success!"

There are a number of biologists who still remember where they were and what they were doing when they heard the news that the interferon gene had been cloned. The news broke on January 16, 1980—"the date on which molecular biology became big business," Nicholas Wade wrote in *Science*—in a Boston hotel, where a hastily arranged press conference by Biogen revealed in public the success that had been anticipated by the patent application and realized in Zurich several weeks earlier. The next day, the story appeared on page one of the *New York Times*, but it was a bittersweet triumph for Weissmann, who returned to Switzerland the night of the press conference and awoke the following day to a swirling controversy about the propriety of conducting industrial research in a university laboratory. "It was all over the newspapers," Weissmann said, rue still fresh in his voice. "The first day it was just about the achievement. The second day, you know, the whole business of the university and this and that and so on. That was particularly disconcerting. And very disappointing."

Whatever ethical odor emanated from this particular project (and it took a long time to disperse), the cloning of the interferon gene was a technological tour de force that rained possibility and optimism on virtually every avenue of biology, not least immunology. Here, finally, was the technology to slay the twin dragons blocking progress in interferon and cytokine research: purity and quantity. Here was a technology that could be applied to other scarce molecules in a vast number of fields: neurology, immunology, endocrinology, embryology, cell biology. And here was a technology, pursued with unabashed commercial intent, that time and again set the world of basic science on its ear.

Within months of Weissmann's triumph, the Genentech team headed by David Goeddel placed a very handsome second in the race. Not only did they obtain the gene in April 1980, but they took only a week to get the gene expressed. The Genentech team went on to identify no less than eight different interferon genes in leukocytes. Clearly this was a gene that evolution liked and used, in all its variations, as a powerful deterrent to infection. It is now known that leukocytes make no less than thirteen different forms of interferon, and studies of the "interferon system" have yielded untold riches in the areas of signal transduction, gene regulation, and carcinogenesis.

Not all the surprises were happy, however. Even as Biogen and Genentech scaled up the production of genetically engineered interferon, ongoing clinical tests of Kari Cantell's natural interferon proved disappointing in test after test. The story seemed always the same: the molecule looked promising in preliminary results but ultimately did not cure disease. These results had become impossible to ignore by the time of the 1981 meeting of the American Society of Clinical Oncology, and despite nearly a decade of promise and hype and the millions of dollars in investment, within two years of its cloning, the molecule looked to have an extremely grim commercial future. And then, just as interest and credibility flagged, a young Texan named "A. J." Goertz got a call from his doctor in Houston. Goertz knew it was serious. "He's about the only person in the world who calls me Arthur," Goertz said in an interview, "and it used to be that when I'd get a call here for 'Arthur,' I knew it was showtime. You know, something's wrong. . . ."

A. J. Goertz was a few days shy of twenty-five years of age when he was first diagnosed with hairy cell leukemia in 1977. Goertz's wife thought her husband took the news of the diagnosis extremely well, but the fact of the matter was that he didn't know what leukemia was. "I thought it was something like pneumonia, something you get a pill or a shot for, and you'd be cured." When the reality set in, he knew he was in for the fight of his life.

Hairy cell is an exceedingly rare form of cancer—the quip in medical circles is that there are more people studying it than people who have it—but it is also exceedingly lethal. "At that point in time," recalls Jorge Quesada, Goertz's doctor at M. D. Anderson Hospital in Houston, "there was not even a definition, a clear-cut definition, for what qualified as a remission in hairy cell leukemia, because nobody could get a remission." Surgeons at M. D. Anderson removed Goertz's swollen spleen, and that bought him a little time, but the disease recurred in less than six months, a bad sign, and they put him on an experimental chemotherapy drug. He remained in the hospital for three months, contracted pneumonia, lost nearly forty pounds, and very nearly died. What pulled him through, he believed, was the intense and constant attention of his physician, the Mexican-born Quesada. Surprisingly, the drug seemed to have worked, and he remained disease free for over two years.

In June 1982, when the call for "Arthur" came, A. J. Goertz went down to Houston for the usual blood and bone marrow tests. By this time Goertz and his wife had a three-year-old son, and his wife was eight months pregnant with their second child, and so it was Quesada's unhappy lot to inform his patient that the leukemia was back. "I recall very well that meeting," Quesada says now. "That was the most difficult time. Arthur was confronting the possibility of dying. He was a very young man, recently married, with a young family. So the poignant point was this: the terrible perspective of a young man that faces death, and the doctor comes in and tells him, 'There's really not much I can do for you, other than going back to the same thing that nearly killed you.' "

They met several more times, discussed several more options. Interferon came up at a later appointment, and the rationale for trying it was slender, born as much of compassion as scientific logic. The growing disappointment over the drug was felt keenly—felt almost personally—at M. D. Anderson, where Quesada's mentor, Jordan Gutterman, had publicly championed its use and presided over many of the disappointing initial trials. Nonetheless, some doctors had reported a hint of effectiveness of the drug against lymphoma and other B cell cancers, and that was in the back of Quesada's mind when he suggested a course of interferon to A. J. "Despite all the theoretical information that we had gathered," Quesada continues, "it was really that human factor that led me to say, 'Let's try it.' It was not something where I sat down at my desk and said, 'Okay, what am I going to do now with interferon? I have all this nice theoretical information. I'm sure we can come up with a model.' And then I decide, okay, hairy cell leukemia? No, that was not the case. It was actually the need to help that particular individual patient that eventually made it click, and I said, 'Why not try it?' "

Desperation and consent are never far removed in patients when an experimental, long-shot therapy is proposed. Arthur, very reserved and not prone to emotion, listened to the news, but didn't say much. "I totally trust you," he remembers telling Quesada. "If you want to try it, let's do it." M. D. Anderson possessed fleeting amounts of the alpha form of human interferon, and most of it was already spoken for in studies against other, more common cancers. Gutterman had trials underway against breast cancer, ovarian cancer, lung cancer, and several others. He was reluctant to squander it on a hairy cell leukemia patient who had already relapsed. "The main drawback," Quesada says, "was that the disease is so rare that, you know, you say, 'Why get involved with such a rare disease?' But after arguing with Jordan for a couple of weeks or so, he eventually authorized me to use some in hairy cell leukemia."

On July 15, 1982, with little reason to expect that it would work, Quesada gave A. J. Goertz his first injection of 3 million units of interferon. Goertz continued to receive shots daily on an outpatient basis. He never suffered side effects, and two weeks later, even before he began to notice any changes, his doctors were shocked to see his blood counts rising toward normal. The rapidity of the response took everyone by surprise. "When you see that this patient is going down, down, down over many months," Quesada says, "and try something new and all of a sudden the trend starts to reverse, you've got to think that it's the drug, and that's what we thought." A. J. Goertz went on to experience a complete remission. "It's kind of gotten to where it's like insulin for a diabetic," he says now. "That's what interferon is for my type of cancer."

That single treatment granted two stays of execution, one for Arthur and one for the drug itself. Given Goertz's remarkable response, Quesada opened up his "study" to six other hairy cell leukemia patients, all of whom enjoyed either partial or complete remissions, and these results, which appeared in the January 4, 1984, *New England Journal of Medicine,* breathed new life into interferon as a potentially useful drug. The following year, the Food and Drug Administration approved the sale of alpha interferon.

A. J. Goertz's recovery occurred at a delicate moment in the commercial development of interferon as the flagship product of the fledgling biotechnology industry, with several companies producing recombinant interferon with little encouragement about its efficacy in the clinic. Quesada says, and many others agree, that if Goertz had not responded, if interferon had not succeeded in hairy cell leukemia, the drug might well have been abandoned—at least temporarily, perhaps for a very long time—by most biotech companies. "Had it been somebody else who took six months to respond, I bet you that the study would have been killed," Quesada says now. "Because if we hadn't seen much in three to four months, we just simply didn't have enough interferon for patients in a no-win situation." "There's no question that hairy cell was really important," says Kari Cantell, who produced the interferon used in A. J. Goertz's treatment. "I think in the field of malignancies, it was a breakthrough, and psychologically meant very much."

The interferon story has been told many times, but it turns out very differently depending on when the story ends. From the point of view of basic science, interferon has more than repaid society's investment and indeed more than lived up to its hype: as work by James Darnell, George Stark, and Ian Kerr has shown, interferon has provided an extraordinarily clear window onto the way signals traffic from the cell surface to the nucleus. From the point of view of immunology, the interferons have been shown to play early and decisive roles in the immune response against viruses and other pathogens; T cells and NK cells, among the most potent agents of self-defense, both produce gamma interferon during an evolving immune response; research published in 1996, nearly forty years after the original discovery, has shown that alpha and beta interferons directly and indirectly activate T cells.

From the point of view of clinical utility, interferon's lesson may lie in the gestation period. It has taken fully a quarter of a century for researchers to *begin* to learn how to use the molecule, first tested in humans in the mid-1960s and declared dead on at least two occasions after that. By 1996 alpha interferon had been approved in the U.S. alone for use against seven diseases, including several forms of hepatitis, genital warts, Kaposi's sarcoma, hairy cell leukemia, and malignant melanoma; the list is even longer in some foreign countries, including major indications like chronic myelogenous leukemia. In addition, several versions of beta interferon—the form originally cloned by Tadatsugu Taniguchi—have been found to reduce the frequency and severity of episodes of multiple sclerosis. In May 1996, for example, Biogen received FDA approval to market its own version of beta interferon, called Avonex, for the treatment of MS, and some Wall Street analysts have predicted annual sales as high as $300 million within a year. "On the learning curve, I'd have to say we're still on the exponential slope," says University of Pittsburgh researcher John Kirkwood, who headed a large, multicenter trial showing that alpha interferon extends the survival of patients with malignant melanoma. "I think we've just scratched the surface."

From the point of view of technology, the interferon story also proves there's more than one way to skin a cat. The British pharmaceutical company Wellcome, eschewing the recombinant DNA and cloning approach, in 1974 opted to produce human interferon in a fashion similar to Cantell's original technique, by harvesting it from a cancerous cell line known as Namalwa. These cells, when infected with virus, produced copious amounts of interferon; perhaps more important, they produced the full spectrum of alpha interferons, about a dozen closely related human molecules. Overcoming prodigious obstacles of production and purification, Wellcome now markets its "Wellferon" internationally, with especially good demand in Asia for use against viral hepatitis B.

From the point of view of business, the moral is perhaps most surprising: to the persevering go the spoils. Despite the widespread perception of interferon as a disappointment, it has in fact become a billion-dollar-a-year molecule, not the first to emerge from the genetic engineering revolution—Amgen's antianemia drug erythropoietin (EPO) earned that honor—but big and continuing to grow. Edmund A. Debler, an analyst at Mehta and Isaly, puts the worldwide sales figures of all interferons at roughly $2.7 billion. No one argues that the drug is perfect against the diseases it currently fights, but some people do argue that the molecule is being used imperfectly. "During evolution, nature created one beta interferon, one omega interferon, one gamma interferon, but there are *thirteen* different alpha interferons," said Kari Cantell. "This is a very rich potential source for clinical use, and it has been very poorly exploited so far. It's a pity that the pharmaceutical companies have focused their efforts on only one type, alpha-2 interferon, so I think there's still a lot to be done."

Finally, from a Dickensian point of view, the story of interferon has evolved into a fascinating tale of biological paupers and princes, of graduate students who became millionaires and visionaries who did not, of profit-driven biotech companies whose research has brought relief to many patients and of beautiful intellectual ideas that haven't saved or extended a single life. Many have benefited from interferon's slow success: both Schering and Hoffmann–La Roche currently gross about $500 million each in annual sales, according to sources. Charles Weissmann received Biogen stock, as did young scientists in the lab who worked on the initial cloning project, as well as a bonus for successfully completing a company goal. Others have had to make do with largely personal satisfactions. Thirty-five years after the discovery of interferon, Jean Lindenmann sat with me in a Zurich coffeehouse and noted in passing, not quite casually enough, that he had not received a penny for his contribution. And asked many years later if compensation was ever an issue in his initial collaboration with Biogen (for whom he later consulted), Kari Cantell paused while slowly and deliberately sipping from a postprandial glass of artic berry liquor before speaking with the uncomplicated candor that has won him so many admirers over the years. "They did not offer," he said, "and I did not ask."

9

"ONE OF MY
BEST KNOWN ACCIDENTS"

*I will proceed with my history telling the story as I
go along of small cities no less than of great. For most of
those which were great once are small today; and
those which used to be small were great in my own time.
Knowing therefore that human prosperity never abides
long in one place, I shall pay attention to both alike.*
—Herodotus

☞ *Standing by a bank of elevators* at Hahnemann University
School of Medicine in Philadelphia to see off a visitor, Doris Morgan
explained her feelings one more time. "I don't bear any grudges," she
said, sounding as if, in some deep and overgrown place, that's exactly
what she did bear. She was referring to research she had done about
twenty years earlier, and of a finding so spectacularly unlikely given
where and how it took place that it took a while for many immunologists to understand its full significance; had it not occurred, however
improbably, in the laboratory of a well-known biologist named Robert
Gallo, she also conceded, it might not even have made the small impact it did at the time of its publication. "If Bob Gallo's name hadn't
been on that paper," Morgan said, still lingering at the elevator, "no
one would have noticed what we did."

In the ceaseless retrospective elbowing and jockeying for credit
that significant observations seem fated to incite, Doris Morgan strug-
gles gamely to make her case but is outgunned. Her two coauthors on
a 1976 paper that sent a slow, transforming wrinkle through cellular
immunology are, in the words of another former lab member, "two of

the most aggressive people on the planet." Doris Morgan does not appear to play by those same rules. She wore a white lab coat over a red sweater, black tights, and small black boots; she had short blond hair, red earrings with plenty of dangle, and large-lensed glasses. She led a visitor back to a small, windowless cinderblock office, a punch-code lock on the door, two blackboards inside, and hardly anything in the way of decoration, save for several illustrations of cats. The usual diagrams of the hematopoietic system hung on the wall; these are posters—distributed by drug companies and rendered obsolete within weeks of their printing, it seems, by new discoveries—which show the networked interaction of white blood cells, what cytokines they make, which cells they influence. Interleukin-2, or IL-2, the factor Morgan and her colleagues discovered in 1976, is represented by arrows that converge, like roads leading to Rome, on what has become the center of the universe to many immunologists: the T cell.

The discovery of IL-2 is a lovely story precisely because the strongest, the smartest, the best-funded, the most focused, the most high-powered laboratories did not win the prize. In fact, the smartest and best-funded scientists didn't even appreciate for a while that it was important. That is why, of all the decorative touches on display in what was otherwise an office almost pathologically bare of ornament, the most poignant and telling memento was a yellowed piece of paper pinned to the bulletin board. It is, Morgan explained, the letter she received when she was laid off from her job in 1978, when she worked in the soon-to-be-famous, and then soon-to-be-infamous, laboratory of Robert Gallo. The letter does not thank Morgan for stubbornly honoring her curiosity in the face of relentless pressure and thus identifying one of the most important factors in twentieth-century immunology. It merely states that "because of the reduced scope of work and cost overruns on contracts in the Department of Cell Biology there no longer will be suitable work available for you within this department."

Morgan clearly felt that, with the increasing historical importance of what she had first found, others were appropriating an increasing share of the credit for the discovery. "Read the book," she urged, referring to Gallo's recently published autobiography, *Virus Hunting*. "The part where he says he 'directed' me [to do the experiments]."

She did not bear any grudges, she repeated. Just read the book and decide for yourself.

⊗

Gallo's book devotes about three pages to the discovery of interleukin-2. He writes that he indeed "directed" Morgan to do several experiments on human bone marrow in the hope of finding a factor that would permit other workers in his laboratory to keep certain leukemia cells alive long enough to study. The real story is, by all other accounts (including an interview with Gallo), more complicated than that.

When Doris Anne Morgan arrived late in 1974 at the National Institutes of Health in Bethesda to work in the Gallo laboratory, her native West Virginia accent still vinegary in her voice, her services were very much needed. Indeed, she walked in on what Gallo, whose taste for larger-than-life hyperbole occasionally invites skepticism of commensurate size, later referred to as "the worst moments of my career."

If nothing else, Gallo and Morgan proved once again that science is often like a square dance, where the most improbable of partners get thrown together and struggle so mightily to keep up with the fast-paced music that they hardly have time to discover how little they have in common; it is difficult to imagine two more opposite personalities. Gallo was handsome and brash, voluble and profane, candid to the point of cruelty, frankly egomaniacal, melodramatic, a shrewd, self-burnishing storyteller, a gifted manipulator of people and personnel with great scientific instincts, and head of a large lab group that he ran, one competitor has said, "like an Italian family—as long as you were faithful to the family, everything was okay." His former office at the NIH, wallpapered with framed photographs of notables posing with its proprietor, reminded one more of a theater district restaurant than of a laboratory (he has since left the NIH to set up the Institute of Human Virology at the University of Maryland). Morgan was by that time reaching her late thirties ("unmarried," Gallo felt obliged to share in an interview), self-described as naive, scrupulously honest, and not a little stubborn, with a circumspect manner and, truth be told, a maddeningly circuitous method of telling a story (scientific or otherwise). Gallo's CV was as thick as a small-town phone book, with hundreds of publications; Morgan has published less than thirty papers over fifteen years. "They almost never saw eye to eye," said a former colleague. "Gallo was a 'let's get on with it' kind of person, let's look at the big picture. And Doris was this cross-every-I, dot-every-T biochemist. They just didn't get along. She had an adamant personality and the system could only be done her way, and she was not very flexible. And with time, she became more and more inflexible."

She had been born and raised in West Virginia, worked as a medical technician in Washington, D.C., and Morgantown, returned to school in 1967 to obtain a Ph.D. in biochemical genetics, and had just finished a three-year postdoctoral fellowship at the M. D. Anderson Hospital and Tumor Institute in Houston. Even now, Gallo gets Morgan's pedigree wrong, describing her as a newly arrived first-year postdoc, "although a little bit older"; she was, he says, "sort of like a super-technician." In fact, she had completed her postdoc and hired on as a senior scientist in the department of cell biology at Litton Bionetics, an NIH contractor whose labs were located on a small dead-end, residential street two blocks off Wisconsin Avenue in Bethesda. And it was to the Litton facility that Gallo's large group had moved en masse in 1971 when its lab space at the NIH was being renovated. The labs were essentially interchangeable; Gallo set the research agenda, did the hiring, ran the lab meetings at Litton. The identities of the two labs had become so intermingled that Morgan, like many others, had trouble figuring out who belonged to which lab.

Every lab has an overriding mission, and Gallo's was no exception. Like many virologists, Gallo had been looking for a human virus believed to cause several forms of human cancer, including leukemias. This particular area of research had recently come to resemble a career graveyard for virologists, and Gallo could be forgiven for whistling as he walked past the graves of colleagues who had preceded him in this endeavor. In the early 1970s, Gallo's lab at first had been extraordinarily lucky. And then, in the blink of an eye, profoundly unlucky.

Leukemia is the umbrella term to denote the unrestricted growth of a number of different blood cells; primitive, "primordial" cells in the bone marrow, cells that normally evolve through a series of steps, in response to a series of cytokine signals, into the various mature cells that make up the blood (red blood cells, white blood cells such as T cells, granulocytes, and so on). Occasionally, some cells get caught in a developmental loop. They fail to mature completely (fail to "differentiate," to use the scientific term) into specialized cells and instead, reiterating a biological loop, churn out greater and greater numbers of themselves until they crowd out normal blood cells and choke the immune system with a kind of gridlock consisting of underdeveloped leukemic cells. According to an attractive hypothesis inspiring a good deal of research (and a good deal of funding) at the time, these leukemias might be caused by a virus that had infected one of these primitive cells, infiltrated the cellular machinery, and tripped some genetic wire, triggering this cancerous tic of repetition. But in order to prove that, researchers like Gallo first had to obtain a supply of these aberrant cells (which, unfortunately, was not difficult given the number of patients) and then keep them growing long enough in lab glassware to study (which, by contrast, was very difficult indeed).

By trial and error, Gallo's group had discovered an essential ingredient, a mysterious "factor," to assist this search. No one else had it, which only made it more precious (and more mysterious). The medical literature of the 1960s and 1970s fairly crackled with reports about these factors—growth factors, transforming factors, inhibitory factors, and all the related "lymphodrek"; "factor" was a catchall term for molecular snipe, evidence for a beast that left footprints in culture flasks but hadn't been isolated or pinned down. Gallo's factor was produced by cells that had been salvaged from a first-trimester human embryo and stored in a freezer. Nobody knew what this factor was exactly, but that in any event was not the main focus of the lab's efforts: the factor allowed them to grow myeloid cells—blood cells that originated in the marrow—long enough that they could paw through them looking for viruses, and that is what really mattered. They had managed to keep more than a dozen different strains of leukemic bone marrow cells growing, and then, in late August 1974, they found strong evidence of what they were looking for in a cell strain named HL-23 (for human leukemia–23). They found the enzymatic footprints of a retrovirus in these cells.

Retroviruses, as the world would learn soon enough with AIDS, are viruses whose genetic material is RNA; these pathogens travel light and pack a gene for a unique enzyme, called reverse transcriptase, which allows them to copy their own genetic text from RNA into DNA, which they then insert into the DNA of the infected cell. Robert Gallagher, the biologist who headed the work on HL-23, hadn't actually wrapped his hands around a leukemia virus per se; rather, he inferred its presence by finding traces of reverse transcriptase in the leukemic cells, traces presumably left by the virus they hoped to isolate. Where giants of the field had stumbled and crashed, Gallo's group had *inferentially* isolated a virus that caused human leukemia. They began to report the result at meetings. They submitted a paper to the journal *Science* in January 1975, announcing a result that newspapers were quick to report and fellow scientists were quick to question.

But months before the *Science* article had even been submitted, someone pulled the plug, quite literally, on Gallo's crowning achievement. Key to the entire mission of the laboratory was this mysterious growth factor (known only as WHE, for whole human embryo factor). Without it, Gallagher and his colleagues couldn't grow the cancer cells long enough to look for virus. The precious WHE cells were stored in a refrigerator, but one morning in late September (or early October) of 1974, workers arrived at the laboratory on Pearl Street to find the refrigerator, sitting outside the lab in the corridor, unplugged. The meltdown did not seem totally calamitous at the time, because a small amount of WHE material did not thaw. But soon, within a matter of months and certainly by the first of the year, it became clear that the surviving embryonic WHE cells produced a factor that had lost its potency. And without the WHE factor, neither the Gallo lab nor any other lab would be able to reproduce what was destined to be a very controversial finding.

In an interview, Gallo recalled a different chronology. "So we published," Gallo said, referring to the *Science* paper. "And then the freezer accident, good-bye embryo factor." Frank Ruscetti, who later investigated the incident at Gallo's request, stated that the freezer accident occurred in the fall of 1974, before the *Science* paper had even been submitted. Regardless of the timing, the implications were ultimately catastrophic. "We were screwed," Gallo said. "You know, we are going to be in the doghouse because this thing has got attention already, by the director of NCI, and we're not going to be able to give it to anybody. We're not going to be able to develop the blood test. We're not going to be able to do anything. We're going to just be able to say, 'Honest, we had it,' and prove it to nobody. So that was terrorizing, as you can imagine." Gallo's paranoia was not ill-placed. About a year later his group attended a meeting in Hershey, Pennsylvania, where critics claimed—in unusually vituperative terms—that the so-called cancer virus was really a laboratory contaminant. "So of course, I've gotta get the myeloid cells going," Gallo continued. "Are you kidding? I could go back to the HL-23 and grow it again because I've got it in the damn freezer and I can't grow it, right? That's the thing; we had almost nothing left. Whatever little we had left, we wanted to grow."

Given Gallo's admittedly emotional worldview, it would not be exaggerating to characterize the mood in the lab at the beginning of 1975 as scientifically based hysteria. And into this dark environment of desperation, despair, suspicion, and frantic scientific scrambling arrived Doris Anne Morgan, about to make what she later referred to as "one of my best known accidents."

∞

Morgan had a feel for cells. There is no other way to put it. She was one of those people who lean over a microscope, peer down on what to less experienced or less patient eyes would appear as a monochromatic flatland cluttered with monotonous shapes, and see instead health or infirmity, growth or stasis, vigor or lassitude, life or death. She had acquired an expertise in the biology of blood cells and the ability to keep those cells alive in culture long enough to study them, and she had come to Bethesda expecting to figure out what was in the embryonic growth factor. That, after all, was the job she'd been hired to do, but of course that job description decomposed as surely as the embryo factor in the unplugged refrigerator.

She arrived in December 1974, just in time to find herself recruited to the frenzied attempt to "rediscover" the magical growth factor. Gallo's mandate to Morgan is crucial: her job was to get those myeloid cells growing again, nothing less and nothing else. As he freely admits, Gallo desperately needed those cells to salvage his reputation. Frank Ruscetti came to the lab several months later, in February 1975, under similar pretenses—false pretenses, he would later claim (both Morgan and Ruscetti have axes to grind, because both were ultimately fired by Gallo, but theirs are not unique sentiments among former lab workers). A small fireplug of a man, with a low voice and hooded eyes, Ruscetti arrived from the University of Pittsburgh thinking he would be working on the famous HL-23 virus, only to discover upon his arrival that there was no virus to study.

So Morgan and Ruscetti, both ostensibly recruited for other tasks, closeted themselves in a laboratory of the Pearl Street facility and struggled to find the cells that would, in effect, save the Gallo lab's scientific bacon. But Morgan worked under Alan Wu, who headed Gallo's blood group, while Ruscetti worked under the head of retrovirology, Robert Gallagher, and only later did they discover that they had both been asked to work two sides of the same street. "Doris had come to do exactly the same thing I was hired to do," Ruscetti said. "And Dr. Gallo had this habit of hiring two people [to work on the same problem], and the one that found it would get the brass ring and the other would find the unemployment line or lose favor. So for the first few months, Doris and I worked independently, until we both found out that we were asked to do the same thing."

Now, as biologists were just beginning to appreciate (molecules like interferon being among the earliest examples), cells are not simply inert islands of cytoplasm, not just static blobs that live for a fixed period of time and then perish. Cells sense the environment around them; they communicate with sibling cells; and they do their talking by making and secreting powerful chemicals into the liquid soup that they need to survive, each cell type speaking a particular chemical dialect, each cell type hearing in turn only a particular set of messages; thus, in the Babel of molecules teeming in the liquid space between cells, separate sects of cells are genetically programmed to receive only certain messages and similarly programmed to deliver only certain messages. The liquid between cells is in fact a blizzard of messages.

When cells are grown in the laboratory, this liquid begins as the nutrient medium; and when white blood cells have floated around long enough in this medium, and have been stimulated by this chemical or that "mitogen" (any substance that provokes a cell to divide), they will have squirted an invisible, colorless pharmacopoeia of molecules into the nutrient liquid. This becomes known as "conditioned medium." And here again, Gallo's group needed luck as much as insight: they hoped to find a factor that would get their myeloid cells growing again, by whatever means. Ruscetti's tack was to test every known bone marrow cell line and as many embryos as possible to see if he could scare up the same factor that had disappeared in the great refrigerator meltdown, and he did a prodigious amount of work. Before he finally stopped looking, he had tested close to 250 cell types, all failures.

Morgan took a different tack. She decided to stick with the existing system of growing leukemic cells and see if she couldn't tinker with the conditions enough to nudge the cells into expanded growth. While working in Houston, she had managed to keep mouse white blood cells known as granulocytes alive in test tubes for up to three weeks, which at the time was considered a phenomenally long out-of-body experience for a blood cell. The leukemic cells Gallo studied were also granulocytic, also of myeloid (or bone marrow) origin. Perhaps Morgan could tweak the system a bit.

At that time, refrigerators humming at minus twenty degrees centigrade lined the basement of the Pearl Street laboratory, and they were crammed full of half-liter bottles containing frozen conditioned medium. A refrigerator was an appropriate place to store the medium, because this material was basically nothing more than the scientific equivalent of leftovers; to generate a source of DNA for other experiments, a laboratory under contract to the NIH named Associated Biomedic Systems in Buffalo shipped gallons and gallons of ordinary lymphocytes—white blood cells—to Bethesda. Well, not quite ordinary. These cells had been "stimulated"—turned on, as it were—by exposure to the substance known as phytohemagglutinin (or PHA). These PHA-stimulated blood cells multiplied briefly; the Gallo lab would run *gallons* of these cells in a centrifuge, collecting the "pellet" of cells for its DNA experiments, pouring off the liquid by-product into plastic containers, and storing them in the basement freezer.

At some point, somebody, it remains unclear who, had the idea to try this conditioned medium. Gallo implies that it was his idea. "Just for no other reason than we were throwing it away did we begin looking in that stuff," Gallo said. "Now I had a reason to look in that stuff, because we found something in it before." Indeed, in December 1974 Gallo's lab had published the discovery that PHA-stimulated lymphocytes produced a factor that stimulated granulocytes. "Maybe there are other things in it, okay?" Morgan states flatly it was her idea. "I can give you a third version," Ruscetti offered when asked about it. "The third one was that I tried it out of desperation, alright? Or Doris and I at the same time tried it out of desperation. We had gone through *everything*."

Morgan certainly doesn't recall being told to "look in that stuff," but she did have excellent scientific reasons of her own to rummage in the basement refrigerators and try the conditioned medium on her leukemia cells. She had used a similar system to keep her cells in Houston alive for a phenomenal length of time. It made perfect sense to revisit the technique that had worked in one setting and try it in this new situation. So she thawed some of the frozen conditioned medium, concentrated it in the lab to make it a little more potent, and started mixing it in 1.5 milliliter cultures—barely more than a few drops' worth—of blood from human patients with myelogenous leukemia (a cancer in which myeloid cells proliferate out of control).

She began at the end of 1974, shortly after arriving in Bethesda, and the experiments foundered for months. A dozen times Morgan tried to coax the leukemic cells into growth; a dozen times nothing happened. Gallo was demanding, as only Gallo could, that they get the myeloid cell line up and growing, and almost from the beginning they began to clash. Gallo found Morgan "very careful, very compulsive, and devoted. But the devotion was not so much to scientific concepts as it was to what she was doing. It was almost like [they were] her babies, the cells." He said this with a certain amount of amused disdain, as if such obsessiveness clearly transgressed the scientific method. But as even he would admit, "I think originally this is what it really had to take."

Morgan's talents may have been a little too eccentric for Gallo to appreciate. She was one of those attentive, intuitive, almost sensually attuned scientists who not only listened carefully to cells but heard what no one else was prepared to hear, saw what no one else had even the patience to look for. In the large community of science, in addition to its justly celebrated theoreticians and its great intellects of breathtaking vision, there are always the journeymen and eccentrics and people who simply keep their noses close to the ground, focused on a small plot of biological real estate, knowing it in a minimalist but profound way. Those people usually do not possess the vision to take in the entire landscape, see the big picture, predict what the world will look like beyond the immediate horizon, but they often have the focus, expertise, and temperament to make contributions every bit as significant as the great theorists. Doris Morgan belongs to that category.

To that may be added personality tics that made her virtually invisible in a high-powered lab like Gallo's, including a humility bordering on self-abnegation. "I was a low-profile person in that group," she admitted many years later, without the slightest bit of irony, "because I was not really doing anything that was of any interest."

In March 1975 Morgan began to notice something of interest. She had taken white blood cells from a patient in "blast crisis," a phase of leukemia during which the patient churns out huge numbers of cancerous white blood cells. She placed these cells in her tiny cuticle-sized wells and gave them a squirt of conditioned medium. Morgan used the fancier biochemical assays that measured the uptake of radioactive chemicals to show if cells were growing, but she really preferred to rely on visual clues. After a day or two, she simply placed a drop from the cell culture on a slide and looked through a microscope.

At first, the scene looked just like the previous dozen failures. About 90 percent of the cells had died—the dying cells all looked "rotten," ragged, full of holes. She came close to pitching them, but decided to wait a few more days. Six days after starting another cell culture, labeled unpromisingly in her notebook as "Experiment # 13," she began to notice changes. The changes were at first so subtle that they almost certainly would have escaped notice of all but the most dedicated and attentive of cell watchers. Morgan placed a drop of liquid on a slide, stained it blue, and maneuvered the slide under the microscope. And she saw something beautiful: that telltale barbell-like shape of a cell straining to pull itself apart to form twins. A cell in the process of dividing and becoming two cells. Mitosis, the textbooks call it. *Growth.*

"When I was looking at a little droplet of the cells, in amongst all of this death and these really rotten-looking cells, I saw cells in mitosis. And anytime—*anytime*—you see cells in mitosis, you know that all is not lost," Morgan explained later. "That tells you that there's still something—*something*—alive and well." Over the next four days, Morgan monitored the culture closely and witnessed what she later termed "an incredible wave of exponential growth." The cells were multiplying at a remarkable pace, changing shape, growing like gangbusters until about day 16; then, when it appeared the growth cycle had peaked at about two weeks, she would spin down the cells, split the cultures (so there wouldn't be so many cells), add more of the conditioned medium, and light a fuse to the cycle all over again. After six days, just as before, the cells suddenly leaped into another explosive phase of growth. The kinetics were remarkable. Within three weeks, she had 220 times the original population of cells; Peoria (pop. 50,000) had become Tokyo (11 million) in less than a month. And as long as she kept the cells fat and happy with her conditioned medium, they just kept growing. It was astonishing. The usual out-of-body experience for blood cells was short-lived, on the order of days if not hours; she had miraculously kept these blood cells going for weeks.

But was it the cell they wanted? *"We* didn't know what it was," Morgan recalled. "Had absolutely no idea what it was, except that a cell was growing." And that uncertainty got Morgan into trouble. She took electron microscope snapshots of their shape; she asked people in the lab what kind of cell it might be. Then, just when it looked as if Morgan's work had turned up something interesting, if not important, she became guilty of what is generally a scientific sin and, in the Gallo lab, amounted to a capital offense. She became too focused on a peripheral issue; she lost sight of the key problem of the lab. She insisted on studying *her* cell instead of *the* cell. Her cells did not appear to be myeloid cells, the cells Gallo desperately wanted. Morgan had discovered the wrong cell! Indeed, in Gallo's mind, the transgression was even worse. Morgan had *rediscovered* the wrong cell.

Morgan's assumption all along had been that she was growing primitive blast cells, the very cells in which leukemia developed and where, Gallo fervently hoped, a retrovirus lurked. But when she presented her preliminary data in a lab meeting, the lack of enthusiasm was thunderous. "When the morphology of the cells was seen by the group," Morgan later recounted, "no one really had any idea for sure what they were. Someone suggested that perhaps they were transformed B cells or even a culture of undifferentiated blood cells, but everyone agreed that they certainly were not the self-renewing leukemic population of cells that had produced the retroviruses in the original culture system." Therefore, she added with admirable equanimity, "The cultures were not received with much enthusiasm."

Especially by Gallo himself. "You know, I tend to be a little monomaniacal, fixed, whatever you want to call it," he said. "You know, you drive for something, you go after it, and once you see the opening, this was clearly—my mind was fixed on finding out what the hell we could do to grow those myeloid cells." Gallo's reputation may have been in virology, but he knew his blood cells, too. Morgan's immediate superior in the lab, Alan Wu, headed a group studying the origins of blood cell lineages. Gallo knew that blood cell cultures often got knocked out of sync by B cells that contained the common and ubiquitous Epstein-Barr virus (EBV); sometimes this virus "woke up" in the B cells, and then these B cells—the immune cells, incidentally, that make antibodies—would proliferate wildly. And that, from Gallo's viewpoint, is exactly what Morgan had spent six months of Gallo's time and money growing—lymphoblasts! Known cells. Contaminated cells. Worthless cells. Not *the* cell. And in his notoriously gentle and mentoring fashion, Gallo said later, he brought this to Morgan's attention.

"I'm looking at Doris, and she's got a lot of these pictures," Gallo remembered in his rapid, colloquial patter. "You know, money is time, and she's always happy. And why is anybody happy growing lymphoblasts?!? Naturally I'm saying, 'Yes, Doris, lymphoblasts can occasionally be grown. But they're EBV-positive transformed cells. You get them a lot of times from [umbilical] cord blood, and sometimes the individual has been activated recently with a cold or something. You'll get them from adult blood! *And of course these are B cells transformed by EBV and have NO SIGNIFICANT INTEREST!!*'" Slightly contrite twenty years after delivering this blast, Gallo added in an interview, "I think I told her what was logic. I mean, the problem is, when you know things, sometimes it's a problem. It's a hindrance. This was a clear example of a hindrance. But to Doris, they were her beautiful things, you know? And there was a pile of electron micrographs, and I'm looking at them and I'm going, 'Nuts, we're spending money on something that's been known for ten years!' "

A lack of enthusiasm from Gallo has, with only slight exaggeration, been likened to the scientific equivalent of sleeping with the fishes, but Morgan probably remained oblivious even to that. She went back to the lab and continued to grow her cells, regardless of what Gallo said. If anything, the story grew more puzzling. No matter where the starter cells came from—patients with myeloid leukemia, patients with lymphocytic leukemia, and, most surprisingly, even healthy people with perfectly normal bone marrow—the same damn thing happened. A single species of cells caught fire, multiplied, overran the culture.

It takes either enormous courage or enormous naïveté (when there is indeed a difference) for a researcher in a high-pressure lab in crisis to willfully swim upstream against the group mission. Morgan knew this was not the cell Gallo wanted and needed, and most scientists interested in advancing their careers would have taken a rain check on their curiosity and gotten back to the common goal. But two senior researchers in the Pearl Street lab—Bob Gallagher and Alan Wu—encouraged Morgan's continuing research, and "prodded" her (her word) to seek advice from experts in immunology to help identify this mysterious cell. Ruscetti had by now joined forces with Morgan, and they systematically began to rule out, one by one, all the different possible species of blood cells. Ruscetti, by most accounts an astute immunologist, did tests to reject granulocytes or monocytes, two other white blood cells that played roles in the immune system. Morgan, through a sophisticated genetic test, determined that they were not leukemic blast cells. And crucially, at Gallagher's insistence and apparently without Gallo's knowledge, the cells were sent to an NIH scientist named Ethan Shevach, who dismissed the possibility that they contained immunoglobulin. This was an especially critical piece of information. Gallo's doubts notwithstanding, if the cells didn't contain immunoglobulin, they couldn't be B cells. And that bit of news, finally, caused even Gallo to perk up. "That was the beginning of the whole thing," he would remark later. "I mean, they were now cells that weren't B cells. They were something different. They might be T cells. Nonetheless, I don't want to give an impression that instantly I thought this was going to be extraordinarily important."

Morgan, meanwhile, kept sending the cells out for more immuno-logical typing. As recently as the 1970s, these tests were amusingly low-tech. If you added T lymphocytes to the red blood cells of sheep, they formed "E rosettes"—distinctive little star-shaped clumps; if you added B lymphocytes, there were no rosettes. Two separate laborato-ries reported back with the same result: the majority of the cells Mor-gan had grown, especially from the normal, nonleukemic human donors, formed rosettes. All the votes weren't in, but by a quick and dirty process of elimination, and against all odds, those little cells Morgan had managed to shepherd for weeks at a stretch looked as if they were T cells, the holy grail of cellular immunology. These were the cells that could identify foreign invaders with numbing specific-ity; more important, these were the cells that, upon encountering a tumor cell or a normal cell harboring a virus within, could summarily execute the cells, purging the body of disease. Gallo's lab had stum-bled upon—indeed, tripped over and nearly tossed away—a method to cultivate and grow the most potent, the most precious, the most *spe-cific* cells of the immune repertoire.

In December 1975 Morgan carried this news to the annual meeting of the American Society for Hematology. The last speaker on the last day, she revealed—to a half-filled auditorium—that she had identified a factor that allowed "blast-like cells," as she put it, to grow on and on and on. Her presentation "most likely made no lasting impression on the few attendees," she noted later.

Just as Isaacs and Lindenmann had to loosen their grip on viruses to discover interferon, Gallo finally began to loosen his intellectual grip on myeloid cells; that exercise in letting go allowed him to focus not on what he wanted, but rather on what they had. The importance of Morgan's discovery truly sank in as he sat through a seminar on an unrelated topic at the NIH. "At that time," Gallo recalled, "I heard a lecture by a clinical professor from Yale saying it was impossible to grow T cells, and that T cells—I remember this vividly—that you had to use young people's blood because T cells had a finite life and they can't grow. So here we were. And when I heard him, we had the obser-vation, and I'm sitting there thinking, 'My God, this is important. But we've gotta be sure we're right . . .'"

It was, in all respects, absolutely the wrong place to make the dis-covery. There were no immunologists in Gallo's laboratory and thus no one to appreciate the true power of the factor they had discovered. Even more improbably, Morgan had managed to coax the growth of T cells out of blood samples from leukemia patients—blood all but over-run with cancer cells. Something, some factor, some molecule, locked in those gallons of frozen media in the basement of Pearl Street conferred an extraordinary ability to cause T cells to grow. They hastily wrote up the results of Morgan's experiments and, in March 1976, submitted a paper to the journal *Science*. (Ruscetti struck a fateful bargain with Morgan on the order of authorship on that first paper, he says; he agreed that Morgan's name would come first, his second, on what became a frequently cited paper, but Morgan insists "there was never any agreement" about authorship.)

Unlike Alick Isaacs, who began using the term "interferon" before the molecule had even been found, Morgan, Ruscetti, and Gallo took an extremely low etymological profile; even with the molecule more or less sighted, if not in hand, they exercised great restraint and did not even propose a name for it in that first paper. The reviewer for *Science* seemed to have exercised restraint, too. "Somebody sat on it for an awful long period of time," Morgan remembered, "and Gallo had to call."

At about the same time (Gallo remembers it being after submitting the *Science* paper, Morgan as early in 1976, before the paper was written), Gallo made a fateful decision. Understanding that he couldn't pursue the implications of T cell growth factor in his own lab, he decided to share the news with several researchers at the National Cancer Institute with expertise in immunology. One was Ronald Herberman, whom he knew as a "good cellular immunologist," and the other was Steven Rosenberg, whom he knew to be "an exciting person" working on immunological approaches to cancer.

"And I can tell you that both expressed great interest," Gallo said, "but Steve almost in a hyper way. I mean, he was excited. He knew what he was going to do right away. That I can vouch for." The timing could not have been better. Quite independently, Rosenberg had reached the conclusion that the best way to attack cancer was to cultivate specific antitumor immune cells; out of the sky fell the elixir that would let him grow those cells. "Steve didn't get this notion from other people," Gallo said. "It was instant. This made him see that there was the potential to go forward, possibly fast. Herberman was more laid-back, more poker-faced. He definitely thought it was important. But I didn't have the clear-cut notion he was going to go off and running. Steve was, you could see, watering at the mouth."

Nobody quite realized it, but the world of T cell biology—which is to say one whole arm of immunology—was about to change in an irrevocable way. Doris Morgan had caught lightning in a bottle; with T cell growth factor (in 1979, the ad hoc scientific committee at Ermatingen changed the name to interleukin-2), still unpurified and largely uncharacterized, immunologists could grow lymphocytes in a test tube. In less than a decade that molecule would be cloned, mass-produced, tested in clinical trials, and on its way to revolutionizing basic studies in immunology. It would even cure a few people of cancer.

But it shed light on something larger even than immunology: the process of discovery. As Morgan herself later put it, interleukin-2 "was discovered in the most unlikely culture environment by a novice in tissue culture technique who had never had a course in immunology. This should tell us something about biased expectations of experiments in our own fields of expertise." She underlined the point in an interview. "Our interest *never, ever* was the molecule that was causing this. It was the cell. . . . It's a bloody miracle that we got it. And I think that it's only because we didn't go after it that we stumbled—absolutely *stumbled*—into it, and had enough help to recognize it, or at least not to throw this down the drain, which we came very close to doing."

It took years for Gallo, obsessed with rediscovering the HL-23 retrovirus that would forever elude him, to understand the significance of the molecule, not the cell. "You can understand why I was emotionally overcome by the other [the retrovirus problem]," he said in his office not long ago. "That's life. I was emotionally overcome by what turned out to be the worst moments of my career at that time. You have to understand, you're fighting against nature almost to say that you could get a human retrovirus. Everybody thinks you're a little crazy to do it. And you do it, and you got it. And then you lose it? What could be worse, okay? *What could be worse?* So I had a very passionate, overwhelming problem. I couldn't get freedom in my mind to do what was the scientifically open field for me, for the lab, to give opportunities to people for their careers, for the field. I had this thing; can I let it go?"

He could, and did. Gallo did not simply let go of T cell growth factor; in 1978, before the ramifications of the discovery had truly sunk in, he let go of the person primarily responsible for the crucial discovery. After the initial observation, Morgan spent the better part of a year simply growing T cells for other scientists—"growing cells until they were coming out of my nose," she groaned, in effect relegated to being the supertechnician she'd been perceived to be. Moreover, she was acutely aware of her shortcomings, according to a colleague in the lab. "She was keenly feeling her ineptitude in immunology," Richard Smith recalled, "and felt it was a real liability to making more of a contribution to the work. Given the personalities of

Ruscetti and Gallo, it's one of those things that happens in human dynamics. They saw the significance of it; it was Nobel Prize stuff. Those personalities never concerned themselves with personal aspects, just with the scientific problem, and Doris got left in the dust." Finally, around January 23, 1978, Phillip Markham of Litton Bionetics informed Morgan in a letter that her job would be terminated.

"To put a scientist out on the street with two months' notice in the spring, when there's not much available, was . . . was not nice." That's about as inflammatory as Doris Morgan's vocabulary gets, but there is no doubt in Morgan's mind that this would not have occurred without Gallo's concurrence, and other lab members support this view, although Gallo vigorously denies this. "Gallo controlled his people's contracts," Morgan said. "Litton did what Gallo recommended."

"It *was* Gallo who fired her," Ruscetti confirmed, "because I tried to talk him out of it." "I remember not being shocked," said Robert Gallagher, "because there was so much frustration between Gallo and Doris. By then she had become marginalized in the whole running of the lab. At a personal level, what a shame. All the stuff that needed doing had been taken out of her hands. . . . She was an incredibly tenacious person. And her basic self-honesty—you'll never find a more honest person than Doris. Her sense of conviction—she's essentially unshakable in her belief. And because of that perspective, she didn't have the capacity to make certain connections and run with it and see the experiments necessary to develop it. And Gallo definitely did." Morgan had an opportunity to plead her case, too, although she found Gallo "not particularly sympathetic. It was just simply, 'Well, this is the way it goes.' And he felt absolutely no particular obligation or responsibility for my staying. I was expendable." Asked if she felt Gallo ever treated her badly, she quickly said no, adding "other than taking credit for the discovery of IL-2, which was a little bit above and beyond." Perhaps the point she made at the elevator was most telling. If it hadn't been for Robert Gallo, no one would have noticed what Doris Morgan found. The flip side, of course, is also true: if not for Doris Morgan, Gallo's lab would never have discovered interleukin-2 or, in all likelihood, the first human retrovirus several years later.

A sympathetic colleague in the lab wrote Morgan a poem upon her leaving that concluded, "Don't despair, Dorothy Anne. There's more to T cell biology than we can guess at."

It took a while for the scientists to catch up with the poets. The *Science* paper made a minor ripple at first. The Gallo lab received hundreds of phone calls from immunologists, but mostly to complain that they couldn't get the growth factor to work. "The immunologic community wouldn't accept it," said Ruscetti, who fielded many of those calls. "Because at the time, the idea was that antigens or lectins drove [T cell] proliferation, and that's what they had in their mindset, alright? And so for four or five years, we always got these calls from the immunologic community saying we're crazy. 'We tried to use your source; we can't grow T cells,' and whatnot. Because they really didn't want to accept it. It was one of those observations that comes out of left field that the community that would be most interested in it would never have found it, because they never would have done those experiments."

In the year following the initial publication, however, the work began to take off in a couple of key laboratories. One was Gallo's lab itself; while Morgan grew her cells, Frank Ruscetti began to take over more and more of the project, showing that these T cells were functional and immunologically active—an important distinction that was reported in 1977 in the *Journal of Immunology.* Meanwhile, in New Hampshire, Dartmouth University researcher Kendall Smith learned of the discovery during a "mini-sabbatical" on the NIH campus with Gallo prior to the September 1976 publication in *Science.* Indeed, he briefly believed he would be collaborating with Gallo's group—until Gallo got wind of it. "Gallo was funny," Ruscetti said. "He was not that interested in it, but he was mad if anybody else would do it. So he told me not to collaborate with Kendall Smith." While Ruscetti and Gallo continued to disagree on the direction the research should take, the Dartmouth group rushed into the breach. The opening allowed Smith to become the leading authority on interleukin-2.

Given a four- or five-year reprieve from competition, Smith's lab virtually extruded a stream of landmark papers—the first generation of T cell clones using the growth factor, the first assay to measure interleukin-2, the first identification of the molecular receptor for the molecule, and the first purification of the molecule to homogeneity, each of which opened up broad and important avenues in cytokine research. "Those years between 1976 and 1980 gave us the window to do our experiments without having to deal with Gallo's big lab," Smith would say later. "He had a huge lab, and if he wanted to, he could have blown us out of the water." ("Yes, Kendall is right," Gallo replied later. "He wouldn't have had a chance.") In 1983 Tadatsugu

Taniguchi, Charles Weissmann's old student, cloned the gene for interleukin-2, opening up fabulous genetic vistas on IL-2 and its relation to the larger cytokine family. The discovery of IL-2 and the follow-up work, Smith noted later, "have had an extraordinary impact on the discipline of immunology, transforming the very nature of immunologic inquiry."

"What I *should have done*," Gallo added in retrospect, "and what I obviously regret not doing—because we can run when we want to run, I don't care what the field is—what I regret is not running with the cell biology of it and the biochemistry of it and the molecular biology of it." In fact James Mier, a postdoc in the Gallo lab, partially purified interleukin-2 in 1980, but both he and Ruscetti maintain that the molecule remained at best of marginal interest to Gallo. "I didn't even think I'd have competition!" Gallo admitted. "I didn't think there would be many people interested. I didn't realize that the entire immunology community was interested and excited by it. I *underestimated* its magnitude of importance to the immunology community."

To hear Ruscetti tell it (with a second from Mier), the final irony is that the molecule Gallo wasn't interested in ended up making his career. He told Ruscetti to stop working on interleukin-2 and get back to trying to grow myeloid cells, but it was precisely that molecule that allowed Bernard Poiesz to grow adult T cell leukemia cells, and it was in these cells that Poiesz and Ruscetti isolated the first human cancer retrovirus, HTLV-1, in 1980. Using IL-2 and the same system on T cells from patients suffering from acquired immune deficiency syndrome, Gallo's lab in 1984 reported the isolation of HTLV-III, now known more commonly as the human immunodeficiency virus, or HIV. "Gallo didn't find a virus in myeloid cells," Ruscetti said, "but he found a virus in T cells, and then he found the AIDS virus, which made him famous." "We knew it was important," Gallo said of interleukin-2 in an interview with the *Wall Street Journal* in 1983. "We just didn't know how important it was."

If nothing else, Gallo made a clean handoff to the right runner. His mouth watering, his vision of the field clear, Steve Rosenberg picked up the molecule discovered by Doris Anne Morgan and ran like hell with interleukin-2 to the end of the next decade.

III

THE RISE
OF THE T CELL
CHAUVINISTS

*It is every immunologist's dream to use the
immune system to cure cancer. It is every
immunologist's nightmare that there is no such thing
as an immune response to cancer in humans.*
—Steven Rosenberg, 1992

10

THE SILK PURSE YEARS

*The term I use is paleo-immunology—we lived in an era where
nothing, really, if you look back, nothing was really understood
at all. And what we did was really hocus pocus, I think,
couched in scientific terms and attempted in good faith, but
really not science. Not science in the way of understanding the
double helix is science. It was trial and error, rife with
empiricism, and really there was no insight into how things work,
save maybe antibodies. And that's true of everything up to ten
years ago, and in some ways maybe even up to five years ago.*
—Michael Lotze, 1992

<> *Pigs, as a general rule,* tend to be too intelligent to willingly play
the role of man's best friend in the milieu of medical research, but this
was especially true of some unsung and unhappy animals at the National Institutes of Health who donated their lymph nodes, and the
cells therein, to the cause of immunology in the late 1970s. No one
can offer a more ruefully expert opinion on this one-sided transaction
than a surgeon named Norman Wolmark. His name appears on no
papers from this period, and his work influenced almost no one (save
the colleagues who fairly fled in horror from the mere sight of it in the
lab). But whenever scientific or medical disciplines experience the jar-
ring, and often unpleasant, temblors of transition, slipping abruptly
like great earthen plates in an earthquake, there is always someone
like Norman Wolmark whose research gets ground up in the friction
of change. Wolmark's contribution to the field of immunotherapy disappeared into the cracks of immunological knowledge, and yet its very
failure exemplified a scientific world shifting from nonspecific immunological therapies like BCG to highly specific ones, from the world of
generalist cells in the immune response like macrophages and NK

committee. "Obviously, we are in desperate need of finding new modalities to treat cancer." Later that morning, Rosenberg suggested that a failure of the committee to vote would have "deleterious effects" on cancer treatment. Each member of the Troika played his appointed role: Blaese laid out the scientific rationale for the experiment, Rosenberg reminded the committee that cancer patients needed new treatments, and Anderson sat back, monitoring the flow. He sensed the tides shifting back and forth. At one point during the midmorning coffee break, one of the reviewers wondered aloud to Anderson, "What's the rush?" This remark served as the irritating grain of sand out of which Anderson created a pearl of a tantrum. Stewing all through the break, although he had been "rehearsing for a week," Anderson rose during a pause to address the committee.

"I was asked at the break, 'What's the rush in trying to get your protocol approved?'" he ranted. "Perhaps the RAC members would like to visit Dr. Rosenberg's cancer service and ask a patient who has only a few weeks to live: 'What's the rush?' A patient dies of cancer every minute in this country. Since we began this discussion 146 minutes ago, 146 patients have died of cancer." "From the look on his face," said one observer at the meeting, "French realized he had overstepped propriety immediately."

Not long after that outburst, the late Bernard Davis, a respected Harvard Medical School microbiologist, raised his hand and asked to be recognized. "Probably *the* most emotional moment," Anderson later said of this intervention. "The reviewers were blasting us and we were pushing back, and it was going this way and that way, and I had used some of my emotional things, which people still criticize me for. And then Bernie Davis raised his hand and said, 'I'm very troubled by what I'm hearing here.' And I just *shrunk*. Because I just knew by the flow that it was at a critical juncture, and most of the undecideds were teetering right on the fence. Bernie Davis was extraordinarily well respected, and when he says he's troubled, whatever he said next was the way things were going to go. And I didn't know what he was going to say because he had kept a poker face the entire time. And the next words out of his mouth were, 'I think we are nitpicking these investigators.' And that was the end. I just . . . just . . .'" Simply reliving the moment, Anderson came close to revisiting the stammer he used to have. "You know, my adrenaline level just warped. I mean, that was it."

Before Davis finished speaking his piece, he chided his colleagues about their wariness toward risk. "It is virtually not possible," he said, down the road, because his boss was an aggressive and restlessly innovative surgeon named Steven Rosenberg.

Wolmark found himself doing this unusual surgery because, when he had finished his residency in general surgery at the University of Pittsburgh in 1976, he'd started looking around for a place to do advanced training. "I was interested in using immunologic therapy for cancer treatment, and Steve Rosenberg had been at the NIH several years. And of the various places at which I had interviewed, this by far seemed to me to be the most innovative. As far as the program, it was perhaps the most countercurrent of them all. It was unconventional. It was almost . . . *abrasive."* Abrasive? "In the way it stood out. It did not blend in. And Steve was very enthusiastic. And I shared his enthusiasm right from the start."

Rosenberg, only thirty-three years old when he took over the Surgery Branch of the National Cancer Institute in 1974, had as a goal nothing less than putting immunologically based cancer treatment on the map. In a hurry. Rosenberg was not alone in this quest (although some readers of his recent book, *The Transformed Cell,* have come away with exactly that impression), and many laboratories throughout the world worked toward a similar goal with similar ideas. But no one pushed as hard, moved as rapidly, and came to dominate the field as much as the team that Rosenberg began to assemble in Bethesda. The mission of this lab combined a belief in immunologically based therapy with an allegiance to brute-force animal experimentation, all fueled by an urgent desire to provide new treatment alternatives to cancer patients, sooner rather than later.

Around the time Wolmark arrived in Bethesda, Rosenberg had been particularly struck with the promise of a procedure called "adoptive transfer." The term referred to the possible use of specific immune cells, harvested from healthy individuals and transfused into patients just like a unit of blood or an intravenous drip of drugs, to fight a specific disease. And that is how Rosenberg came to ask Norman Wolmark to dedicate his considerable surgical skills, honed during eight years of education at the finest medical schools, to test this radical approach. In pigs.

The reluctant collaborators were mini-pigs, actually, a specially bred population of pig used at the NIH for experiments in organ transplant surgery. Indeed, transplantation surgery had always represented the flip side of tumor immunology; in transplants, you wish to dampen the immune response (or "induce tolerance") to prevent the rejection of tissue that appears foreign, whereas in cancer you wish to enhance the immune response ("break tolerance") to a tumor that appears only marginally foreign, if at all, to the immune system. Because of the intense interest in manipulating and dampening the immune system to prevent the rejection of transplanted organs like kidneys, livers, and hearts, the NIH maintained a large facility for animals on its Bethesda campus to facilitate the study of transplantation immunology. Ultimately, Wolmark and Rosenberg turned to large animals like pigs because they were going to need a lot of cells—billions of them per patient, in fact. "When you needed those numbers of cells," Wolmark said, "you were limited to livestock, to farm animals—horses, sheep, pigs."

Several years prior to Wolmark's arrival, Rosenberg and William D. Terry, head of the Immunology Branch, had embarked on an exhaustive survey of the literature of immunotherapy. "I went way back," Rosenberg would later recall. "I went to Coley and back beyond Coley. I mean, I spent well over a year in the library, intensively, weekends and nights, trying to find out everything that had been known before I decided on what I thought would be a reasonable approach." At the time, nonspecific bacterial approaches were still being pursued, and a flamboyant French hematologist named Georges Mathé had created a sensation in the early 1970s by claiming cures of childhood leukemias with the use of bacillus Calmette-Guérin. BCG is an attenuated (or weakened) strain of the tuberculosis bacterium that, since the 1940s, has been safely used in over 2 billion people throughout the world as a vaccine against TB. It is known as a "nonspecific" form of immunotherapy because the microbe causes an indiscriminate commotion in the blood, not specific to any particular cancer (or cancer antigen).

Terry and Rosenberg rejected nonspecific approaches like BCG (Mathé was so miffed at Terry's disinterest that his 1985 novel, *The Man Who Would Be Cured*, is facetiously dedicated "To Bill"). They also rejected the notion of cancer vaccines, believing that attempts to immunize patients against their disease would not be effective. But they had been intrigued by several reports suggesting that immune cells known as lymphocytes might be transfused into cancer patients to attack their tumors. The technique was generally known as "adoptive immunotherapy."

The idea had gained some momentum with the 1973 report of

M. O. Symes, a British urologist. Symes had implanted bits of human bladder tumors into pigs to incite an immunological reaction, harvested lymphocytes from their abdominal lymph nodes seven days later, and then injected these supposedly sensitized cells into twenty-five patients with bladder cancer, some of whom received supplemental radiotherapy. There wasn't a great deal to show for so arduous a procedure, but Symes reported that several patients had tumor shrinkage and also claimed some evidence of clinical improvement. Rosenberg had a somewhat paradoxical reaction to Symes, who presented his research at an NIH seminar in the mid-1970s. He found Symes to be "excitable," his research unconvincing. At the same time, he felt intrigued enough to try the same thing himself. "We *knew* that this was sort of a para-scientific method experiment," Wolmark recalled. "*But*—you couldn't argue with the x ray showing a decrease in tumor size."

Transplants of tissue from species to species, be it of organs or cells, are fraught with complications and unlikely to succeed, but with certain intractable diseases like cancer, the degree of scientific rationale considered sufficient for experimental medicine is inversely proportional to the enormity of the problem and the desperation of both doctors and patients. In 1975 the desperation over cancer treatment justified, in the minds of at least some oncologists, proceeding on the basis of such slender hints of efficacy. So Rosenberg asked Wolmark to prepare and study pig lymphocytes for possible use in humans. "We were going to do it not the way Symes had done it," Wolmark explained later. "We were going to bring modern immunologic techniques to this model. Which by today's standards, of course, was barbarous."

The sheer surgical contortions defy belief. In pigs, as in humans, a thin, veillike sheet of tissue known as the mesentery attaches the intestines and other viscera to the rear of the abdomen, and this tissue is studded—like spangles in some scarves—with small, pea-sized lymph nodes; as in human biology, the immune cells known as macrophages reconnoiter and roam in the immediate vicinity, returning to (or "draining" into) these nodes rather like detectives returning with evidence to the forensics lab of a local precinct station, in this case bringing with them biochemical evidence of any foreign antigen—bits of bugs or toxins or aberrant cells—into which they might have bumped. If the antigen represents something sufficiently novel and threatening, a large number of identical lymphocytes—identical in that they all specifically react to this particular antigen—begin to proliferate in the nodes. Since each of these reacting lymphocytes is the same in the sense that they all respond to one very specific molecular provocation, they are known as a "clonal" population, and since many T cells respond, the overall response is termed "polyconal."

Wolmark's task, therefore, might be likened to lymphocyte farming: surgically implant little pellets of human tumor near the many lymph nodes that crowd around the small bowel of the pigs, sew the animals back up, and then hope that the immune response of the pigs will generate clonal populations of lymphocytes specifically targeted to attack and kill the transplanted human tumor cells. Then, two weeks later, Wolmark would operate again on these very same, and now very sore, pigs, removing the lymph nodes and the spleen to harvest the cells within. There were several excellent reasons to think this approach would not work, beginning with the fact that the pigs would attack human tumors not so much as cancerous tissue but as simply human tissue—foreign, transplanted tissue. Or, as Wolmark wryly put it, "Of course it was to be demonstrated that pigs developed *any* kind of immune response to human tumor that wasn't related to its natural enmity to the human." Pause. "Both immunologically and psychologically." Wolmark experienced, to borrow the immunological term, both forms of intolerance.

"When we were really in high gear, there were pigs everywhere," Wolmark continued. His twice-weekly expeditions to the animal facility for surgery became an ordeal of ear-splitting, odoriferous mayhem. "Getting the pigs corralled was not an easy experience," he recalled. "I mean, you would get the pig out of his run, and the pig *would* run, wherever he desired. I think it took three of us to corral the pig, in sort of a triangular crate, so that he would be wedged in, and then we would inject him as he was . . . as he was apprising you of his displeasure. Which included, you know, high-pitched screeches and efforts to bite off an appendage. And finally, just to add a bit of poetry, he'd give you an immediate critique. And if you weren't careful, you'd end up slipping in it. Which we often did." A few surefooted colleagues undertook the treacherous pilgrimage to witness the future of immunotherapy and emerged absolutely incredulous. "The chorus of 'This will never work' and 'You're wasting your time' and other words of encouragement were a daily process," Wolmark said. Did he ever pause to ask himself how someone with years of medical training and high-powered mentors ended up in an animal facility operating on resentful pigs? "If I had ever *forgotten* to ask myself that question,"

Wolmark replied, "there was a host of people who would ask it for me."

At the end of the laborious operations, Wolmark would trudge back through the tunnel to Rosenberg's small tenth-floor lab. "We'd be walking through the tunnels with this vast array of tissues, different parts of the pig, wheeling them in carts," Wolmark recalled, "to the great chagrin of the people who saw this group of invaders, basically, some occasionally covered with blood, marching along. And then the work just started."

It was gory science. Living tissues do not thrive for long outside the body, so once back in the lab, Wolmark hastily squeezed fresh spleens and lymph nodes through large, metallurgic strainers to break down the tissue. One after another, these huge strainers, about a foot in diameter, piled up next to the sink, blood and fluids oozing down the mounting tower, no time to pause, no time to clean up. They had to work quickly to harvest their cells. They treated the broken-down tissue with enzymes that dissolved the connective tissue, essentially liquefying it, and then spun this porcine puree in a centrifuge to separate the cells, break up and remove the red cells, and isolate the bland, almost colorless band in the test tube that represented the harvest of white blood cells—the lymphocytes. As morbid backdrop to this precious colorless fluid, the rest of the lab "was just *covered* with blood," Wolmark recalled. People would walk into the lab, survey the scene, and cry, "My God, what the hell are you doing?!?"

The short answer—the answer they could always use to silence friend and foe alike—was that they were trying to cure cancer. They learned how to grow more of the cells in flasks; they studied the ability of these cells to be "cytotoxic," to kill tumor cells; they tested the system in mice and rats, and finally, in November 1977, they prepared to take the great leap: they were ready to inject pig cells into a human being. "I don't think they *ever* would let us do that experiment today," Wolmark said in 1993. "Nobody, probably, thought that we'd have the unmitigated audacity to inject these cells into human beings. Which we quickly did."

As he would later admit, Wolmark's boss never actually believed it would work, either. But then, that wasn't exactly the point. "I wasn't *expecting* us to inject pig cells and watch tumors disappear," Norman Wolmark explained later. "But on the other hand, I thought that it was a *start* of something that could possibly have merit." It was indeed the start of something: the beginning—at least in intent—of specific immunotherapy in humans.

❧

The first patient was a woman from Pennsylvania; Rosenberg refers to her as "Linda Karpaulis" (in this as in other patient names, the same pseudonym has been used as in Rosenberg's book to avoid confusion). Only twenty-four years old, she had been diagnosed with an aggressive sarcoma in one thigh, and early in 1976, as part of another clinical trial at the National Cancer Institute, underwent a "hemi-pelvectomy"—an amputation of the leg, including part of the hipbone—followed by chemotherapy. A year and a half later, she came back, and follow-up x rays revealed that the cancer had spread to her lungs. They attempted to remove these nodules surgically, but when the tumors proved to be hopelessly numerous, Rosenberg instructed Wolmark to implant bits of Karpaulis's sarcoma tissue into a pig. They applied for permission to infuse pig cells into Karpaulis at the end of October 1977.

Even before the lung surgery, Rosenberg had explained to the woman that "there is something else very experimental that we've never done but that we can try, if you are interested." He outlined the pig cell transfer to her and asked if she would be willing to try. These are moments of such extraordinary vulnerability on the part of the patient that it is fair to wonder if it is possible for any person, looking up from a hospital bed, up from an ever-deepening well of terminal illness, to assess such a question objectively, or whether objectivity even matters. Like many terminally ill patients to whom similar questions have been posed, Linda Karpaulis consented to a highly experimental procedure with, her doctors knew, little chance of success.

On November 15, 1977, they gathered in Linda Karpaulis's room for the historic transfusion. First they gave Karpaulis a small dose—5 cc, about one teaspoon—of the pig lymphocytes. When there was no adverse reaction to this first sip of cells, they went on to infuse 5 billion pig cells intravenously; Wolmark manipulated the IV tubing and controlled the flow as the pig cells seeped into the woman's bloodstream. The physicians who crowded the room, Wolmark and Rosenberg included, didn't dare hope for a clinical response; they just wondered whether the thick slurry of white blood cells, harvested from pigs and hanging now from a transfusion bag, could be safely transferred into a human being. "We didn't know *what* would happen," Wolmark said later. "Including the possibility of these cells embolizing [forming blood clots] and causing sudden death. Which happened in mice very frequently. And we stood there at the bedside with a cart outside, as I recall, a resuscitation cart, and slowly injected the cells." To everyone's great relief, there was no dangerous reaction.

"Of course I knew that these cells would get rejected," Rosenberg said later. "The only hope was that they would mediate some anticancer effect before they got rejected." Unfortunately, not even that occurred. The woman developed chills, and her temperature shot up, but that was about it. Linda Karpaulis was discharged forty-eight hours later with no change in her condition, and she later died at home. Five other patients received transfusions of pig lymphocytes at the NIH in the fall of 1977; none experienced any benefit from this unusual treatment.

It was Rosenberg's first, crude attempt to exploit the remarkable specificity of lymphocytes as a form of human therapy against cancer. The NIH researchers had demonstrated that massive amounts of lymphocytes could be safely transferred between patients—and indeed, between species—but no one seemed particularly impressed. Not one mention of the work, two arduous years of experiments in animals and trials in humans, has ever seen the light of day in the medical literature. Rosenberg and Wolmark sent out papers describing every phase of the work, and all four papers came flying back from astonished and indignant peer reviewers as if shot out of a cannon. "It was considered just a little too offbeat," Rosenberg admitted. "People thought it was ridiculous. People thought the idea that you'd transfer T cells and a tumor would go away, especially if you took them from another species. . . ." He shrugged at the memory. "After the raving reviews that these initial efforts got," Wolmark added, "I think we were probably too embarrassed to submit the experiments in which we had actually done adoptive transfer itself."

Even as Norman Wolmark shoved pig spleens through strainers, the coming revolution in molecular immunology made itself painfully, immaculately apparent. With the discovery of a molecule that allowed one to grow millions of identical T cells in a lab dish and with the suddenly reasonable prospect that genetic engineering might provide unlimited amounts of that molecule, the days of silk purse immunotherapy were clearly numbered. "We were very much aware of all that was going on there," Wolmark said, not without remorse. Indeed, he knew Rosenberg had already assigned someone else in the lab to work on T cell growth factor (soon to be renamed interleukin-2). Wolmark could only watch from a distance, not unaware that he was among the first immunologists caught standing on the wrong side of the molecular divide.

"These experiments—I was locked into them," he said plaintively. "So I basically was looking very longingly and with some degree of perhaps jealousy at the younger individuals who had arrived, just arrived, in the lab, who were now starting to grow human lymphocytes with T cell growth factor! In a pristine surrounding! With incubators, with white coats on that never seemed to turn red! And you know, they came in at a reasonable time. They didn't have to go to the abattoir. They didn't have to soil their hands. And they didn't have to risk injury by animals who had a very excellent recollection of what you had done to them last time, and were not entirely pleased by the prospect of having it done again. It made the pig work all the more difficult, because here somebody was doing exactly what we had *hoped* to be doing, only doing it in the human. And doing it with techniques that were really a log ahead of what *we* were doing. So when I say that the technology was barbaric, it wasn't only barbaric in retrospect, it was almost barbaric on sight. Because here, you know, one saw one era closing and another beginning. And I was committed. I couldn't straddle the ship leaving the wharf because I knew I'd fall into the water."

Barbaric though those first experiments might have seemed, they anticipated by twenty years an exploding interest in the transplantation of cells and tissues between species, a field now known as xenotransplantation. The most publicized example in recent years has been the 1995 experiment in San Francisco, when researchers from the University of Pittsburgh and the University of California–San Francisco transplanted bone marrow cells from a baboon into an AIDS patient in the hopes that they would reconstitute the man's immune system (by February 1996 it appeared that cells from this species, which is resistant to HIV infection, had failed to take, although the man's health has reportedly improved). Nonetheless, a bustling cottage industry in xenotissues has developed, and a number of major pharmaceutical companies, including Sandoz and Baxter Health Care Corp., have made significant investments in the field.

Even in failure—perhaps *especially* in failure—the Rosenberg lab had embarked on a rather pugnacious approach to biological therapy, as aggressive philosophically as it was scientifically. They did not wait to do safe experiments. They were not afraid to bang their heads against the wall. They did not wait for perfect knowledge before trying new approaches in humans. And all those qualities could be traced back to the laboratory chief. "He was very persistent," Wolmark said. "And unbending, as far as scientific endpoints. I think that was all to his credit. From a scientific standpoint, undertaking a new methodology that was ridiculed by some—I think that Steve Rosenberg was exactly the kind of person who needed to take it on."

"You know, I've always felt scientifically, just to paraphrase Pasteur's motto, that if chance favors the prepared mind, then chance favors a prepared mind only when the mind is at work," Rosenberg later said of these experiments. "And it's always been my feeling that you have to be *doing* things, even if they're not the right things right then, even if they're not the perfect things right then, so that you'll be thinking about them, and when the right things come along, you'll be ready to leap upon them as opposed to having to start from the beginning. And so, even though I had very little hope that those were the ultimate answers, at least it was a way to get started and begin to look at immune reactions and begin to study patients, with the idea that as we learned, we then would already have a big head start." If science is often a process of learning from failure, Rosenberg had the sheer energy and faith and confidence to fail more often than anybody else, as long as it got him closer to the success he believed was sure to come.

11

THE RISE OF THE
T CELL CHAUVINISTS

I'm probably the most impatient person you've ever met.
—Steve Rosenberg to television interviewer, 1996

❧ *There are countless ways* to document scientific single-mindedness, but Steve Rosenberg tells a revealing anecdote about himself that gives special resonance to the phrase "need to know." He had just attended a small scientific meeting in England in September 1977 and joined several colleagues on a flight to a second meeting in Japan. It was, to his mind and many others, a unique moment in immunology. The theoretical importance of T cells—the stalker cells, the effector cells, the memory cells of the immune system—had only recently been matched by the practical ability to cultivate and study those cells in the laboratory. As usual, several enterprising bloodhounds had been sniffing around the edges of this black box, their research (or intuition) too far ahead of the curve to attract much notice. But the discovery of interleukin-2—*the* potion for growing T cells—changed all that.

On the flight to Japan, even as Norman Wolmark was harvesting pig lymphocytes by hand, Rosenberg discussed these emerging molecular possibilities with the immunologist Richard Hodes, currently director of the National Institute on Aging, who was describing some pioneering research by Edward Boyse and Lloyd Old in the 1960s that identified different antigen-specific cell subpopulations. "I got real excited about it," Rosenberg recalled in an interview, "and then we got into this turbulence, and it was putting on seat belts and all the rest, and I remember thinking at that point—and then being amused by the fact—that I wanted him to talk a little faster because it looked like we were going to crash, and I wanted to find out what it was all about. I wanted to hear it fast in case we crashed first and I would never find out." Here, in short, was a man so intent on learning a scientific fact useful to him that it didn't matter whether the plane he was in was about to go down or not, only that he acquired the fact before it did.

Since those first pig lymphocyte experiments in 1977, Steve Rosenberg has left an indelible mark on the field of cancer immunotherapy. To some, he may have pushed too far and too fast; to others, he was just the right person to do the heavy lifting in a field with a reputation for sacrificing far too many mice in order to save far too few human lives. He has become a lightning rod to an exceptional amount of criticism, in part because he tries to do so much and in part because one gets the impression that stepping on the toes of colleagues, like riding in airplanes about to fall out of the sky, doesn't much distract him as he pursues his oft-professed goal, which is to cure cancer.

Despite having worked very hard to enjoy the credentials and perks of the quintessential insider, Rosenberg has never quite ceased being the striver, the outsider with something to prove. The youngest of three children, Rosenberg was born in August 1940 into a deeply religious, "very lower middle class" Jewish family in New York that overcame classic immigrant deprivations. Rosenberg's father Abraham arrived at Ellis Island in 1919 at age twenty-two from his native Poland, parentless and penniless (his mother having died in his arms in one of countless anti-Semitic pogroms). The elder Rosenberg found lodging in the Lower East Side and work in a dress factory. His wife-to-be arrived with her family in 1920, and the couple married in 1927. In 1929 Abraham Rosenberg lost his job but bought a candy store, the first of several small businesses he would run until his retirement. The family lived off the Grand Concourse and 173rd Street in the Bronx during the war years, and Rosenberg has written movingly about the postcards that would arrive at the Bronx household after World War II, bearing news of yet another relative in Poland who been executed, who had disappeared, who remained ominously unaccounted for. In a passage from his book, memorable both for its passion and its transference, he would later equate cancer with the Holocaust: "Indirectly, my family history accounts for my choosing to work on cancer. To me the disease resembles a holocaust. It is a disease one can hate. Other diseases, including heart disease, tend to attack older people, but cancer kills randomly, and it kills the young. Among people between the ages of fifteen and thirty-five cancer kills more than any other disease. In cancer, one's own cells turn alien and grow out of control. Tumors thicken, grow, spread, and eat away at the body. The way cancer gradually takes over one's body and forces its victims and their families to watch impotently as it grows and spreads makes it hateful. Cancer murders innocents. It is a holocaust."

Not coincidentally, he has prosecuted his campaign against cancer with an almost religious fervor. Raised in an Orthodox household, Rosenberg never dared switch on an electric light on the Sabbath, never tasted nonkosher food until he went to college. His work ethic was of a very old-fashioned sort. Each and every day during his school years, he spelled his father at the New Madison Luncheonette, the Manhattan coffee shop owned by his father on East Thirty-second Street; Abraham Rosenberg would open the restaurant at 5 A.M., and the teenage Rosenberg would show up after school to relieve him and run the place until closing time. As a student at the Bronx High School of Science, Rosenberg was bright, but not blindingly so. His older brother Jerry, now a surgeon in Michigan, used to chide him about his laziness ("I think that was just his way of goading me to do better," the younger brother explained).

The sibling indictment of laziness seems inconceivable to anyone who has encountered Rosenberg since roughly 1957, when he began his four-year pre-med education with a scholarship to Johns Hopkins University. He describes himself as one of those lucky people who always knew what he wanted to be. "I recall my first ambition, aside from becoming a cowboy, was to become a doctor and a scientist," Rosenberg writes of his youth. His very first scientific paper, coauthored with his brother in 1961, had to do with malignant melanoma (in hamsters), and he is nothing if not consistent; his 687th paper, published in 1996, also had to do with malignant melanoma (in humans).

The single-mindedness, the focus, was always there. During medical school training at Hopkins between 1961 and 1964, he recalls how he would "allow nothing to deflect me even momentarily from pursuing my learning," not even the eminently distracting Cuban missile crisis; and when he served his internship and residency as a surgeon at Peter Bent Brigham Hospital in Boston, one of the teaching hospitals for Harvard Medical School, he recounts with admirable but revealing candor the anxiety he felt at becoming romantically involved with an emergency room nurse named Alice O'Connell. Trying to extricate himself from the relationship, he told her, "You're getting in my way. I'm very young, have a lot of work ahead of me, and cannot be distracted. I cannot allow myself to get involved in a relationship that takes my mind off my work." (Happily, they married five years later). He brought that same singleness of purpose to medicine.

Rosenberg perceived early on the need to bring the rigor of basic research to the inherent messiness of human biology. With that in mind, he interrupted his surgical training for four years to obtain a Ph.D. in biophysics from Harvard in 1968, specializing in the biochemistry of red blood cells. But medical training only goes so far in explaining Rosenberg's accomplishments; his is as much a story of personality and temperament as intellect. Like many surgeons, Rosenberg is immensely confident and makes no attempt to hide his ego. His training at Harvard seems to have imbued him not only with terrific medical instincts and superb surgical skills (the most gripping parts of his autobiography, *The Transformed Cell,* may be about surgery, not immunotherapy), but also the sometimes self-aggrandizing advertisements for self that might better have been left to other voices (of which there are many). He will remind an interviewer that he has published "more than 500 papers" in the literature; he will pause in mid-interview to count the number of references in a review article to establish the full citational sprawl of his scholarship; he will recall that he "probably made 10,000 egg creams" at his father's luncheonette. With Rosenberg, one gets the impression that quantity matters— amount of data, number of references, number of honors, and certainly the most important number among scientists who live near the leading edge of research: number one, being first.

And yet, though he probably wouldn't be flattered by the comparison, Rosenberg has a good deal in common with William Coley. Both trained as surgeons, both were exemplary clinicians, both had a reputation for innovation, both were perceived by at least some colleagues as self-promoting, both insisted on the propriety of offering imperfect but nonetheless promising treatments, and, perhaps most revealing, both were drawn to immunological medicine by a single exceptional patient early in their careers.

In Rosenberg's case, it happened in 1968 when, having finished his Ph.D. at Harvard, he resumed his surgical training, which included a rotation at the West Roxbury VA Hospital. One day in June of that year, a man Rosenberg referred to pseudonymously as "James DeAngelo" walked into the emergency room, complaining of severe abdominal pain. Tests revealed a gall bladder in need of excision, and

Rosenberg was assigned to handle the surgical honors. But while examining DeAngelo, he could not help but notice the large autograph of a scalpel across his abdomen. When he asked about the surgical scar, the sixty-three-year-old DeAngelo recited a remarkable story.

It began in July 1956, when Rosenberg was still just a fifteen-year-old slinging egg creams at his father's Murray Hill luncheonette. James DeAngelo wandered into the West Roxbury VA Hospital in Boston, complaining of intense abdominal pain. Medical records indicate he hadn't taken good care of himself: he consumed two packs of cigarettes a day and three or four fifths of booze a week, but it wasn't weed or whiskey that was causing the pain. X rays suggested a mass in the stomach, and exploratory surgery revealed a fist-sized tumor as well as several enlarged and hardened lymph nodes nearby and, more ominously, three cherry-sized tumors that had spread to the man's liver. Surgeons at the VA Hospital removed 60 percent of the man's stomach but had no illusions; once stomach cancer hops the body's fire line and spreads to the liver, the game is usually up. DeAngelo's recovery, as doctors like to say, was "uneventful," except for one thing: he developed a postoperative bacterial infection so severe that it required follow-up surgery two weeks later just to clean the wound.

Sent home to die, DeAngelo steadily improved. He gained twenty pounds and even returned to work. He said he felt fine when he came back to the hospital five months later, and he looked it. Three years later, in 1959, he returned with a lump just below his left ear, which doctors believed to be a metastasis. "No diagnostic or therapeutic efforts were instituted," they noted, and yet two years later, the lump had disappeared. Even though he had been essentially given up for dead in the summer of 1956, his cancer had mysteriously been held in check. The only clue, possibly, was that when pathologists looked at the stomach tumor that had been removed in the 1956 operation, they noticed a "dense infiltration" by lymphocytes, plasma cells, and eosinophils—three types of white blood cells that were vaguely thought in those days to be related to inflammation or immune function. But in a 1956 pathology report, their appearance would barely raise an eyebrow.

Twelve years later, Rosenberg could scarcely believe his eyes as he leafed through DeAngelo's medical records. The man's story was true: biopsy slides and test results confirmed that an aggressive, metastatic cancer had spread to his liver in 1956. Rosenberg wondered if any traces of the cancer could be found when he removed the gall bladder. With the permission of senior colleagues, he closely examined De-

Angelo's abdomen, probed his liver, felt under it and behind it, and found no tumor. It had disappeared. "The cause of such regression is unknown," Rosenberg and colleagues concluded in a 1972 paper describing the rare case. The man had cured himself of cancer, and a single question nagged at Rosenberg: "How?" The question was so open-ended that in a sense Rosenberg has spent the last quarter of a century trying to answer it.

There was another question Rosenberg couldn't wait to answer, however. With the same urgency of the man in the bouncing airplane a decade later, Rosenberg wanted to know if James DeAngelo's blood might contain some mysterious antitumor component that could confer health to an otherwise terminal cancer patient. In the first in a career-long series of bold, controversial, and in some respects wishful experiments, Rosenberg drew a sample of DeAngelo's blood in 1968 and transfused it into an elderly man with stomach cancer who had failed all available treatments. The treatment dabbled in ancient humors; it was barely more sophisticated than BCG or Coley's toxins. Not surprisingly, the patient in this experiment died within two months. "In retrospect," Rosenberg would write later, "it was a naive experiment—perhaps even then I knew this—almost embarrassing in its simplicity. But I had to try it."

I had to try . . . The remark hides in plain sight all the qualities that explain both Rosenberg's great accomplishments and the equally great exasperation expressed by many of his peers about his style of achieving them. Aggressive, compelled to help, impatient with (and sometimes contemptuous of) medical bureaucrats who snarl medical progress in red tape—that is what cancer patients love in Rosenberg the physician, as well they should. But it is those same qualities that have made some fellow scientists uncomfortable. Whatever one's view, he launched his immunotherapy crusade, with the puzzle of James DeAngelo's remarkable recovery fresh in mind, at a unique juncture in American biomedical research. The rise of Steve Rosenberg and what have been called the "T cell chauvinists" turned out to benefit greatly from two wars (both losing efforts) and one revolution.

The first war, ironically, was a world away, in Vietnam. Up until the war's end in 1975, many of America's best and brightest young physicians satisfied their obligation to military service by working for the government's Public Health Service (PHS). Instead of doing postdoctoral research in the nation's premier university medical centers, an unusually talented cohort of young scientists gravitated to the campus of the National Institutes of Health in Bethesda, Maryland. This influx did not occur overnight, but neither did its impact cease with the fall of Saigon. Steve Rosenberg himself joined the PHS in 1970 and spent two years at the National Cancer Institute as a clinical associate in immunology before returning to Boston to finish his surgical residency at Harvard Medical School. The Vietnam War created a kind of positive brain drain to Bethesda of scientists who were doing exciting work, so that by the time Rosenberg returned to the NIH on July 1, 1974, to take over the Surgery Branch of the National Cancer Institute, a month shy of thirty-four years, he rode the crest of an incredible wave of talent.

His arrival closely coincided with a second war, more amorphous, closer to home, equally unwinnable, and declared by the same man many associated with the debacle of Vietnam: the War on Cancer. Hostilities were formally declared in December 1971 when Richard Nixon signed into law the National Cancer Act of 1971. Books have described the intense lobbying and behind-the-scenes politicking that shaped this historic legislation; and although there are a number of ways to interpret the results, the bottom line—an especially apt metaphor, given the historical impulse to throw money at the problem of cancer—was that the federal budget for the National Cancer Institute began to skyrocket. The government spent $200 million on cancer research in 1970; by fiscal year 1972, spending had jumped to $337 million, and now it is roughly $2 billion a year.

These riches didn't necessarily stay at home: the NIH disbursed funds both to deserving researchers at universities throughout the country (the extramural program) and to researchers on its own Bethesda campus (the intramural program), and arguably the funding bore precious, if unintended, fruit; a large (and largely unsuccessful) effort to pin down viral causes of cancer led to the coincidental discovery of oncogenes, which in turn opened up research into molecular biology, signal transduction pathways, cell cycle events, and the genetics of tumor suppression and carcinogenesis that dominate basic cancer research today. The important thing for Rosenberg was that the budget virtually doubled overnight. Said to be the best-funded cancer researcher in the world, Rosenberg could afford to try a number of approaches at once; moreover, the resources gave him the space and personnel to pounce upon and run with new developments as soon as they were reported by other labs. In addition, researchers based at the

NIH had the added advantage of receiving funding without having to ask for it in the normal bureaucratic fashion—intramural scientists did not have to write the elaborate grant applications, the form of rationalized begging that by some accounts now takes up approximately one-third of a working scientist's time. The NIH scientists could simply do science. They had access to patients from all over the country; they did research and medicine in close approximation; they had a mandate, unspoken but clearly understood by people like Rosenberg, to develop new approaches to wearisomely familiar problems—to shoot for the moon, medically speaking, because the War on Cancer had in fact been conceived as the biological equivalent of the moon program.

Finally, the war coincided with a distressing reality in cancer research. Aside from several well-publicized exceptions, the limitations of most cancer treatments in the 1990s could be depressingly foretold in limitations apparent in the 1970s, and in some respects were almost identical to the problems bemoaned by William B. Coley and James Ewing in the 1920s and 1930s. When Rosenberg embarked on his crusade, surgery, chemotherapy, and radiation represented the three main arms of treatment, but their efficacy always represents a shrinking pie of hope. As recently as 1996, for example, statistics indicate that of all cancer cases (the annual number of new cases in the United States is about 1.3 million), only 50 percent will survive for five years. Of that fortunate cohort of survivors, the vast majority benefit from surgery, and what was true at the beginning of this century is for the most part true at the end—the best hope, short of prevention, is to detect a cancer early and, if possible, cut it all out. Some statistics indicate that only 10 percent of all cases respond to radiation, and no more than 5 percent to "chemotherapy and/or radiation." "The misperception in cancer therapy over the past twenty years," says Michael Lotze, associate director of the Pittsburgh Cancer Institute, "is that chemotherapy does very much for solid tumors. If you look at much of what we've done in the last twenty years in oncology, one of the problems is that we continue to test notions that we should have disposed of years ago. Single-agent chemotherapy and multiple-agent chemotherapy just do not work in many settings! And yet, because of the lack of other ways of treating patients, and since those patients demand treatment of some sort, doctors keep giving these drugs in more complex, and often more toxic, regimens."

There are noteworthy and gratifying exceptions to this trend, of course. Childhood lymphocytic leukemias (currently about 11,000 new cases a year), the relatively rare Hodgkin's lymphoma (7,500), and testicular cancer (7,400) all represent satisfying exceptions to the general lack of success of chemotherapy, but these account for only about 2 percent of new cases of cancer in the United States each year. The big killers—prostate cancer (317,000 new cases), breast cancer (186,000), lung cancer (177,000), colorectal cancer (133,000) and so on—tell a rather more sobering story, largely because these cancers often cannot be contained, and spread to other organs. "The reality is that chemotherapy doesn't work very often," said Phillip Frost, an oncologist and drug industry executive. "If you're lucky and you have one of the very rare cancers that respond well to chemotherapy, like carcinoembryonic seminoma of the testes, which has a 90 percent survival rate, you're in good shape. If you have lung cancer, it's essentially zero. It's just not working."

It was in the context of this second lost war that Rosenberg sought something better. Immunotherapy, in theory, represented the perfect, microscopic complement to surgery. Just as surgeons cut away the gross evidence of disease, the immune system was uniquely designed to find and cut away microscopic remnants of disease, which if left untreated accounted for the fatal spread of local disease. Indeed, it might best be thought of not so much as a panacea as the second of an unusually effective one-two punch, where the patient would undergo surgery or radiation to "debulk," or remove most of the tumor, and then receive a course of immune modulation to stimulate the immunological elimination of any surviving malignant cells.

As he began to shape the Surgery Branch as a research-driven, innovative, and exploratory band of physicians, Rosenberg remained partly inspired by the case of James DeAngelo back in West Roxbury, but also by an example from transplantation immunology. He learned of the case of a man at Brigham Hospital who had undergone a kidney transplant. Unbeknownst to doctors, the donor kidney contained a tumor, and when this patient received immunosuppressive drugs to prevent rejection of the transplanted organ, metastatic tumors spreading from the kidney began to sprout up all over his body. However, when doctors removed the tainted kidney and discontinued the immune-suppressing drugs, the man's own immune system mopped up the remaining metastatic nodules and eliminated them entirely. "That, it seemed to me, was what could make huge kidneys and livers get rejected, and that's what I wanted to do with cancer," Rosenberg said. "So it was always a search for a *specific* immunologic response against a known antigen."

His timing could not have been better, because these aspirations coincided with the unfolding technological revolution in biology: the discovery and application of recombinant DNA techniques to study immunological processes. Though it would take several years for the technologies to evolve, its progress slowed in part by a self-imposed scientific moratorium in the mid-1970s, genetic engineering would change the way everyone—immunologists, molecular biologists, Steve Rosenberg—would go about their business.

Once at the NIH and once committed to immunological approaches, Rosenberg single-handedly recapitulated the long and rather problematic history of immunotherapy up to that time. His publications during the 1970s reflect an eclectic mix of interests, a compass needle spinning in every direction: nonspecific therapies like BCG, swine transplantation experiments, studies of chemotherapeutic agents alone and in combination, even a randomized trial of the active agent in marijuana to see if it eased the side effects of chemotherapy. Many approaches were quickly tried and quickly abandoned, including the use of BCG against melanoma and another bacterium called *Corynebacterium parvum* in patients with sarcoma. "That was something that I started doing right off the bat when I got here," he said, "but I don't believe that any of it ever worked." There had to be something better. Rosenberg began looking for it.

He was not, it goes without saying, the only pilgrim traversing this Slough of Despond. The 1970s, just before recombinant DNA technology made its mark, saw a last explosion of nonspecific immunological research, along with the beginnings of more specific therapies. Because most of these reports have not held up over time, it is sometimes easy to overlook the few that have. In 1976, for example, "when interest in BCG was waning fast," according to an editorial in the *Lancet,* Alvaro Morales of Queen's College in Ontario, Canada, reported that BCG proved unusually effective against superficial bladder cancer; after extensive testing against chemotherapy in randomized trials that took more than a decade, BCG is now part of the regular armamentarium in bladder cancer. Philip O. Livingston, Herbert Oettgen, and Lloyd Old at Memorial Sloan-Kettering began a decades-long campaign to develop a melanoma vaccine based on surface markers known as gangliosides, which are recognized by antibodies: variations on that vaccine have been tested in several hundred melanoma patients, seem to be associated with prolonged survival, and are currently the focus of a large, multicenter study. Meanwhile, Sanford

Kempin and colleagues at Sloan-Kettering even saw promising hints of efficacy in the trial of a mixed bacterial vaccine based on Coley's toxins.

Rosenberg lit out in a different direction. "What I decided as I began to look into this was that attempts to *immunize* patients against their disease would not be effective," he said, in effect ruling out vaccines that incorporated either microbial agents or tumor antigens (nearly twenty years later, this would be back in vogue, Rosenberg chasing it as hard as anyone). "And that what one had to do was transfer to a patient *with* cancer immune substances that were capable of destroying their cancer. And so I spent well over a year looking *exhaustively* at the world literature on what I call 'passive immunotherapy.' " "Passive" in this case meant that researchers passively transferred immunologically active agents, in the form of either cells or antibodies (in the form of serum), to patients rather than actively stimulating the patient's own immune system with vaccines. All this reading resulted in a detailed review published in 1977 in *Advances in Cancer Research*, and it was clear that Terry and Rosenberg ended up unimpressed by all the immunotherapeutic research that preceded them.

"It was just a mass of people who were giving nonspecific immune stimulants," he recalled. "With virtually no basis in animal data. I mean, that's one of the things that struck me, that there were these enormous numbers of clinical trials being done with nonspecific stimulants *when none of them worked in animals!* And yet there were all of these human trials, an entire focus of human trials, on things almost none of which worked in animals as *therapy.* And early on, I realized that one of the major problems was that when you immunize an animal against a tumor and keep him from getting a tumor, that's easy. There are dozens of manipulations that will do that. But there were virtually *no* manipulations that would make an established tumor disappear. And it seemed to me that people were really confusing the two of these issues. And so what I set out to look at were true therapy models, as opposed to preventive models, which represent a whole different kind of problem."

That, in effect, was Rosenberg's one-man self-imposed mandate: Therapy, not prevention. Treatment, not vaccines. Specific, not nonspecific. Like a decision tree, each choice carried him further along toward the use of immune cells. After much study and discussion with people like Bill Terry, then head of the Immunology Branch at NCI, and David Sachs, a leading investigator in transplantation immunology, Rosenberg finally reached an almost inevitable conclusion.

"Cells," he decided, "had a lot more likelihood of being effective" than transfers of serum, the liquid portion of the blood. And it was beginning to look as if lymphocytes were uniquely qualified for the job. They not only could find cancer cells but could kill those malignant cells once found. There were, he recalled, "a few examples that I thought were quite suggestive, that such an approach might be effective." Peter Alexander at the Institute of Cancer Research in England had done some work during the previous decade, he recalled; Herb Rapp at the NCI had done some work with cells, too, as had Herbert Oettgen and colleagues at Sloan-Kettering in New York. And then there was another researcher whose name popped up repeatedly in the references of the 1977 review article, as it would repeatedly pop up as a counterpoint to Rosenberg's work over the next few years. If scientists, in Newton's memorable phrase, are always standing upon the shoulders of giants, there are some immunologists who suggest that a good deal of Steve Rosenberg's visibility in the 1980s came because he stood on the shoulders of a five-foot-five man at the University of Washington named Alexander Fefer.

It is hard to imagine two men with so much in common turning out to be so different. Both of Eastern European Jewish ancestry, both of families traumatized by the Holocaust, both overachievers who earned educations at the finest schools, Rosenberg and Fefer bet their research careers on the immunology of cancer at a time when that looked like a very bad bet. Perhaps the single greatest difference between the two men lies in their personal temperaments, which goes a long way toward explaining why Rosenberg ended up on the cover of *Newsweek* a few years later while Fefer would be apologizing for stammering through an important talk at a large cancer meeting—a talk intended to critique the work that made Rosenberg cover-boy material in the first place. Revealingly, they both like to cite Pasteur's famous dictum—"Chance favors the prepared mind"—but choose to place the accent on different parts of it. Where Rosenberg sees his story as a case of the prepared mind, Fefer views his career as a tale of chance.

Born in Tarnopol, Poland, in January 1938, one year before the Nazi invasion, Fefer became separated from his parents during World War II, lived in "various and sundry places" by his wits, and had basically become a displaced person before he could talk in complete sentences.

Although his parents survived the Holocaust, he lost all his other blood relatives—grandparents, cousins, sixteen uncles and aunts. After the war, the family made its way to a camp for displaced persons in West Berlin, and then to the United States. On June 16, 1949—he remembers the date as automatically as his birthday—Fefer arrived with his displaced family in Brooklyn, where they lived in an apartment above a garage in the Williamsburg section. Fefer's father worked the night shift in a paper factory; his mother made bras and girdles.

Against his parents' wishes, Fefer attended public high school in the Brownsville section of Brooklyn, where his work earned a scholarship to Harvard as a liberal arts student. Then it was on to Stanford University School of Medicine in 1959. Upon arrival at Stanford, in an effort to make money, he found work—"by accident, complete and total accident"—as an animal cage cleaner in the department headed by Joshua Lederberg, a man whose name meant nothing to Fefer (although a few months earlier Lederberg had received a Nobel Prize in medicine). Nor did he know much about the fellow for whom he cleaned cages, a visiting Australian named Gustav Nossal. "I didn't know who he was, either," Fefer remembered. Soon he learned that Nossal was the intellectual and institutional heir to Macfarlane Burnet, one of the most insightful immunological theoreticians in the world; soon thereafter he was not simply cleaning cages, but doing experiments and getting a crash on-the-job course in immunology. "From that point on—this happened about two months before I started medical school, when I was twenty-one—from that point on, I basically have never stopped doing research," he said.

Almost everywhere he turned, he bumped into estimable mentors—a postdoctoral fellowship with Georg Klein at the Karolinska Institute in Stockholm, a research job at the NIH, where he did important early work with the Moloney sarcoma virus, a virus that causes tumors in mice. "I set up a whole series of experiments right off the bat," he said, "in which I put murine sarcoma virus into adult mice of various strains, and indeed what happened was they *did* develop tumors. And you could document it. And the tumors all went away. But when I put them into newborns or irradiated adults"—animals, that is, incapable of generating a mature immune response—"lo and behold, they induced tumors and the tumors grew, and then the cancer killed them."

Fefer was on his way. Within two years, he pumped out a dozen papers showing the interaction between these tumor viruses and the immune systems of mice. He traced this immune protection to a certain class of cells, namely the lymphocytes, and soon thereafter proved that when he transferred these cells from immunized animals to newborn or irradiated adult mice, these animals did not develop tumors, even when injected with the cancer-causing tumor cells. "I used the most egregiously ridiculous, exaggerated system in order to test a principle," he said. "But once it was tested, I knew it could be applied in situations akin to clinical conditions." So did Rosenberg, who cited this work in his 1977 review.

Fefer and Rosenberg never overlapped at the NIH. In 1968, E. Donnall Thomas, the University of Washington physician who pioneered the modern bone marrow transplant (for which he was awarded a Nobel Prize in 1990), recruited Fefer to Seattle as a hematology/oncology fellow to work on the transplantation immunology team. Even as he assisted at the first bone marrow transplant to treat leukemia in March 1969, Fefer committed himself to the long-term study of cellular immunity to cancer. In some respects, he had a tougher row to hoe than Rosenberg; he had to convince peer reviewers that the transfer of lymphocytes was a worthwhile intellectual and medical enterprise before he received grant money. "I used to get reviews of my grants," he said later, "which I kept getting funded, and every time I'd get comments like, 'These are fascinating, but it's basically a voice in the wilderness. None of these are going to apply to humans. After all, if it takes 10 million lymphocytes to cure a 20-gram mouse, how many truckloads of lymphocytes will be necessary to cure a 70-kilogram human?' They didn't put it exactly that way, but between the lines, that was sort of the essential message. And then of course, after all there's no evidence that there's an immunological target in human cells. And at the time, indeed, there wasn't."

In the problems they hoped to overcome, Fefer and Rosenberg were very much soul mates. They knew T cells could ferret out tumor cells, although no one understood exactly how. They knew T cells could see something, some target, on those cells, but they didn't know what. And they knew that T cells, sufficiently stimulated and treated gently enough, could kill established tumors. T cells could also secrete powerful chemical messages, further orchestrating and modulating the immune response, and they could respond to chemical messages from other cells, like the incitements of the growth factor discovered by Doris Morgan, Frank Ruscetti, and Robert Gallo. The emergence of T cell biology and genetic engineering at roughly this same juncture suddenly promised answers to intractable, century-old problems. But these two streams hadn't entirely merged in 1977, and perhaps more important than the differences between Fefer and Rosenberg was that they were very much brethren occupying a fringe of scientific respectability: tumor immunology was not a hot field where young people flocked to make their reputation. Nonetheless, just as Rosenberg rallied his increasingly large army of researchers to tackle tumor immunology head-on, Fefer committed to that same intellectual path with a rather smaller retinue and ultimately turned the pursuit over to two talented younger researchers, Philip Greenberg and Martin Cheever.

It was just like Fefer to mull the ramifications of his animal experiments, ponder the next step, weigh with great caution and deliberation the ultimate treatment of humans with lymphocytes. And it was just like Rosenberg not to wait—not for perfect knowledge, and certainly not for twenty years. As he had shown with the pig lymphocytes, he was ready to try whatever was at hand, even if it was naive or imperfect. All they needed was patients, and there was never a shortage of those.

12

TO BE IN MOTION . . .

In research, the usefulness of error is that it leads to more
research, and this is what the word tells us. To err doesn't
really mean getting things wrong; its etymology derives from
the Indo-European root ers, *signifying simply "to be in motion";*
it comes into Latin as errare, *meaning "to wander," but the*
same root emerges in Old Norse as ras, *rushing about looking*
for something, from which we get the English word race.
In order to get anything right, we are obliged first
to get a great many things wrong.
—Lewis Thomas, *The Youngest Science*, 1983

ॐ *The Warren G. Magnuson Clinical Center,* known less euphoni-
ously as Building 10 of the National Institutes of Health, rises as an
imposing multistory wall of red brick in the midst of the leafy, gently
rolling NIH campus. Viewed from the rear, with its thin eyebrow of
windows running like a band across the top of a slightly convex wall,
the structure can sometimes resemble the bridge of a massive brick
ship; not only can visitors imagine it noiselessly plowing through the
Maryland turf, but cynics might view the illusion of motion as instructive, its progress against a disease like cancer so incremental that
indeed it appears not to be moving forward at all. That was the impression of many critics looking at the federal cancer effort in the late
1970s, and that perception of unsatisfactory progress became a static
frieze against which Steve Rosenberg appeared as a one-of-a-kind,
white-coated dervish.

For anyone with feet in the worlds of both research and medicine,
the NIH was about the best place in the world to be, and one reason
was as simple and unique as the interior architecture of Building 10:
there are many laboratories as well equipped as the facilities that Ro-

senberg created at the Surgery Branch of the National Cancer Institute, but nowhere, literally, is the distance between basic research and the patient's bedside shorter. In Building 10, Rosenberg's labs line a crowded, linoleum-tiled corridor narrowed, like an atherosclerotic artery, by the steady accretion over time of freezers, filing cabinets, mailboxes, and the inevitable overflow of laboratory hardware onto the shoulder of the hallway; pass through a double door at the end of the corridor, however, and you suddenly find yourself in a carpeted hallway running between rooms filled with cancer patients. In Building 10, technology transfer from the lab to the patient is literally a matter of fifty or so steps. In Building 10, those double doors serve as the thinnest of membranes separating theory and practice. In the drama of medical research, the threshold might even be thought of as a proscenium that separates the researcher as playwright, feverishly penning lines behind the curtain and passing them out, from the doctor onstage, who immediately tests them on a small, desperate audience of patients. Every drama is real, every line important (including the ones that don't work), every revision precious and potentially life-saving. No theater is more experimental, more gripping, more for keeps.

"We're always working in animals and people, in parallel," Rosenberg explained, and in the years between 1977 and 1985, Rosenberg and his coworkers wore a path between lab and patients. The group tested ideas in mice, tried them on patients, fine-tuned them in the mice, went back again to the patients; and in this back-and-forth dialogue between rodents and humans, between the clean geometry of petri dishes and genetically identical mice on the one hand and the splendidly messy, richly informed complexity of human immunology on the other, they advanced toward, circled around, and finally converged upon a "protocol"—an immunotherapeutic design to treat cancer patients. And in this complicated evolution of ideas and approaches, the lab unwittingly recapitulated all the lessons of research, the harsh quotidian ones as well as the occasionally gratifying ones. Like other labs, they experienced failure, but because they tried to do so much so quickly, they failed more often. Because they tried things without perfect knowledge, they failed more spectacularly. Because they worked at a large, well-funded federal research institution, they failed more publicly. And because patients died when they failed, each failure was deep, significant, and unforgiving.

Almost by sheer force of personality, Rosenberg persevered, main-

"Her death came to me as a great shock.": John D. Rockefeller, Jr., and Elizabeth Dashiell, the friend who inspired his lifelong interest in cancer research *(Courtesy of the Rockefeller Archive Center)*

"Nature often gives us hints to her profoundest secrets.": William B. Coley, around 1888, when he began his medical career
(Archives, Cancer Research Institute)

"Apparently he had only a few weeks to live.": Signor Zola, who was treated by Coley in May 1891 and survived another eight years
(Archives, Cancer Research Institute)

"I determined to try inoculations in the first suitable case.":
William Coley, one month before first use of Coley's toxins,
December 1892 *(Archives, Cancer Research Institute)*

"He had original views about *everything*.": Alick Isaacs, co-discoverer of interferon, ca. 1957
(National Institute for Medical Research, London)

"In the course of a true collaboration, one 'plays' the role of the enthusiast and the other the role of the skeptic.": Jean Lindenmann, co-discoverer of interferon, ca. 1957
(National Institute for Medical Research, London)

"We felt we'd got him back again. And then, then he had another hemorrhage.": Alick Isaacs with twins Stephen and David *(Courtesy of J. Lindenmann)*

"Many visitors were surprised when they came here . . . seeing that it was so small and modest.": Kari Cantell in his Helsinki office *(Photo by author)*

LEFT: "Chance favors a prepared mind only when the mind is at work.": Steven A. Rosenberg, Surgery Branch, National Cancer Institute *(NCI)*

BELOW: Georges Mathé, Hôpital Suisse, Paris, 1994. In 1969, he and colleagues claimed cures of childhood leukemias with BCG. *(Photo by author)*

"My critics say I'm a manipulator, but it's committee work at its finest.": French Anderson *(center)*, flanked by Mike Blaese *(left)* and Ken Culver *(right)* after a first attempt at gene therapy in 1989 *(National Heart, Lung and Blood Institute)*

"For the first time, we can select as candidates for therapy those patients who have a chance of benefiting from immunization.": Thierry Boon *(left)* and C. M. Melief *(Cancer Research Institute)*

"Human tumor immunology was reborn" with the 1991 *Science* paper on melanoma antigens, discovered in the cancer cells of Alexander Knuth's patient "Frau H" *(Cancer Research Institute)*

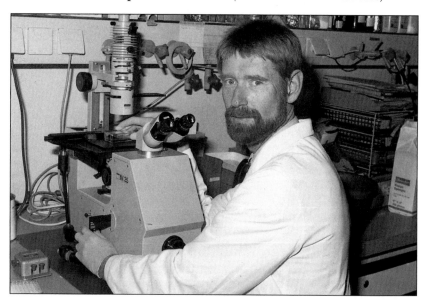

taining confidence in himself and retaining the confidence of his patients. He is an average-sized man with owlish, wire-rimmed glasses. The smile, almost boyish, is so sincere and becalming, the eyes genuinely moist with concern, the compassion almost palpable, that you get the feeling that this man would run through a brick wall, jump off a tall building, strangle a tumor with his bare hands, and *then* pluck every lingering metastatic cancer cell out with tweezers, if that's what it took to save a life. *Your* life. If you had cancer, this is the doctor you would want to come walking into your hospital room. This is the guy in whom you would want to invest your last savings, which for patients with advanced cancer is a resource never measured in dollars, always in hope. Whatever critics have said about Steve Rosenberg, it is impossible to dismiss or diminish the effort he makes on behalf of his patients, which is to spend every waking moment of his life and every erg of his considerable energy trying to rewrite the tragic ending of every bedridden story in his ward.

In his battles with critics, administrators, and regulators, Rosenberg has always justified the speed at which he moves by positioning himself on high moral ground, the better to fire down on his detractors. "Every day I take care of cancer patients," he said in one interview, during a time when he felt under particular attack. He sat in his small office, munching M&Ms and drinking a soda, speaking in a surprisingly low, somewhat gravelly voice that admitted no uncertainty. "It's rare that I ever have less than fifteen patients in the hospital," he continued. "All of the patients I take care of have advanced cancers that have failed other treatments. Very often they're young, with young families, children, husbands and wives. The disease represents an incredible tragedy, and every case has its own human drama associated with it. It's a horrible thing to watch. And it's hard not to be affected by the fact that innocent people and their lives are being destroyed by something that's intrinsically not their fault. It's unfair. How could anybody confront the situation and *not* be personally horrified at the toll that this disease takes, and want to do something about it?"

Rosenberg has heard all the criticism, knows all his detractors. When you ask him about his critics, however, he gestures behind him toward a plaque on the wall of his office in Building 10, which frames an aphorism by the Belgian philosopher and writer Maurice Maeterlinck. "At every crossway on the road that leads to the future," it reads, "tradition has placed against each of us ten thousand men to guard the past."

ॐ

As they move briskly on their morning rounds through Building 10, the doctors of the Surgery Branch sweep into and out of rooms wearing white coats with a trademark green patch over the left breast. The patch depicts a muscular Greek god ramming a sword through the heart of a crab, and that emblem at some level reflects the special sense of mission that possessed the young doctors gathered by Rosenberg in Bethesda. Cancer, the crab, was in fact a dragon, and they were dragon slayers; moreover, they were determined to use those most invisible of scepters—cells and molecules—to bring the dragon down.

Even as Norman Wolmark bid adieu and good riddance to his pigs and headed off to Pittsburgh to practice a less adversarial form of surgery, other young researchers were drawn to the Surgery Branch, most trained to wear the two hats of medicine and science: Maury Rosenstein, Elizabeth Grimm, Michael Lotze, James Mulé, Suzanne Topalian, Jim Yang, Yutaka Kawakami, to name but a few of many stellar talents who joined the crusade. Mulé, whose animal experiments laid the groundwork for many clinical approaches, claimed the collection of talent reminded him of the 1927 New York Yankees, arguably the greatest team in baseball history, and fellow Douglas Fraker added, "You couldn't *help* but succeed here."

It *was* a crusade, nourished as much by faith as science in those early days, because when Rosenberg set out in the late 1970s, they were united by the disbelief of others. "We didn't know if there were cancer antigens in the human, and we didn't know whether immunotherapy would ever work!" Rosenberg said. "And so there was an enormous amount of doubt. In fact, tumor immunology was not in the mainstream of immunologic research. It was sort of considered something that one didn't credibly pursue, because after all, it had been talked about for a century and nobody had ever been able to do much." Eventually they would have timing and luck on their side, too, for recombinant DNA was about to change the way all immunologists would go about their business. For Rosenberg and many other researchers, the dawn of this new world began to break with Doris Morgan's accidental discovery of interleukin-2.

IL-2—or "T cell growth factor," as it was still called in 1977—at first caused just a slight rustle of interest among many immunologists. The discovery had emerged, improbably, from a nonimmunological laboratory, and even though Rosenberg had benefited from the scientific equivalent of insider trading when he learned of the new factor from Robert Gallo in 1976, he not only did not run with it at first, but almost refused to accept the handoff. The findings "seemed significant to me," he noted later, "yet I did not pick up on them immediately." What ultimately sparked Rosenberg's interest, what would soon convert his little nook of Building 10 into something resembling an immunological microbrewery, with vats and columns percolating with milligram amounts of this precious growth factor, was the paper by Steven Gillis and Kendall Smith in *Nature* in July 1977.

Smith ran a lab at Dartmouth and had learned about "T cell growth factor" while on a sabbatical at the NIH—so early in its history, in fact, that he possessed a vial of it in his lab refrigerator labeled "Frankie factor," in honor of one of its codiscoverers, Frank Ruscetti. Yet he recognized at once that the factor was of enormous immunological importance. Indeed, a first tentative turn of the key in the lock opened a door onto almost unlimited promise: Gillis and Smith reported that they could take killer T cells—the T cells that kill cancer cells—and grow them with interleukin-2 in lab glassware for up to four months while still maintaining their cancer-killing capability. Rosenberg, having just published his long review article with William Terry on the adoptive transfer of cells, could hardly have missed the implications of the *Nature* article's conclusion. "The ability to propagate large populations of [T cells]," Gillis and Smith wrote, "makes possible the testing of these lymphocytes in adoptive immunotherapy." Those words "registered strongly" on Rosenberg, as well they should have. If Gillis and Smith were right, if physicians could use this factor to grow cancer-specific T cells, the game would suddenly have changed. Smith and Gillis rescued interleukin-2 from potential obscurity and spent the better part of a decade defining its basic biology, but Rosenberg's group put IL-2 on the map as a therapeutic agent—not, however, until after a very circuitous route.

"We were looking for cells that reacted against tumor," Rosenberg recalled, "but we had no way to get them. You couldn't get them from patients because you had no way to study them or grow them. So when T cell growth factor came along for the first time, we could get them from patients, expand them, and study them. So that made a big difference to us. And I remember really becoming entranced with it at this meeting in Oxford."

In September 1977, about a year after Morgan and Ruscetti published their first paper in *Science* and just a few weeks after the Gillis and Smith paper in *Nature*, Rosenberg attended a small meeting in Oxford, England. William Terry of the NIH brought a group of NIH researchers over to meet with a group of British scientists gathered by Avrion Mitchison, a prominent British immunologist, to "rethink" tumor immunology. It came at a time when cell-mediated immunity had matured into a separate, equal arm with antibodies in the immune response. And now the T cell, *the* crucial cell that lay at the epicenter of that tremendous biological power, could be kept alive and studied in the laboratory.

The T cell is a small and rather unremarkable-looking package for so much immunological power. In micrographs, its surface sometimes appears like a macaroon, a sphere almost plush with folds and pleats. Like many of the clinical immunologists studying it, the T cell also wears two hats: one subset acts as a superb impresario among immune cells, while another subset performs the role of a discriminating, cold-blooded assassin. Once these cells pass through the thymus (from whence the "T" of their name), T cells are biologically certified to protect the body from assault, external or subversive. Certain T cells, known as helper T cells, coordinate the response to an incursion, secreting cytokines like interleukin-2 and ordering other cells, including the antibody-spewing B cells, into the fray; killer T cells recognize very specific antigens and eliminate the cells that display them. There are an estimated 1 trillion T cells in the body, and they might be thought of as a single, fluid, almost omniscient organ of self-defense in the sense that they are genetically programmed to "recognize" intruders they have never encountered before.

The discovery of T cell growth factor—and the ability to keep T cells alive and kicking for months at a time—meant to some the possibility of laboratory experiments on the nature of the T cell and its receptor, and to others, like Rosenberg, the more remote but daring possibility of treatment. "It was in the course of those discussions that I became more interested in the possibility that this could have," he explained. "Don't forget—the fact that you could grow this T cell didn't mean the T cell, once it expanded, had any activity of any kind. And that was the first thing we had to demonstrate." As he would so often do, Rosenberg pounced. He decided "to shift some of the things that I was doing to explore whether or not this really had something to offer."

Now, finally, Rosenberg was ready to *run*.

As soon as he returned to Washington, after an eighteen-hour layover in Japan for a meeting, Rosenberg began the series of experiments in test tubes, animals, and humans that would, many years later, lead to a first tentative treatment. "I came back and decided that I myself—not anybody else in my lab, but I myself—would actually see whether there was anything here. And then I hired Paul Spiess [as a technician], and we just did it ourselves. Nobody else did it. And it started to work. And that's why I started putting other people on it." Ultimately, it turned out that they'd ripped a dog-eared page out of the scientific playbook, namely the old misdirection play. Intent on working with T cells, they reversed field several years later and became distracted by another type of cell. Using interleukin-2 merely as an ingredient in a recipe to make cells grow, they later discovered it served quite nicely as a drug in its own right. It took nearly ten years for them to circle back to where they started.

Rosenberg quickly set to work on several questions: Could they grow animal and human T cells in the test tube? Could they demonstrate in mice that these cells did the things that T cells are supposed to do? And could they, finally, show that these cells could attack experimentally induced cancers in mice? Behind these three lurked a fourth question, never enunciated but always hovering in the air: Could they do this before anybody else?

During the course of 1978, the NCI lab repeated and confirmed that T cells could be grown in culture and maintain their killing ability—Paul Spiess and Sue Schwarz showed it in mice, John Strausser with human cells. Next, they had to show that these cells maintained immune function in animals. This took nearly two years of frustrating labor by Maury Rosenstein, who went through hundreds of mice a week trying to show that activated T cells accelerated the rejection of skin grafts in animals, publishing the results in the *Journal of Immunology* in 1981. This allowed the NCI group to move on to a rather more pertinent question. Would T cells do the same thing to tumors?

In 1980, even as Rosenberg's group embarked on this uncertain mission, the noted Australian immunologist Gustav Nossal circulated a paper at the International Congress of Immunology in Paris with the puckish title, "The Case History of Mr. T. I.: Terminal Patient or Still Curable?" The patient in question, Nossal wrote, "is about twenty-five years old, and at present appears to be in a rather serious state." The symptoms included "fatigue and confusion, which have followed a period of unusually intense activity," and also "periods of euphoria alternating with depression, and close observers have noted a certain malaise throughout." Mr. T. I.'s full name, Nossal continued, was Tumor Immunology, and he presented a "diagnostic and prognostic puzzle."

From the time of William Coley and even earlier, there had been hints that at least some tumors were immunogenic, but the field of tumor immunology had become something of a scientific red-light district: a seedy intellectual neighborhood of fantasy and wishful thinking, a landscape littered with the hulks of abandoned hypotheses and charred reputations. Several distinguished researchers had traversed this biological minefield with reputations intact, however, including Ludwik Gross, Edward Foley, the team of Richmond Prehn and Joan Main, Georg Klein, Peter Gorer, and Lloyd Old. Solid, imaginative, even heroic work emerged from their laboratories, but their collective effort also delivered a lesson that scientists do not often advertise: a single new insight can render decades worth of prior research irrelevant.

In the field of tumor immunology, that single devastating fact was summed up most succinctly in 1937 by Peter Gorer, a British scientist interested in the biology of transplantation. In a pioneering set of experiments at the Lister Institute in London, Gorer showed that the immune system is responsible for rejecting foreign, transplanted tissue during experimental transplants, and is able to do so by first recognizing the incompatibility of tissues (or "histocompatibility") from markings on the surface of transplanted cells, and then attacking those incompatible, foreign cells. This one insight doomed years of research purporting to demonstrate how the immune system of experimental animals rejected transplanted tumors; unless the animals were genetically identical, and almost none were, the rejection often had to do with histocompatibility, not cancer. Perhaps just as important, Gorer's work suggested that the body's ability to distinguish Self from Non-Self lay in its ability to read the surface of cells as if the organism's life depended on it, which it did.

In the average adult human body, there are something on the order of 100 trillion (10^{14}) cells. Most are fixed in place; some circulate. Some come in the standard spherical fuzzy golfball shape, an average 20 micrometers across, 10,000 to the head of a pin; others, like the nerve cells called neurons, send out appendages one to two feet in length. Significantly, all cells are chemically social in the sense that their behavior is affected by the touch and taste of nearby cells. All trace their ancestry back to the embryo, and there are several hundred different lineages—the beta cells in the pancreas, for example, which make insulin, and the smooth muscle cells in the heart that keep that most crucial organ pumping. The reason there are so many different kinds of cancers, which take so many different courses, is that each cancer begins as a genetic disruption within a single cell, and each different cell type, along with each possible genetic lesion within that afflicted cell, can evolve differently, making cancer an almost uniquely personal kind of disease.

In the context of immune responses, the most critical feature of the cell is its surface. The surface of a cell is not smooth and flat, like a ball bearing; it is more like a microscopic garden tended in darkness, bathed by warm salty fluid, a rounded and shaggy convex landscape with cellular vegetation waving like seaweed above the cell membrane. This vegetation represents the aboveground tendrils of proteins and other molecules, which are anchored in the membrane. These are the eyes, the ears, the taste buds, the nerve endings of the cell. At any given time, the surface of a typical cell may be studded with thousands of different tendrils. But perhaps the hardest concept to bear in mind, sometimes even for scientists, is that this moist, shaggy landscape is dynamic. It is constantly changing. New vegetation shoots up and old vegetation collapses as in a time-lapse film. In real time, these proteins bloom and shrivel, unfurl and fold up, in the course of hours, sometimes minutes, depending on what's going on in the immediate neighborhood, because cells are exquisitely sensitive, and constantly reacting, to their local ecology. Certain receptors, like perennial flowers, grow on the surface, so permanent and unchanging a fixture of the cellular landscape that they can in essence serve as a reliable molecular landmark, or fingerprint, that reveals the identity of the cell itself. Indeed, they are known as clusters of differentiation (CD), or "markers," and were first reported in mice, by Lloyd Old in the early 1960s.

Different cells in the body, like different soils, support different kinds of vegetation. Lymphocytes as a group all bear the CD3 marker, but the classification gets more specialized than that. The lymphocytes that decline among AIDS patients are called T helper cells (or CD4+) because these cells not only flash the CD3 marker but, alone among the family of lymphocytes, also have a permanent bit of vegetation on the cell surface known as the CD4 marker. Cytotoxic, or killer, T cells carry the CD8 marker. Natural killer cells carry the CD56 marker. Markers can even reveal what a cell is doing—if a killer T cell possesses the CD44 marker, it means the cell has been "activated," or bumped into some threat or another. Put more colloquially, immunologists began in the late 1970s and early 1980s to judge a cell by its cover—that is, by the kind of vegetation that grows on the surface—and indeed, another immunologically based technology, the creation of monoclonal antibodies (antibodies specifically engineered to recognize these markers, among other things), has allowed unprecedented precision in the identification of functionally distinct types of lymphocytes.

Those same cells, however, can "grow" new, complementary molecules when genes inside the cell turn on, an event that occurs as a result of the constant chemical conversation and communication that goes on between a cell, its neighbors, and the surrounding liquid environment, a watery Rialto full of chemical whispers and tidal rumors, of proteins lapping up against the vegetation with news. Such a gene will become activated when a molecule on the cell's surface, the receptor, receives a signal; receptors are a special class of vegetation that sits on the surface of the cell and interacts with specific signaling molecules out and about in the local environment. If a virus is in the neighborhood, for example, the news will lap up in the form of interferon, and once this signal molecule docks with the interferon receptor, messages travel to the cell nucleus telling it to turn on several genes that will abort viral replication.

Like all proteins in the cell, a receptor molecule is created only when the gene encoding its amino acid sequence is turned on in the nucleus and expressed; when the gene for the receptor for interleukin-2, for example, is turned on, the cell makes the molecule, ships it to the surface for planting, and then can "receive" signals from and respond to this cytokine and grow. So the cell surface is in part a garden of softly undulating molecular antennae, constantly being erected, dismantled, or left permanently in place depending on the instructions of genes inside the cell; those genes, in turn, can be turned on and off depending on signals picked up by the receptors and conveyed to the nucleus of the cell. Biologists speak of receptors as being "up-regulated" or "down-regulated" in response to signals. One might even think of them as cues like *andante* or *lento* in a musical score. It is a daily exercise in molecular theme and variation, chemical point and counterpoint, a garden whose growth is as syncopated, as fluid and cyclical and logical, as a Bach cantata.

Buried in that thick underbrush of gently undulating molecules is the central question of tumor immunology, the crucial premise upon which all of immunotherapy finds either rationale or ruination: What would make a cancer cell appear outwardly different enough from a normal cell to attract the notice of the immune system? Do tumor cells betray any signs—any surface vegetation, in this extended metaphor; a surface marker, or "antigen," in biological argot—of being different from normal cells? Why should a cancer cell, which after all is a cell with the same DNA, the same genes, the same lineage and architecture as all the other cells in the body, betray this common ancestry and suddenly appear foreign? Why, in short, should the Self suddenly appear Non-Self?

In the 1970s, when Steve Rosenberg began treating patients at the National Cancer Institute with pig cells, immunologists presumed that tumor cells *were* different, although molecular evidence of those differences was just beginning to be uncovered in other laboratories. Indeed, the fact that answers were wanting was critically clear as early as 1931, when one researcher lamented that published results in the field of tumor immunology were "conflicting to a surprising and bewildering degree." A giant step toward clarity occurred in the 1940s, when a more rigorous approach to tumor experimentation in animals allowed researchers to perform meaningful experiments in this extraordinarily tricky region of immunology.

It began with a landmark paper in 1943 by Ludwik Gross, a Jewish émigré who had fled his native Poland in advance of the Nazis and ended up in the laboratory of Christ Hospital in Cincinnati. There Gross demonstrated the crucial importance of using inbred mice in tumor experiments. He chemically induced a sarcoma in a strain of mice known as C3H, allowed the cancer to develop for a while, and then transplanted a standard dose of cancer cells from this tumor into other C3H mice, which rejected the tumors. The significance of using C3H mice is that they had been inbred for twenty years, until every animal was genetically identical; this neatly skirted the histocompatibility problem identified by Peter Gorer and allowed Gross to conclude, with far greater authority than researchers before him, that rejection of transplanted tumor cells in his experiments occurred because of something specific and intrinsic to the tumor, not simply because of genetic differences in foreign tissue of the sort that lead to rejection in normal transplantation procedures. Furthermore, just before the war, Gross had conducted a series of transplantation experiments in 1938 at the Pasteur Institute in Paris showing that "humoral antibodies are not necessarily responsible for the presence of such immunity." Here was one of the earliest hints that some immune agent other than antibodies accounted for tumor rejection.

Gross's work didn't exactly dispel the gloom. As Richmond Prehn and Joan Main at the National Cancer Institute admitted a decade later, "The history of attempts to immunize against cancer is one of long frustration. As a result of apparent failure during the past half century, it is the current consensus that immune mechanisms probably will be of little use in the control of this disease." However, Prehn and Main had been intrigued by experiments reported several years earlier by Edward J. Foley of the Schering Corp. in New Jersey. Foley injected methylcholanthrene (coal tar) under the skin of mice to cause the formation of a tumor, which could then be passed on serially to identical inbred mice. What caught Foley's attention was that if one of the transplanted tumors was surgically removed from a mouse, the animal managed to mount dramatic resistance to a second tumor transplant, almost as if it had been vaccinated. It seemed to have acquired a certain immunity to the tumor. Those experiments, according to Herbert Oettgen of Memorial Sloan-Kettering, "put tumor immunology on the map."

Prehn and Main repeated and extended Foley's work. They used coal tar to induce sarcomas in inbred mice, transplanted these tumors into naive animals of the same strain, and surgically removed them after a period of time; then they rechallenged these animals with tumor again, and found that most had become immunized against developing the cancers. Significantly, as a control, Prehn and Main also challenged these same animals with a different, spontaneous tumor, which was not rejected, suggesting that the mice recognized antigens specific only to the tumor they had already seen. With hindsight, we can see both the blind spots and the flashes of brilliance in these early experiments. Prehn and Main still assumed—incorrectly—that antibodies lay at the root of this protection. On the other hand, they shrewdly noted that one tumor among the six they studied seemed to lose its antigenicity in the course of the experiments—a hint of a biological phenomenon that would later merit distinction as "tumor escape mechanisms."

Prehn and Main restored credibility to the notion that cancers might possess tumor-specific antigens. Georg Klein, a gifted essayist as well as immunologist, and his coworkers at the Karolinska Institute pushed the field along even further by establishing that such antitumor immunity could be traced to a commotion in a tiny fraction of the blood. In a complicated but artful series of experiments, they too used chemicals to induce experimental sarcomas in mice. In conditions even more rigorous than those imposed by Prehn and Main, they removed the tumors surgically, irradiated the tumor cells (thus killing them), and used these cells as a kind of vaccine to immunize each of the mice to their own particular tumor. Upon subsequent challenge with live tumor cells, vaccinated animals showed increased resistance to the cancers, and often remained disease free until the dosage of tumor cells rose beyond a certain threshold.

Then Klein and his colleagues did a clever thing—they collected blood and lymph nodes from the mice that had mounted a successful immune response and divided the blood into two parts, the cells and the liquid, or serum, portion; they then showed that immunological protection mounted by these mice against cancers resided in the lymphocytes, not in the serum. Like others before them, only reluctantly did the Swedish group relax their grip on antibodies as the likely explanation for this resistance: in a bemused tone, they conceded that immunological protection appeared to be "mediated through cells rather than humoral antibodies." "These findings," Lloyd Old has remarked, "placed the field of tumor immunology on a firm experimental basis for the first time and lifted the sense of unrelieved gloom that had come to be associated with immunological approaches to cancer over the preceding decades." Gloom, alas, would soon settle again on the field.

Klein and his colleagues, including his wife, Eva, and Karl-Erik Hellström, had correctly noted that their experiments raised "several fascinating questions," preeminent among them whether chemically induced tumors possessed the same telltale markers, or antigens, on their surface as spontaneous tumors. The answer seemed to come fifteen years later, and it loomed before Rosenberg, and everyone else in the field, like a washed-out bridge. "There had been this article by this fellow Hewitt," as Rosenberg recalled, "who had claimed there was no such thing as tumor antigens, and there was an enormous pessimism about whether or not such a thing really existed."

Working at Mount Vernon Hospital outside London, Harold B. Hewitt summarized an extraordinarily exhaustive series of experiments undertaken over a period of twenty years that appeared to undermine the very idea that tumors could be immunogenic. Most of the animal tumors that researchers studied were caused either by viruses or by high doses of chemical carcinogens (such as coal tar); these experimental cancers were then "seeded" by direct injection of cancer cells into related inbred animals. In many cases, researchers could show that these "man-made" tumors were visible to the immune system, and their experiments often resulted in impressive immunological recognition of cancer. Hewitt argued that such tumors were merely "seductive artefacts" that had no relevance to real life.

The cancers that develop in human beings are of course never transplanted and only rarely result from viruses or massive chemical insults of the sort laboratory mice routinely suffer. Much more commonly, they arise spontaneously, due—it is increasingly clear now—to a succession of genetic disruptions. Hewitt and his colleagues patiently identified and studied twenty-seven different types of spontaneously occurring tumors through some 20,000 transplants, all in genetically identical mice, and claimed that not a single one of these more "natural" tumors was blunted by immune attack. According to Hewitt, spontaneously arising tumors, the kind that most resemble the cancers that afflict human patients, were *not* immunogenic.

Tumor immunology had always been a field built of shaky, makeshift materials; Hewitt's tendentiously authoritative 1976 paper describing these results blew through the field as destructively as a tornado cutting through a tent city. But it was also something of a fresh wind: falling just short of accusing the field of intellectual dishonesty, Hewitt invited greater rigor in a discipline that may have set its scientific standards too low.

Tumor immunology was ultimately redeemed by basic research in what has rapidly become one of the cornerstones of modern immunology: the story of antigen presentation. These complex transactions between the immune system and its constituents, namely the body's cells, are among the most sophisticated and beautiful in nature. They have been millions of years in the making from the evolutionary point of view, but attracted the notice of immunologists only in the last half-century. Indeed, the 1996 Nobel Prize in Physiology or Medicine recognized the scientists who first explained in the 1970s how T cells receive news that something foreign is on the loose.

The immune system has evolved a variety of ingenious surveillance mechanisms to distinguish Self from Non-Self—and perhaps, according to a new theory, the Endangered Self from the Normal Self. The easiest way is to spot something foreign on the surface of the uninvited guest; bacteria have antigens on their surface that make them giveaway targets for immune elimination, and antibodies are particularly effective at picking off bacteria and their toxins floating through the bloodstream. But viruses do their dirty work behind closed doors, as it were, inside the cell and therefore beyond the usual scrutiny of antibodies. And so the immune system has devised a remarkable way to see through the cell wall and spot the intruder within. This ability is connoted by three initials—MHC.

MHC stands for major histocompatibility complex, and it is a form of self-testing, self-tasting machinery. The pioneering insight again belongs to Peter Gorer, who had studied the phenomenon of tissue rejection in transplants for decades and discovered in the 1950s that the "foreignness" of any tissue was given away by molecules on the surface of cells. George Snell, who studied the same phenomenon in human biology at the Jackson laboratory, called them "histocompatibility" molecules and showed that all the cells in an individual share the same set of MHC markers; when transplant surgeons speak of a "match," they are in fact talking about MHC compatibility.

The first hint of an explanation came in 1974, when Rolf Zinkernagel and Peter Doherty, both at the John Curtin School of Medical Research in Canberra, Australia, showed that these MHC molecules are essential to healthy immunological function. In order for killer T cells to spot a cell infected by a virus, they need to see a viral antigen, of course, but they also need to see a family of MHC molecules (called Class I) on the surface of the cell. Why? Part of the answer came about a decade later, when Alain Townsend and colleagues at John Radcliffe Hospital in Oxford, England, demonstrated that these T cells in fact recognized very small bits of viral protein known as peptides as well as the MHC molecules. It turns out that on a constant basis, easily millions of times as you read this sentence, your immune system is scanning the surface of normal cells, looking for trouble. Inside each and every cell, there exists a separate and continuous cycle of activity whose purpose is to chop up every protein inside the cell into bite-sized chunks of amino acids known as peptides and load them up on one of the MHC molecules, which snakes through special corridors of the cell known as the endoplasmic reticulum and Golgi stack and then, as if emerging from a manhole onto the street, delivers this short protein fragment to the surface of the cell.

Packaged just so, cradled within the MHC molecule as if posed by a photographer in such a way as to catch the light most revealingly, the peptide is held up in such a way to become visible to the immune system. Without the MHC molecule, T cells are blind to antigen, kittens whose eyes have yet to open; cued by MHC presentation, they become tigers with a predator's hunger only for a specific target—cytomegalovirus in the case of one subset of T cells, perhaps melanoma cells in the case of another. Thus even in our normal cells, the innards are routinely cannibalized, the internal proteins sampled and inspected to make sure everything belongs. Through a crash course in highly personalized immunological education at birth, each of our bodies permanently retires T cells that recognize—and therefore might attack—peptides derived from normal Self proteins, in a process known as immunological tolerance; we literally "tolerate" every peptide snippet typical of our Self. Intolerance, on the other hand, is reserved for cells that contain viruses, bacteria, other parasites, and perhaps cancers, as well as cells from another organism (such as a transplanted organ or transfused blood).

The beauty of evolution is, as always, in the details. What makes T cells so special? The way they handle this inspection, this system of self-monitoring, is known as MHC restriction. It takes as few as twenty copies of a given peptide on the cell surface in the context of these MHC molecules to trigger a T cell response, but it is even more precise than that. The peptides delivered to the cell surface are short segments of amino acids, not quite long enough to fold into the three-dimensional pretzels of a full-fledged protein. The business end of the T cell, its receptor, can instantly recognize this snub-nosed, nine-peptide shape as homegrown or foreign, friend or foe. The precision of the recognition mechanism is breathtaking. There are 20 possible amino acids that can be used to build each bead in a chain of protein, so for each 9-amino-acid chain, the number of possible combinations is staggeringly high: $20 \times 20 \times 20$ and so on, to 20^9, or more than 500 billion possible combinations. Despite this astronomically rich possibility for variation, at least one subtype of T cell will normally detect a single amino acid difference, a one-in-500-billion aberration that changes a normal Self peptide into a novel foreign peptide. If it is foreign, if it doesn't taste right, doesn't look like Self, the entire cell that it now marks, like an enemy flag inadvertently waving from the parapets of the membrane, is destroyed. If a virus lurks inside a cell, laboring away at replication, the MHC machinery will routinely find it, chop it up, present bits and pieces of viral protein to the immune system. When all is working correctly, the cell will be marked with this scarlet A (the "A," in this case, standing for antigen) and terminated with extreme prejudice by killer T cells, which sidle up to a target cell and within five minutes or so blow fatal holes in it. Pamela Bjorkman, now at Caltech, working with Don Wiley and Jack Strominger of Harvard and their colleagues, by virtue of x-ray crystallography, were the first to provide spectacularly intimate views of a peptide nestling in the groove of a MHC molecule—"like a hot dog in a bun," according to one immunologist.

The MHC system represents the body's version of quality control at the cellular level, a biological way of seeing through cell walls and inspecting the merchandise. In human physiology, the work is divided between professionals and amateurs. The Class I MHC molecules operate in the amateur, lumpen, everyday cells that undergo routine self-examination. But a certain class of cells of the immune system are professionals in this line of work—they are known as "antigen presenting cells," and they do the same thing, only they take their catch straight to headquarters, as it were, in the immune system. One such professional cell is the macrophage. When he first identified these "wandering cells" a century ago, Elie Metchnikoff called them phagocytes because they are able to encircle and swallow bacteria and viruses in one gulp; what Metchnikoff did not appreciate was how important digestion was to immunity. Macrophages digest foreign material by chopping up proteins into peptides, packaging them in the second but related Class II family of MHC molecules, and "professionally" presenting the antigens of viruses and bacteria to a different set of inspectors: the helper T cells. "Present" is a verb pregnant with immunological import. Unlike killer T cells, helper T cells are not assassins; what they do, extremely well, is initiate, coordinate, and oversee an overall immune response by the judicious secretion of cytokines such as interleukin-2, which in turn activate and summon in specific order a host of other immune cells and molecules. Squirting cytokines and arousing immune cells, the helper T cell raises a true molecular commotion in the blood when the organism is at risk.

There are two main classes of these professional presenting cells in the body: some, like macrophages, are mobile, swimming through the blood and bodily fluids, following chemical signals, roaming the vasculature, while others remain more or less stationary and embedded throughout tissues, sending out slender tendrils like little snares laying in ambush. These latter are known as dendritic cells, because their branching architecture resembles the sprawling nerve cells known as dendrites. James Allison of the University of California at Berkeley recently provided a more complicated view of these and other cells at work by showing that certain antigen presentations require a secondary signal for stimulation. Like the grounding prong of an electrical plug, these so-called costimulatory molecules make contact with T cells and turn them on more powerfully, thus eliciting a more enhanced immune response.

In 1980, as all these pieces began to fall into place, the specificity of the immune response started to become clear in molecular terms, and with that clarity, the prospect of more precise, molecular manipulation of the immune system became, if not a reality, at least plausible and pursuable. It would take a decade for the dust to settle, but the future was clear enough by 1980 that Georg Klein felt obliged to revise Gustav Nossal's grim prognosis for Mr. Tumor Immunology. "TI is not a patient at all," Klein told colleagues, with a characteristic twinkle in his eyes. "He's still a youngster who had a very complicated childhood."

"A lot of people have said: Why would anybody think that immunology would have anything to do with cancer?" asks Michael Lotze. "Why is that? Since we know that the immune system evolved primarily to protect people of reproductive age from viral illnesses, it was not evolved to try to protect people who developed cancers, which often occur late in life. But in actual fact, maybe those two things *are* true. Maybe the reason why you usually don't get tumors until late in life is because you've had some dissolution or decrease in the ability of your immune system to respond to the internal environment."

The $64 million question for cancer biologists, therefore, has always been, Do cancer cells appear to the immune system as Self or Non-Self?

Tumor immunologists began to see a very fuzzy answer in the early 1980s, and interleukin-2 helped them see it—the molecule caused proliferation in only the T cells that had been activated, or seen antigen. That was why Steve Rosenberg wanted to know if IL-2-activated T cells would perform their immunological duties against established cancers in mice. Finally, they got their answer. Unfortunately for Rosenberg's group, the answer came from Seattle.

Alexander Fefer's group had worked with lymphocytes and mouse cancers for years, especially a mouse tumor known as FBL-3. Using this model, Fefer had demonstrated that one could harvest T cells from one tumor-immunized mouse, transfuse them into a second inbred (and thus identical) mouse with cancer, and cure it of the malignancy. The next step, made imperative by the discovery of interleukin-2 and obvious to Fefer, Rosenberg, and everyone else in the field, was to see if one could remove T cells from an animal, expand their number to tens of millions with the growth factor, and reinject them to cure cancers that were already established and spreading. This was in effect the litmus test for the idea of adoptive transfer of lymphocytes, because it was analogous to the kind of treatment doctors might ultimately attempt in cancer patients: remove their cells, find the T cells that reacted specifically with the tumor, expand them in the lab with IL-2, and give them back as medicine.

Rosenberg's group repeated Fefer's work with a few variations. Fefer had injected both cancer cells and T cells "intraperitoneally," or directly into the abdomen. In his experiments, Rosenberg decided to inject the cancer cells into the footpads of the mice, then injected the T cells intravenously (that is, directly into the bloodstream). Over such miniscule variations are claims of priority made, and such was the case with the Seattle and Bethesda groups. Rosenberg considered this to be the first convincing demonstration of the adoptive transfer of T cells; the Seattle group saw it as confirmation of *their* work.

"He goes ahead and makes the statement that he was the first one to cure a mouse with *disseminated* disease," Fefer said. "Well, the problem with that is, number one, we also gave cells intravenously, and could cure the mouse. And this is known and we did it, and we didn't make a big deal out of it. It wasn't surprising. Number two, much more importantly, is that there was an article in 1980, two years before this one, by Phil Greenberg of our group showing clearly that if you give our mice the tumor intraperitoneally and then you do a bioassay—you take the spleen cells from this mouse and lymph nodes from this mouse and blood from this mouse—by day five, which is when we usually started our treatment, the leukemia has spread and is everywhere. And Steve *knew* that. This is published work, okay?"

Claims of priority notwithstanding, the ground had been cleared from the immunological point of view for the idea of treating humans with T cells, and Rosenberg was already looking ahead at human applications. Indeed, he performed what he called the "Cinderella experiment" around this time. It was not a real experiment. Half thought-experiment, half-fantasy, it was based on what they had learned so far and where, ideally, they thought it would lead.

∽

In Rosenberg's imagination, the Cinderella experiment worked like this: researchers would give huge numbers of lymphocytes to animals with cancers, and then large amounts of interleukin-2, which would arouse the lymphocytes until their cancers were cured. What made it a fantasy was that they couldn't get their hands on much IL-2, and they hadn't yet identified the best lymphocyte to give. If anything, the slipper, so to speak, seemed to fit the wrong cell. Ilana Yron, a visiting Israeli biologist, had fished an interesting cell out of a pool of mouse lymphocytes with the use of IL-2, and Michael Lotze identified its counterpart in human blood. This cell killed cancer cells, but it wasn't a T cell and it wasn't antigen-specific, which meant it worked outside the MHC system and really wasn't what Rosenberg had in mind. That did not stop him from trying it on three patients, none of whom benefited.

It was another clinical associate named Elizabeth Grimm who ultimately characterized this curious cell. Grimm, a staff scientist in the laboratory, had been trying to reproduce work published by researchers in Europe that suggested that when you mixed T cells and cancer cells in a test tube, a subset of those T cells—a small fraction of the total population, known as T cell subtypes or clones—specifically killed the cancer cells. *This* was the cell Rosenberg had been looking for. Grimm couldn't find any T cells, but she did not come up empty-handed. She kept finding a different species of lymphocyte, a nonspecific killer cell that was not as powerful a killer as the T cell but still killed cancer cells—weakly, but nonetheless consistently. It turned out, upon further study, to be essentially the same cell that Yron had found. Known informally at first as "Grimm's reapers," these lymphocytes seemed to grow in the test tube when she added interleukin-2 (a lymphokine), so Rosenberg ultimately decided to call them "lymphokine-activated killer cells"—LAK cells, for short.

"We tried to throw out LAK cells for years and years and years," Michael Lotze remembered of those early experiments. "And it was only out of a real sense of frustration in terms of our inability to get the T cells that we wanted that we felt we had to study LAK cells and their ability to mediate antitumor effects." What exactly was a LAK cell? That was an issue of vivid immunological debate at the time, and remains so today. In the early 1980s, the biological tests one could run to determine the traits of a cell were so crude that not a lot could be said about the LAK cell, except that it was a lymphocyte and killed "nonspecifically." The Rosenberg lab upped the ante, however, by suggesting they had discovered an entirely new class of immune cell, virtually a third arm of the immune system. It was not a T lymphocyte and, they claimed, not a natural killer, or NK, cell.

One respected immunologist, Jerome Ritz of the Dana-Farber Cancer Institute in Boston, has suggested that the LAK cell was misidentified all along, that it was nothing other than an ordinary natural killer (NK) cell enjoying a new and sudden and short-lived celebrity under an alias, although anyone with the temerity to point that out within earshot of the Surgery Branch got a chilly reception. "The original papers from Steve Rosenberg and those people maintained very strongly that these were a different set of cells, that this was a new cell, and it wasn't clear how it was related to NK cells or T cells," said Ritz. "I maintained, and got sort of roundly booed a few times [for doing so], that these were really just NK cells. Subsequently, we, in a number of papers, now have clearly shown that it's the NK cell that is the activated effector cell, so a LAK cell is really an NK cell." ("Lymphokine activated killer cells kill fresh cancer cells," Rosenberg said in response to some of this criticism. "It's a totally unique observation. Whether the LAK cells are related to NK cells as precursors is another issue.")

No one disagreed, however, that when fresh cancer cells were added to a culture of these lymphocytes along with high doses of IL-2, certain immune cells became activated and killed the cancer cells. Regardless of whether they were related to NK cells, they were, unlike T cells, nonspecific cells, and therefore the LAK cells behaviorally belonged to other nonspecific treatments like BCG and other bacterial vaccines. To use a military analogy, NK cells were to T cells as a mortar is to a radar-guided missile—less accurate, less discriminating, and therefore less effective.

From Rosenberg's point of view, any ordnance was better than nothing, and here in particular is where the hare in Bethesda left the tortoises in Seattle and elsewhere behind. While Greenberg in particular stuck with T cells, in a methodical march of research that only bore fruit in the 1990s, Rosenberg decided by May 1981 to try Grimm's LAK cells in humans. As he later admitted, there was barely any animal data to support this leap, and good reason to consider it risky. The experience of Norman Wolmark's pig cell infusions back in 1977 provided at least some comfort here; those experiments showed it was feasible and theoretically safe to infuse billions upon billions of nonhuman immune cells into humans without gross evidence of harm. After weighing this bold request for two months, the National Cancer Institute gave Rosenberg the green light to treat cancer patients with LAK cells. In no time at all, Rosenberg covered those few footsteps between the lab and the ward. Between May 1981 and November 1984, Rosenberg tried every variation on the Cinderella theme—LAK cells alone, IL-2 alone, at different doses, before hitting the right combination.

Rosenberg's first patient was a young woman who had developed a sarcoma in her thigh that had spread to her lungs. On May 13, 1981, the Surgery Branch team removed some of her blood and culled out her white blood cells. Since lab technicians couldn't make enough human IL-2 in their impromptu distilleries, they settled for an admittedly inferior stimulant: they turned on these lymphocytes with another, much weaker class of activating agent known as a lectin. Some 10 billion of these lectin-activated cells silently dripped back into the woman during the first infusion, and she registered receipt of them with fever, chills, rigors, and a headache. They waited a few days, and repeated the process. Over a period of three weeks, this patient received approximately 72 billion activated lymphocytes, but upon follow-up examination it became clear that the tumors in her lung had continued to spread.

Rosenberg treated 10 patients, all with advanced cancers, with these so-called LAK cells; none benefited, and all 10 died. The next 11 patients received a combination of LAK cells and cyclophosphamide, a well-known chemotherapy drug that was thought to dull the arm of the immune system that suppresses activity; all 11 died without showing evidence of positive benefit. Twenty-three patients, 23 failures. Including the 6 patients treated with pig lymphocytes, the tally for the adoptive transfer of cells was 0 for 29.

Rosenberg had a hard time dealing with the uninterrupted string of deaths. "Failures in lab experiments," he wrote, "were discouraging but I could always make a change and try again. Failures in these clinical trials yielded no second chances for the patients. This thought, although rarely expressed, was constantly in my mind as I spoke with the patients and their families. Time was running out for them. Patients place extraordinary trust in NIH doctors; they think that, enveloped as we are with science, we can cure anyone. I felt I could cure no one."

Back to the lab. With the apparent failure of LAK cells alone, Rosenberg began to think more and more about interleukin-2 as an anticancer agent in and of itself. The molecule posed mammoth problems in terms of production (starting out with two gallons of cultured cells, they could manage to make only 100 milliliters—20 teaspoons—of human IL-2). Nonetheless, they had published a dozen IL-2 papers; they were "running" with it now.

Rosenberg's rising interest in IL-2 coincided with the return on July 1, 1982, of Michael Lotze as a clinical associate. Born in Pasadena in 1952, Lotze had trained as a surgeon and shared with Rosenberg a keen interest in exploring immunologic alternatives to traditional cancer treatment; indeed, when he heard about Rosenberg's work, while serving a fellowship at M. D. Anderson Hospital in Houston, he immediately applied for a position and spent two years, between 1978 and 1980, as a visiting fellow. Of all the fellows who passed through the Surgery Branch, Lotze enjoyed a particularly close and complicated relationship with Rosenberg; he was at once star pupil, budding rival, clinical confidant. Like Rosenberg, Lotze liked to push open doors and didn't mind stepping on toes; unlike Rosenberg, he seemed a little more self-aware as he was doing it. Lotze "had the ability to do large things," Rosenberg wrote of his return in 1982. "When I first went back to the Surgery Branch in 1982," Lotze said, "what I decided was really important was to get sufficient IL-2 for clinical trials."

Lotze was likably ambitious—he had voracious intellectual curiosities, but sometimes his eyes were bigger than his stomach. He quickly gained a reputation among colleagues for spreading himself too thin. "Michael would always set up a thousand experiments," Maury Rosenstein recalled. "He always set up more than he could possibly do. And then he'd have to decide at the end which experiments he was going to complete and which experiments he *couldn't* complete, and that's just the way he was, and is. He always set out to do more than was humanly possible." Lotze, to his credit, knew this about himself. "In very Mike Lotze fashion," he said of his return to the Surgery Branch, "I started about two or three projects all at the same time."

He initiated what would become an extraordinary ad hoc effort to produce enough human interleukin-2 to use in patients. In those days, prior to the cloning of the gene, you did this by stimulating cells in laboratory flasks, collecting the biochemicals they sweated into the medium, and trying to purify out the desired molecule—essentially the same approach used by Kari Cantell to produce interferon nearly two decades earlier. Key to this effort were a few aberrant human cells that made the rounds of the world's laboratories, called cell lines. Each usually derived from some forlorn human case history, and each was distinguished by some quirk of nature that enabled it to make prodigious amounts of one molecule or another. Lotze auditioned several cell lines for the job: one called MLA-144, another from monkeys, yet another originally from Germany, from a patient called Jurkat. The Jurkat line produced unusually large amounts of IL-2, and it seemed that no flask, no test tube, no thimble was too small to contribute to its cultivation and collection. "I had huge vats all over the lab making IL-2," Lotze recalled. "And then motors with stirrers in them, trying to make enough of this stuff. I'd filled up all the incubators, I'd filled up the cold room full of these huge vats of supernatants—and it became clear to me: this is crazy! Especially with what was a burgeoning biotech capability of making these things, it became clear to me that we really could not do it, that we could not be running an industry inside a laboratory."

When Rosenberg and Lotze learned in the fall of 1982 that researchers at Du Pont were using the same Jurkat cell line to produce industrial-sized amounts of interleukin-2, they persuaded Du Pont to supply human IL-2 for clinical trials. Even before Du Pont agreed to provide clinical amounts of the molecule, Rosenberg had drawn up an experimental protocol for IL-2 use in humans, and the NCI approved its use. Du Pont at first blanched at Rosenberg's request—100 milligrams for clinical trials would take a task force to purify—but after some difficult negotiations, Du Pont finally agreed to do it for free.

Du Pont could read the handwriting on the wall even before they began, and they could read it in the journals within a few months. In February 1983 the Japanese cloner extraordinaire Tadatsugu Taniguchi announced at a meeting in Philadelphia that he had cloned the gene for interleukin-2; not only did this promise unlimited amounts of the molecule, but it meant that rights to the drug would be controlled by a small biotech company in Seattle called Immunex, a not inconsequential fact because one of the company founders, Steve Gillis, had recently left the University of Washington and maintained close ties to Alex Fefer's group. In the chess match to get rare reagents, Rosenberg began in the spring of 1983 to court a rival biotech company in California known as Cetus, whose scientists were still trying to clone the interleukin-2 gene. In June 1983 Du Pont delivered their first 7.5 milligrams of clinical grade interleukin-2, ready for use in patients; that same month, Rosenberg received his first vial of recombinant IL-2 produced by Cetus's genetic engineers for animal testing. And then, just as he planned to rush through the double doors with these precious vials of IL-2, he was scooped again.

This time his thunder had been stolen by a group in Australia. The very first humans to receive interleukin-2 therapy were not in Bethesda but in Sydney. Using an approach similar to Lotze's, but distilling their IL-2 from the white blood cells of volunteers, a team headed by a cancer researcher named Peter Hersey treated two desperately ill melanoma patients with tiny amounts of partially purified drug in 1982. Hersey's rationale, however, was slightly different: the Australian group interpreted a burst of experiments around 1980 as suggesting that physicians didn't need to remove blood from a patient and culture lymphocytes outside the body, as Rosenberg was doing. Rather, they believed they could simply inject IL-2 to trigger the growth of cancer-specific T cells already present in the bloodstream. The low doses of impure material not only failed to help the two patients but lulled researchers into a false sense of security. "The side effects noted were relatively mild," the Australians wrote. No one would be saying that after Rosenberg and Lotze took their turn with the drug.

Back to the clinic. Rosenberg's first IL-2 patient, a sixty-nine-year-old woman he refers to as "Sheila Hopeland," suffered from metastatic melanoma. She checked into the clinical center at the NIH on July 20, 1983, to receive the Du Pont version of interleukin-2. As a precaution, she was placed in the intensive care unit, and both Lotze and Rosenberg waited expectantly as she received the first dose of 14 micrograms—14 millionths of a gram, less even than the weight-adjusted dosages that mice were receiving on the other side of the double doors. They had so little firm information to go on that they did not know if it was best to drip the drug slowly into a blood vein as a continuous infusion or give one single "bolus" injection. They opted to try bolus injections, once a week for four weeks. Rosenberg's summation is succinct: "There were no effects of any kind."

Over the next several months, 10 patients received interleukin-2 for their cancers; 5 of them, suffering from AIDS, had the cancer known as Kaposi's sarcoma. As before, none of them responded. The Surgery Branch was running through the precious IL-2 as if it were bottled water, and for that matter, its therapeutic efficacy looked to be about the same as seltzer. The tally for immunotherapy in the Surgery Branch became 0 for 39.

Back to the lab. Rosenberg, as usual, pursued an almost dizzying array of experimental leads all at once. So even as they were trying IL-2 alone, even as supplies of the drug dwindled in September 1983, Rosenberg and his longtime technical associate Sue Schwarz attempted one more set of experiments with interleukin-2 and LAK cells. Mice were deliberately injected with tumor cells that caused metastases in the lung. Rosenberg and Schwarz let the tumors develop for a few days, then treated the mice with either interleukin-2 alone, LAK cells alone, or the combination of IL-2 and LAK cells. Within two weeks, they had what at the time seemed like an unambiguous answer: animals treated with the combination had half as many metastatic lung tumors as any other group. "Thunderbolt moments do not happen often in research—sometimes they come years apart and many scientists never experience even one—but when they occur they make any amount of previous failure seem trivial," Rosenberg wrote later. "And one knows such a moment instantly." The combination of LAK cells and IL-2, he had no doubt, was a "thunderbolt moment."

Nonetheless, it also looked very different from what Rosenberg had set out to find. Ever since the 1977 meeting in Oxford, he'd been convinced that T cells held the key to tumor regressions; seven years later, he'd ended up with a less lethal white blood cell, the nonspecific LAK cell. Committed to a cellular therapy, he now entertained the idea of a combination therapy with the addition of a cytokine, IL-2, to the treatment. Somewhat tattered and barely recognizable, this was the Cinderella experiment in the flesh. More important, it seemed to work—in mice.

Back to the clinic. Sixteen more terminal patients received courses of IL-2, with no responses noted and deaths in every case. The genetically engineered version of the drug made by Cetus looked purer and better, and it also worked in combination with LAK cells in mice. But the tally stood at 0 for 55. By October 4, 1984, when the FDA approved the experimental use of LAK cells with recombinant IL-2, Rosenberg had reached one of the lowest moments in his career. He counted 75 straight failures in all, stretching back to Norman Wolmark and his pig cells. Even the promising new combination therapy disappointed. They treated 4 patients with the combination of LAK cells and IL-2, and got 4 more failures.

Rosenberg was running out of options, and the final crisis of confidence came in the spring of 1984. "Perhaps for the first time," he wrote later in his autobiography, "at least a part of me began to doubt the path I had chosen to follow." Of that time, Rosenberg said, "I had put almost ten years of my life directly into the effort of developing immunotherapy, and indirectly more than double that many years. In all that time, not a single person outside my own laboratory had ever done more than express interest in what we were doing. Critics had never ceased arguing that it was quite possible that immunotherapy could not work in man." This may overstate circumstances; others were working the same side of the street, and many critics—Fefer's group, for one—did not question the general principle so much as its specific execution in Rosenberg's hands. Yet Rosenberg was exposed as no one else, and few onlookers would dispute his own assessment that the Surgery Branch's efforts had been "flailing and futile."

Certain scientists listen intently to the message of their experiments, even when the message is negative; others refuse to take no for an answer from nature. His ears ringing with negative replies—"No! in thunder," as the critic Leslie Fiedler once put it—Rosenberg decided to go for broke by doubling the dose of interleukin-2. "You know, we'd been inching up and inching up and inching up," he said in an interview. "So finally I decided, 'Look, it was either going to work or not work. I would give it exactly as we gave it to mice, three times a day.' I hadn't done that before, ever."

And just when Rosenberg decided he needed to be more aggressive, in walked a thirty-three-year-old army brat and career naval officer named Linda Taylor. She had been commanding a U.S. military base in Guam until November 1983, when she felt marble-like bumps under her skin and knew her melanoma had come back. Just about the time she showed up in Bethesda, Steve Rosenberg had decided he would push his patients as he never had before, to see once and for all if the Cinderella experiment that cured mice could be made to cure humans. He gave himself one more trip through the door that separated the lab from the patients, and he came within minutes, literally, of losing the patient, the cellular treatment, perhaps the entire prize of immunotherapy.

13

THE END OF
THE BEGINNING?

He who saves a single life,
it is as if he saved the world.
—Talmud

⚭ *She had first been diagnosed* with cancer in 1982, when a mole "bubbled over"—stopped looking like an ordinary mole and started growing beyond its margins, a thick, chocolate extrusion of dark cells. Doctors surgically excised the mole, which proved to be melanoma, and told her she had a 50 percent chance that it wouldn't come back. But it came back by the end of 1983, and after a year of going from one clinical trial of biological agents to another, Linda Taylor—the patient known by the pseudonym "Linda Granger" in Rosenberg's book—reached the last station of the cross on November 14, 1984, when she was referred to the Surgery Branch of the NCI. Less than two weeks later, she checked into Building 10 at the NIH. Fate chose her to be the first patient to receive treatment under the newly aggressive Rosenberg protocol of LAK cells and high-dose interleukin-2.

It was a season for bad prognoses, not just for Linda Taylor but for Steve Rosenberg's dreams of immunotherapy as well. With seventy-five straight clinical failures behind him and nothing even resembling a partial success, Rosenberg faced his own personal, internal deadline, which is why he had decided to push the next patients to the utmost. Working with his senior associate, Michael Lotze, and a large team of Surgery Branch colleagues, Rosenberg repeated with Taylor essentially the same procedure the NCI team had tried before. They removed her white blood cells, stimulated them with a bracing out-of-body dose of genetically engineered interleukin-2, and then, on a Thursday morning in November, exactly one week after Thanksgiving, infused a slurry of 3.4 billion revved-up immune cells into her bloodstream, topping it off with 72,000 units of IL-2. She developed chills, little else.

The following week, Rosenberg tripled the dosage of IL-2. "Dose escalation," as it is called, is not always done in individual patients during the initial phase of clinical testing, but there was nothing customary about the handling of this patient. Eleven years of research, an emotional commitment and intellectual effort involving dozens of patients, a score of researchers, and countless thousands of mice, hung in the balance, and the man who had on another occasion stated that "experiments do not work; investigators make them work" now found himself trying to force arguably the most important experiment in his career to work—a human experiment, no less. As Rosenberg upped the dosage of interleukin-2, Taylor sunk slowly, soggily, into a bedridden bog of side effects: first came waves of nausea, then vomiting and swelling. Her joints throbbed with pain. Open sores developed in her mouth. She began to bloat with treatment-related weight. Still Rosenberg pushed. He realized that if the combination of LAK cells and IL-2 did not work against this patient's disease, it would probably mark the end of these efforts. "Everything was on the line," he admitted in his written account of the episode. "And so I had been determined not to stop pushing her until I was made to stop." Until, that is, the toxicity became too great.

Although they did not realize it at the time, high doses of this particular cytokine drug provoked an alarmingly dangerous complication: Taylor's blood vessels became porous, a condition known as "vascular leak syndrome," so that the liquid portion of the blood simply oozed through the vessel wall and began to accumulate in the tissues, in the space between cells, a flood filling the basement under the skin. From 122 pounds at the beginning of the treatment, Linda Taylor put on 35 pounds in about three weeks, and it wasn't from the hospital food. During one hellish twenty-four-hour stretch, she gained 13 pounds, all fluid.

And then one Thursday afternoon in December, exactly three weeks into the treatment, Taylor nearly died a paradoxical death: landlocked and lying in a hospital bed, she nearly drowned. Her lungs had become entombed with fluid, so much so that she struggled just to draw a breath; her mind filled with panic, as if someone were holding her head under water. Her face and skin began to turn blue, a penultimate warning sign, just before death, that the tissues were not receiving enough oxygen. She was minutes from dying—not from cancer but from her cancer treatment. Then she stopped breathing and slipped into unconsciousness.

Fortunately for Linda Taylor, for Steve Rosenberg, and for the field of immunotherapy in general, a nurse noticed Taylor's desperate condition, and a code for medical emergency summoned an army of white-coats to her bedside. The first physician to reach her, Steve Ettinghausen, inserted a breathing tube; out spurted a pink, frothing foam of liquid, fluid tinted with blood, and in rushed life-sustaining oxygen. After another two days Linda Taylor was out of the woods, but she received no more treatment. The crisis added yet another harrowing side effect, this one psychological, to an already rugged treatment: Taylor refused to sleep lying down, fearing she might drown in her sleep. It was a close call, and Rosenberg knew it. "I was made to stop," he wrote later.

In retrospect, it might be seen as a more pivotal moment than even Rosenberg has allowed. Some years later, he downplayed the significance that a treatment-related death would have had on his program at that critical juncture. "I think it would have been a problem," he said in an interview, "but I think we would have seen responses in other patients." In point of fact, no other patients had shown a response, no other patients were receiving the combination treatment at the time, and a long string of failures, punctuated by a death directly related to overly aggressive treatment, might well have marked the stroke of midnight for the "Cinderella protocol." The NCI doctors sent Linda Taylor home for Christmas. At the time of her departure, there was no obvious change in the status of her disease, and little reason to think there ever would be.

But one of the lessons of the immune system, dating back to Jenner and Coley, is that just because you don't immediately see gross evidence of change doesn't mean that changes—momentous changes— are not underfoot at a microscopic level. They had removed a snippet of melanoma tumor for a biopsy before Linda Taylor went home for the holidays, and the NCI pathologist noticed several subtle irregularities. Several spots in the midst of the tumor tissue appeared dead, or necrotic. Moreover, immune cells, namely lymphocytes, had poked their noses into the margins of the tumor—had infiltrated it, according to the jargon. Also curious. When Taylor returned to the clinical center on January 9, 1985, for more tests, her melanoma tumors again appeared unchanged, but a second biopsy confirmed the initial observation—some of the tumor cells had become ghosts, inert shadows of their rapacious selves.

There is perhaps no greater indication of Rosenberg's desperation than that, despite the close call with Linda Taylor, he not only initiated the same treatment in a second patient but vowed to push even harder. The man's pseudonym was "James Jensen," and he suffered from a cancer of the colon that had already spread to his lungs. Rosenberg began treating this patient on January 5, 1985, and decided privately that this would be his make-or-break case; he reasoned that Jensen was stronger than Linda Taylor, could tolerate the ravages of treatment better than she. "And I felt desperate," he wrote later, "as desperate as I have ever felt in my life." For three weeks, they pumped Jensen full of LAK cells and interleukin-2. He gained twenty-four pounds, all fluid; the number of platelets in his blood, crucial to the ability to form clots, dropped off the charts, and at one point he was in danger of bleeding to death. On January 29, after three weeks of treatment, they sent him off for a chest x ray.

A colleague brought the x-ray films back to the second floor. There, on a small lightbox in Rosenberg's office, right next to the quote from Maurice Maeterlinck and opposite a wall of framed photographs of all the good soldiers who'd marched through the double doors between lab bench and bedside, the two doctors compared that day's x ray to the pretreatment films, and Rosenberg must have viewed the difference as if it were magnified by a telescope ten years long. Before treatment: one large tumor and many small ones in the lung. After: the large tumor had shrunk to more than half its size, and the others had disappeared. "My God!" he thought. "All the years I have waited for a result like this. Can it be real? Can it have finally happened?"

Linda Taylor provided an even more definitive answer two weeks later. When she returned for a checkup on February 13, 1985, Rosenberg asked her how she was doing, and he was startled to hear her say she thought her tumors were going away. Quickly they measured the nodules with calipers, and it was clear that they had indeed shrunk. A biopsy showed the tumor cells were all dead. A month later, she returned for a more thorough workup. Where once marble-sized lumps had rolled beneath Rosenberg's probing fingertips, there was nothing left to feel now. Nothing at all. No tumor. The treatment had worked, at least in two patients, at least for a short period of time. "That for me was a watershed," Rosenberg said later, "because it demonstrated, 'Hey, this can happen!' "

Once something worked, once he got his foot in the door, Rosenberg was back on terra firma; he could tinker, he could "optimize." By the end of summer, they had treated twenty-five patients with a combination of interleukin-2 and LAK cells; nearly half of them experienced "objective" responses, meaning that their tumors retreated at least partially. With time, all the limitations of the approach would be duly noted—the toxicity, the small number of responses, the short duration of most partial responses, and the relatively small number of complete responders. But oncologists rarely if ever detect advanced tumors suddenly in retreat, whether on x rays or under probing fingers. In the span of two weeks, Rosenberg had seen it twice.

Years later, sitting in his office, Rosenberg was asked what it was like to peer at an x ray and see metastatic disease disappear. "It's always a wonder," he replied in his low, gravelly voice. "An incredible thrill is the best word [to describe] when you see a tumor regress. But I can tell you, it doesn't last long, because three minutes later you walk into the next patient's room or you see the next patient in clinic or the next x rays, and there is an equally viable, innocent person who's *not* responding, and then you have to sit down and say, 'I'm sorry, it didn't work.' If anything, the failures are harder to take now that we're getting responses than before we had ever seen a response."

From the moment that Linda Taylor and James Jensen enjoyed surprising recoveries from their advanced cancers, immunotherapy entered a different orbit—a wobbly one, it would turn out, but launched along a trajectory that escaped no one's notice in the scientific or lay community. Like interferon before it, the Rosenberg protocol was about to experience all the hype, all the disappointment, all the backlash—all the exaggerations of both optimism and disillusionment—mustered by a society that seizes upon the secular miracles of modern medicine and blows them up into phenomena of monstrous dimension. Add to this high-profile scrutiny the desperation of cancer patients, more than a million strong in the United States alone, their numbers (and despair) amplified through family and friends, and perhaps the best metaphor to describe public disclosure in this kind of environment is combustion: patient hopes are as dry as kindling, media attention acts as an accelerant, and all it takes is a bolt of lightning, which indeed struck the landscape of cancer research in late 1985.

Cancer researchers smelled a storm in the wind a few months earlier, when rumors of remarkable remissions at the NCI began to circulate; one oncologist recalls hearing the news in the fall of 1985 from a friend at the Cetus Corp., which provided the IL-2 for Rosenberg's trials. But implicit confirmation seemed to arrive around the end of November 1985 when Vincent DeVita, director of the National Cancer Institute, issued an "RFP"—a "Request for Proposals"—to the several dozen comprehensive cancer centers throughout the United States, soliciting applications on very short notice to be one of six sites in a trial that would attempt to confirm what would be informally described thereafter as the "Rosenberg protocol"—the combined use of interleukin-2 and LAK cells in the treatment of kidney cancer. No one had seen a publication; a lot of researchers, to say nothing of clinicians, didn't know a LAK cell from a jail cell. But the feeling was that if the NCI had sufficient interest in the approach to push it into the provinces, it must be pretty damn good.

The Surgery Branch was about to embark on the most treacherous part of the treatment: public disclosure. Rosenberg had gone to DeVita after the first ten patients had received the treatment, and DeVita needed little convincing that something significant had occurred. "It was very straightforward," DeVita said later in an interview. "The first two patients that were treated had significant responses, including a complete response. I've always felt very strongly that when you look at the introduction of a new agent, you can do two things with it. If it produces partial responses, that's of interest, because you can develop the agent further or you can develop analogues to the agent or you just in the general area go ahead. But in fact there's far less interest than in any kind of agent or any new therapy that produces complete responses, and especially if you're in a tumor where you don't often see complete responses. So when that happened, we said to Steve, 'Look, treat a number of patients, and if you consistently see more than an occasional complete response, then I think that it's time to make note of it and report it.' And he asked me when I thought it would be publishable, and we stopped at 10, and 15, and 20, and I think the final number was 25 [patients], with at that time I think more than a 40 percent response rate. So that was a reportable observation that was worthy of note, and that was all that really counted."

Two other momentous events occurred that summer that would forever complicate perceptions of the Surgery Branch work. In July 1985 Rosenberg had been asked to be one of the surgeons who participated in the most intensely monitored operation of the year: the removal of a portion of President Ronald Reagan's cancerous colon. Moreover, after the operation, he had come across as the most candid and forthcoming member of the medical team in describing the president's illness. At a press conference, on the record and in front of every species of audio and visual recording device known to human-kind, he had stated ("totally off the cuff," he said later), "The President has cancer."

The remark caused a furor. Nancy Reagan was livid, according to DeVita, who recalled that Rosenberg got into "a lot of trouble" for uttering that simple, accurate statement and "was immediately cut out of that case." The *Los Angeles Times* conducted a poll on whether he should have been so blunt. Two hundred interview requests poured into the NCI; Rosenberg refused to address another press conference. "It was the first time I had been on the inside of a story, just to see how things started to get muddled," Rosenberg said later. "And what happened is the press then turned to experts who didn't have the facts. A lot of things they were saying were wrong; I *knew* they were wrong. And I knew that I could have cleared up a lot of the confusion very quickly, but I wasn't willing to do that, because it just did not seem appropriate that as a doctor I would talk about a patient's problem to the press." Whatever the rationale—and this was by no means an average patient—Rosenberg undoubtedly emerged from the experience a more guarded personality when it came to dealing with the press, which in turn created an awkward paradox: he still curried its favor on the one hand, but may also have attempted to exert greater control on the flow of information.

Then Rosenberg took his family on an extended vacation the following month, an absence that became a crucial ingredient for the other, quietly profound event that summer. Earlier in the year, Michael Lotze had been heading a clinical trial of interleukin-2 alone, which Rosenberg had suspended—against Lotze's wishes—because of inadequate supplies of IL-2. While Rosenberg was away, Lotze took it upon himself to skirt the Rosenberg protocol and treat a melanoma patient with high doses of interleukin-2 alone—no LAK cells, just the cytokine. There was medical justification for this departure—the man was considered so close to death that infusions of cells would be impractical. As Lotze recalled, "He was one cell-doubling away from dying." The patient, a middle-aged and desperately ill man whose melanoma had spread to vital organs (including the spleen, liver, and lungs), responded spectacularly. Month by month, hundreds of tumors melted away.

This spectacular piece of good news, however, came with a cloud attached—it immediately raised questions about the other 25 patients. Were their responses due to LAK cells *and* IL-2, or just IL-2 alone? It was an absolutely crucial question, with important scientific and medical and even commercial ramifications, but before it could even be properly framed, much less answered in the clinic, the inexorable momentum of the first round of treatment and its imminent publication assumed a life of its own. Not least, it turns out, because of the efforts of the NCI itself to publicize the success.

DeVita, in particular, avidly spread the word. As director of the National Cancer Institute, he had paid his dues as ceremonial whipping boy before Congress; he had been grilled about the lack of progress in the War on Cancer, heard the plaintive wail of congressmen lamenting the billions of dollars spent on cancer research with so little to show, and had even gone out on a limb the year before by publicly stating that cutting the number of cancer deaths in half by the year 2000 was "achievable." In the summer of 1985, after 4 of the first 12 patients treated at the National Cancer Institute responded to the combination of LAK cells and interleukin-2, DeVita encouraged Rosenberg to submit a brief report to the *New England Journal of Medicine*, which is published by the Massachusetts Medical Society and is closely associated with Harvard Medical School and its flagship teaching institution, Massachusetts General Hospital. The deliberations of a medical journal are confidential, but the *Journal*'s editors undoubtedly wrestled with a complicated and perhaps controversial decision. The editors asked for more cases. Rosenberg was only too happy to oblige; he'd treated a few more patients in the interim and had seen a few more partial responses. Everyone understood the caution; purporting to herald a new form of cancer treatment would inevitably cause a huge stir. Even when Rosenberg submitted a revised paper, the *Journal*'s acceptance was not automatic. Dr. Arnold Relman, who then edited the journal, said in an interview, "It was clear to everyone, including Rosenberg, that the final answer wasn't in. It's a judgment call. . . . I concede it's an arguable case. I would not concede that we shouldn't have published it."

Events ultimately overtook everyone's caution. In late October, a panel of medical experts met in New York to discuss candidates for the General Motors Cancer Research Foundation's annual award for excellence in cancer research, one of the most coveted prizes in the field. After a daylong session behind closed doors, the panel members met with members of the press during a dinner at the World Trade Center, and several researchers excitedly disclosed some of the recent results coming out of the Surgery Branch of the National Cancer Institute—results duly noted by one of the attendees, William Rukeyser, the editor of *Fortune*, who passed the information along to the magazine's science writer, Gene Bylinsky. This tip kicked off one of the more memorably manic (and now generally lamented) jags in the long manic-depressive history of immunotherapy, beginning with a magazine story that asserted, among other things, that "many clinicians believe the odds in the struggle against cancer will soon be tipped in favor of the patient."

While the editors at the *New England Journal* had ruminated long and hard on the implications of the research, editors at *Fortune* had already reached a decision: "CANCER BREAKTHROUGH," its cover of November 25, 1985, proclaimed, and the accompanying story reported that Rosenberg's results were "unheard of in cancer drugs." Cetus, the California-based biotechnology company that supplied the Surgery Branch with its interleukin-2, found its name handsomely plastered on the cover of the nation's premier business magazine. Rosenberg professed to be furious, convinced that someone had leaked his data to the magazine ("Totally blew me away," he said of the article). *Fortune* reported that "even the most careful researchers are having trouble containing their excitement" about the approach. One careful physician, who spoke at length with *Fortune* to explain why the results weren't nearly as exciting as they appeared to be, later noted that his cautionary remarks weren't included, perhaps because they didn't fit the upbeat tone of the final story. "Enough data have leaked out around the cancer research circuit, however, to suggest not only that Rosenberg's results are striking," the article continued, "but also that they mark a milestone in medicine—the first successful and reliable enhancement of a major part of the human immune system."

It was almost anticlimactic when, two weeks later, a "Special Report" from the Surgery Branch appeared in the December 5, 1985, edition of the *New England Journal of Medicine*. Almost, but not quite.

For the first time since the BCG craze of the 1970s, an immune-based medical approach seemed to have worked—indeed, in a minority of cases seemed to have worked sensationally. Patients with extremely advanced cancers, their tumor burden large and widespread, people who had reached cancer's endgame—people who like Linda Taylor had failed all the other standard regimens of surgery or radiation or chemotherapy, some of whom had additionally failed other experimental immunotherapeutic medicines like interferon and BCG—had responded.

As the New England Journal article documented, 11 of 25 patients—44 percent—had experienced responses to the treatment, including the one complete regression. The 11 responders represented a broad range of malignancies, including cases of melanoma, kidney cancer, and colorectal cancer, and inevitably the report caused a firestorm of public interest. As Rosenberg remembers it, "When the [New England Journal] article came out, I faced the decision of letting the article either speak for itself or clarifying things when I was asked questions about it. And I decided that, look, this was science, it was now published. And I had in mind all this confusion after the Reagan thing, and [thought] maybe I ought to try and clear it up. But it just exploded."

As so often happens in mass reporting, the initial print story held a spotlight on the breaking news, the better to help others flock to the story. It didn't hurt that Rosenberg's face was known to millions of Americans for his "articulate and reassuring role," as Newsweek put it, as Ronald Reagan's surgeon. But at some point, as the dynamic unfolded, the LAK/IL-2 treatment ceased to be a medical story and began to be a medicosociological phenomenon. The New York Times described the "Special Report" on page one; Tom Brokaw of NBC News paid a visit to Rosenberg's lab in Bethesda; Rosenberg appeared on the MacNeil-Lehrer News Hour, answered Leslie Stahl's questions on Face the Nation, posed for Newsweek's December 16 cover story, "The Search for a Cure." He told interviewers on the Today show that the treatment represented "the first new kind of approach to cancer in perhaps twenty or thirty years," and he shortly thereafter received the $100,000 Armand Hammer Cancer Prize in 1985, along with Tadatsugu Taniguchi. Rosenberg even used the "B" word when he told People magazine, "It's not a cancer cure in 1985, but it is a breakthrough in that it marks the first successful approach to using the body's own immune system to reject a cancer." Time later described him as "a researcher of near celebrity status." Switchboards at the National Cancer Institute lit up, logging more than a thousand calls for each of the two days following the initial press reports—one of the heaviest volumes of calls ever recorded. Before long, there were even references to a "black market" in IL-2, although no such thing developed. "The whole thing just got totally out of hand," Rosenberg said. "And I think raised expectations to incredibly unrealistic levels."

Some of those expectations, unfortunately, had been aroused in other physicians. "A lot of people were under the impression that this was going to be *it*, so to speak—that this was going to cure cancer," recalled James W. Mier, an oncologist at Tufts University–New England Medical Center. In response to De Vita's request, many of the comprehensive cancer centers across the country enthusiastically prepared their applications, on very short notice, for the honor of being one of six sites in the NCI's hastily arranged follow-up study. Soon thereafter, testing of the Rosenberg protocol began in earnest at Tufts University in Boston, Loyola of Chicago, University of Texas–San Antonio, City of Hope in Duarte, California, the University of California at San Francisco, and the Albert Einstein School of Medicine in the Bronx, New York. At the same time, a private cancer clinic based in Franklin, Tennessee, and run by two former NCI physicians, William West and Robert Oldham, initiated an IL-2 and LAK cell treatment program, at a cost to patients of up to $30,000 per treatment. All this even though the initial study awaited confirmation.

Alex Fefer, reduced by circumstances to a kind of pesky one-man Greek chorus, watched with bemusement from Seattle, bristling at the speed with which events were unfolding—and not a little disappointed that his group had not been selected for follow-up studies. "It was the height of chutzpah, in my opinion, for the NCI to select six centers, to give several million dollars, to try and specifically repeat what Steve had done," he said. "That is not the way to do science. If his results were 50 percent, 80 percent, you might make a case. When the results are 30 percent *and* very toxic, then it seemed to me more appropriate for the government to entertain proposals for what you would like to do that might improve the therapeutic ratio—either decrease toxicity, increase therapeutic effect, try it on other diseases, whatever." There were similar sentiments even within the NCI. "It's an example of the precipitousness with which this was developed," said one senior researcher. "It was done as much for, 'Oh my God, we really need something that's new and important and working in cancer, and we need to be able to convince Congress and the American people that we're really making progress, and we by God need to confirm whether this is real or not as quickly as possible.' When the very first and most important question is whether the LAK cells *added* anything. You know, there were like four steps that somehow got lost in the process of propelling forward, in one's anxiety, to find an effective cancer therapy."

It all unfolded so quickly that many doctors and patients didn't know what had happened, either. "I think Steve was taken a little bit unawares," Mike Lotze said later. "I don't think he recognized the kind of impact that this had on physicians. You've got to recognize that a million people a year get cancer, and half of them will die of it, so at any one time in the country, there are 500,000 patients *dying* of cancer, each one of them connected to a physician or a hospital or whatever, and when you hear in the newspaper that there's a new cancer treatment, you immediately go to your doctor and say, 'What do you know about this?' And for something as exotic at that time as IL-2, which had only recently been cloned and expressed and much of the immunology was not part of the lexicon of the conventional medical oncologist of the time, you either say, 'I don't know anything about it (and therefore I am uninformed and must not be a very good physician),' or 'People report these kinds of things all the time, and they almost never work out.' And the appropriate response is to caution your patients against the next thing that appears in the newspapers. So there was a pretty good reaction that was building up against the IL-2 LAK reports, primarily in the medical community, because one, people didn't understand it, and two, felt that it had far too big a play. And because it had the full support and authority of the NCI, with Vince pushing it *very* hard."

By year's end, Steve Rosenberg had been selected as one of *People* magazine's "25 Most Intriguing People of 1985." Rosenberg would later write in his book that he disliked the publicity intensely, that he found the reaction "disturbing, even frightening." There is indeed ample evidence that Rosenberg made an effort to dampen the reaction; on a number of occasions he was quoted as saying (as he told *Time* magazine), "I am really anxious that this be kept in perspective. This is a promising first step in a new approach to use the body's own immune system against cancer. It is certainly not a cancer cure in 1985." Rosenberg's boss at the time, Vincent DeVita, also defends Rosenberg, suggesting that the big push came from the NCI. "In Steve's case," DeVita said, "it was me who noticed it and called it to the attention of Armand Hammer, the chairman of the president's cancer panel at the time. And you know, every time he got near a microphone, he started talking about Steve Rosenberg and what he had done. And there was nothing I could do about it after that. So it was really me, not Steve, who got the first attention to it, and so I don't think he should get blamed for that."

But at the same time, there is a part of Rosenberg that courts the very limelight he finds "disturbing and frightening," and his peers are aware of it. "He was on the cover of *Newsweek*. Now no one put a gun to his head to do that, I don't believe," said one former colleague, speaking anonymously but reflecting a widely held sentiment. "Rosenberg grew up in a world where you pass out your press release as you walk into the room to treat your first patient," said another physician. Nor was he shy about claiming his rightful due. "He transposed the field from witchcraftery to one judged according to the standards of the scientific method," says a former colleague. "And he's not reticent about stepping forward and taking credit for it." Indeed, throughout his tenure at the NCI, Rosenberg has opted to put his name first in the list of authorship on many important papers (while this is not exactly rare, the general practice in basic research is for the laboratory chief's name to come last), and he has aroused considerable enmity among fellow researchers for, in their view, not sufficiently crediting the work of others. "There's a general sense *outside* the NCI of a great deal of discomfort with Steve," said Mike Lotze. "Much of it probably has to do with jealousy, I think. Just the fact that he *is* so accomplished. But it's not just that. Why should Steve's success cause so much acrimony among people who do this business? I think in part it's because he tells the story, as he did in his book, without telling about all the important work that preceded it or was going on concurrently, or gave it just minimal lip service." In any event, everything from Rosenberg's method of describing Ronald Reagan's cancer to his appearance as the cover boy on national magazines attached further controversy to an experimental biological therapy that had enrolled many more detractors than patients.

Perhaps more significant, the *New England Journal* report offers a perfect example of one of the most troubling trends in the mass-media coverage of medical news. Many physicians (including experts in a given field) may first learn the details of a new experimental therapy from a newspaper story or, worse, a reporter's phone call seeking their reaction, while researchers on the other hand may first encounter critical remarks from colleagues, a kind of alfresco peer review, on the front page of national newspapers like the *New York Times*. The journal article, while not yet an afterthought, has become an after-the-fact accessory to public disclosure—and public marketing—of new medical developments, and given the increasingly competitive funding environment in the research community, this conflation of biomedical science and institutional publicity is destined to raise unrealistic expectations again and again in the future.

ॐ

Not long after the *New England Journal of Medicine* report appeared, a well-known immunologist began to pepper his talks with a slide of a 1986 Gary Larson cartoon showing a cowboy, gun smoking, standing over a dead body and hurling questions like "What's the circumference of the Earth? . . . Who wrote 'The Odyssey' and 'The Iliad?' . . . What's the average rainfall of the Amazon Basin?" to which a bystander replies, "Burt, you fool! You can't shoot first and ask questions later." The reference, everyone understood, was to Steve Rosenberg. The LAK/IL-2 episode represented the first time that many people in the lay public had heard of immunotherapy and what they heard was confusing. Given the enormous amount of publicity that swirled around this one single clinical report, the hopes it raised as well as the hopes it dashed, it is worth taking a moment to revisit this landmark study, to see exactly what it said and what was left unsaid, what has held up and what has not.

First of all, the section of the paper entitled "Lymphocyte Harvest and Culture" would give pause to any doctor who toiled in a regional cancer center or private clinic instead of a $2-billion-a-year national institute. Collecting white blood cells from patients (a procedure known as leukopheresis), plucking out the right subset of lymphocytes, cultivating them in the lab for weeks at a time, and then reinfusing them brought the ardor, and odor, of industrial production into the private laboratory, to say nothing of worrisome problems with quality control, safety considerations, and federal regulation.

Then there was the issue of toxicity. The side effects shocked almost everyone who ventured beyond the summary of results and read the fine print. Virtually every patient treated according to the NIH regimen suffered fever, malaise, nausea or vomiting, diarrhea, sharp drops in blood pressure, skin rashes, difficulty breathing, liver abnormalities, chills, and irregularities in blood chemistry (all but one developed anemias requiring transfusions of blood). Fluid retention alone caused patients to balloon 10 percent above their normal weight; a patient weighing 150 pounds at the start of the treatment, for example, would predictably gain 15 pounds in a week or two. And to counter those side effects, many patients received drugs ranging from acetaminophen (for the flulike fevers and chills) to an antihistamine for skin rashes to ranitidine (for gastrointestinal bleeding) to doxepin (a sleeping pill). One of the paradoxes of immunotherapy is that, while the immune system is viewed as "natural," and therefore presumably as gentle and knowing as a mother's caress, its potent molecules and remorseless cells are in fact designed to wreak biological havoc; they pack considerable physiological wallop, especially at the high doses Rosenberg recommended.

Because the treatment was so grueling, patient selection became very important. None of Rosenberg's first twenty-five patients were over sixty years of age; many were in their thirties and forties, obviously better able to withstand the rigors, the wear and tear, of such toxic treatment. In selecting patients so carefully, Rosenberg invited subsequent criticism for enrolling only "Olympic athletes," as Robert Oldham once put it, into these trials—again, not a trivial criticism, since the vast majority of people develop cancer after age sixty.

Coming nearly a century after Coley's toxins were first reported, the Rosenberg report also offers some striking, and revealing, parallels. Like Coley's, Rosenberg's very first patient treated on the new protocol enjoyed a complete and long-lasting regression; as with Coley's toxins, the mechanism of action of LAK cells and IL-2 was (and remains to this day) somewhat unclear, although cytokines clearly played an important role in both; like Coley, Rosenberg was perceived by peers as mongering an unseemly amount of attention; and as with Coley, other researchers would claim they were unable to replicate Rosenberg's reported rate of responses. What distinguished Rosenberg's work from Coley's, as indeed what distinguished immunotherapy at the end of the century from immunotherapy at its beginning, was the precision of the science, of the diagnoses, of the molecular rationale for attempting such treatment and the improved rigor in assessing it, the extensive animal studies that preceded the work, and the willingness of the community at large to put it to the test in larger trials. It is a truism in cancer treatment, however, that unless you hit a home run, it's very hard to tell why something works in one patient and not in another, and how to make it better; and just as Coley was at a loss to explain why some patients responded when so many didn't, so was Rosenberg. Indeed, they both became painfully familiar with another truism about immunotherapy in cancer: there are virtually no home runs.

There was nothing the paper could say, coming so soon after treatment, about the duration of these remarkable responses. From the time of William Coley to the present, duration is always the other penny in cancer treatment; getting a tumor to respond is one thing, but if the response doesn't last, you've merely bought a little time for the patient, and sometimes at great cost, in medical expense and quality of life. One unhappy clarification on that point came from James Jensen, whose x ray provided the first exhilarating evidence of success in January 1985. James Jensen's cancer returned, and he died several years after the *New England Journal* article came out. He is forever immortalized in the medical literature, sealed in amber as a dramatic partial response, but his remission lasted less than five years. Despite the impressive 44 percent initial response rate, NCI researchers believe that Linda Taylor remains the only one of the first twenty-five patients who received the LAK/IL-2 treatment still alive at this writing.

But of all the reactions to the LAK/IL-2 approach, two of the most revealing came from the Surgery Branch itself. First, rather than devoting all his time to optimizing the technique that had attracted so much attention, Rosenberg delegated this task to others and virtually abandoned LAK cells; as he would do so often, he lit out in another direction, tantalized by an even more promising kind of cell, a different cytokine, a different combination. Second, Michael Lotze continued to treat patients on high doses of interleukin-2 alone—no LAK cells at all—and began to achieve a comparable response rate. Which begged a very important scientific question: were LAK cells even necessary?

Rosenberg understood the importance of the question. He immediately launched a randomized trial at the NIH to test interleukin-2 treatment alone against the combination of IL-2 and LAK cells. But the quiet buzz of this work was soon drowned out by the deafening roar of backlash. Toward the end of 1986, the Rosenberg lab learned of a devastating editorial that was about to appear in the *Journal of the American Medical Association*. No one was heartened by the news that it had been written by a physician known informally at the NCI as "Dr. No."

The occasion for this famously vituperative editorial was a follow-up study by the NCI group that appeared in the December 12, 1986, issue of the *Journal of the American Medical Association*. Michael Lotze, the study's lead author, had treated a small group of ten patients with high-dose interleukin-2 alone, including his detour from the Surgery Branch protocol in August 1985. Ultimately, three of the first ten patients Lotze treated had partial responses; in fact, all three responses occurred in the six patients with melanoma in the study. One of them continued to recover after the study was published and, alive and well eight years later, is now considered a complete responder.

When thirty-six-year-old "Nancy Burson," as they called her, appeared for evaluation at the NIH in October 1985, she was, regrettably, a perfect candidate for IL-2. Several months earlier, the woman had gone to her doctor complaining of pain in her right hip; tests revealed she had a bone fracture caused by melanoma, and although they never learned where the cancer had started, it had already spread so rapidly that her right leg was removed at the pelvis in June 1985. Even as she recuperated from that major surgery, the cancer popped up again. "She had a huge tumor on her forehead," Lotze recalled, "and in the lymph node in the groin, and also pulmonary metastases." In the geography of metastases, the more widespread and distant the wildfires, the more pessimistic the likely outcome. "It was the sense of the service," Lotze said, "that she wasn't going to respond." Indeed, they deemed her condition to be so terminal that they didn't even bother to take photographs to document her advanced condition.

At first, their clinical instincts were absolutely correct; Mrs. Burson didn't respond. She began receiving high doses of IL-2 on October 8, 1985—two months before the *New England Journal* article appeared— and continued to receive them three times a day for nearly three weeks. By the end of week one, the tumors on her skin had decreased markedly; regression continued until, twelve months later, the location of the tumors could not be detected at all. But the tumors in her lung, though they shrunk by half in a month's time, started increasing in size again. "This was right before Christmas," Lotze recalled, "and I remember having a big discussion with Steve. I really wanted to push and treat her again because I was worried that I was going to lose her as a response! And his argument at the time was that this might be her last Christmas, and didn't I really think the likelihood was that she was going to die of her tumor, and wouldn't it be better for her to be with her family and so on. And I pushed very hard and ultimately won out and treated her again, and she went on to have an additional response, almost complete." Lotze, in other words, out-pushed Rosenberg.

Lotze initiated this second round of treatment on November 29, 1985. The lung tumors continued to diminish, and with a booster dose the following May, the two remaining lung nodules shrank even further. By September 1986, about a year after she began treatment with IL-2, Mrs. Burson's disease had virtually disappeared, except for a single pea-sized nodule in one lung (it has remained quiescent as of this writing, the fall of 1996). "She has finally been called a complete response," Lotze said. "She's now nine years out."

Such results are often more difficult to interpret than so startling and improbable a recovery might initially suggest. Melanoma has an unusual natural history—"one of these fishy diseases that goes into hibernation," according to one oncologist. The disease takes unexpected twists and turns in its behavior; it can spread and then stabilize, metastasize and then grow quiescent. It has long been acknowledged to be among the more responsive malignancies to immunological treatment, but its unpredictable course often makes it difficult to determine if responses, however modest or dramatic, are due to treatment or to the curious twists and turns the disease takes on its own.

In any event, the Burson case and nine others formed the modest database for an article Lotze prepared for the *Journal of the American Medical Association*. It was, Lotze later reckoned, "a very cautious paper, not in a flashy journal. Putting it in *JAMA*, it's not a big deal. I just wanted to let people know that it was something that we were studying, and that we could see responses with IL-2 alone." The paper didn't skimp on any gory details. In even greater detail than the *New England Journal* paper, the horrendous side effects of interleukin-2 got a full, sobering airing; eight of ten patients ended up in the intensive care unit as the drug took its usual toll in fluid retention, nausea, fever, and chills. The treatment required extraordinarily invasive procedures: cardiac catheters to monitor blood pressure of both arteries and veins, transfusions, oxygen; no less than twenty-two secondary drugs, not including antibiotics, were at times employed to control the side effects of the primary, therapeutic drug. For all that, Lotze had his three responders, and all three had such advanced disease that their predicted five-year survival rate was only 5 percent. Lotze felt vindicated. And then he read the editorial accompanying the *JAMA* article. "One of the things that bothered me," he would say later, "is that I got the backlash." In point of fact, he just happened to be standing in the path of a hail of bullets aimed at Rosenberg.

Usually, editorials accompanying journal articles are cautiously favorable and contextual, explaining why the research is significant; *JAMA* invited Charles G. Moertel, a well-known and respected oncologist at the Mayo Clinic, to deliver an opinion on the use of interleukin-2 as an experimental cancer drug, and in a departure from the usual collegial back-patting, he delivered a preemptive nuclear strike. As even Moertel admitted, his anger had been festering since the boom of publicity accompanying the initial National Cancer Institute report in December 1985; the Lotze report merely lanced the boil. Known as "Dr. No" by some at the NCI, Moertel had clashed in public frequently with Vince DeVita. "He was the debunker, the person who kept saying, 'No, this doesn't work, and all it does is enrich the oncologists who are applying it,'" said one NCI insider. "Vince thought he was too much of a naysayer."

He certainly could always be counted on to be an outspoken critic of novel, especially alternative, cancer treatments such as laetrile, vitamin C—indeed, any cancer treatment that was ineffective, expensive, or toxic—and in that ignominious company he reserved a special place for interleukin-2. Moertel took exception to the treatment's side effects, its cost, and, interestingly, its attendant publicity. "Following their first publication," Moertel wrote, "these authors described their work as a 'breakthrough'; since this claim bore the imprimatur of the National Cancer Institute, it is scarcely surprising that the impact on patients with advanced cancers, their families, and their physicians was extraordinary." In unusually blunt language, Moertel noted that the treatment itself was "an awesome experience," that it usually "requires weeks of hospitalization, much of which must be spent in intensive care units if the patient is to survive the devastating toxic reactions," and that the costs were exorbitant. "In short," he concluded, "IL-2 therapy as administered in these studies is associated with unacceptably severe toxicity and astronomical costs. These are not balanced by any persuasive evidence of true net therapeutic gain. This specific treatment approach would not seem to merit further application in the compassionate management of patients with cancer."

Devastating, awesome, astronomical, uncompassionate—these were the adjectives of movie blurbs, not the medical literature. At a certain level, however, the sheer exasperation of the editorial went well beyond purely medical and scientific cavils. In griping about the widespread press coverage, its characterization as a "breakthrough," and Rosenberg's enthusiastic endorsement of the therapy, during one of many television appearances (this one on the *Today* show), as the "first new kind of approach to cancer in perhaps 20–30 years," Moertel had ventured into sociological criticism, and had done so on an equally national stage. There was even a little dig at the integrity of the data. "Response rates are fragile commodities," Moertel predicted. "If oncological precedent is sustained, response rates in confirmatory studies will not even come close to equaling the response rates claimed by initial enthusiasts in small and early studies." Here was a spear hurled right at the castle keep of the NCI, at DeVita and Rosenberg; the attack had been picked up and reiterated on national television and on the front page of the *New York Times*. The worst part, for Rosenberg and the NCI, is that Moertel turned out to be right about that last point. Rosenberg's "thunderbolt moment" had only presaged stormy weather, and Moertel was the first to rain on everybody's parade.

The term "backlash" often connotes envy and enmity, and certainly professional jealousy can never be minimized as a potential motivation for criticism; some researchers in the cancer community, struggling to get funding, reacted to the Surgery Branch's annual multimillion-dollar budgets with a form of sour grapes that had aged and been bottled like a fine wine. In truth, substantive scientific and medical issues divided the community, including concerns about the extreme toxicities associated with immunotherapy, the speed with which some of these experimental therapies had been implemented, the cost of reproducing the results, and the prediction in some circles that doctors in the provinces would be hard-pressed to obtain the same response rates. But perhaps the most surprising aspect of the Moertel editorial, obvious now at the distance of a decade, is that it distracted from a larger truth. In one of several surpassing ironies surrounding this remarkable episode, it turns out—surprisingly—that the Surgery Branch's *JAMA* article was probably closer to the scientific truth than their *New England Journal* article a year earlier. If current practice is any indication, virtually no oncologists use LAK cells today, simply because they don't seem to work. IL-2, on the other hand, enjoys modest but continued use throughout the world.

"See, the problem is that IL-2 became the afterthought," Lotze complained later. "And then subsequently, with the test of time, it was shown to be *the* major active agent." Indeed, it took many years, but the NCI published a randomized study in 1993, eight years after the initial firestorm, that answered the question raised by the controversial *JAMA* paper: LAK cells plus IL-2 versus IL-2 alone made no difference in the outcome of kidney cancer patients, and gave only a slight—and ambiguous—survival advantage to melanoma patients. Every time LAK cells have been studied in randomized trials, they seem not to make a difference. "There have been three studies," said Nicholas Vogelzang of the University of Chicago, whose group published the most recent in 1996, "all of them negative." Given those results, we can in hindsight see the back and forth of the debate that took place in the interim in an altogether different light.

The editors of the *New England Journal of Medicine* replied—indirectly, of course—to Moertel's blistering editorial with a one-two-three punch of their own several months later. In the April 9, 1987, edition of the journal, the Surgery Branch group weighed in with the first counterpunch. Rosenberg's group updated its original, December 1985 "preliminary report" with data from an expanded trial that seemed to confirm the earlier conclusions. In 106 patients with metastatic cancer who received the combination of interleukin-2 plus LAK cells, they reported, 23 responded (21.7 percent), including 8 complete responses (7.5%) where all discernible tumor disappeared; by contrast, 46 patients received IL-2 alone, and 6 responded (13 percent), only 1 completely (2.2 percent). The combination of IL-2 and LAK cells, judging from this data, appeared to be the superior treatment. The NCI authors took the occasion to state, "This immunotherapeutic approach can result in marked tumor regression in some patients for whom no other effective therapy is available at present. Determining its ultimate role in cancer therapy awaits further attempts to increase the therapeutic efficacy of treatment and decrease its toxicity and complexity." Translation (with a nod to Moertel): It's experimental, it's dangerous, and it's imperfect, but it can also achieve regressions where everything else we have ever tried has failed. That was exactly the case; nothing more, but nothing less.

A second, semiconfirming report appeared in the same issue of the *New England Journal*. It came from William H. West and colleagues at the Biological Therapy Institute and Biotherapeutics, Inc., the controversial for-profit cancer treatment center located in Tennessee. In an attempt to reduce the toxicity, West and colleagues had administered their IL-2 in a slow, constant infusion rather than in one large dollop every eight hours, with the result that side effects were reduced. The good news was that 13 responses were seen in 40 patients with advanced cancer who received both IL-2 and LAK cells, a response rate of 32.5 percent; the less heartening news was that none of the responses were complete. The study was hailed by some as confirmatory of the original Surgery Branch report when it might more profitably have been mined for another, more illuminating lesson implicit in the data from both studies: with cytokines, physicians had to be prepared to accept (at least initially) the rather extraordinary toxicities at higher doses in order to achieve more complete and durable remissions.

Taken together, these two studies suggested a response rate of between 20 and 30 percent when using the combination of LAK cells and IL-2. That prompted the third "punch," a redeeming—and, to Rosenberg, vindicating—editorial accompanying the two research articles entitled "Immunotherapy of Cancer: The End of the Beginning?" It was written by John R. Durant, then at the Fox Chase Cancer Center of Philadelphia, in an attempt to provide "an oncologic perspective on the meaning of the results" for both the medical profession and, by extension, an eager and confused public that was hearing both "Yes" and "Dr. No."

Perhaps the most significant thing about Durant's editorial is the punctuation in its title: a question mark. Only in retrospect can one appreciate how carefully worded, how conditionally enthusiastic, how elegantly noncommittal this seeming endorsement of the Rosenberg protocol really is; it is lyrically elusive in its use of qualifications, its optimism perched artfully on a nearly invisible superstructure of ifs and perhapses and maybes. To Durant, the key medical unknown— and this would become an issue of considerable significance in the industrial as well as the medical sphere—was whether the two studies conclusively showed that cultured human immune cells could be manipulated to "restore or enhance" normal immune function rendered ineffective by the spread of disease. "If LAK cells are essential to the best response rate," Durant wrote, "the therapeutic activity described in the current studies reflects successful cellular immune manipulation and represents much more than the effects of another new drug in the traditional therapeutic sense. Instead, the results become the basis for a whole series of important biological questions. . . . If they reflect, as seems possible, the successful manipulation of the cellular immune system, then we may be near the end of our search for a meaningful direction in the immunotherapy of cancer." Here was an antidote to Moertel's bitter emetic. Amid a host of shrewd questions, Durant posed an essential one: Did the LAK cells really add anything? And if so, how exactly did they subdue the uncontrollable growth of tumors? No one—not Rosenberg, not Lotze, not West or Oldham—could provide a satisfactory answer to that essential scientific question.

Rosenberg still cites the editorial today as a repudiation of Moertel's broadside. "To me, that was a demonstration that in fact it's . . . *real*," he said in an interview. To Lotze, it was perplexing, because the 1986 *JAMA* paper and its follow-up had implicitly under-mined the role of LAK cells in cancer treatment—IL-2 alone was capa-ble of achieving complete responses. Indeed, when asked if the results of the study published in the 1987 *New England Journal,* of which he was an author, were misleading, Lotze replied yes! To interested skep-tics like Alexander Fefer, it merely highlighted the obvious: a prospec-tive, randomized trial had not yet been designed in such a way as to answer Durant's question. The uncertainty—and the testy egos on either side of the question—placed inordinate pressure on the six phy-sicians who had been asked by the NCI to do the follow-up study. Everyone expected them to provide at least some guidance.

Their moment came one month later, in May 1987, in a plenary session at the annual meeting of the American Society of Clinical Oncology. It fell to Richard Fisher of Loyola University, head of the multicenter study testing the Rosenberg protocol, to disclose the dis-appointing data; as Moertel had predicted, fourteen excellent physi-cians at six leading cancer centers had failed to replicate the National Cancer Institute's initial 44 percent success rate. In fact, they didn't even come close. They only managed a 16 percent response rate, less than half of what the Surgery Branch doctors had claimed. The failure to reproduce the results caused great consternation inside and outside the NCI; according to one NIH insider, "There was a great hue and cry within the Cancer Institute with other extramural investigators who complained both to Steve and the Institute. 'Well, you know, is there some hidden step in the way you're doing this thing, because we're just not getting the same results.' Those sorts of things tend to make people wary." "No one has reported data like that first *New England Journal* paper, including Steve Rosenberg himself," said another physi-cian who participated in the follow-up trial.

To this day, the medical community is still puzzling over these results. Samuel Hellman of the University of Chicago, writing in the *Journal of the American Medical Association* in 1994, said the use of IL-2 "represents a significant advance in cancer therapy because of the novel mechanism of action of IL-2 and the long duration of complete responses." "In some cases," says James Mier of Tufts, "the clinical results are miraculous. I mean, when you see these tumors melt away, you just shake your head in awe. This *has* happened. We have proba-bly fifty to a hundred cases in Boston where within a span of a month a person who is literally terminally ill, lungs just *littered* with pulmo-nary metastases, will come back feeling better, having gained fifteen pounds because their appetite has returned, and who has a clean chest x ray and then goes on to have a complete eradication of the disease with another cycle. This has happened many times."

"Whatever else you can say about IL-2—good, bad, or indifferent— despite its checkered history and so on, to date it's the single biologic agent that achieves purely through immunologic effects clinically evident, often complete regressions of some cancers," said Alexander Fefer. "And that is the *only* biologic agent you can actually say that about. There's no question that it works in kidney cancer. There's no question that it works in acute myelocytic leukemia in some cases. There's no question that it works in melanomas in some patients. There's no question that it works in lymphomas in some patients." But because of the way clinical studies have been conducted so far, Fefer continued, many essential questions remain unanswered about how best to use it. "We don't know what the optimal regimen is. We don't know how to differentiate those people who will respond, and therefore should be pushed, from those patients who won't respond and who therefore should not be exposed to this sometimes horrible and toxic procedure."

The utility of IL-2 as a stand-alone drug remains an issue of ongoing exploration. Michael Lotze, known as "Mr. IL-2" during his years at the NIH, now claims a "solid 20 to 30 percent" response rate in melanoma and kidney cancer patients at the University of Pittsburgh. Other researchers have begun to explore the usefulness of this undeniably potent immune stimulator in chronic infectious diseases, where it might ultimately have broader impact than against cancer. Two NIH researchers, H. Clifford Lane and Joseph Kovacs, have demonstrated that a combination of high-dose IL-2 and antiviral drugs like AZT significantly boosted T cell counts in a small group of AIDS patients, and in a 1996 publication, they suggested that IL-2 may be combined with a new class of drugs known as protease inhibitors to reconstitute the devastated immune systems of people with AIDS. And Kendall Smith at New York Hospital recently published work showing that AIDS patients treated for six months with low-dose IL-2 experienced significant reconstitution of their immune systems with minimal toxicity.

There is only one thing virtually everyone agrees on, if clinical practice is any indication. The LAK cells, hailed in newspaper headlines in 1985 and in the *New England Journal of Medicine* editorial in 1987, have quickly fallen into total disuse. "It did not remain a mystery for any great length of time," said Mier, "that the cells added very little, except a lot of expense and a certain risk of infection, and that the real benefit was from IL-2."

∾

To this day, the "Rosenberg protocol" story—the quick rise and sudden fall of the combined LAK IL-2 treatment—leaves a sour taste in many mouths. Moertel's skepticism apart, a number of researchers—especially in Europe—found the data premature, the results hyped, the rush to confirm an inefficient use of time and money. The utility of IL-2 as a stand-alone drug has emerged relatively intact, but its toxicity has so far limited its application. By 1992, when no group had confirmed the high initial response rate of the LAK IL-2 treatment, some shared the feeling of Kendall Smith, who flatly refers to the high-dose protocol and the frenzy it touched off as a "fiasco."

To those who believe the research interests of society are best served by a more methodical, regrettably slower, and undoubtedly more painstaking approach, there existed widespread uneasiness about the sudden impact of, and swift backlash against, Rosenberg's work in immunotherapy. Ronald Herberman, director of the Pittsburgh Cancer Institute and an exceedingly cautious proponent of immune-based medicines, credits Rosenberg for moving the field along more rapidly than it might otherwise have gone, but attaches a price to that speed.

"Because of either his own optimism or the degree of attention that got focused for one reason or another by the press, it did lead to this overexpectation, in each case," said Herberman. "And that sort of set up the whole field for disappointment. And then for others of us who are in the same general area, it's made it harder for us to persist in it, because we get hit with the backlash. And we're trying to really keep moving forward and proceed with fine-tuning a number of these questions along the lines that we think are as sound and promising as possible, at a time when negativity or skepticism has entered into it. That makes it very difficult, particularly because when there's that kind of skepticism around, I can tell you it's very hard to get grants funded to do this kind of work."

Similar caveats were expressed by another researcher in the field, Philip Greenberg of the University of Washington. "That's actually been one of our problems with the field of immunotherapy," Greenberg said, though specifically not referring to the Surgery Branch. "Commonly what happens is that people move so quickly through things that ultimately knowing why things work poorly or only work in a small fraction of people is often left behind. And, you know, solving these problems really takes enormous commitment. It doesn't have the same pizzazz that being the first one to give something has. But ultimately it may be the much more important work. Sometimes being the first one to do something has to do mostly with time and place, not necessarily insight. So I think really getting down to the hard science *is* actually the hard part. It's going to be what's necessary to really take us to the next level."

At the same time that the Surgery Branch geared up to treat the first twenty-five patients with LAK cell therapy. Rosenberg's ever restless eye began to track a different immunological cell—indeed, came full circle, back to what he'd started out looking for in the first place. Around February 1985, just as the first startling responses to interleukin-2 and LAK cells appeared on Linda Taylor's x rays, Rosenberg and his valued long-time technician Paul Spiess returned to the lab to begin an important new series of experiments. Spiess assumed that there was an immune system response—perhaps futile and underwhelming, but a response nonetheless—to the experimental sarcomas that grew in mice. He also assumed that at least some immune cells attacked the tumors. So he removed these tumors and poked through them looking for cells of immune lineage that had infiltrated and attacked the cancers.

Here was an instance where interleukin-2 proved to be immensely useful as a lab tool, the chemical equivalent of a centrifuge, for it separated out a select group of cells from its cousins by the power of biological stimulation. IL-2 stimulated only lymphocytes that had already been aroused (or "activated") by cancer antigens; in other words, only those lymphocytes that had already bumped into mouse cancer cells and recognized them would be primed to receive IL-2's chemical imperative to expand. By adding IL-2 to the usual motley array of lymphocytes in a test tube, Spiess reasoned that only those cancer-specific lymphocytes would multiply and come to dominate the culture in which they grew. Remarkably, they did. In short, Spiess had developed a technique to pluck out tumor-specific lymphocytes. In finding those cells, the lab had crossed an important theoretical threshold, from the nonspecific to the specific. Rosenberg called them tumor-infiltrating lymphocytes (TIL).

The NCI group was not the first to land on the specific side of immunology. Bijay Muhkerji and Thomas MacAlister at the University of Connecticut reported in 1983 the ability to generate killer T cells against human melanoma cells, and in 1984 Alexander Knuth, a guest researcher in Lloyd Old's group, used IL-2 to isolate killer T cells from a patient that recognized his own melanoma tumor cells.

Unlike LAK cells, TIL cells seemed to be more potent, more re-fined, and more specific killers. Indeed, they turned out to be killer T cells—the cells that carry specific immune memory, right down to the peptide pattern of antigens. By March 1985, with the euphoria over Linda Taylor's response still thick in the air on the other side of the double doors, Spiess quietly compared the performance of LAK cells versus TIL cells against mouse tumors. The preliminary numbers were small but tantalizing: TIL cells appeared to be twice as effective as LAK cells in killing cancer cells in the test tube. Over the next year, Rosenberg's group cranked out a bushel of data, sending a paper about the activity of TIL cells in mice to the journal *Science* on July 10, 1986. Rosenberg's paper, which appeared about two months later, laid out the argument for what it called "a new approach" to immune-based cellular therapy against cancer. This was quintessential Rosen-berg: while the world was still reeling from last year's news, he had already announced next year's model.

The detection of T cells that could recognize tumor antigens on cancer cells gave new impetus to tumor immunology. In terms of basic research, it gave immunologists an avenue to tumor antigens—a way of identifying what these T cells saw on cancer cells, although a number of researchers had already embarked on this path. In terms of therapy, it meant that these patients had mounted an immune attack, however weak and futile, against their own tumor, thus opening the door to enhancing and optimizing an existing immune response: need-less to say, Rosenberg didn't stick around to hold the door while oth-ers walked through. Less than a year after the *New England Journal* "Special Report" had caused such a stir, the Surgery Branch had read-ied a protocol to treat patients with TIL cells and IL-2, had indeed generated T cells from several patients in preparation for infusions. The operation required cellular horticulture of an unprecedented na-ture, and the pilot project was entrusted to Suzanne Topalian. Topalian excised bits of tumor from her patients for preparation; easily accessible tumors made for the least invasive approach, so many of the first patients had melanomas, where the tumors lie at or just be-neath the surface of the skin.

Once excised, the knotty tumors would be mixed with enzymes that digested away all the gristly connective tissue, until all that was left was a soup of cells, mostly tumor cells but presumably also the immune cells that had reacted to the tumors. These two very different cell populations could be separated by placing the soup in a centrifuge and spinning them. At the end of the process, Topalian would have a thin layer of white blood cells, a few of which would be T cells, even fewer of which had been "turned on" by their encounters with the tumor. Since interleukin-2 spurred the growth only of immune cells that had already been turned on, however, the addition of IL-2 would preferentially goose exactly the cells they wanted. The process was slow: it took four to six weeks to grow enough T cells, and Rosenberg kept raising the bar in terms of the requirements for treatment—he wanted 10 billion cells per patient at first, but soon requested 200 billion TIL cells per treatment. The plan called for these cells to be infused, followed by a blast of high-dose interleukin-2 and also cyclophosphamide, an anticancer drug that seemed to enhance the killing effect of the TIL cells in animal tests.

In the beginning of December 1986, just before Charles Moertel's editorial appeared, Rosenberg revealed that he had begun treating patients with TIL cells, and the initial results tantalized everyone—the first five patients, all with melanoma, responded either partially or completely. Rosenberg pronounced himself "extraordinarily excited," although his enthusiasm had a somewhat different ring to it after the LAK cell story. Later, at a scientific parley in March 1988, Rosenberg reported "substantial tumor regression" in eight of nine melanoma patients. It seemed too good to be true. It was.

As they treated more and more patients, as some of the cases became more difficult, the response rate settled to a nonetheless impressive 55 percent. (As always, readers must regard this figure as an incomplete promissory against cancer until the duration of such responses can be known: on the other hand, these results were singularly impressive because no other treatment, accepted or experimental, had similarly made such a dent against these advanced cancers). Rosenberg published the results of the first TIL study in 20 patients in the December 22, 1988, issue of the *New England Journal of Medicine*. Only one patient enjoyed a complete response.

Subsequent studies have yielded much lower response rates. Rosenberg possessed the resources not only to continue trials with LAK cells and TIL cells but to reinvent himself once again. "The pressure to improve what we had was intense," Rosenberg later wrote. "We knew our approach would work. *But the treatment wasn't working well enough.* For all the progress we had achieved, I felt frustrated and stalemated. I could see only two ways to end the stalemate." One was to fling troops from his now quite large research group into the fray of basic research, trying to tease out details of the immune response by studying the role of other cytokines like interferon, interleukin-4, interleukin-6, interleukin-7, and tumor necrosis factor, all of which had been genetically engineered in the previous few years, all of which had demonstrated some antitumor effects in the laboratory.

But truth be told, Rosenberg had become seduced by an even more exotic approach. As even he admits, the basic science approach produced only "incremental improvements," and to his thinking, no one hits home runs in increments. Rosenberg still wanted to swing for the fences, or in his words "to leap a chasm" into new territory. The leap was to take the gene for a molecule known to attack tumors, a molecule like tumor necrosis factor, and smuggle it into the T cells that they had already learned to grow by the billions. Such an approach would require an entirely novel, totally unproven, and socially sensitive technique known as gene therapy. It had never been legally attempted in human beings before; indeed, it had never been *allowed* by federal regulators in human beings.

New, unproven, controversial, first—all the siren buzzwords of breakthrough science circled like flies around the idea of gene therapy, as did a growing number of ambitious scientists in the mid-1980s. "About that time," Rosenberg writes, "Michael Blaese and French Anderson approached me about using another avenue to gene therapy." After meeting to discuss the idea, these three NIH scientists decided to collaborate and would go on to achieve the world's first attempt at gene therapy. The collaboration was indeed historic, but it didn't quite happen that way.

14

"THERE'S JUST SO MUCH
YOU CAN LEARN FROM A MOUSE"

Scientific research, like literature, has two objectives—
the egoistic satisfaction of the investigator and other interested
individuals, and the betterment of mankind. The former
we call basic research, and the latter applied research.
—Chester Southam, "Applications of Immunology to Clinical
Cancer: Past Attempts and Future Possibilities," 1961

∞ *Ever since the 1960s,* when biologists cracked the genetic code, and the 1970s, when researchers acquired the tools to manipulate DNA, a few visionary scientists began to predict that this lofty biological knowledge, coupled with the biochemical equivalent of carpentry skills, could be harnessed to correct hereditary diseases in humans. Steve Rosenberg was not one of those visionaries, which is why it is ironic that he found himself in the thick of the race to be the first to attempt the uncertain new technology of gene therapy in humans. Although he will go down in history as one of the three pioneers, along with two other NIH scientists, W. French Anderson and R. Michael Blaese, their short-lived and somewhat unhappy collaboration as the "Troika" may live on as a case study not so much of visionary, cutting-edge science as of a marriage of convenience that says a lot—too much, perhaps—about the pressures of high-stakes, high-profile biomedical research.

The birth of gene therapy flowered not in the Surgery Branch but within another branch of immunology, on another floor of the National Cancer Institute, with a different and unrelated set of goals. Fittingly, however, it came to fruition because of a disease afflicting the cell so beloved of cellular immunologists: the T lymphocyte.

In the early 1980s, pediatricians became tantalized by the possibility of correcting inherited genetic flaws, especially in children with severe hematologic defects like sickle cell anemia or thalassemia. Indeed, in 1980 a UCLA scientist named Martin Cline traveled overseas to perform unauthorized gene therapy experiments in two individuals; he returned to face scientific censure, cancellation of his federal research grants, and public disgrace for jumping the gun on a technology that was not ready for human application. The Cline episode gave gene therapy the odor of a field that attracted glory seekers who, with the rationale of treating desperately ill patients, were willing to cut ethical corners. It also helped prove the point that doctors still had a lot to learn: it turned out that, because of the complex regulation of genes, hematologic disorders were too difficult to lend themselves to easy genetic correction.

But immunological disorders offered different possibilities. One of the nastiest of a bad lot was known as ADA (for adenosine deaminase) deficiency, which arose from a flaw in the gene for the enzyme adenosine deaminase; this tiny genetic glitch produced T cells that lacked the ability to make this enzyme, and the enzyme was crucial to the T cell's ability to clear a toxic chemical that built up in these immune cells alone. The result, in children unlucky enough to inherit ADA deficiency, was that their T cells basically committed suicide, poisoning themselves to death. The children became the human equivalent of nude mice, the laboratory animals specially bred to have no lymphocytes, and were helpless against a broad range of normally modest but now suddenly monstrous pathogens.

One of the experts in these virtually untreatable disorders was a tall, stolid, and big-boned Minnesotan, R. Michael Blaese. Since his earliest days as a medical student at the University of Minnesota, Mike Blaese had specialized in the rare and heartbreaking immune disorders of young children: Wiskott-Aldrich syndrome, SCID (for Severe Combined Immunodeficiency Disease), ADA deficiency, the kinds of diseases sadly exemplified and popularized (if that is the right term) by the plight of David, the "Bubble Boy" of Houston, who lived in an antiseptic, germ-free bubble throughout his childhood, awaiting a treatment that never came. Trained as a pediatric immunologist, Blaese had a long-standing interest in developing new treatments for these children and their otherwise fatal genetic diseases, and during the early 1980s he joined forces with W. French Anderson to explore the possibilities of restoring the missing or damaged genes to these children to treat their illnesses.

Anderson and Blaese made for an odd couple, even before Rosenberg entered the picture. Anderson, a Harvard-educated biologist with a black belt in tae kwon do and a reputation of quirky brilliance, had staked his intellectual claim on gene therapy as early as the 1960s, when it seemed like a mirage to everyone else. Indeed, he mentioned genetic medicine in an essay when applying to Harvard for undergraduate studies in the early 1950s and enlarged upon this vision in the 1960s and 1970s, after completing his training in the NIH laboratory of Marshall Nirenberg, who would win a Nobel Prize for helping to crack the genetic code. An expert in molecular hematology, Anderson originally bet that gene therapy would be tried in a blood disorder like thalassemia, but when it looked as if immunology was the place to be, Anderson needed to seek an alliance outside his area of expertise to keep his eyes on the prize. That is what brought him to Blaese. And he recognized, correctly, that in order to win approval for so radical a form of medicine, especially after the controversy aroused by the Cline affair, scientists would have to devote considerable effort to social and ethical as well as scientific arguments. By the early 1980s Anderson had groomed himself as a kind of good shepherd who would selflessly nurture the technology until it was ready to cross the threshold into clinical reality.

"I think French saw the big picture as well or better than anyone else in the field," said one former collaborator; another former colleague at the NIH remembers him as being a "wonderfully wily character," especially adept at playing the politics of science. But what seemed like a shrewd visionary intellect to some struck others as shallow opportunism; for a senior scientist at a major institution like the NIH, Anderson seems dogged, as he almost cheerfully acknowledges (and dismisses), by doubts about his grasp of the very field he sought to champion. Indeed, Anderson has managed to claim an impressive number of excellent scientists as detractors. One prominent Nobel Prize–winning biologist familiar with his work dismisses it without qualification as "a joke"; says another physician active in gene therapy: "If you look historically, I don't think he'd quite make it to the category of lightweight." These may be unusually ungenerous characterizations, but they became part of a larger problem. "French has had a history with a lot of people in the field," Blaese would say later, "and for me, the hardest part for a lot of this was that I didn't know the history, and I *still* don't know the history. But I found out that when I would go out to make presentations, I'd be met with incredible hostility." The reason, in part, is that Anderson seemed more at home with the public relations and politicking of gene therapy than with the science, and his confrontational gamesmanship often riled his fellow scientists.

This might all be reasonably dismissed as distracting and forgettable scientific gossip, except for one important regulatory ramification: many of Anderson's peers, and his severest scientific detractors, formed part of the overnight apparatus whose approval he would ultimately have to win. Two impeccably credentialed critics—Richard Mulligan of the Whitehead Institute for Biomedical Research in Massachusetts and William Kelley, then of the University of Michigan and now at the University of Pennsylvania—served as scientific members of a powerful NIH subcommittee that represented one of the most important hurdles to regulatory approval of gene therapy. Remarkably, even Anderson's partner in the venture seemed to acknowledge doubts about his grasp of the science of gene therapy. "I think there's a feeling that French had not made any scientific contributions," Blaese said. "You know, he sort of had organized and had taken advantage of things, but hadn't really made any scientific advances, or contributions, to advance the field. They sort of viewed him as an opportunist. . . . I mean, he certainly played the role of someone who had this vision and whatever it was going to take to get it done. But I don't think he understood the scientific issues particularly well. I think he still doesn't particularly understand much about the lymphocyte story and what was done and actually even why it may have been done."

Where Anderson was slender and fidgety, full of gesticulatory charm, a bit of a scientific loner, calculating down to the sound bites he would privately rehearse for every big interview and the strategic "spontaneous" tantrums he planned in advance to throw at public meetings, Blaese was large, soft-spoken, maddeningly cautious, a big bear with a gentle manner whose science was as respected by colleagues in the field of immunology as was his disinclination to attract attention to it. He describes himself with the prose of beige paint: "a relatively easygoing guy," a person that "most people don't have problems with."

Most colleagues would agree with Blaese's self-assessment. "I think when the history is written," said Nelson Wivel, who as former executive director of the RAC had extensive dealings with the Troika, "French will be given due credit for staying in there in the trenches and fighting for this until it became a reality. But in terms of the science and the scientists, I think you could poll a hundred people, and among those three, Mike Blaese is clearly the most respected one, no question about it. . . . And on top of it, he's just a genuinely nice guy, the kind of guy you'd like to have as your pediatrician." There is no doubt that Anderson saw himself as the conceptual godfather of gene therapy, but if one understands both the scientific strategy and the undercurrent of distrust among the researchers on this historic project, Blaese arguably emerges as the key figure in shepherding gene therapy over the threshold to reality, in part because he proposed and reduced to practice the scientific strategy that ultimately won regulatory acceptance (although not without a lot of help) and in part because he seemed to be the psychosocial glue that held together a highly friable amalgam of egos. New technologies move ahead or career off track on the basis of just such behind-the-scenes sociologies. Powerful groups in Seattle, San Diego, and Boston were also advancing on the same prize. Anderson, Blaese, and Rosenberg were certainly not the only biologists pushing to be the first gene therapists.

"ADA was the disease that French and I had selected to go after, and actually there were a number of breakthroughs that let us get going," Blaese explained. During the winter of 1984–85, Blaese asked colleagues treating patients with ADA to send blood samples to the NIH for experimental studies before the children received bone marrow transplants. In laboratory experiments, Blaese's postdoc Donald Kohn managed to slip a virus with the gene that corrected the defect into the T cells of these children, transforming them and allowing them to grow normally. "And that for the first time," Blaese recalled, "gave us a target for really looking at gene therapy that was relevant to the disease."

The principle of gene therapy is straightforward, though the mechanics are devilishly tricky, and not without theoretical risk. Molecular biologists—Richard Mulligan was the first to do this as a postdoc at Stanford—took advantage of a tiny pathogen known as a retrovirus, which has the ability not only to infect human cells but also to insinuate its slender volume of genetic information into the cell's own library of DNA. Through genetic engineering, one could gouge out the virus's genes for replication and, using it like a smuggler's bible, insert a desired gene that the virus could ferry into the cell. By the summer of 1985 Donald Kohn and Phillip Kantoff, a postdoc from Anderson's lab, had a retrovirus—known generically as a "vector"—and demonstrated the approach in principle. They popped the ADA gene like a cassette into the reengineered virus, mixed the virus with T cells from children whose ADA gene had been damaged or missing, and watched with tremendous excitement as, just as planned, the virus smuggled an RNA version of the gene into the T cells, copied it into DNA, and inserted it into the cell's DNA. More important, these cells could read the newly inserted DNA and "express," or manufacture, the missing protein; when the gene turned on, the protein—as it was supposed to do—cleared the toxins from the T cells, restoring them to healthy immune competence. All this had occurred, of course, in the friendly confines of laboratory glassware. Could they pull off the same trick in animals? If so, they would be one step closer to making a very persuasive argument for attempting gene therapy in human beings—an attempt, they were acutely aware, that would have to be done in young children because of the nature of the disease. Human experimentation is parlous enough; no one wanted to argue for child experimentation in public without compelling data from animal experiments.

Then Blaese and Anderson got a little greedy. They weren't satisfied with sneaking the gene into any old cell; Anderson in particular wanted to hit a home run the very first time out by inserting the ADA gene into the grandmother cell of the entire blood system, a scarce and elusive archival cell that resided in the bone marrow and was known, almost mythically, as the "stem cell." As precious as a jewel, and placed for safekeeping in as close as the human body has to a safe-deposit box, deep in the marrow and garrisoned by bone, this was the cell from which all other cells in the blood derived—not just the lymphoid lineage with its T cells and B cells and natural killer cells, but also the myeloid lineage with its macrophages and eosinophils and platelets. The entire repertoire of blood cells derived from this one queen bee of a cell. When the body needs to replenish its blood cells, as it does on a regular basis, the stem cells receive a chemical signal that activates them and initiates the process of blood cell formation. Thus, if one performed gene therapy on lymphocytes, the blood cells circulating through the arteries and veins, the effect would be helpful but short-lived, lasting only as long as these cells, many of which perish after about 90 to 180 days of service whipping through the bloodstream, each cell completing a lap every 14 seconds or so. If one managed to slip a gene into a few stem cells, however, every descendant cell, every freshly made cell of blood would now contain the full, and hopefully functional, ADA gene.

There was a problem with this approach, however. Nobody knew how to isolate a stem cell in 1984: and, as it turns out, nobody realized that retroviruses would in any event be uniquely ill suited to slipping a gene into stem cells, for what would turn out to be a very simple and compelling biological reason—these viruses insert their DNA into cells only as cells divide, and stem cells hardly ever do. They couldn't get the approach to work in mice, so they moved on to primates. Joining forces with Richard O'Reilly's group at Memorial Sloan-Kettering in New York, Anderson and Blaese by the end of the summer of 1985 attempted to work out the kinks in the procedure in monkeys. They removed the bone marrow (where the stem cell was believed to lurk) from rhesus monkeys, mixed in the virus containing the ADA gene, transplanted the marrow back in, and looked to see signs of a successful smuggling operation. "The bottom line of the whole study," Blaese said, "was that we could get human ADA produced in monkey lymphocytes—that was the good news. The bad news was that we didn't get very much, and it didn't last very long. It would last for a few weeks and disappear." For two years, they kept swinging for the fences. They kept trying to enhance the possibility of getting into stem cells, and they kept banging their heads against the cell wall. Getting the ADA gene into a stem cell, Jeff Lyon and Peter Gorner have written, was "like trying to send a postcard to the Flying Dutchman."

By the spring of 1987 Blaese felt intensely frustrated by the lack of progress. Then one day, reviewing the literature in the field while writing a book chapter at home, he realized that at least one way around the problem would be simply to scrap the stem cells and go back to his "roots," as he later put it. Go back to the T cells. Slip the ADA gene into circulating T cells. They had already done that in the lab, just to prove the principle. Why not for practice, too? Why not for treatment?

About this time, on July 1, 1987, a congenitally cheerful young buck named Kenneth Culver arrived in Blaese's tiny sixth-floor lab. Trained as a pediatric immunologist at the University of California in San Francisco, immensely confident and, beneath the gregarious warmth of his native Iowa, an aggressive and not publicity-shy fellow, Culver lost a little of his bounce when Blaese immediately put him to work inserting genes into T cells. There was nothing intrinsically difficult about the task. Culver removed a sample of his own blood, separated the T cells by what had become a standard and straightforward technique, then kept them happy in labware long enough to infect them with a retrovirus; this vector carried a marker gene into the cells it infected, and the T cells performed just fine with this new gene aboard. Culver quickly showed that T cells from mice could similarly be transformed. Indeed, it was all so quick and easy that Culver refused to believe it could be significant. In the vocabulary of scientific greatness, "quick," "easy," and "important" usually don't go together, and Culver indignantly complained to his boss that the work wasn't sufficiently challenging.

"I remember I went to Mike one day," Culver would say later, "and I said, 'Mike, you know, I'm kind of frustrated because it seems to me like anybody could do this.' And that it wasn't all that novel, I guess, or that dramatic. And Mike basically said, you know, 'You'll see. Just shut up and go back and do it.' That wasn't the wording, but that's how I remember that. And that was enough just to keep me energized and enthusiastic about doing the work. And little did I know—and I don't think Mike fully knew then, either—the dramatic consequences it would have." They both got a strong whiff, however, several months later during the weekly joint meeting of the Blaese and Anderson laboratories. By then Culver had repeated the experiments and assembled enough data to make a convincing argument that a foreign gene like ADA could be inserted into a human T cell with rather high efficiency, and that the inserted genes performed well. "I presented it at the meeting," Culver recalled, "and one of French's Ph.D.s, named Martin Eglitis, threw himself back in his chair when he saw the data and said, 'If this is true, this will change *everything!*' " Culver, who occasionally inflects his speech with the equivalent of Shakespearean stage directions, added, *[Alarum]* "Little did any of us realize how prophetic a statement that would be."

Even as the idea of inserting a gene into T cells looked more feasible and attractive to Blaese and Culver, French Anderson dug in his heels. He'd been indoctrinated in the high church of stem cells; anything else was cheap, vulgar, a blasphemy. "I'd been talking to French, and sort of telling him about this data that Ken and I had been developing," Blaese said. "And French was singularly uninterested. Because he's a molecular hematologist, and he comes from the perspective of hematology, his holy grail had always been stem cells." But it was not just that Anderson wasn't interested in T cells. While Blaese and Culver were in the lab doing what scientists are presumed to do, Anderson was doing what scientists spend increasing amounts of their time doing: laying the bureaucratic groundwork, working the committees that in this case would ultimately have to approve any human experiment in gene therapy, and perhaps, as one observer has suggested, "marking the bushes" to establish that his group had been there first in seeking approval for what clearly would be a historic experiment in human medicine. He took it upon himself to prepare for what he later called the "early rock 'em, sock 'em meetings" with regulators.

To gain approval for human gene therapy, Anderson and Blaese knew they would have to surmount regulatory hurdles the likes of which were probably unique to postwar American medicine—"the most formidable bureaucracy ever assembled to ponder a clinical trial," according to one account. Enemies galore lined every gauntlet they needed to traverse. They had to submit their detailed protocol for the approval of NIH investigative review boards and then, in succession, the human gene therapy subcommittee of the NIH's Recombinant DNA Advisory Committee (known informally as the RAC), which was stacked not only with prickly molecular biologists, but—as Anderson never tired of pointing out—with competitors; then the full RAC committee; and finally the FDA, to say nothing of winning over a dubious jury in the court of public opinion. Reluctant to abandon the stem cell strategy, even though they couldn't get it to work in monkeys, Anderson decided to submit a 500-page "preclinical data document" that became affectionately known as the "Phonebook" in April 1987. Enamored of political strategizing, Anderson admitted later that he wanted to smoke out the opposition by tipping all his cards in advance and inviting the gene therapy subcommittee and outside reviewers to enumerate all their objections.

"French had decided that he wanted to put in this ADA protocol," Blaese recalled. "And the background, at least as it was presented to me, was that this was sort of a dry run, not a totally serious protocol, to give us experience and give the RAC something to play with." Virtually every prominent researcher in the field got a chance to review the document, and it bombed in Boston, Seattle, and just about everywhere else because the science of gene transfer into stem cells still seemed too iffy, too premature, potentially even dangerous or unsafe. Indeed, by the end of the year, when Anderson polled members of his own lab on whether they would allow the dry-run protocol to be performed on their own children, only a third said yes. As even Anderson admitted, "We had to provide convincing objective evidence that our protocol would help and not harm. That we could not do." Convinced by Anderson that an upcoming December 7, 1987, gene therapy subcommittee meeting would be an informal discussion of no consequence, Blaese showed he was a political naïf when it came to committee work: he didn't bother to attend the meeting. He went goose hunting one day a year: December 7 happened to be the day.

Unbeknownst to Blaese, the December 7 meeting at the NIH very quickly turned into French Anderson's own little Pearl Harbor. As many subcommittee members pointed out, an acceptable ADA protocol would have to establish safety, technical efficacy in terms of gene transfer, and therapeutic efficacy, and the stem cell strategy under consideration had yet to satisfy any of those criteria. If anything, the requirements grew stiffer; William Kelley went so far as to propose an ironclad rule that no protocol be approved for human use until an animal model for the procedure had been established. There was also in the air at the meeting a subtext of contempt for Anderson's scientific arguments, especially from Richard Mulligan, who may have stepped over the line of propriety by grilling Anderson with a series of questions about very basic procedures in molecular biology, according to others at the meeting; Anderson patiently answered the questions, although aware that his very scientific competence was under attack. Despite all the questions, he kept angling for an opening, floating the notion that a child might qualify for gene therapy as a "last hope." Unfortunately, even this was demonstrably not true in the case of ADA children, and the subcommittee knew it; a new experimental drug called PEG-ADA, hardly perfect but nonetheless promising, had just been approved for these children, and there was no evidence that it would stop working. One subcommittee member warned of the "danger of exploitation" associated with last-hope cases.

In retrospect, it's probably just as well that Blaese didn't participate. The subcommittee's clear discomfort with Anderson's last-hope gambit would have been even more acute and quizzical if Blaese had been present; Anderson's chief collaborator would have felt compelled, he says now, to dissociate himself publicly from a last-hope experiment. "I would not have done that protocol in patients," Blaese stated recently. "Since I was the clinician, I didn't think it was worthwhile worrying French about that. But I just wasn't going to do it. I didn't think that there was any chance that it could work."

Since Anderson had not formally submitted a protocol, no formal vote was taken, but there was no mistaking the subcommittee's mood: the technology was not nearly sophisticated enough, and too many scientific questions remained unanswered, for stem cell gene therapy to proceed. Indeed, the general sense following the meeting was that Anderson's persistence, his attempt to smoke out the opposition, had only stirred up the bees more, and that prospects for the entire field of gene therapy had suffered a setback. As even Anderson admitted, "We weren't dead in the water, but we were *close* to dead in the water"—this from the man who once said, "I have always had the habit of seeing success where others see failure." Some observers even suggest now that if the NIH group couldn't get a protocol approved, nobody else would, either. "Gene therapy research is so inordinately expensive," said Nelson Wivel, "that it is no accident that its genesis was at the NIH in the Clinical Center. That's one of the few places, if not the only place, on earth where there was sufficient support to carry out those kinds of clinical trials."

That's where matters stood at the end of 1987. By digging a political hole for themselves, Anderson and Blaese had managed to make the scientific mountains even taller; indeed, there were now two huge peaks to surmount. First, as even Anderson admitted, the "science" of gene transfer into stem cells wasn't reproducible and wasn't predictable; stated bluntly, it didn't rate as science at all. Second, no government committee, with an anxious and concerned public peering over its shoulder, could be expected to approve such a scientifically borderline therapy for young children unable to give informed consent, even on a compassionate, last-ditch basis. To this day, Anderson insists that all the opposition was "cultural," not scientific—that society, and its proxies on the NIH subcommittee, was simply not psychologically ready to make the leap. "We already knew that we needed to do something to break the emotional logjam," he explained later, "because it really was a cultural problem." And this interpretation may explain why Anderson has managed to provoke so many scientific detractors: his group couldn't find stem cells and couldn't get inserted genes to work reliably in mice or monkeys, and when a committee entrusted with public oversight frowned upon letting him inflict demonstratively unpredictable techniques on young children, he viewed the problem as "cultural."

One can imagine Blaese coming home from his lone day of goose hunting in 1987 to find his clinical dreams full of bureaucratic buckshot. Now more than ever, Blaese believed, they had to forget about stem cells. He again tried to win over Anderson with the idea of inserting a gene into T cells rather than stem cells. "I started working on him," Blaese remembered. "I said, 'You know, we could do this. We know we can do it now in an animal, and get long-term survival. Let's forget trying to hit the home run. Let's just let the camel get its nose in the tent, okay? Let's do a safety study, not even asking a gene to do anything. Let's just use it as a label. And I know somebody in the Cancer Institute who's treating patients with T cells. Maybe we can approach him about combining efforts?'"

That "somebody," of course, was Steve Rosenberg.

∞

In the late 1980s, few phrases could be expected to win a cancer immunologist's ear as quickly as talk of T cells, no more so than at the Surgery Branch of the NCI. Since December 1986 Rosenberg's group had been removing T cells from patients, stimulating them with interleukin-2 until they had copied themselves billions of times, and then giving them back as medicine, claiming an objective response rate greater than 50 percent. Blaese kept trying to bring the idea of gene therapy to Rosenberg's attention, but Rosenberg was always too busy. "I could never get in to see him," Blaese remembered, "and Mike Lotze and I had been friends for a long time, and Mike was sort of second in command in the Surgery Branch, so I went and talked to Mike. This was probably in November or December of 1987. *Before,* actually, the RAC meeting. I sort of ran it by Mike, and he was quite enthusiastic."

"Mike Blaese and I sat down," Lotze confirmed, "and started chatting about the possibility of using TIL cells as vehicles to deliver genes. Actually, we set up everything, and talked through what the issues would be. And he gave me a bunch of stuff to read, and I read it, and then we went and talked to Steve about it as a possibility." There was a fatal flaw to Lotze's enthusiasm; it was his belief that Rosenberg would view him as a coequal in developing the technology, just as Anderson had originally involved Blaese as a partner. "Mike, very simply, believed that he was at the same level that Steve is, and could run his own show," said one lab member. "And it got to the point where they outgrew each other. Once you reach that point here, there's only one conductor. Period. And no backup."

Lotze made it to one preliminary meeting in March 1988 where Blaese and Culver described their results to Rosenberg. "At that first meeting where we met with Steve," Culver said later, "it didn't quite all sink in. You know, that here Mike Blaese and I were presenting this data that we could genetically engineer T cells, and it was then at the second meeting where I presented more of my in vivo data, which showed that indeed these lymphocytes would survive for *months* in animals. And I remember it's as if the lightbulbs went off in Steve's mind, because now that meant that they could probably survive long enough to give some meaningful information about the biology of TIL cells." That first meeting was noteworthy not so much for Rosenberg's perceived disinterest as for a memorable interruption. "As we sat there," Culver remembered, "Mike Lotze got paged not too long after the start of the meeting." Lotze answered the page, and that was the last anyone saw of him on the planning team for gene therapy. The history according to Rosenberg's book is somewhat simpler. "About that time," Rosenberg writes, "Michael Blaese and French Anderson approached me about using another avenue to gene therapy."

It is an omission that, to this day, seems to rankle all the other collaborators. "What Steve had left out of that description," French Anderson said in an interview, "is really the key person who got us together, and that's Mike Lotze. It was Lotze and Blaese who kept pushing and pushing until Blaese finally convinced me that it *was* reasonable to at least sit down and talk with Steve. But Steve did not have Lotze there. . . . There were four critical people, and unfortunately Steve left out one of them." Culver: "Mike Lotze and I commonly get left out of that story, and we were there." Blaese: "I always tried to include Mike because he was the person I went to first; he's a person that I've known for many years. And Steve just didn't seem to want to have him involved." (Asked about this later, Rosenberg insisted that Lotze "certainly wasn't one of the key people" and that he was "just peripherally involved in the gene therapy work.")

"The reason I left the NIH, in part, was because of my discomfort with the gene therapy story," Lotze said later. "It became quite clear to me that I could not continue on there and be a major player, that Steve's feeling was that there should only be a senior member from each group that was nominally identified with the effort. And that excluded me. And in fact, if the truth be known, he said, 'I would just as soon do it all on our own, and not have these other guys involved.' And I think that was pretty clear to Mike Blaese and French."

So began the historic collaboration.

On March 17, 1988, Rosenberg, Anderson, Blaese, and Culver sat down in Rosenberg's office to plot in detail their joint approach to gene therapy. "Rosenberg and I hit it off the first two seconds," Anderson gushed. "In a little over an hour," Rosenberg wrote, "we plotted a detailed path we would follow virtually without deviation for the next three years." Blaese: "Everybody agreed that it was sort of logical and something we could probably pull off. We had the technology to do it. And we made the commitment to each other at that point to go with this." Anderson would later call it "an absolutely seminal meeting." Since 1985 Anderson and Blaese had held joint lab meetings; after

March 1988 there were meetings every Monday afternoon in the Surgery Branch of the Anderson, Blaese, and Rosenberg labs.

The plan they roughed out that March afternoon addressed political and social criteria as much as scientific concerns; indeed, as Anderson explained, "The *whole reason* for hooking up with Rosenberg was social, cultural; it wasn't the science." Blaese again pushed the idea that they had to table the dream of using stem cells and instead insert a gene into T cells; Rosenberg had experience handling T cells, in the form of TIL, to treat cancer patients, and he explained what his lab could do. But perhaps the crucial stratagem was that they wouldn't even propose using a *therapeutic* gene at all in cancer patients. (That would almost certainly bite off more than any self-respecting oversight committee would care to chew.) Rather, Blaese argued that the simplest, cleanest demonstration of the new technology, and the easiest way over the cultural hump, would be in essence a scientific show-and-tell project to establish that the technology could work as an idea, not as medicine. They would propose to insert a harmless marker gene known as *neo* into TIL cells; this gene allowed bacteria to neutralize an antibiotic known as neomycin, but had been purloined by molecular biologists as a marker to allow researchers to follow cells in experiments. By giving these genetically altered TIL cells to cancer patients, NIH doctors could more easily establish if the cells were indeed homing in on tumors. Following infusion, they could biopsy a tumor, analyze the T cells attacking it, and test them to see if they contained the *neo* gene.

This so-called marker study would not benefit patients very much, if at all, over existing therapy, but it would benefit the field of gene transfer a great deal. It would help the NIH team establish that gene therapy was reasonable and safe, and once the general idea had been nudged across that important psychological threshold, they could move on to the main attractions. Anderson and Blaese could seek approval for what they really wanted to do, which was to put the ADA gene into T cells and treat young children with it, and Rosenberg could seek approval to do what he really wanted to do, which was to insert the gene for tumor necrosis factor into his TIL cells and then let these cells carry that powerful anticancer molecule directly into tumors. As Blaese admits, using cancer patients rather than children for the initial study was a "real part of the strategy"; it was what had made him think of Rosenberg's group in the first place.

Rosenberg may not have appreciated how lucky he was to marry into a family of gene therapists with such extensive hands-on experience with the technology. But the beauty of collaborating with Rosenberg was not lost on Anderson and Blaese; they knew they now had access to the perfect patient population for winning regulatory approval. "You're talking about a terminal cancer patient who can give an informed consent!" Anderson pointed out. Not that Rosenberg didn't bring something special to the group. "Steve had the persona and the ability to overcome obstacles that would have been impossible, I think, for even people as talented as French and Mike Blaese collectively to do," Lotze believed. "Steve really provided the *oomph* that made it possible." But he did not yet possess much scientific know-how about gene therapy; that came primarily from the Blaese lab. "Mike was the man with the ideas behind the scenes for the science," Culver says, "and I don't know that there are *any* ideas fundamentally that were used in that TIL trial that didn't come out of Mike's lab, either stuff that he thought of or stuff that I came up with through actually doing it."

"We knew what we wanted to do," Blaese said. "We knew there was a lot of preliminary stuff to get out of the way first, but the basic concept about *what* we wanted to do [was clear]—that is, to put a gene in, use it as a marker, which is sort of what my proposal had been in the beginning, and to forget the idea of doing something therapeutic. Let's just do a marker study to address safety issues. This was a patient population that was going to be dead in three months. It was going to be, from my perspective, a Phase I trial in cancer patients. It was done all the time, and this was the *only* way that we were going to get permission to get started. Particularly after the response of the RAC to that trial proposal, which was so negative."

Not all the obstacles were regulatory. While Anderson and Blaese were delighted to have Rosenberg's patients, they were rather more wary about getting Rosenberg himself. They worried about Rosenberg's well-known penchant for wearing his impatience on his sleeve; Rosenberg viewed the regulatory network of RAC committees and review boards and FDA, Anderson recalled, as "nuisances at best and obstructionists at worst." They worried that his appeals to emotion might strike the committees in charge of approval as self-righteous and self-serving. Finally, they worried about his forceful personality, what one collaborator termed "his need to control." Publicly Anderson professed immense admiration for Rosenberg. "I think he's a *superb* physician. I think he's a superb human being. Now, he has an ego. He's a surgeon and, you know, he has an enormous ego. And he runs his own shop. I have been told by others I am the *only* person that he carried off a long, successful collaboration with. Because other people just get eaten up by him. But of course other people weren't as obsessed as I am. I mean, I'm just as obsessed in what I do as Steve is in what he does, and so we understood each other—understood and *understand* each other—completely." That was for public consumption.

"I used to do psychotherapy on French all the time," Blaese recalled with a laugh. "Because I was sort of between them. Surgeons very much *have to have* control, and Steve is very much that way. And French was going crazy, because he always thought Steve was trying to take over, was trying to sabotage him. And then I'd go meet with Steve, and Steve was upset because French was upset, and it was one of those things where I was trying to keep the thing together. I think there was a lot of suspicion." Anderson states with great pride that he went on rounds with Rosenberg's team "every morning for three and a half years"; part of the reason, less eagerly advertised, is that he was afraid Rosenberg might ease him out of the project. After staking his reputation, his scientific prestige, his entire career since the late 1960s on the prospect of gene therapy, now, when the moment seemed finally within grasp, French Anderson wasn't about to let his baby out of his sight, not even for one day.

Differences in scientific style also created some tension, although perhaps of a more creative sort. In a collaboration that would ultimately number dozens of scientists, keeping everyone moving toward the same goal had aspects of a cattle drive at times, and Rosenberg did not hesitate to crack an arbitrary whip to keep things moving along. As even he admits in his book, these meetings would occasionally become rancorous and "hot." To the chagrin of some senior scientists, Rosenberg had the habit of declaring success on a particular scientific problem when others felt the experimental data remained ambiguous. "That's how he ran the meetings," Ken Culver recalled. "In fact, his words were, 'I declare success' or 'Let's declare success and move on.' And that really didn't sit well with everybody in the group. Sometimes it was the Ph.D.s, because they felt from their laboratory training that it deserved more of a basic science evaluation. And sometimes it was just from the group as a whole saying, 'Wait a minute now, we're putting the cart in front of the horse.' "

For Rosenberg, the overriding concern—as always—was to test a promising idea in patients as soon as possible, and he could run a pretty good race even with the horse behind the cart. In June 1988, a mere three months after the "seminal" planning meeting and even before the labs had learned how to insert the *neo* gene consistently into TIL cells, Rosenberg went to the in-house Institutional Biosafety Committee at the NIH seeking approval for the historic gene transfer experiment. They were getting ready to walk onstage, and by the time the Troika went before the regulators—the gene therapy subcommittee in July 1988, the full RAC the following October—the whole world would be watching.

Once the group entered the public arena, French Anderson was in his element. He prided himself on the ability "to *feel* the flow" of meetings. "A harsh word for it would be to manipulate it," he conceded. "My critics say I'm a manipulator, but I mean, it's committee work at its finest. Some of my emotional outbreaks are planned *weeks* in advance." Rosenberg was less calculating, more prone to defaulting directly to moral indignation. "I think one of the reasons why Steve is difficult with groups," Anderson would say in his defense, "is that he's had so many people blast him, and every time that there is a real investigation, the conclusion is—and *has* to be—absolutely superb data management, *every* conclusion absolutely justified. And so Steve almost has a righteous indignation. He *knows* everything he does is right. People keep badgering him. They've badgered him for years, and he just can't be bothered with it anymore."

There were some real badgers on the gene therapy subcommittee of the RAC, and the Troika's first audition in public was rocky at its July 29, 1988, meeting. The TIL strategy clearly threw the NIH subcommittee off balance, however. As the minutes of their meeting later noted, "The Subcommittee wrestled with the issue of whether the protocol was, in reality, human gene therapy and whether it fell under the jurisdiction of the Recombinant DNA Advisory Committee and its Subcommittee. After deliberation, it was decided the protocol was 'very similar to human gene therapy,' and that the Subcommittee should review it."

They reviewed, but declined to approve. After lengthy discussion of the TIL-marker protocol, the subcommittee unanimously agreed to defer action until more data were provided on five specific scientific questions, including further animal testing on gene transfer efficacy and safety. Anderson felt the demands were unreasonable—and so, up to a point, does one of the scientific experts on the subcommittee. "That's a pretty stiff list," admitted Scott McIvor of the University of

Minnesota eight years later, "except for the safety issue. The protocol submitted in July, and the data, was not there—and mind you, this was the very first human gene therapy protocol." The demand to provide animal data appeared to be a reasonable request, but Anderson and Blaese understood it to be a deliberate obstruction, because an animal model of TIL transfer would be impossible.

Anderson attributed the subcommittee decision not to vote, then and now, to sour grapes, and at least some observers agree that Kelley's insistence on an animal model was, in the words of one observer, "disingenuous and intended much more to block a protocol." But the opposition put the TIL protocol in a desperate bind. The subcommittee demanded data that Anderson, Blaese, and Rosenberg insisted they could never provide. How to proceed? The Troika made a deliberate and strategic decision to "bypass" the subcommittee, as Anderson later admitted; because of the perceived intransigence of the molecular biologists on the subcommittee, Anderson, Blaese, and Rosenberg chose to answer only questions they could, ignored questions they couldn't answer, and, as it turned out, withheld yet other forms of data from the subcommittee, which held a telephone conference on September 29, 1988, and once again deferred a vote on the protocol. "It wasn't all noble and altruistic," Anderson conceded later. "There wasn't *some* maneuvering; there was a *lot* of maneuvering around the subcommittee." The fact that the three investigators suddenly presented new data when they appeared four days later before the full Recombinant DNA Advisory Committee, on October 3, 1988, seemed exactly like an end-around to subcommittee members, too. Anderson admitted as much in an interview later when he said, "We skipped the subcommittee because we would *never* get to the subcommittee. They were competitors, and everybody wanted to be first, and the only way to make sure that you're first is to make sure that nobody else gets there first."

But the stakes were high at the October meeting, too: a rebuff here, they felt, would effectively kill the notion of gene therapy for years— for themselves, and perhaps for everyone else. "We felt that logic and persuasive arguments, although vital, might not be enough," Anderson later admitted. "Our trump card, if needed, would be a carefully planned emotional appeal."

Early in the tense, day-long meeting, Rosenberg made his customary appeal. "Four hundred and eighty-five thousand Americans died of cancer last year, and one out of every six Americans now alive will die of cancer if no new treatment modalities are developed," he told the committee. "Obviously, we are in desperate need of finding new modalities to treat cancer." Later that morning, Rosenberg suggested that a failure of the committee to vote would have "deleterious effects" on cancer treatment. Each member of the Troika played his appointed role: Blaese laid out the scientific rationale for the experiment, Rosenberg reminded the committee that cancer patients needed new treatments, and Anderson sat back, monitoring the flow. He sensed the tides shifting back and forth. At one point during the midmorning coffee break, one of the reviewers wondered aloud to Anderson, "What's the rush?" This remark served as the irritating grain of sand out of which Anderson created a pearl of a tantrum. Stewing all through the break, although he had been "rehearsing for a week," Anderson rose during a pause to address the committee.

"I was asked at the break, 'What's the rush in trying to get your protocol approved?' " he ranted. "Perhaps the RAC members would like to visit Dr. Rosenberg's cancer service and ask a patient who has only a few weeks to live: 'What's the rush?' A patient dies of cancer every minute in this country. Since we began this discussion 146 minutes ago, 146 patients have died of cancer." "From the look on his face," said one observer at the meeting, "French realized he had overstepped propriety immediately."

Not long after that outburst, the late Bernard Davis, a respected Harvard Medical School microbiologist, raised his hand and asked to be recognized. "Probably *the* most emotional moment," Anderson later said of this intervention. "The reviewers were blasting us and we were pushing back, and it was going this way and that way, and I had used some of my emotional things, which people still criticize me for. And then Bernie Davis raised his hand and said, 'I'm very troubled by what I'm hearing here.' And I just *shrunk*. Because I just knew by the flow that it was at a critical juncture, and most of the undecideds were teetering right on the fence. Bernie Davis was extraordinarily well respected, and when he says he's troubled, whatever he said next was the way things were going to go. And I didn't know what he was going to say because he had kept a poker face the entire time. And the next words out of his mouth were, 'I think we are nitpicking these investigators.' And that was the end. I just . . . just . . ." Simply reliving the moment, Anderson came close to revisiting the stammer he used to have. "You know, my adrenaline level just warped. I mean, that was it."

Before Davis finished speaking his piece, he chided his colleagues about their wariness toward risk. "It is virtually not possible," he said, "to have more risk than certain death." Another fifteen minutes of debate ensued, but soon after the vote was called, and the committee voted 16–5—all the nays, significantly, were scientists—to approve the TIL protocol. The decision, historic as it was, literally made no sense, as Anderson has cleverly pointed out. As submitted by Rosenberg, the protocol required patients to have a life expectancy of more than ninety days to allow enough time to grow TIL cells; as amended by the committee, the protocol limited the experiment to ten patients who had a prognosis of less than ninety days. "So there is no patient that could qualify for the protocol!" Anderson laughed.

In public, Anderson and Blaese and Rosenberg celebrated the decision; there was a dinner at a Chinese restaurant in Bethesda, humorous skits, a temporary lull in hostilities, surely not least because these three men now knew their names would attach to that most coveted of scientific sobriquets: The First. "I remember being *so high* because we had gotten permission," Blaese remembered. "But I was a little uneasy, because I also realized that it wasn't unanimous and that the subcommittee hadn't voted. And I figured there was going to be something coming out of that."

He figured correctly. Someone on the RAC committee complained to the then director of the NIH, James Wyngaarden, about the fact that the subcommittee had never approved the protocol. Wyngaarden was bothered that all the scientists on the RAC committee had voted against the experiment, according to one observer, and also by Anderson's emotional outburst. He wanted unanimity on an issue of such broad social import, so he blocked formal NIH approval until he could poll each and every member of the committee by mail. By January 1989 he had elicited twenty favorable votes (with one abstention), and he formally granted approval on January 19, 1989. Finally, the Troika had the green light. Rosenberg, with the added shoulders of Anderson and Blaese, had knocked down another door.

And promptly bumped into several others. First, biotech gadfly Jeremy Rifkin filed suit to block the experiment, arguing that the RAC's vote to approve did not occur, as required by law, in a public meeting. A federal judge approved a settlement of that dispute four months later, but the timing worked out just as well for the researchers; once gene therapy retreated from the public stage to the laboratory bench, hurdles of a more unforgiving scientific sort quickly stopped the giddy momentum of the scientists. They almost couldn't get the technology to work. It took another four months before they could manipulate the marker gene successfully into TIL cells and then grow enough TIL cells to infuse back into a patient. After two false starts in other patients, after dealing with contamination problems and finicky cells, the historic first infusion was finally scheduled for May 22, 1989. Given all the constraints that had been placed on the experiment, and all the talk that had preceded it, the first approved use of gene therapy—or "gene transfer," as the committee put it—turned out to be exactly the kind of nonevent one might have expected.

Rosenberg arose early, as usual, checked on the cells, made sure everything was ready. Anderson, having strategically insinuated himself into every round and seen every patient, attended this culminating event, as did Blaese; significantly, Culver did not. The patient, a dignified and good-natured fifty-two-year-old Indiana man with metastatic melanoma later identified as Maurice Kuntz, received 200 billion TIL cells that contained the *neo* marker gene, and the entire experiment—like most human endeavors shaped by committees—was anticlimactic. Five minutes into the experiment, as the researchers stood around hoping nothing would go wrong, Kuntz cracked them up by saying, "Well, I haven't grown a tail yet."

From a therapeutic point of view, the first use of gene therapy proved utterly useless; the experiment did not save Mr. Kuntz (he died eleven months later)—but it did revive the once moribund technology of gene therapy as surely as a miracle drug. Intended to demonstrate that gene transfer could be performed safely in humans, the first trial did just that. Within several weeks, tests showed that the marker gene had made it back into his body. Perhaps under the glare of such intense public scrutiny and such cautious prosecution, it would be unfair to expect anything more. Not one to stint on the historic significance, however, Rosenberg later remarked that the experiment "would open a door in medicine" and that gene therapy created "an entirely new and almost unlimited way to deal with disease that could change the way medicine is practiced in the twenty-first century."

Rosenberg was absolutely right. The wobbly collaboration had maintained its balance long enough to career through that door, and once the threshold had been breached, dozens of would-be and wannabe gene therapists crashed through with plans to treat cancer, AIDS, cystic fibrosis, and a host of other disorders, genetic and otherwise. Anderson, Blaese, and Rosenberg led the pack back to the RAC committee. To all intents and public purposes, they moved forward together as a team, still the same Troika, still one happy family seeking to treat both ADA-deficient children and cancer-ridden adults. Public appearances deceived, however. The race against each other, according to most of the principals, had now become open and unpleasant.

All three scientists had agreed from the very beginning, at that "seminal" March 1988 meeting, to a long-term commitment, according to Blaese; they would jointly submit the first two genuinely therapeutic applications of the technology after the marker experiment: ADA gene therapy for children and then TIL cell gene therapy for cancer patients. But as one of the collaborators recalled, there was "a very intense competition evolving underneath all of that, who was going to get into the patient first with a therapeutic protocol."

"We were going to continue working together. That was always our view," Blaese remembered. "And the plan was, had always been, that the first gene therapy was going to be ADA. . . . So the plan was then to continue on and to expand on what we were learning from ADA and just immediately put in a TIL protocol. But suddenly Steve was writing the TIL protocol. He had done two or three experiments with tumor necrosis factor, and all of a sudden he had this protocol written. And in the meantime, we had been working with ADA for four or five years, and probably knew a lot more about it."

Blaese is being diplomatic here; they knew a *lot* more about gene therapy than Rosenberg's group, had studied the safety issues much more carefully, were handling the gene for a much less toxic molecule than TNF, and had worked out the molecular mechanics of expressing the ADA gene in T cells to a much more sophisticated degree than Rosenberg's group had with the TNF gene in TIL cells. Indeed, Ken Culver later claimed to know that the TIL strategy was probably doomed to failure—knew it even before the famous March 1988 meeting, because he had conducted experiments inserting cytokine genes like TNF and interferon into T cells and discovered that the cells quickly turned off the genes. It is unclear if Rosenberg knew this, but it's also uncertain if it would have stopped him.

"See, where it started to fall apart was when Steve had gone out and gotten somebody else to make him a retrovirus vector," Blaese said, his voice still weary with the memory. Rosenberg wasn't satisfied with the retroviral vectors being developed by a private company Anderson had helped to found, Genetic Therapy Inc. And so he decided to get his own from the Cetus Corp. in California and go it alone. "He went to Cetus to have them make a TNF vector for us. As surgeons do, he had to have control. And he felt that he needed French to get him the vectors, and he didn't like that feeling of having to be dependent on somebody else. But that was taken immediately as a sign that he was becoming a competitor rather than a collaborator." Rosenberg had taken this step in the middle of 1988 and now Anderson had to calm Blaese. "It is true that things got competitive, and that Steve and Mike were not speaking to each other," said Anderson, who claims to have been aware of the Cetus involvement all along. "But I never had problems with what Steve did."

Blaese, however, was "viscerally upset," according to others, because of the internal tensions. Rarely does a mild-mannered scientist like Blaese express as much public remorse about a collaboration gone sour as he did in a later interview. "Suddenly it seemed like now we had degenerated from a cooperative group into really competitors to get going," Blaese said. "And I never did understand that part of it. At that point, the cooperative relationship between us sort of fell apart. French rounded regularly with Steve's group while the marker study was going on, and I never really did. I just didn't want to get involved in it. There were too many things to do to get worried about the politics, and I decided to just sort of withdraw a little bit and get the ADA thing going. And if we could develop anything that was useful for the TIL, fine. But I didn't want to sort of keep tightly related to that, because it was distressing." "Eventually Mike Blaese was kind of left out," Culver recalled, "and then eventually French was left out. That was kind of how the process evolved with time. That was part of Steve then taking over and running the show." Asked about this later on, Rosenberg sounded genuinely surprised that his coworkers felt this way, and this social insensitivity may help explain why others find this often genial man to be infuriating. "It was the smoothest collaboration I've ever been involved with," he said, "and I never sensed any competition at all. We worked incredibly closely together, as close as I've ever worked with anyone." He wasn't dissembling; he simply remained oblivious to the discomfort he had caused his colleagues.

"I hadn't really been intent on being first," Culver remembered. "But when I saw that whole thing happen and saw how I'd poured lots and lots of effort into it, trying to make it work, and then, you know, it kind of became, 'We don't need you anymore . . .' That allowed me to focus with all the more determination." Culver worked at generating the kind of animal data that would persuade regulators that real gene therapy for ADA was ready for clinical testing. "And I remember coming in, in late 1989, in the fall, and telling both Mike and French that I had all the data we needed." Not quite, but certainly enough for Culver to sketch out a plan of action, a rough draft of a gene therapy trial he wrote in November 1989 that formed the basis of the protocol submitted for government approval by the RAC in 1990. The race was on.

Both the Blaese-Anderson team and the Rosenberg group submitted gene therapy proposals for consideration at the July 1990 meeting of the RAC, and both won approval. As Jeff Lyon and Pete Gorner relate in their book *Altered Fates,* Anderson and Blaese needed to fly in an Italian scientist, Claudio Bordignon, as a last-minute "collaborator" in order to include data from his animal experiments to win regulatory approval. As for Rosenberg's plan to insert the gene for tumor necrosis factor into TIL cells, Michael Lotze likes to refer to government approval of that protocol at the same RAC meeting as "the day there was a crack in the fabric of the universe," not least because preclinical data on the procedure were so sketchy and incomplete.

Although both the cancer therapy and the ADA therapy received RAC approval on the same day, Rosenberg's lab got bogged down for months with questions from the FDA, and his erstwhile collaborators privately savored the experience of leaping ahead. In September 1990, using the technology they had first attempted in the lab in 1987, Anderson, Blaese, and Ken Culver inserted the human ADA gene into the T cells of Ashanti DeSilva, a four-year-old girl from suburban Cleveland, and infused those cells back into her in the first genuine test of gene therapy. The procedure received global attention.

The first therapeutic attempt of the technique by Anderson and Blaese may also be its first success, although there is hardly scientific unanamity about this. Ashanti DeSilva regained more or less normal immune function and has been able to attend school for the first time (the therapeutic effect of the transfused cells continues to be supplemented by drug treatments, although the drug treatment has been halved). A second patient treated according to the NIH protocol, Cynthia Cutshall, has also responded well to gene therapy, although not all critics believe the results establish the success of the technique. Interestingly, because of the historic nature of the treatment and the need for public disclosure, the DeSilva and Cutshall cases were described to reporters and the real names ultimately revealed, even though the first peer-reviewed paper describing the experiment would not appear in the literature until 1995.

Rosenberg did not get his gene therapy program with TIL cells underway until January 1991, four months after Blaese, Culver, and Anderson. Since then, he has treated a total of seventeen patients with gene therapy, with less impressive results. One patient in the first TIL-marker study had a spectacular response that has not recurred; but although the Rosenberg lab has published its gene therapy work in the best journals, the protocols have not stirred much excitement in the medical community. "Unfortunately, he picked the wrong molecule, the wrong cell, and the wrong delivery system," said one scientist familiar with the work. Rosenberg said recently that the gene therapy protocols had been "superseded" by more promising work in tumor vaccines.

If, as French Anderson has maintained, all the opposition to gene therapy was cultural, it in part reflected the age-old clash of laboratory science, with its insistence on unambiguous results, and clinical experimentation, which is more the art of doing some good while doing no harm in a setting of desperate medical need. But in winning approval with emotional appeals and political stratagems, the NIH researchers may have helped to raise unrealistic expectations for gene therapy, and to that is attached a social price.

Long after the front-page stories had ceased to appear, the collaboration of NIH scientists continued to be a model of dysfunction. Anderson and Rosenberg had a major blowup over the order of authorship for the first paper describing the initial gene transfer experiment, according to Culver; Rosenberg prevailed, and his name appears first on the paper. In April 1991, meanwhile, Mike Blaese inadvertently saw a copy of a letter from Claude Lenfant, director of the National Heart, Lung and Blood Institute, nominating French Anderson—and Anderson alone—for an Albert P. Lasker Award for his work in gene therapy. The letter, as reported by Lyon and Gorner, "amounted to a major revision of the facts, and a clear injustice." Blaese was furious, and the nomination amended, but it was one more reminder that when so much prestige, grant money, and reputation attaches to pioneering work, a disproportionate amount of energy tends to get expended on issues decidedly less lofty than science.

In March 1995, Genetic Therapy Inc., the company founded by French Anderson in 1986, received a broad U.S. patent covering any gene therapy procedure in which cells are removed from the body, genetically manipulated, and then reinfused. However, only the names of Anderson, Blaese, and Rosenberg appear on the patent; omitted were Ken Culver, who claims to have performed many of the experiments that reduced the idea to practice, and A. Dusty Miller of the University of Washington, who provided the retroviral vector eventually adopted for use in the first human experiments. Almost immediately, in a fracas delightedly followed by the journal *Science* and others in the press, former collaborators and outside observers immediately challenged the legal, not to mention scientific and ethical, integrity of the patent. The most galling fact, according to many, was that Anderson's own gene therapy company, Genetic Therapy Inc., had to rely on technology developed by Miller, who sits on the board of a rival company, Targeted Genetics of Seattle, in order to get a retroviral vector that would work in their pioneering experiments.

Nonetheless, a number of researchers have rushed through the door opened by Rosenberg and his NCI colleagues, testing several different gene therapy approaches to cancer. The team of Gary and Elizabeth Nabel at the University of Michigan, for example, has come up with an approach where they inject "naked DNA"—a gene, essentially, disrobed of its swaddling viral cloth—directly into the tumors of melanoma patients; the gene instructs cancer cells to make a surface marker called HLA B-7, designed to instigate immune assault upon the tumors, and in initial safety testing in five patients, the Michigan researchers reported one encouraging response.

In collaboration with Richard Mulligan's group at MIT, Drew Pardoll and his colleagues at Johns Hopkins University have taken a different approach. In a rare randomized and double-blinded Phase I safety study in metastatic kidney cancer, they remove cancer cells from patients, insert the gene for an immunologic cytokine (granulocyte macrophage–colony stimulating factor, or GM-CSF) into the tumor, and return these cancer cells into patients. The irradiated cells are no longer cancerous but are functional enough to manufacture GM-CSF, which chemically recruits antigen presenting cells and ultimately helper T cells to the site of the tumor. In animals, the strategy results in substantial tumor regressions, and the Hopkins group has seen one partial response in eight patients treated with the vaccine in a preliminary trial. In the treatment of brain tumors, Ken Culver, Edward Oldfield, and their colleagues at the NIH have pursued a strategy that creates a little suicide bomb, in the form of a viral gene called thymidine kinase (abbreviated TK) that can be smuggled via a retrovirus into mouse cells, which are then inserted into the brain; later treatment with ganciclovir, an antiviral drug that acts specifically on thymidine kinase, in effect triggers suicide in all the cells that contain the TK gene. And Michael Lotze, thwarted in his attempt to participate in the NCI gene therapy trials, has launched several gene therapy trials at the University of Pittsburgh. His group has seen preliminary responses in several cancers after inserting the gene for another cytokine, interleukin-12, into cells known as fibroblasts and then reinjecting these cells just below the skin. In all, some 160 trials have been approved for test in patients since the initial TIL marker study.

Like virtually every previous "new" approach to immunotherapy, however, gene therapy has already ridden its obligatory wave of hype and is now entering its first backlash period. There is plenty of blame to go around, according to researcher Theodore Friedmann of the University of California–San Diego. In a 1996 editorial in *Nature Medicine*, Friedmann wrote that the gene therapy genie is out of the bottle, but is still so young and unformed that any steps it takes are bound to be staggering. "As is true in all rapidly moving, competitive and important areas of research," Friedmann wrote, "the enormous potential payoffs for definitive therapies combined with forces such as competition for scientific eminence and diminishing federal support for biomedical research have inadvertently opened the doors a bit to touches of exaggeration, scientific self-delusion and possibly even self-promotion. Some gene therapy papers, manuscripts and funding applications have in the past sent, and at times still send, overstated messages of successful *therapy*, even in the obvious absence of convincing evidence for clinical benefit."

The rigor of the science has also come under fire. Committees like the RAC, for example, had a mandate only to assess if a proposed clinical experiment was safe, not to rank it against other similar proposals in terms of scientific merit, as is done in the NIH grant-awarding procedures. "I think there was a very real cognizance that there were shortcomings in the science," said Nelson Wivel, who served as executive director of the RAC when many gene therapy proposals were considered, "but it wasn't the purview of that committee to reject it solely on the basis of the science. . . . As we used to say, the committee voted with one hand holding their nose while the other hand was up in the air." A more formally chastising assessment came from a special committee chartered by NIH director Harold Varmus to assess the NIH's investment in gene therapy research. The so-called Orkin-Motulsky report, issued in December 1995, criticized the field of gene therapy for "overselling" the science and creating the impression that "gene therapy is further developed and more successful than it actually is."

Eventually that backlash even caught up with Steve Rosenberg. His gene therapy program based on TIL cells, according to erstwhile collaborator Anderson, "stalled right on the tracks," and ultimately became the focus of a highly publicized funding dispute at the National Cancer Institute. Indeed, all the doubts, expectations, jealousies, and concerns aroused by the Surgery Branch work since the mid-1980s came to a head in the fall of 1992, when a committee of expert immunologists serving on an NIH oversight committee for in-house researchers did something no one had managed to do since Bob Gallo handed Steve Rosenberg the IL-2 ball back in 1977 and watched him run with it. They tackled him and stopped his forward progress. Not surprisingly, it took gang-tackling by an entire committee to bring him down.

The problem, ironically, grew indirectly out of the ruptured collaboration on gene therapy. For two years Rosenberg's group at the Surgery Branch had struggled to get the technology to work in animal experiments. Part of the problem was bad luck: gene therapy experiments in humans require extensive pretesting in animals, and no one—not Rosenberg or anyone else—has had much success inserting the gene for tumor necrosis factor into mouse TIL cells. Gearing up for a large-scale gene therapy effort in humans, Rosenberg nonetheless requested approval for a $3.9 million contract with an outside lab to provide raw materials for his protocols ("That's a lot of dough!" said one extramural scientist, putting the amount in perspective. "That'd keep a lab going for years!"). The request, however, required approval by the board of scientific counselors to the NCI's Division of Cancer Treatment, an expert panel of outside cancer researchers that functions as a peer review committee, assessing the work of NCI researchers and advising the government institute about approving such fund requests. In October 1992, the panel stunned everyone in the biomedical community by refusing to approve Rosenberg's request. The surprisingly exasperated tone of comments by panel members, which soon came to public attention, made it clear that this represented an unusually candid referendum on Steve Rosenberg's science by his most knowledgeable peers.

The dispute, which first came to light in an article in *Nature* by

Christopher Anderson, reflected the panel's growing frustration that Rosenberg would not, or could not, furnish crucial data to the over-sight committee about his stalled gene therapy experiments. The is-sue evolved around technical details: Rosenberg was still trying to insert the gene for tumor necrosis factor into his TIL cells, and the committee wanted to know where these cells were going in the body and how much TNF they made when they got there—perfectly legiti-mate questions about a molecule capable of bona fide, life-threatening toxicity. The degree of frustration expressed by a number of expert cancer researchers, including Rosenberg's onetime collaborator, Ron-ald Levy of Stanford University, and Philip Greenberg of the Univer-sity of Washington, made the dispute an especially significant public spectacle within the community of immunotherapists.

"I just think that it is premature and unscientific to proceed with a clinical trial based on the pre-clinical or preliminary data that has been developed," *Nature* quoted Levy as saying. Greenberg said of a site visit to the Rosenberg lab, "There was a lack of preclinical data for virtually every one of the issues that [Rosenberg] was propos[ing], and it would help if Steve provided the preclinical data to us. What we got instead were a series of preprints, none of which really answers those questions. His response to the site visit was not adequate." Even an unresolved issue left over from the landmark 1985 *New England Jour-nal of Medicine* article, Rosenberg's ongoing study comparing LAK cells plus IL-2 to IL-2 alone, came under fire. "There were a series of— I don't want to be misquoted—not particularly successful trials that he was going to continue," Ralph R. Weichselbaum of the University of Chicago Hospitals told *Nature*. The journal noted that "one former collaborator (and now critic) believes that in the TNF TIL trials Ro-senberg has 'crossed the ethical line' in portraying research as ther-apy." A member of the panel later said, not for attribution: "I don't think he's ever been held to the same scientific standard the rest of us are held to. By getting his way, he hurts himself. And he *could* do good experiments. . . . He's just driven to *do*. It's a surgeon's mentality. The problem is, it's a very expensive and very inefficient way to do science."

My first lengthy conversation with Steve Rosenberg came the very same week that *Nature* reported on the grant dispute. In his office, surrounded by photographs of fellows in the lab and respected sur-geons, of Rosenberg posing with Nancy and Ronald Reagan, I brought up the *Nature* article and the fact that it contained some unusually harsh criticism from colleagues. Seeming unusually subdued, Rosenberg agreed. "It was pretty tough," he admitted. Then he suggested that the *Nature* reporter didn't understand the science, and predicted that the NCI committee would reverse itself at its next meeting, scheduled two months later; in truth, it took another eighteen months to obtain the necessary preclinical mouse data.

"You know," he added, "there's no question I try to move fast. I mean, I'm trying to solve problems that afflict people I watch die every day. I don't think I'm going fast enough."

It has always been Rosenberg's style to push—push the technology, push his patients, push his colleagues, push himself, push back at critics. He opened—and in some cases simply knocked down—a number of doors in the 1980s that led into promising new applications of immune-based medicines, which in turn seeded an enormous bloom of work. Several years earlier, when I first spoke with Steve Rosenberg, I asked him if there had ever been an occasion when he felt he had pushed too hard. He paused for a moment, which in itself was unusual.

Finally, in his low voice, he said, "See, you have to understand that as we were applying these things for the first time, we did not know what to expect. We had animal data. But there's just so much you can learn from a mouse."

The stalled gene therapy approach, the internal NIH dispute, lingering resentments over the publicity for the early work—all served to slow Rosenberg down and take a little steam out of the momentum he enjoyed in the 1980s. "Steve has never published anything in a big way out of any of those gene therapy experiments," said Nelson Wivel, formerly head of the NIH's RAC committee. "He refuses to close those second and third trials though he hasn't enrolled any patients for four, four and a half years, and it's fairly clear that he doesn't plan to do any more because it didn't work. And I think the scientific community is very well aware of that." Wivel added, "If you promote yourself and you don't meet the expectations you've set out for yourself, then I think people tend to pay less attention to you, and I think that's probably why people are wary of Steve."

But just as the accolades may at first have been exaggerated, so too may be the brickbats. Basic science needs someone willing to break through the inertia of mouse experiments, to move beyond the elegant reductionist exercises in test tubes to real human biology. Although he would probably much prefer being remembered as a brilliant medical innovator, Rosenberg's greatest value to date—to his peers and to medicine—may have been opening doors that other, more methodical researchers have poured through.

Every major conceptual avenue pursued by immunotherapists in the 1990s, including many that continue to show enormous promise, were experimentally advanced most famously, if not first, at the Surgery Branch of the National Cancer Institute during the 1980s. These include the treatment of cancer by cytokines like interleukin-2, by immune cells like LAK and TIL cells, by combinations of cells and cytokines, by gene therapy, and most recently by a broad campaign in antigen-specific cancer vaccines. With a large group of researchers, endless resources, an excellent nose for what's important, and that single-minded zeal to exploit new information as soon as possible, Rosenberg is in excellent position to come up with major clinical insights and observations in the future, not least because the knocks he has taken in recent years probably feed that part of him that remains the outsider with something to prove.

"You know, these are *new* things," he said in his office one day, when asked to put his twenty-year immunotherapy crusade at the NCI in perspective. "And I think [when] you try to do things new, there are going to be different responses to them. But I think there's also a difference in medical research, when one often has to proceed without having *total* information at one's disposal. This is what medicine is like. You're often making clinical decisions with imperfect information, and when one's attempting to do this kind of very directed research, you often have to do that as well.

"What I've seen happen over the years now is that in fact tumor immunology is very mainstream immunology," he continued, "because we've demonstrated unequivocally that there are immune responses against cancers that patients naturally react against, and it was our findings with TIL cells, I think, that were instrumental there. Even before that, we demonstrated that in fact there was a cell population, LAK cells, which we described, which were previously unrecognized, which could distinguish normal from malignant cells. So there were two different kinds of immune reactions against cancer, a non-MHC-restricted and the MHC-restricted, that we were able to find. And then our studies with interleukin-2 to *my* thinking were crucial, because here for the first time we had a purely immunologic maneuver that could mediate tumor regression. After all, you can bathe tumor cells in high concentrations of interleukin-2 and it doesn't bother them at all. Interleukin-2 has *no impact* on a cancer cell at all directly: it mediates all of its impact through the immune system. And yet with treating with interleukin-2, you can see cancers disappear. So it showed that it was *possible*, that was the first time that it was *possible*, to mediate antitumor effects with a strictly immunologic maneuver. . . . I mean, to me that was a demonstration that in fact it's . . . *real*. And the challenge became: Could you exploit it to make it more common and extend it to other cancers? But it changed it from the realm of: Is this the case? Can this ever work? to: This *is* the case and this *can* work, and of course there's been a tremendous outpouring of work in this field."

In the 1990s immunotherapy has become more molecular, more elegant, more precise, and for all those reasons more promising. Rosenberg's lab is very much in the thick of things, especially in the race to develop tumor vaccines. But the field is ascendant in many other places as well: Brussels, where the first T cell–specific antigens have been discovered, and Frankfurt, where early tests of a cancer vaccine based on those antigens have been promising; Seattle, where the adoptive transfer of cloned T cells has been shown to protect immunodeficient humans against disease; Stanford, where the use of monoclonal antibodies has, after nearly two decades of disappointment, appeared finally to prove of enduring worth against certain cancers; Lausanne, where a unique combination of surgery and cytokine therapy is saving both lives and limbs; and Pittsburgh, where the latest "cytokine du jour," interleukin-12, is being tested, alone and in gene therapy. Perhaps more important, these are but a few representative dots on a map where the advances of the last two decades in the laboratory are now being tested against what Lloyd Old has shrewdly described as "the unforgiving yardstick" of human benefit.

IV

IN VIVO
VERITAS

In a sense, the human is the best experimental animal
we have in which to test these new medicines.
—Steve Rosenberg, 1996

15

"ONE PLUS ONE
EQUALS TEN"

*It is ironic that even though we now know so much
more about the molecular nature of tumor antigens and
the mechanisms by which synthetic and recombinant vaccines
generate immunity, our clinical results are not yet nearly
so impressive as Jenner's original smallpox vaccines.*
—Drew Pardoll, 1996

❧ *In December 1982,* a young German physician named Alexander
Knuth climbed into his car and set off on the three-and-a-half-hour
drive from Frankfurt to Brussels. Carefully tucked into his shirt
pocket, and kept warm by his own body heat, was a flat plastic flask
known as a cell culture bottle. The flask contained cancer cells isolated from one of Knuth's patients, a German postal worker we will
call "Frau H," who at the time was losing ground to a rapidly advancing form of skin cancer and was not expected to live much longer.
Researchers traffic in biological materials like this all the time, but
the exchanges between Frankfurt and Brussels that began with
Knuth's car trip would have an unusually significant impact on the
field of cancer immunology. Several months later, in a smaller con-
tainer packed in dry ice, Knuth sent another consignment of cells to
Brussels. This one contained a small amount of Frau H's white blood
cells—the smaller fraction of blood that includes the most potent cells
and molecules of the immune system. There, in their most reductionist forms, were the cellular protagonists in the battle between the
immune system and cancer.

Working in a small, two-person laboratory at the University of
Mainz, Knuth had noticed that Frau H's white blood cells, and by

implication her immune system, recognized and killed her cancer cells with exquisite specificity. Knuth did not yet know the full extent of the interaction, nor even exactly what it was that those immune cells recognized. But he was among a handful of researchers in the world who could find and extract these cells from the blood, and he also knew of a researcher who had painstakingly been developing the tools and the technology to figure out what these cells saw. Thus the trip to Brussels.

Alex Knuth's initial journey to Brussels might have been viewed—indeed, was viewed by some of his colleagues in Germany—as the latest in a long succession of fool's errands in the field of tumor immunology, because of the long-standing difficulty researchers in that field had in identifying meaningful antigens that were truly specific to tumors. Once in Brussels, Knuth passed on Frau H's cancer cells to Thierry Boon, a Flemish-born, French-speaking, American-trained onetime medical school dropout who headed a large laboratory funded by the Ludwig Institute for Cancer Research and located at the Université Catholique de Louvain. Boon had become fascinated—almost against his will, to hear him tell it—with the immunology of cancer, which, to a younger generation of scientists who had cut their teeth on molecular biology, was a field in considerable disrepute. Boon did not set out to find a treatment for cancer. "When you start out," he said some years later, "you have no clues how long it's going to take. You have a certain curiosity. You try to move to the next step." In such stepwise fashion Boon, Knuth, and their colleagues in Brussels and Frankfurt set out to understand the relationship between the immune system and cancer cells at the molecular level—what the immune system sees, the context in which it sees, how it is able to distinguish the cancer cell from a healthy cell, and what the tumor does when it realizes it is under such intense and potentially lethal scrutiny.

These are puzzles that many researchers in many labs have helped to clarify, but crucially important clues to this remarkable interaction lay buried in the cells that traveled from Frankfurt to Brussels. Using them, Boon and Knuth set out to do nothing less than to teach the immune system to fight cancer. The first classroom was Frau H's cancer-riddled body.

Frau H was a dark-haired, somewhat plump woman, thirty-seven years old and mother of two teenagers, who intensely disliked her job sorting mail during the early-morning shift at the local post office in a small provincial town located southwest of Frankfurt. The tragedy that ultimately led to Frau H's exalted status in the world of tumor immunology announced itself in a whisper of minor symptoms: fatigue, vague abdominal distress, cramps, pallor. As her health declined, she spent entire days lying in bed. Finally, after Frau H consulted her regular physician in March 1982, doctors at her community hospital suspected she had developed several large cysts, probably cancerous in origin, and sent her to the department of gynecology at the nearby University of Mainz for surgery. The "cysts" turned out to be three tumors, one ovarian and two in the adrenals, that were later estimated by her doctors to be about the size of grapefruits.

Like a postmark on an envelope, cancer cells often carry the unmistakable stamp of the tissue where they originated, and Frau H's several masses had actually spread, or metastasized, from another location. Her doctors never located this original tumor, known as "the primary," but laboratory tests eventually confirmed that all the trouble started in a pigmented cell, or melanocyte, in the skin. She in fact had the deadly skin cancer known as melanoma, a disease that will strike an estimated 40,000 Americans this year (and which has seen a near-epidemic increase throughout the world in recent years due to social customs like sunbathing and, perhaps, environmental conditions like a thinning ozone layer). Frau H's melanoma was especially aggressive. At the time of this first surgery, doctors discovered a bad omen: the cancer had already spread to other places, including a lymph node beneath the right armpit.

Medical history, like any other form of history, often unfolds along meridians of chance and happenstance; in this case, Frau H's case intersected with the right set of researchers at precisely the right time. Frau H's surgeons referred their patient to the University of Mainz's new oncology clinic, headed by a thirty-four-year-old physician who had just returned to Germany in 1981 after a three-year research fellowship at Memorial Sloan-Kettering Cancer Center in New York. Tall and rail-thin, Alexander Knuth considered himself a product of the United States despite his very rigorous Germanic upbringing in Hamburg. He learned excellent English during two extended stints in the United States—once as a foreign exchange student at Berkeley High School in California in 1965, later as a visiting researcher. While at Sloan-Kettering, he had worked with Lloyd Old's team on an experimental melanoma vaccine; fresh in his memory was the very special patient known as "A.V.," who in 1976 experienced a rare and remarkable event. This man had been injected with his own tumor cells as a vaccine, and the treatment triggered an immune response that helped him to recover from an extensive metastatic melanoma (the patient is still alive today, about twenty years after being diagnosed). Knuth had helped to show that this patient had made killer T cells against his melanoma, so the idea that a patient could be "vaccinated" against an ongoing, metastatic cancer was very much on Knuth's mind.

Even in desperate cases, however, treatment begins with the conventional. "We put her on chemotherapy, first choice," Dr. Knuth recalled some years later, sitting beside his long, sleek, and impeccably neat desk in his office at the Krankenhaus Nordwest hospital in Frankfurt, "because we needed to do something that was an established therapy at the time." When drugs provided no benefit over the course of a year, they moved on to a second strategy. In a bold attempt to arrest the spread of her aggressive cancer, Knuth decided to repeat the crude tumor vaccine experiment from Sloan-Kettering: he took chunks of Frau H's excised tumor, broke them down into their component cancer cells, and started growing these melanoma cells in laboratory culture flasks. This cell line, christened with the name "MZ2," was destined to become one of the most productive in recent cancer research.

The tumor is only one partner in the tango of melanoma; the other partner, it is now becoming clear, is the immune system, and it was to that relationship that Knuth next turned his attention—specifically, the relationship between the subset of immune cells known as killer T lymphocytes and Frau H's melanoma cells. As in team sports, the immune response has its role players and its stars, and T lymphocytes are a particularly vital component of immunity. Like antibodies, killer T cells retain a memory for the viral and bacterial antigens they see, but they are unusually effective killers, more so than antibodies, and can even identify cells within which a virus—and a cancer—lurks. To their surprise, as the German researchers peered into their lab dishes, they witnessed a remarkable scenario. A small number of Frau H's white blood cells—specifically, certain of her cytolytic T lymphocytes, or CTL—not only recognized the melanoma cells from her body but killed them in the test tube. In other words, her immune system not only had learned to pick out the cancer cells but went on to eliminate them.

With thoughts of patient A. V. much on his mind, Knuth was "antsy" to try to enhance Frau H's response by vaccinating her in similar fashion. This would require altering, or mutagenizing, her cancer cells prior to treatment and at the suggestion of Lloyd Old, his former mentor in New York, Knuth contacted Thierry Boon. After receiving Frau H's cancer cells, Boon exposed them to powerful chemicals to create mutations; Boon knew that when you inject such mutated tumor cells back into the body, sometimes the unmutated tumor suddenly becomes visible to the immune system and then succumbs to an attack by T cells. Boon had demonstrated this before, but only in mice, never in human beings. He and Knuth almost didn't get the chance with Frau H.

"The chemotherapy was not working," Knuth said. "And since the mass in the lymph node in the armpit was growing so fast, we then asked the surgeons to remove it, because it appeared at the time that that was just a single, isolated, additional site." It was not. Frau H's disease had advanced far more than they had suspected; it not only had already spread to her ovaries and lymph nodes but by now had invaded other organs. In July 1983 surgeons removed one cancerous kidney and her spleen but did not manage to cut out all the affected tissue—did not achieve, in Knuth's words, "clean surgical margins," which made her already grim prognosis even worse. "It was questionable at that time if she was really free of disease," he said. "We definitely thought she would die. With such a vastly metastatic tumor, it would be very unusual for someone not to die." Frau H could no longer tolerate chemotherapy by the fall of 1983, and so Knuth began injections of the vaccine in February 1984. Roughly 100 million of Frau H's own cancer cells, altered by Boon's chemically induced mutations and then killed by irradiation in Frankfurt shortly before each treatment, were injected at intervals of four to six weeks.

If you dig behind the dry data of medical records, you almost always discover an unexpected plot twist that makes each case study unique. In Frau H's case, the "key situation," according to Knuth, occurred soon after the vaccinations began, when the German doctors detected yet another tumor mass; it appeared, not surprisingly, in the region where surgeons feared they had not been able to remove all the cancerous tissue. Monitoring the mass with ultrasound and CAT scans, they watched with apprehension as it grew and ultimately reached a diameter of eight centimeters, about the size of a large plum. "I remember it very well," Knuth said, "because I wanted the surgeon to operate on her in the summertime. It was in August, and he wanted to go on vacation, and he couldn't do it before his vacation. She was almost scheduled, and then it didn't work for time reasons. So we continued to give her the vaccination. And during the time that the surgeon was away, that tumor started to shrink. It later disappeared, and the surgery was never performed."

That was in September 1984, and Frau H's cancer has never come back. Depending on one's point of view, Frau H was cured by her doctors or, with a little help from her friends in Frankfurt and Brussels, she had somehow managed to cure herself of a highly malignant and undoubtedly terminal cancer. Passing as through a revolving door from medical miracle back to quotidian banality, Frau H walked out of the hospital and back to sorting mail in a provincial post office.

I first heard a detailed account of Frau H's remarkable remission from Thierry Boon, who recalled how Knuth had begun to inject Frau H in the spring of 1984, how her condition waxed and waned over the following months, how she began to improve steadily. "Now that was in the fall of 1984," Boon continued, relating this story nearly ten years later, in early June 1994, sitting under a little gazebo atop one of the hills that overlook Cold Spring Harbor Laboratory on Long Island, where he had come to give a symposium talk. "And then, you know, we continued injecting her every three months, and then every six months, and now it's more irregularly. But she is alive and tumor-free. And," he paused to add, mustering as much irony as the pause would support, "we'll never know if our injections have anything to do with it."

If truth be told, anecdotes like Frau H's miraculous cure (and popular accounts of them) are heartwarming eyesores on the landscape of medical progress; with the subtlety of billboards, they promise so much, and yet rarely deliver useful scientific information that will ultimately benefit other patients. Frau H's remarkable case history has never been published in the medical literature, primarily because her German doctors failed to gather, painfully on purpose, one crucial piece of evidence. When that last plum-sized tumor began to shrink and disappear in the fall of 1984, Alex Knuth wrestled with the kind of dilemma that illustrates how good doctoring and good research often work at cross purposes. He knew they could make an airtight case that the vaccine had worked if they performed a biopsy on the shrinking tumor and confirmed that it was indeed a recurrence of the original cancer. He also knew, however, that by snipping away even a small piece of malignant tissue, they could endanger Frau H; they might accidentally disseminate tumor cells and promote the spread of what had already proven itself to be a highly malignant cancer. Wisely, they chose not to tamper with the shrinking tumor. Sadly, all that they were left with was one very happy patient, one very pleasing anecdote, and no proof that their improvised vaccine had made a difference.

What rescues this story from mere anecdote is what happened afterward with the cells that traveled to Brussels. The adversarial relationship between Frau H's T cells and her cancer cells in the laboratory mimicked in miniature the complex interaction between her immune system and the tumor, and using those two precious raw materials, along with some backbreaking molecular genetics refined over two decades of basic research, Boon, Knuth, and their colleagues in Brussels and Frankfurt set out to define, at a molecular level, what Frau H's T cells saw. They set out to find something that had never been found before: a tumor-specific antigen recognized by the most potent avenging angels of the immune choir.

"I must tell you," Thierry Boon apologized when I visited him in his Brussels laboratory, "I'm by training a molecular geneticist. I'm not an M.D. I'm not an immunologist. And I was certainly not a tumor immunologist to start with. I stumbled into this whole thing by accident."

"Stumble" seems far too blundering a verb to describe anything that Boon (pronounced "bone") might do. Tall, thin, elegantly dressed, and charmingly outspoken, with a close-cropped, graying beard, dark eyebrows, and a long and slightly hooked nose, Boon wears his two predominant attitudes—modesty and arrogance—like a reversible coat. "I really grew up on this principle of molecular biologists—you know, we are intelligent, and the less we know of what other people are doing, the better," he explained, and in almost the same breath confessed an unfamiliarity with the literature in his own field, adding with an apologetic shake of the head, "I am particularly illiterate." He is also something of a scientific loner, one of those increasingly rare researchers, manifestly intelligent and sufficiently self-confident, who thrive outside the scientific mainstream. "He is not a man of fashion," Knuth remarked. "He's really a man of science. And that is something that is very rare these days, because people are out for immediate rewards."

Born in 1944, Boon grew up in the great university town of Leuven, located about ten miles north of Brussels and known also as Louvain when, in the sixteenth century, Vesalius and Mercator walked its narrow, curving streets, one reconfiguring the atlas of the body and the other of the planet. Boon's mother came from a French-speaking background, and his father, who for many years was the chief executive at a Belgian brewery, came from a Flemish background. "And I did my secondary studies in Flemish, because by the time I got to secondary school, it was already forbidden to have a French-speaking secondary school. You know, we have our tribal problems here. Fifty years ago, if you wanted to climb in society, you had to learn French." His father would have preferred that he pursue a career in business—"money and all that," Boon noted dryly—but he chose research instead.

When he decided to study biology at the Catholic University of Louvain, he received a shrewd bit of counsel from a friend of his father. This friend, Boon recalled, "gave me the advice that if I wanted to study biology, even for research purposes, I should study medicine, because the quality of the medical faculty was better. And I think he was damn right." This comes as no surprise; the source of the advice was Christian de Duve, a biochemist who would win a Nobel Prize in 1974 for work on the internal organization of the cell. Boon studied medicine for three years, just long enough to get turned off by the prospect of clinical practice, just long enough to get excited by the kind of research he was doing in de Duve's laboratory.

Then he decided to quit medical school and pursue basic research. This established a pattern that, under the right circumstances, can result in an especially potent combination of scientific traits: excellent training by terrific mentors on the one hand, an independent spirit willing to defy them on the other (or, as Boon put it, "to be choosy about who you disagree with"). "De Duve told me I was taking a big risk," Boon recalled, "but I told him I didn't care. And so I managed to convince him that, whether he would help me or not, I was not going to do these clinical years [of medical training]." Here, truly, mentorship made a difference. In 1965, technically a medical school dropout, twenty years old and thumbing that arched, patrician nose in the face of his sponsors, Boon landed—with de Duve's assistance—as a graduate student at the Rockefeller University in New York. After spending two years in his room studying mathematics and physics, he ended up in the laboratory of one of the central figures in the early history of molecular biology, Norton Zinder.

"Before he worked for me," Zinder recalled recently, "he went and read all my reprints and told me which ones were wrong." Boon spent six years at the Rockefeller, and he was by his own appraisal not a very good guest, "not particularly friendly or helpful because I wanted to work all the time." He did a piece of research on bacterial toxins that attracted a lot of attention, however, and that allowed him to "publish very little for many years without being fired." The research was for the most part esoteric. But the *ethic* of the research—reductionist, rigorous, supremely confident in the power of genetics—was not.

Boon, like Knuth, considers himself "a child of the States intellectually," but it was more than that; he had coupled his loner's sensibility to a powerful scientific approach. "If I had now to really drop every semblance of modesty and say why we were successful in doing what we did, it was really two points. First, like almost every area of modern biology, we introduced genetics into tumor immunology. And second, by illiteracy and also by psychology I would say, I've never listened to or trusted one bit of what the tumor immunologists were saying. We believed only what we saw ourselves." Molecular genetics gave him a way to see with exceptional clarity and rigor. "His approach to science is infinitely, scrupulously detailed and analytic," Zinder said. "He is as obsessive about detail as anyone I've ever seen. His experimental protocols were worked out to the nth degree, and were done in such a way that any ambiguity was at a minimum."

At the same time Boon's career path had all the jagged coherence of random collisions, a combination of the pilgrim's progress and Brownian motion. When Jacques Monod, the brilliant French molecular biologist, gave a seminar at the Rockefeller in the early 1970s, he visited Zinder's lab, heard about Boon's work, and invited Boon, who was finishing up his Ph.D., to join the staff of the Pasteur Institute in Paris. But Monod had recently become director of the Pasteur; he traveled frequently, assumed more and more administrative duties, and had less time for laboratory inquiry, so Boon transferred to the lab of François Jacob. Jacob, who had shared the Nobel Prize with Monod for their seminal hypothesis about how genes are turned on and off in normal cell biology, had by now focused his attention on differentiation (formerly known as embryology), the study of the genetic and molecular processes that orchestrate the orderly development of a single fertilized cell into the more specialized cells and tissues of a mature organism. Against Jacob's advice—yet another example, Boon admitted, of "getting advice from first-rate people and never following it"—he decided to study this process in mice. "And there was a tumor," Boon remembered, "a very strange type of tumor. . . ."

Teratocarcinoma is a cancer that derives its name from the Greek word *teras,* meaning monster. It is a grotesque aberration of normal embryological development. The tumors arise in the embryos of many animals, including humans, during the earliest stages of development, and usually result in spontaneous abortion. In these dysfunctional embryos, the genetic wiring has short-circuited in the midst of the process that forms different, specialized tissues; as a result the tumor gets shuttled to a particular genetic pathway and in essence repeats the same anatomical phrase over and over again. Hence these tumors have been known to generate the most peculiar special effects: teeth, hair, eyeballs, bone, muscle, cartilage, almost any type of tissue.

Boon had no particular interest in cancer. For many years researchers had tinkered with these tumors and then transplanted them into mice as a way of studying the process of normal embryological development. In the course of studying these strange tumors, Boon and his colleague, Odile Kellermann, hoped to tease out details about the genetic blueprint that governed the embryological development of specialized tissues. "And I thought my contribution to this thing," Boon would say with a small laugh, "was to really stick to the classical genetic approach, which is to try to make mutants, hoping to get mutants that would make tumors that were unable to differentiate into certain directions, and that maybe the *pattern* that I would see would make sense."

Their very first experiments produced a surprise. Beginning in 1972, they exposed teratocarcinoma cells to a very strong chemical agent that caused mutations and then injected these mutagenized cancer cells into mice to see if there was a genetic pattern to the weird variations that grew out of the tumor; as a control in these experiments, they injected other mice with the regular, nonmutated teratocarcinoma cells. The control animals invariably developed rampaging cancers, but in roughly 30 percent of the other animals, the mutated cancer cells never took hold. As a card-carrying geneticist, Boon instantly realized he'd fiddled with the genes of the tumor cells in an interesting and possibly important way. Quite by accident, he had created a mutation in cancer cells that genetically changed them in so fundamental a way that they were no longer cancerous.

At that point Boon knew nothing about cancer and less about immunology, but he felt "impelled" to understand why the mutated tumor cells no longer formed tumors. Unknowingly, he was about to abandon embryology forever and venture onto a scientific land bridge leading to a tangled—and in the view of many, rather discredited—wilderness of biological investigation known as tumor immunology. Why didn't the mice develop tumors? The short answer, Boon's group learned, was not because the researchers had altered the tumor's malignant traits but because they had artificially changed the cancer cells, marked them with a badge of mutation, so that the cells suddenly appeared visible to the mouse immune system—indeed, appeared somewhat like a mismatched organ that might be rejected following transplant surgery. The mutated cancer cells appeared foreign, something Non-Self and something therefore to reject. Unwittingly and still only by inference, Boon and Kellermann had created man-made tumor antigens, markers that appeared on the surface of cells as a kind of identifying biochemical flag. They had genetically tricked tumor cells into waving such a flag for the immune system, and immune cells not only came and saw but, in the unforgiving idiom of Caesar, conquered as well, for once activated by the sight of antigen, cytolytic T cells turn into ferocious killers.

To decommission a tumor, to genetically pull the plug on its malignancy and make it disappear, was a pretty unusual stunt, even in the artificial environment of inbred mice. Boon at least knew that much, although not much more. Sometimes, however, ignorance can be a scientist's greatest ally, and in order to understand the delicious beauty of what transpired next in Paris and later Brussels, it is useful to revisit a bit of scientific sociology from that era. "What I sort of learned *afterwards*," Boon said with a little laugh, was that the field of tumor immunology was "an absolute morass," that he had committed himself to the "untouchables of immunology."

Just as Boon launched himself into the study of tumor immunology, practical applications of immunology against cancer were going up in smoke. Georges Mathé, the flamboyant French hematologist, had electrified the medical community in 1969 by claiming that children with leukemia could be cured by stimulating their immune systems with injections of bacillus Calmette-Guérin (BCG), an attenuated live bacterium, and cancer cells. That BCG stimulates the immune system there can be no doubt; it has by now been used in over 2 billion people worldwide as a vaccine for tuberculosis. But Mathé's claims touched off a frenzy of clinical trials of BCG against a variety of cancers, with almost total failure. Other researchers, with similar empirical fervor, attempted to inject blood serums and antibodies and other experimental therapies without success, and by the mid-1970s the BCG story in particular—and immunotherapy in general—came crashing and burning back to earth after an extraordinary burst of publicity. But the most devastating blow came from England and the work of Harold Hewitt. As mentioned in an earlier chapter, Hewitt published a widely read (and hotly debated) paper asserting in no uncertain terms that spontaneously arising tumors, the kind that most resemble the cancers that afflict human patients, were *not* immunogenic; in fact, he

strongly suggested that any results suggesting otherwise were laboratory artifacts. The effect of Hewitt's broadside was twofold: to rob the
field of its conceptual foundation and to impugn its scientific integrity.

Happily, Thierry Boon remained blissfully "illiterate," as he put it,
of Hewitt's indictment. In November 1975, just as Hewitt's paper was
being accepted for publication in the *British Journal of Cancer,* Boon
performed what he now considers to be the key experiment of his
entire career ("the last experiment in Paris," he calls it, because he
was about to leave the Pasteur to return to Brussels). Still working
with the teratocarcinoma, he once again stumbled upon a remarkable
aspect of the immune response to cancer. As before, he injected his
mice with mutated cancer cells; as before, many of the mice did not
develop cancers. Then Boon took the experiment one step further. He
later injected these same mice with the normal, *unmutated* cancer-
causing cells, the ones that invariably produced tumors. The animals
were caged in a corner of a garage at the Pasteur; to Boon, the moment
he went to check on the animals remains frozen in time.

"I still see myself on a Saturday afternoon," he said, voice full of
unexpected wonder for one who prides himself on scientific rigor,
"alone in the animal house at the Pasteur Institute in Paris, and I
thought I would faint. Because all the control animals had the tumor
and more than three-quarters of the [other] animals had rejected the
tumor. There was no tumor at all!" Once the mouse immune system
had recognized and rejected a mutated version of the cancer cells, it
gained the ability to "see" the previously invisible, nonmutated par-
ent tumor. In mice at least, the immune system had learned to see
things that had previously escaped its notice, had learned to see antigens it hadn't seen before. "That is certainly the day when I became
totally excited about this," Boon said, "and sort of envisioned that this
could lead to some breakthrough in cancer immunology."

After completing the move to Brussels, and with unusually ample
and long-term support from the Ludwig Institute for Cancer Research
to tackle risky work, Boon and his key collaborator, Aline Van Pel,
tried to attack this puzzling result in a classic genetic way. Geneticists study genes, so they set out to develop a technology for finding
the gene for the antigen that made these mutated tumors visible to the
immune systems of mice. This line of research, which stretched over
the better part of a decade, was not for the faint of heart (nor, some
American scientists point out ruefully, for researchers who need to
show progress every three years in order to get their NIH grants re-

newed). Boon ultimately realized that mutations in these so-called tumor-minus variants created tiny differences in small, bite-sized parts of proteins known as peptides, and that T lymphocytes recognized these tiny changes. The scientific challenge, therefore, came in two parts: to isolate and amplify (or clone) populations of T cells from mice that recognized specific antigens, and to then use those T cells as a kind of molecular magnet to pick out the needle in a genetic haystack, namely the gene responsible for the antigen.

There was no easy way to find those genes. From the moment Boon stood breathless in the animal room of the Pasteur Institute to the time his group finally got the techniques to work in mice, fourteen years transpired. But they had good reason to persevere. In 1983 Aline Van Pel repeated Harold Hewitt's experiments with several spontaneous tumors, and when her mice were "vaccinated" with mutated versions of Hewitt's spontaneous and invariably aggressive cancers, they too acquired the ability to reject the unmutated "natural" cancers.

If there was any criticism to be made of this slow, painstaking approach, it was that Boon's lab became almost too focused on mice. Just about this time Lloyd Old suggested that Boon give Alex Knuth a call. Soon thereafter, the cells of Frau H arrived in Brussels.

Like all tumor immunologists, Boon was working at a boundary as much philosophical as scientific: Where does the Self end and Non-Self begin? In practical terms, the distinction is crucial to survival: the immune system is designed to identify and root out foreign pathogens like viruses. Indeed, the phenomenal rapacity with which T lymphocytes in particular can attack and utterly ravage a transplanted kidney or liver or heart—large organs weighing several pounds—has always tantalized immunologists with the possibility of unleashing the same violence upon tumors.

But cancer cells are not nearly so obliging a target. They teeter on the very edge between Self and Non-Self. Cancer, after all, begins as a single aberrant event in an otherwise normal cell, and that cell contains the exact same genes and the same homely pulp as a normal cell. In a cellular sense, it is still the Self, albeit slightly deranged. The cancer cell eventually breaches every convention of civilized cellular etiquette: it drives through cellular stop signs, as it were, in its relentless cycles of replication; it sabotages internal checkpoints that normally arrest such aberrant behavior; and as the tumor grows, crowding neighboring cells, it actually evolves, apparently reactivating genes used in fetal development, for example, to metastasize. Rather than seeing a dangerous stranger, as it does when it encounters a flu virus, however, the immune system sees a wayward uncle—behaving a bit odd, perhaps, but nonetheless a blood relative. As Boon and Van Pel learned in their work with teratocarcinomas, tumor antigens *do* exist, but they are weak and don't normally arouse much of a response. Thus, just as a family tends to indulge the erratic behavior of kin, the immune system is said to be "tolerant" of the slight differences in cancer cells; immunologists therefore speak of "breaking tolerance," of overriding this forgiving relationship and nudging the immune system into recognizing cancer as the threat it is. And lurking behind that issue is a question that has remained unanswered for years: is this Self sufficiently deranged to look, finally, like Non-Self, sufficiently different to elicit an intolerant immune attack?

In order to answer that question, Boon's group typically recast it as a reductionist question: What did Frau H's immune cells see, and how were they able to arrest her cancer? Here, Boon's work began to intersect with the MHC story—that is, the discovery in the mid-1980s, principally by Alain Townsend and colleagues at Oxford University, that killer T cells can only recognize antigens when peptides are "presented" to them, as if held up to the light, by the Class I major histocompatibility (MHC) molecules. If the proteins are normal, T cells won't even notice. If a flu virus has infected a cell, for example, some of its viral protein will inevitably be chopped up and shipped to the cell surface along with normal peptides, and T cells will recognize it as foreign instantly. Indeed, the term "recognition" in an immune context is laden with alarm and violence. First, the T cells will *lyse,* or kill, the marked cell (hence "cytolytic" T cells), literally perforating its cell wall with tiny pores until, only minutes later, the cancer or virus-infected cell swells and bursts; at the same time, other T cells— the "helpers"—will blare a chemical alarm if they become involved. They will gush molecules called cytokines, some of which recruit other immune cells to the site and some of which act like hormones, causing the very subset of T cells that recognized the foreign marker in the first place to multiply.

Frau H's dramatic recovery suggested to Thierry Boon that her miracle lay in the molecules; her cancer cells had invited immune annihilation because they possessed some telltale molecular marker that aroused a T cell response, the very T cells that Knuth and Boon had identified in her blood. While simultaneously continuing its search for the mouse gene as well, the Brussels team began searching for this human melanoma antigen in 1983. As is often the case in biology, Boon's lab lacked the technology to look directly for what was the object of their search, namely the antigens, the bits of protein on the surface of tumor cells that served as red flags to the immune system. There existed neither a microscope that could provide a visual glimpse of these surface markings nor a technology for, in effect, taking cuttings of a cancer cell's surface vegetation and identifying a single peptide as uniquely antigenic (though such a technology has been developed recently by Donald Hunt and his colleagues at the University of Virginia). So the team looked simultaneously in mice and humans for genes that contained the instructions for making cancer-specific antigens. And they did possess an extraordinarily powerful tool, more precise than an electron microscope, to feel out the three-dimensional shape of the human cancer antigens. This tool, too, had traveled from Frankfurt to Brussels. It was Frau H's blood. By 1987 the researchers had stable clones of her T cells and, one by one, they would screen every gene in Frau H's cancer cells until they found the one that made the marker recognized by her T cells.

Only fellow geneticists can appreciate the Jobian enormity of this task and the cleverness with which Boon, Knuth, and their colleagues stalked this elusive molecular prey. They knew only in 1989 that such an approach might even succeed, when the work with mouse tumors finally resulted in the discovery of a tumor-specific antigen in mice called P1A. Using the same basic technology, they tackled Frau H's genome. Each human cell—cancer cells included—possesses up to 100,000 genes. Each gene contains instructions, encoded in DNA, for a different protein, which is a large, three-dimensional chain composed of dozens or hundreds of amino acids. The large gene for, say, dystrophin, the protein in muscle tissue linked to muscular dystrophy, contains about 2 million base pairs (or letters) of DNA; the hormone insulin, by contrast, has fewer than 2,000 letters, which code for a protein fifty-one amino acids in length. But because T cells can see only what fits onto an MHC molecule, the immune system typically recognizes much smaller units, selected outcrops of these larger globular proteins typically measuring about nine amino acids in length. A human antigen, if one existed, would therefore represent at most a tiny, 30-letter portion of a gene up to 2 million DNA letters long. To put this in perspective, searching for this antigen would be a little like looking for a 5-word sentence in a 300,000-word novel, buried in a 3 billion–letter text of uniformly fine print.

"This is always very difficult work," Boon would say later, "because you're using a method of which you are unsure, for a goal that you don't know is achievable, even if the method is good. So it's only if and when you *get* your thing that you know that the goal was achievable and that the method was appropriate." Buoyed by the success of finding a tumor antigen in mice, the Brussels lab proceeded to chop up the entire DNA of Frau H's melanoma cells—all 3 billion base pairs, all 100,000 or so genes—and parceled tiny bits of the DNA into approximately 700,000 "cosmids," which might be thought of as separate and temporary biological slipcases for these smaller man-made volumes of the total DNA library (the technique's blunderbuss philosophy may be inferred from the term by which it is colloquially known: "shotgunning"). Each of these 700,000 volumes, each containing a slender excerpt from Frau H's genetic text, could then be inserted into recipient cells; each of these special cells would form colonies, read the instructions, make the corresponding bit of protein, and ship it to the cell surface; and thus smaller segments of Frau H's DNA library could be scanned to see if they contained the gene for the antigen. That was the beastly part of the Brussels experiment.

Now comes the beauty. Frau H's T cells provided a kind of immunological radar to lock onto any cell that made an antigen. And something like biological radar was sorely needed: in one series of experiments, some 29,000 colonies had successfully taken up a volume of Frau H's DNA. Pierre van der Bruggen of Boon's laboratory, who did the brunt of this work, used her cancer-specific T cells to screen large pools of clones in search of promising candidates. He ultimately identified a mere 5 colonies that contained the gene encoding the antigen; in another experiment, her T cells pinpointed 2 more positives out of 13,000 colonies.

Early in 1991, about eight years after Alex Knuth's first visit with Thierry Boon, Frau H's white blood cells homed in on the prize. Her T cells were able to distinguish a tiny molecular bump on the surface of one of the cell colonies, a bow-backed peptide bump arching out from the surface of the cell and measuring exactly nine amino acids long. Using standard recombinant DNA techniques, the Belgian scientists cracked open this cell, identified the gene, and produced its complete sequence. Boon called this gene MAGE (for "melanoma antigen"), and the group published the results in the journal *Science* at the end of 1991.

"Human tumor immunology up to that point represented a very empiric, observational, nonmolecular kind of discipline that was held in incredibly low esteem by the scientific community, and probably rightly so," said Drew Pardoll, an immunologist at Johns Hopkins School of Medicine. "All of a sudden, here was a hard-core molecular approach that represented a legitimization of human cancer immunology. There were genes and antigens that had obvious and clear-cut implications for therapy, and could be crafted into antigen-specific vaccines." Added Alex Knuth, "The December *Science* paper was the turning point, because then human tumor immunology was reborn."

Despite all the dispatches about newly discovered genes that appear almost weekly on the front pages of newspapers, be they related to obesity or breast cancer, it cannot be stressed enough that the discovery of a new gene does not so much end a story as punctuate a very long preface to what is usually an even longer, more complicated biological tale. Researchers need to understand what protein the gene makes, when it makes it, how that protein fits into cell commerce, and whether the protein plays a central, irreplaceable role or is part of a redundant biochemical chain of events. What MAGE had to do with cancer—and, more important, with cancer treatment—still belongs to that larger, more uncertain, and still unfolding story.

The Brussels group discovered that MAGE was a normal gene in melanocytes. It sat on its hands, one small part of the X chromosome, doing nothing, remaining mute. In about 30 percent of melanoma tumors, however, the gene had become active; it also appeared to be active in a lesser proportion of several more common tumors, including breast cancer and lung cancer. Boon soon had reason to append the numeral 1 (MAGE-1) to their discovery, because it quickly became clear that this first gene had many siblings—belonged, in fact, to a large family of related genes, all previously unknown to biologists, all permanently inactive and silent in normal, healthy, mature cells, with one exception: the testes. In 1994 Boon's group reported that the related gene called MAGE-3; it has turned out to be present on the surface of a great many tumors—not just in 70 percent of metastatic melanomas but also in lung cancers and bladder cancers. At last count, Boon's group has identified three different genes in the MAGE family, and three other related genes, BAGE and GAGE and PRAME, which encode tumor antigens. With the exception of PRAME, every one of these genes came to light first in Frau H's cancer cells, and all were ferreted out using Frau H's white blood cells.

As to why a tumor would avail itself of a normally silent gene, Boon speculates that once a cell is "transformed," or becomes cancerous, it reaches into its own genetic archive and resurrects dormant genes that served the organism during its earliest embryological moments, genes that allow it to do something that advanced tumors do very well: move around, or metastasize. "Clearly, genes have not been created to be expressed in tumors," Boon explained. "I believe that *all* types of genes, like MAGE and BAGE and GAGE, must be expressed in the embryo sometime or maybe in some cells of the placenta. And the function that these genes have in the embryo—say, perhaps, for allowing tissues to move or cells to move within tissues—might be reused by tumor cells when they want to metastasize."

Publication of the MAGE gene broke the long-standing theoretical bottleneck in tumor immunology, and not coincidentally, the field has blossomed. Since the discovery of MAGE-1 in 1991, a growing number of tumor-associated antigens have been identified. Boon's group is responsible for the MAGE, BAGE, and GAGE genes; they have also identified two so-called differentiation antigens, which have complicated the immunology of cancer in an interesting way. Using the same technique that uncovered MAGE, Boon discovered these antigens in the cancer cells of A. V., the same patient Alex Knuth studied at Sloan-Kettering in the 1970s. Called Melan-A and tyrosinase, they are specific not to melanoma per se but to the type of differentiated cell, melanocytes, in which the cancer arises; for some reason certain melanoma patients develop T cells that recognize their own normal proteins, a situation that verges on an auto-immune response. Steve Rosenberg's group at the National Cancer Institute has also reported two differentiation antigens, called gp100 and MART-1 (the latter is identical to Boon's Melan-A), and a team at the University of Virginia independently identified gp100. In addition, there are several somewhat less specific antigens under study, including an unusual glycoprotein called mucin in breast cancer, a well-known protein called CEA (for carcinoembryonic antigen) in colon cancer, and a protein in ovarian cancers. According to reports at a recent meeting, possibly another dozen antigens are in the process of being characterized.

The discovery of T-cell-specific antigens has, according to Lloyd Old, who heads the Ludwig Institute and is an elder statesman in the field, "inaugurated a new era in cancer immunology." If they did not exactly scream "non-self," these antigens at least whispered "not-quite-self," and did so loudly enough to attract notice. In light of this and subsequent discoveries, Alan Houghton of Sloan-Kettering has come to refer to cancer cells not as Non-Self so much as Altered Self. By identifying tumor antigen genes, Boon and many other cancer immunologists throughout the world could begin to test some ideas about deliberate manipulations of the immune system to react more aggressively to these weak signals from the Altered Self. Not surprisingly, several considerable hurdles remained.

If tumors indeed wave red flags at the immune system, why don't T cells charge like a herd of aroused bulls? The answer might have been suggested by Charles Darwin: during the development of a malignant disease, tumor cells must evolve and adapt in order to survive, and one adaptation especially favored for survival is an ability to evade the immune system.

Joost Oppenheim, a researcher at the National Cancer Institute, has observed that viruses know a lot more about the human immune system than immunologists do, and the same could be said of tumors. Far from the mindlessly replicating drones of popular description, tumors are exceedingly clever about thwarting the immune system; Lloyd Old has referred to this ability as the "Houdini phenomenon." If the immune system spots a tumor antigen, for example, then a tumor cell that shuts off the gene that encodes the antigen—muffles the cry, as it were, of the gene that gives away its location—may be selected for survival. Tumors also tend to "upregulate," or churn out, molecules that dampen the immune response, and to "downregulate," or silence, genes that make the molecules that promote communication between cells. In extreme cases the tumor may shut down any of several genes that tell the cell how to make and use those crucial major histocompatibility complex (MHC) molecules, thus short-circuiting the body's security system altogether.

Taken as a whole, these evasive strategies go by the name "tumor escape mechanisms," and it has been one of the revelations of cancer research over the past decade or so that tumors are not dumb, but exceedingly crafty. Sir Walter Bodmer, director of the Imperial Cancer Research Fund in London, has specialized in the study of tumor evasion mechanisms, and never tires of reminding colleagues to "think evolution, evolution, evolution" when pondering the mutations common to cancer cells. Playing by the same Darwinian rules of survival as penicillin-resistant bacteria, for example, tumors constantly evolve so that the heartiest, most evasive, and, alas, most malignant tumor cells survive. "There is constant competition among the tumor cells," Boon explains, "for the one that is going to be a little more malignant than the others, multiply a bit better, invade [surrounding tissues] a bit better."

In the context of such evasive behavior, Boon has used the MAGE gene to show just how individual and personal a disease cancer ultimately is. Melanoma at least, and probably many other cancers, is inextricably entwined with not only how the tumor makes use of its genetic archive but which genes the patient has at his or her disposal to locate and flag the tumor. Once the cancer begins to grow, it does so in an immune ecology that is unique to each individual, because of genetics and personal immunology. Personal immunology? As Boon and Knuth discovered with the MAGE genes, one reason that many melanoma tumors remain invisible to the immune system has as much to do with the genes of the patient as with the tumor. Again, killer T cells are incapable of seeing an antigen—be it a snippet of flu virus or a scrap of MAGE-1—unless it is presented on the surface of the cell by the MHC molecules. But humans possess enormous genetic variation in the MHC system, much more so than in hair or eye color. In this regard, the O. J. Simpson trial may be cited to make, for a change, a useful point.

Each human being inherits six different "sizes" of MHC transport molecules that "fit" with killer T cells. Known as HLA markers and much discussed in the context of DNA blood analysis performed for the Simpson trial, these molecules are inherited in combinations fairly unique to each individual; just as each of us inherits either A, B, AB, or O blood type, we also inherit six HLA subtypes, such as HLA A-1, A-2, B-6, C-44, and so on (hundreds have been identified to date). In the context of the Simpson trial, these subtypes serve as very specific blood markers, similar to but much more precise than blood type; in the context of immunology, HLA subtypes are like shoe sizes into which only certain peptide antigens, like certain feet, will fit. Thus, for the patient with melanoma, it is not enough to have a tumor expressing the MAGE-1 gene; one must inherit from one's parents the proper HLA molecule, with the right-sized molecular groove, into which the relevant peptides from MAGE-1 or any other antigen will fit. If the antigen doesn't fit, to paraphrase Johnnie Cochran, the immune system must acquit and let it pass.

Boon's work with Frau H's blood demonstrated, for example, that MAGE-1 fits snugly only into the molecule known as HLA-A1. Approximately 26 percent of Caucasians, including Frau H, are A1. In practical terms, whether a melanoma is immunogenic or not depends on heredity and therefore, like all of heredity, becomes an exercise in genetic probability. If 30 percent of melanoma patients produce MAGE-1 antigens (not all do) and 26 percent of Caucasians possess the HLA-A1 protein, a simple statistical calculation predicts that only about 8 to 10 percent of Caucasians with melanoma will possess the right combination of tumor antigen and MHC molecules to allow the immune system to see the tumor. That number sounds rather low. But each additional antigen discovered improves the odds, and this has been especially true since the Brussels laboratory reported the discovery of MAGE-3. This antigen, presented by A1 and A2 molecules, appears not only in 70 percent of metastatic melanoma cases but also on many other common cancers, including lung tumors, bladder tumors, and head and neck squamous cell carcinomas.

What this means, as Boon correctly realized, is that "for the first time, we can select as candidates for therapy those patients who have a chance of benefiting from immunization." And that is exactly what the identification of these very specific antigens has led to: cancer vaccines and immunization.

The notion of vaccinating people against their cancer has been in and out of fashion since Coley's toxins, and a number of "nonspecific" melanoma vaccines have been tested since the early 1970s, including work by Michael Mastrangelo and David Berd at Jefferson Medical College in Philadelphia; Malcolm Mitchell, now at the University of California, San Diego; Donald Morton at the John Wayne Cancer Institute in Santa Monica, California; and Jean-Claude Bystryn at New York University, among others. In most cases, these vaccines have been created by grinding up or irradiating tumor cells and then injecting them into patients, somewhat as Boon and Knuth originally did in the Frau H case. But these approaches have typically lacked any precise molecular knowledge about what antigens the vaccine contained, what sort of immune response it might inspire, and who might be genetically disposed to respond. Although many physicians feel that most randomized clinical trials of these vaccines have fallen short of clear-cut demonstrations of efficacy, there has been consistent evidence that between 10 and 20 percent of patients, at a minimum, have favorable responses; indeed, Frau H probably responded to just such a crude, almost desperate intervention. What separates the MAGE story from these previous attempts is that Boon and Knuth successfully tackled the very substantial and basic problem of explaining on the molecular level how and why an immune response occurred. A physician might now reasonably predict ahead of time, with a simple blood test and an analysis of the tumor, which patients are likely to respond to vaccination.

Each antigen provides the starting point for a different cancer vaccine. A number of them are being prepared and tested, and although the results are extremely preliminary, it would be fair to say that researchers are quite optimistic. Boon's group, for example, in conjunction with Alex Knuth in Germany and doctors in Italy, France, Belgium, and the Netherlands, has prepared a bare-bones vaccine using a small immunogenic fragment, the MAGE-3 peptide, simply to see if it provokes a measurable rise in T cells in genetically compatible patients and if it is safe.

In December 1995 Boon, Knuth, and their colleagues published a report on their MAGE-3 vaccine in patients selected for their compatible HLA types. Because these were patients with rapidly advancing cancers, only 6 of the 16 candidates received the complete course of treatment—injections of the MAGE-3 peptide once a month for three months. Nonetheless, 3 of the 6 patients with advanced disease experienced what were termed "very significant" tumor regressions, including 2 complete responses, and the side effects were reportedly negligible. After three injections, a Belgian woman with about 100 small melanoma nodules on her left leg began to experience a remission, and four months later, all the tumors appeared to have been attacked by immune cells and have disappeared. A Dutch woman whose melanoma had spread to her lungs also enjoyed a complete remission, with five lung tumors shrinking and then disappearing about three months after her third injection. Updating these results at a meeting in the fall of 1996, Boon reported that of 14 melanoma patients who have received the entire course of vaccinations, 5 experienced significant antitumor responses. The results are both preliminary and puzzling, since regressions are slow to develop and doctors cannot find MAGE-specific T cells in the responders, as would be expected. "We often have to wait more than a month after the last injection to see anything," Boon said. "You kind of start a bandwagon effect that takes a long time to develop."

Alex Knuth, meanwhile, has begun testing a melanoma vaccine using the Melan-A and tyrosinase antigens and has produced preliminary data from two provocative studies in humans, one encouraging and the other sobering. In Phase I safety and dosage tests, patients in Frankfurt received injections of this peptide antigen along with a five-day course of the blood hormone granulocyte macrophage–colony stimulating factor (GM-CSF), which, as its name implies, stimulates the formation of macrophages and granulocytes, two immune cells critical to shaping the early immune response. As of October 1996, when he described some preliminary results at the "Cancer Vaccines 1996" meeting in New York, Knuth reported that the first four patients who received the treatment all responded, including one complete response—"which we consider encouraging," Knuth added, "without overstating the results." Optimism is tempered, however, by the other study led by Knuth's colleague Elke Jäger, which suggests that the use of these vaccines may trigger the Houdini phenomenon in certain cases. In one patient treated with the Melan-A peptide, metastatic tumors that spread after vaccination no longer expressed the Melan-A antigen, the message being that, in certain circumstances, a tumor vaccine may apply pressure on a tumor to evolve and evade the immune system, possibly even accelerating the course of the disease.

Nonetheless, many other researchers are pursuing similar approaches in melanoma—indeed, some are testing the very peptides first identified by Boon's group. Steven Rosenberg at the National Cancer Institute has no fewer than nine trials with melanoma antigens, including the MART-1 antigen administered in several regimens. The Surgery Branch group has seen "sporadic" responses, but not of the same magnitude as those seen by Boon. A group at the University of Virginia headed by Craig Slingluff and Victor Engelhard recently began testing a vaccine based on the gp100 antigen in melanoma patients. In several cases, some of these vaccines include the use of an adjuvant, a somewhat mysterious concoction of oil, water, and detergent that stimulates a generalized immune response.

Once antigens have been identified, they may be presented to the immune system in a variety of different tumor vaccine strategies. Ronald Levy of Stanford University has been testing a vaccine against non-Hodgkin's lymphomas since 1988; antigens in this setting are unique to each patient, but Levy's group has introduced these antigens into dendritic cells and then given these cells as vaccine to induce a particularly strong immune response. A group headed by Drew Pardoll of Johns Hopkins School of Medicine has begun treating patients with advanced kidney cancer with a novel vaccine approach using gene therapy; they have inserted the gene for the molecule GM-CSF into irradiated tumor cells and injected these disabled cells back into patients, where they recruit helper T cells and other immune cells to the tumor; Pardoll says they have seen "some evidence of clinical response." And Michael Lotze of the University of Pittsburgh, using a similar gene therapy approach with a different cytokine, interleukin-12, has seen three "interesting" responses in early trials.

Second-generation approaches are already underway. Both Thierry Boon and Steve Rosenberg have been testing vaccines in which tumor antigens have been introduced into recombinant viruses, which are then injected into patients—with or without adjuvants—in the hopes of triggering a more vigorous immune response. One of the most unusual approaches has been developed by Pramod Srivastava and his colleagues at Fordham University in New York, who have discovered that clean-up crews inside each cell known as "heat shock proteins" tend to gather tumor antigens in cancer cells and thus can be used in the preparation of vaccines. With a sample of tumor, Srivastava explains, a customized vaccine unique to each patient can be prepared in less than a day; a preliminary clinical trial was scheduled to begin in New York by the spring of 1997. "I think we're at the point where it's time to pay the piper, one way or another," said Victor Engelhard. "We've got five to six different proteins in the case of melanoma, at least one good one in the case of ovarian cancer, and one very unusual antigen in breast cancer. So in the next year or two, we're probably going to know if they're going to work or not."

When asked to put all this in perspective, Thierry Boon immediately referred to the experiments done by Aline Van Pel shortly after he returned to Brussels, where animals had indeed "learned" to attack their tumor. These were mice that routinely succumbed to "spontaneous" tumors; but when they were treated first with one of the mutant strains of tumor, they acquired the ability to reject the nonmutated, transplanted form of cancer as well when exposed to it later on. The message, he said, is that when tolerance is broken, when the immune system can be induced to see a little of the tumor, it suddenly sees a lot more. "If you have only one antigen," he said, "you can pass the control. If you have two or three or four antigens, then all of a sudden you really become a target and you really elicit a response. In immunology," the accidental immunologist explained, "one plus one equals ten."

Indeed, with each report of a new tumor-associated antigen, Boon is optimistic that a variety of strategies of immune modulation may shift the odds in favor of the patient. "I believe that if we're able to attack a tumor through two or three or four antigens as opposed to one," Boon said, "we are going to do infinitely better. And the good news is that our analysis indicates that up to now there *are* going to be many antigens we can pursue in every tumor. Now in what sense are we going to do better? First, in one obvious sense, we have shown that a way for tumors to escape the immune system is just to stop making some antigen, deleting the gene or inactivating the gene. And clearly, if we attack them simultaneously through five antigens, it's going to be very difficult for them to delete them all. These are multiplicative percentages of chances, or probability, so that it becomes very unlikely that you can escape all of them. So that becomes one important concept.

"But besides that, I *do* believe that if the immune system can recognize two or three antigens in a tumor, even forgetting about the problem of escaping, the primary attack on the tumor by the immune system is going to be enormously facilitated. That is based on what we have seen [in mice]. You have a series of tumors that don't elicit the slightest response. Now you add one antigen on that, and not only is that antigen recognized, but all of a sudden the others are recognized, too. And in the present-day view of immunology, this is very easy to explain, because the lymphocyte that recognizes one antigen is going to secrete interleukins, interferons, so that every antigen is going to be expressed better, so you clearly see some sort of positive feedback. The more you start being recognized, the more you're going to be recognized further."

Definitive answers about these new vaccines will take several years to sort out, in part because the value of any cancer treatment cannot truly be assessed until the status of patients is known five years after treatment. There is, however, at least one indisputably complete and durable remission that has grown out of the MAGE work, and as she sat in a functional black chair in Alex Knuth's office one day, listening to my questions with a wary, almost grave expression on her long face, reluctant even to make eye contact, I realized that if you saw Frau H on a street corner, at the market, on a subway or train, you would never give her a second glance. When she agreed to meet with me in November 1994, I didn't know quite what to expect, or what she could tell me that her doctors could not. I think now that I mostly wanted to satisfy the notion that extraordinary scientific discoveries can begin in the cells of very ordinary people. And Knuth had forewarned me that

Frau H was an unassuming and down-to-earth woman who "had a rather simplified understanding of the circumstances." Her personality, he added, did not comport with New Age fabulists who sing the praises of cantankerous, angry, and feisty patients. "Very compliant," Knuth said. "She did everything we asked of her."

Frau H arrived in Knuth's office promptly at 11 A.M., looking nothing like the woman whose traumas and tribulations Knuth had spent the previous hour detailing (an ordeal that, in addition to metastatic melanoma, included open heart surgery, episodes of psychosis requiring hospitalization, and the death of her husband from lung cancer). She wore black jeans and a white pullover blouse with lace trim, and perhaps the most remarkable thing about her was simply how healthy she appeared. Her life had blossomed following her remission, Knuth had told me. I took it as an almost unquantifiable measure of good health that she had even lived long enough to quit her job at the post office.

She had very little understanding of what they had done with her blood, only that they seemed to have poked an inordinate number of needles into her over the years in order to get it. Did she know that her cells had been sent to laboratories throughout the world, that important genes had been discovered inside them, that vaccines were being created based upon them? She was vaguely aware of this, but mostly because it involved "many, many injections" and much taking of blood. Was there ever a moment when she felt she might not survive? "No, not really," she said with a shrug. "I never had the feeling that there was an end to it, that it was a terminal situation. There were moments of distress," she continued with admirable understatement. "But I never felt that it was the end." And then she appended a throwaway observation that betrayed the innocence that lay at the heart of this remarkable story. "I never thought it was so dramatic," she said.

A fellow scientist has remarked that Thierry Boon is the "quintessential purist," having made a career trying to understand with utmost clarity an aspect of biology in a single patient. Interestingly, Boon has still never met Frau H in person. Her recovery has now lasted twelve years; her cancer has never recurred, and she remains an anecdote whose medical story will never be told because her doctors did not obtain that conclusive biopsy sample. I asked Thierry Boon if he ever planned to write up her case in the literature. "No," Boon replied, wearing on this occasion his coat of modesty. "No, because this is not the kind of thing you write in a paper."

16

"BEAUTIFUL LIVING THINGS"

*Lymphocytes circulate and recirculate, so that the
cells present in the blood at any time are like the chorus
of soldiers in a provincial production of the opera* Faust—
*they make a brief public appearance and then disappear
behind the scenes only to reenter by the same route.*
—P. B. Medawar and J. S. Medawar, *The Life Sciences*

℮ *On an overcast September morning* in 1993, shortly after 9 A.M.,
a young man arrived at an outpatient clinic of the Fred Hutchinson
Cancer Research Center in Seattle for a seemingly routine infusion.
Several floors above, a technician named Kathe Watanabe, wrapped in
a papery gown a bit like an amaretto cookie, her hair lovingly and
tightly purled into a magnificent two-foot braid, rushed back and forth
across a laboratory, laundering hundreds of millions of cells for a treatment scheduled to begin in less than an hour. These were human
lymphocytes, incubated for weeks in a glass-doored incubator, and if T
cells could see, their view out the lab window would have taken in the
quiescent volcano known as Mount Rainier, looming to the south. We
have been conditioned to think of the eruption that coughed up Rainier as among the most powerful forces of nature on the planet, but
the very same thing could be said about the collective biological
power of the white blood cells gently washed by Watanabe. They were
specifically groomed to protect an immunodeficient patient against a
frequently lethal virus, and they would be performing exactly that job
by lunchtime.

A physician stuck her head in the lab door long enough to ask, "Are
you going to make it in time for the infusion?"

"Oh, yeah," Kathe (pronounced KAY-ta) promised, moving imper-
ceptibly faster, checking her watch. It was 9:10.

People who watch cells for a living—observe them, nurture them, coax them to grow, learn from them—may be the wisest of biologists, not because they know so much (although there is a holistic quality to their knowledge that is not quite so true of the more reductionist wings of biology), but because they have been fully and productively humbled by just how much there still remains to learn, and how much the art of growing cells remains just that: an art, not a science. There are textbooks thick as a brick with molecular knowledge about cells, and yet the vast majority of scientists who attempt to keep lymphocytes alive and thriving in a laboratory dish are, for all their knowledge, like chefs who suddenly can't boil water. The most remarkable thing about the preparations undertaken this morning by Kathe Watanabe was that enough wisdom had been won about this particular lymphocyte, at least in this particular laboratory in Seattle, for it to be grown reliably and to schedule, handled safely and without contamination according to a reproducible technology, and used to treat patients (like the one downstairs in the clinic, just settling into his chair in the waiting room) to prevent illness and, arguably, to save lives.

A little more than four weeks earlier, Watanabe and her coworkers had started with a single cell. It was a single T lymphocyte, but well chosen, with a very specific mission in the complex array of immunological function. Known formally as a cytotoxic T cell (and more colloquially as a "killer T cell") because of its ability to sidle up to and selectively dissolve an unwanted cell, this particular lymphocyte possessed the ability to identify normal cells within the body that had been infected with cytomegalovirus (or CMV), a common and usually mild pathogen to all except people with compromised immune function, such as patients with AIDS or, like the young man downstairs in the clinic, leukemia patients who have recently had their immune systems completely destroyed during bone marrow transplants but whose new immune systems have not totally kicked in. Indeed, one reason for the high mortality of bone marrow transplants is the vulnerability of patients to posttransplant viral infections such as CMV. Over many years, this Seattle laboratory, headed by Philip D. Greenberg, had developed the technology to expand populations of these cells outside the body to therapeutically useful numbers. And the numbers are enormous: like the conductor of an invisible symphony, Watanabe had nurtured, coaxed, and cajoled this one lymphocyte into a rampage of exponential cell divisions, bathing the burgeoning population in ersatz human blood serum to keep the cells happy, adding the cell growth factor interleukin-2 at precise intervals to stimulate the clones to expand, knowing when to let the cells rest, knowing when to restimulate them at exactly the optimal time, poking and prodding them biochemically until, in the space of a month or so, she had created half a billion identical copies of this one lymphocyte—a clonal population, as it is known—lying in milky white clumps at the bottom of six fat test tubes.

Her face pressed to the glass of a sterile laboratory hood and hands moving as deftly as a dealer at a Vegas casino, Watanabe smoothly unscrewed the tops of the six thick, blue-capped tubes with her left hand and maneuvered the pipette in her right to suction off an excess of reddish fluid. Watanabe and Stanley Riddell, the scientist who developed the step-by-step methodology for the care and feeding of these lymphocytes, discovered through a woefully long process of trial and error that T cells are extremely finicky about their environment. In their natural habitat, the blood and immune organs like the spleen and lymph nodes, they are bathed and nourished by serum (the liquid portion of the blood); in these stubby plastic tubes, they seem to flourish better in the presence of serum concocted in the laboratory.

"We wash them twice," Watanabe explained, rinsing the cells a second time, suctioning them in and out of the pipette like a child playing with a straw, washing away whatever cellular excretions might have found their way into the liquid. To the untrained eye, the "washing" occurred in a dilute red fluid, like a watered-down mouthwash. After one rinse, they went into a centrifuge for ten minutes, spun at 1,200 revolutions a minute until all the cells formed a small gray pellet at the bottom of the test tube. She suctioned off the excess reddish fluid, added a rinsing solution, and repeated the process. It was easy to forget that the white cloud bubbling into and squirting out of the pipette was composed of living things.

It is an unfortunate stereotype in the world of molecular biology that those who indulge their fascination with cells are seen to reside on the fringe of biology—eccentrics or mother hens (they seem to be predominantly women), who describe these living creatures with the same fitful and reverent incoherence that one often associates with artists struggling to explain their own work. "I remember in biology class, the first time I was exposed to the inside of cells, when another student pointed out that they were undergoing mitosis," Watanabe had mentioned a day earlier, speaking of the event in which one cell pinches and pulls itself into two daughter cells. "I was stunned. I just *love* cells. That's something Stan and I have in common; we just both love looking at beautifully growing cells. There's some sort of intuition or instinct that goes with being a good cell culture person. A common sense, maybe." At a loss for a more scientific explanation, she added, "Maybe you just need to think of cells as beautiful living things."

At 9:25 A.M., after ten minutes in the centrifuge, Watanabe washed her beautiful living things once again, in and out of the pipette, the tube cloudy with lymphocytes, the liquid now the faint pastel of pink lemonade. "The cells look very good," she said. "They obviously have been proliferating very rapidly." After washing them a few more minutes, she combined all the cells into two tubes, which went back into the centrifuge for a final spin. She consulted a piece of paper, which confirmed that the patient downstairs was scheduled to receive upwards of 540 million of these T cells.

The technology for growing T cells to order is still in its infancy. "If it were less black box than it is," Kathe had said the day before, "everything would be so much easier. We're not there yet. We still sweat it. We still don't know if these cells will be ready for infusion until we get to know them. Because there is tremendous variation in terms of proliferation rate, even within one individual's cells. Among clones that are virtually indistinguishable, one cell may grow three times as fast as the other." It would overstate the biology to suggest that cells have a mind of their own, but their behavior is variable and unpredictable, making flexibility and suppleness an essential part of the job description for a cell horticulturalist. Rarely does medicine get more individual and customized than this.

By 9:45 A.M., as the patient was ushered into a room downstairs in the clinic and offered his arm for the insertion of an IV line, Kathe raced back and forth upstairs in the lab, preparing a 100 cc infusion bag. Then she took a quick census of the cells, staining a sample with blue dye and peering through a microscope. A hand counter clicked quietly in her left hand like tinny castinets, and she pronounced the cells healthy. "We have exactly what we want," she said. "We have 600 million cells." Holding a huge, 60 cc syringe, which looked just about the right size to deliver a dose of insulin to a diabetic elephant, she slowly pushed the milky white cells, thick as viscous pudding, into the infusion bag. She passed the pudding-soft bag to a waiting colleague, who whisked it off to the outpatient clinic.

If this form of cellular farming sounds arduous, it is. If it seems revolutionary, it is that, too. Since treating its first patient with laboratory-grown T cells on July 16, 1991, the Seattle group has infused 26 patients at high risk of CMV infection following bone marrow transplants; only one of the 26 has contracted an infection during the most dangerous window of vulnerability, and all but one are alive today.

The Seattle team has come a long way from the summer of 1991, when the entire lab team anxiously gathered at the patient's bedside to watch the very first infusion of cells. "Phil insisted that we all go," Kathe recalled, tidying up a bit after the bustle. "He said it was an historic event, at least for the lab. It was something we had worked on for so long that he wanted all of us to be there." That first patient apparently didn't share their sense of history. "Basically, he fell asleep," explained Philip Greenberg, the scientist who has provided the medical leadership for this effort. "We had to wake him to say good-bye. Clearly it was a bigger event for us than it was for him."

When I first showed up at the Fred Hutchinson Cancer Research Center to meet Greenberg, I thought an aging hippie orderly had been sent down to the lobby to fetch me. A stocky, wild-haired man, who only moments earlier may have stuck his hand in a socket, shuffled toward me in a blue flannel shirt, gray Dockers, and suede moccasins that looked dangerously close to slippers. The beard and rimless glasses gave him a vaguely middle-aged Aquarian look, but the truly distinguishing feature was that curly bulb of salt-and-pepper hair that silvered out like beams of light; his barber could have been Gino Severini, the Italian Futurist. Greenberg had left his native Brooklyn after high school in the 1960s, but Brooklyn had not left the voice, his accent streetwise and his style of speaking rather like a pot boiling over—urgent, agitated, frothy, *enthusiastic*. Eager to establish his credentials as a denizen of the great Pacific Northwest, he immediately escorted me across the street to the espresso bar in Swedish Hospital. There, for the first of many times that day, he urgently canvassed colleagues to see if they were "going to see Nolan tonight." My first thought was that an eminent immunologist named Nolan was in town to lecture, but it turned out Nolan Ryan would be pitching in Seattle that evening for the last time in his twenty-seven-year baseball career.

Baseball in fact provides the operating metaphor that sums up the philosophy of the Greenberg lab, and one that might easily sum up the record for the entire field of immunotherapy so far—there are no home runs in cancer research. While Greenberg may have the look of a hedonist, he possesses the scientific soul of a penitent. The word that recurs again and again in his conversation is "commitment"; since 1976 he has committed himself and his laboratory to a very methodical, systematic effort to develop techniques for adoptive immunotherapy—"adoptive" meaning that the patient receives cells that have been grown or otherwise immunologically manipulated outside the body, as opposed to active immunization, where the patient receives some stimulant that tweaks his or her own immune system. Whereas in the 1980s Thierry Boon's group successfully used cytolytic T cells as a laboratory tool to identify tumor antigens that then could be developed into a vaccine, Greenberg's group cultivated them as a form of medicine unto themselves—and, to many minds, has been the first to prove that they can reliably provide therapeutic benefit in humans.

It is sometimes said that the smartest thing a young scientist can do is gravitate toward a field where everything is up for grabs, including the field's intellectual integrity, and tumor immunology certainly qualified in the mid-1970s. Greenberg studied biology at Washington University in Saint Louis, obtained his medical degree at the State University of New York at Brooklyn, and then became fascinated during postgraduate training at the University of California at San Diego with medical applications of immunology. At the time cancer immunology was mired in one of its periodic depressions, but one attractive exception was the program at the University of Washington, where E. Donnall Thomas had pioneered the practical use of bone marrow transplantation in the treatment of leukemias. It was also where Alexander Fefer had been one of the conceptual pioneers of cellular therapy, demonstrating in the late 1960s and early 1970s that one could cure experimental cancers in mice by injecting them with specific lymphocytes. "Maybe naively," Greenberg reflected, "you're young enough then to think that overcoming great challenges is not an issue. That all it takes is commitment. So I think my perception was that I would get training in basic immunology and then try to bring that to bear on the state of cancer biology. And in the end I think probably we're doing that. But it sure took a lot of time."

Greenberg joined Fefer's group in 1976 and worked with him in perfecting cellular treatments in animals. Later on, when Fefer tabled his research activities between 1979 and 1981 to help organize the Biological Response Modifiers Program at the National Institutes of Health, Greenberg and Martin Cheever essentially took over the laboratory effort in immunotherapy and published two key papers in 1980 and 1982 showing that infusions of carefully selected lymphocytes could cure mice of the experimental equivalent of metastatic cancers. When, in the mid-1980s, Steve Rosenberg at the National Cancer Institute made the leap to human patients and cured at least one patient with advanced cancer with a combination of cells (the so-called lymphokine activated killer cells, or LAK cells) and cytokines, the Seattle group doggedly stuck to their animal models, trying to understand every aspect of the transfer of lymphocytes before attempting it in people. Indeed, they took what appeared to be a step backward. In order to understand the T cell immune response against cancer, they decided to study a viral disease.

Although Greenberg is reluctant to discuss it, one can easily imagine that this must have been a particularly difficult period for the Seattle lab. Fellow researchers say Greenberg felt Rosenberg had not sufficiently credited Fefer and himself for their earlier work, which they believed laid the groundwork for cellular therapies, and to all the world it appeared in 1985 that Greenberg's philosophy of slow, steady, painstaking investigation had been superseded by a home run by the Surgery Branch sluggers at NIH with their LAK cell therapy. To add insult to injury, a talented and heavily recruited young doctor named Stanley Riddell arrived in the lab in 1986 and, initially dazzled by the work coming out of the Surgery Branch, promptly lobbied Greenberg about shifting the lab's focus to LAK cells in hopes of extending the NIH's apparent success.

A graduate of the University of Manitoba School of Medicine, Riddell came to the Hutch for training in oncology and had allotted but one year for laboratory research. "When I started in the lab," Riddell admitted, "I tried to convince Phil that I should do a LAK project. I wanted a quick hit." After a little reeducation from Greenberg, Riddell realized that using high doses of interleukin-2 produced too much toxicity and left too many questions unanswered. "My problem with it," Riddell said later, "is that you just don't know *what* you're doing. It's created a situation where, because you got a few patients that respond, it continues the mythology." It turned out to be a good decision: virtually no one is currently treating patients with LAK cells.

Greenberg and Riddell set out on a different tack. In the mid-1980s, no one had definitively identified a tumor-specific antigen, so they resolved to work with a well-known virus like cytomegalovirus; the antigens could be identified precisely and could be used to induce subpopulations of CMV-specific T cells. They would capture those T cells and then, just as a zoologist or naturalist would study a rare species, they would examine this particular cell intensely—track it in its natural habitat (the blood), breed it in captivity, feed it, learn its habits, learn everything about its behavior and how to manipulate it, first in mice, ultimately in humans. Only then would they move on to the T cells they ostensibly set out to find, the ones that recognize and attack cancer cells.

Such a mission would indeed take commitment. "It wasn't like there was a method that you could take out of a previous paper or book," Riddell noted. Several groups of very able viral immunologists had looked for the same beast, and failed.

By all accounts, Stan Riddell is an exemplary cell watcher, a conjurer who can keep lymphocytes alive and coax them to grow. Yet it took him more than two years to find the type of cell he needed, another three years to learn how to grow them properly, yet another three to test them tentatively in human patients. The effort has been rewarded by being widely perceived as one of the most elegant and thoughtful manipulations of the immune system yet attempted for the treatment of human disease.

The first order of business was to find the right T cell. In the pecking order of immune cells, the cytotoxic T cell sits atop the pile for a very persuasive reason. These cells have the molecular equipment to detect a virus hiding inside a cell and the power to selectively destroy only the cells acting as safe houses to pathogens. Just as there are millions of distinct and unique individuals who belong to the Caucasian subset of *Homo sapiens*, each endowed with special skills and abilities, there are millions of distinct and unique individual cells that belong to the thymus-derived (or T) subset of lymphocytes; they are unique because of a small appendage on the surface of the cell known as the T cell receptor. There are approximately 10^{12} T cells in the human body—roughly a trillion, or ten times the number of stars in the Milky Way. Each one is antigen-specific, meaning that each one is designed to identify a single, short, distinct molecular bump, usually about nine amino acids long, that dimples the surface of any cell that contains a virus, a bacterium, a parasite, or even—it is now beginning to emerge—normal proteins in a cell undergoing some sort of stress, including the stress of cancer. Like sculpted tongs, the T cell receptor reaches out and "feels" the topography of these small bumps literally at the level of atoms, as if detecting the three-dimensional whorls and grooves of a molecular fingerprint. Among those one trillion cells are a few, for example, with molecular tongs able to recognize various antigens associated with cytomegalovirus, a common viral infection.

It was Stan Riddell's job to figure out how to find those cells attuned specifically to CMV, to grow them outside the body, and then see if massive infusions of them could tip the balance to prevent viral infections in bone marrow transplant recipients. If the approach was methodical, the interest was hardly academic: up to 80 percent of transplant recipients reactivate their latent CMV infections, and up to 40 percent develop disease. Pneumonias caused by reactivated CMVs are especially lethal in this setting.

In a sense, cytomegalovirus presents an enormous target for the immune system. CMV is the largest of the DNA viruses that afflict humans, with about 200 genes (the human immunodeficiency virus, by contrast, possesses only 11). CMV is also fairly ubiquitous; about 90 percent of us have been infected by it at one time or another, which means virtually all of us have a few memory T cells in permanent circulation in the blood that remember what CMV looks like. But once they bump into the antigen they are genetically programmed to recognize, these T cells begin a galloping replication, copying themselves, spewing chemicals that recruit other cells to the scene, and creating a proliferating wave of investigators all looking for the same suspects: cells that display the viral antigen on their surface. In the body, the emergence of these expanding T cell populations usually takes about seven to nine days if the body is seeing an antigen for the first time, so that a week or two after an infection, a noticeably elevated number of these specific white blood cells can be detected in the blood; once immune memory has been established, however, a second encounter with a virus or bacterium can trigger a T cell response in one or two days. A practical demonstration of this latter response time is the common tuberculosis skin test—the angry welt that may appear on the skin in a day or two reflects the surge of reactivated T cells to the scene and their intense proliferation in response to an antigen they have seen before. This is technically referred to as the delayed-type hypersensitivity response.

It is one thing to identify such a lymphocyte, quite another to grow (or clone) it. "We literally spent eighteen-hour days, seven days a week, to get these cells to grow," Riddell recalled. "In the beginning, you had to look at the cells every day. You had to make decisions every day. And there were a lot of times where you made decisions that were wrong. The cells died." Indeed, to grow cells in culture, they learned, was a lot like gardening: it wasn't simply a matter of having the right plant for the right location, but also choosing the right soil, how much to feed it, when to water, and perhaps most important, realizing that cells are living things, as fickle and unpredictable, as individual and subject to biochemical whims, as the scientists nurturing them.

Since T cells flourish in the blood and in lymph nodes, the Seattle researchers made a conscious effort to emulate the milieu of blood and lymph in their sterile plastic flasks. "It was the philosophy from the very beginning," Greenberg said. "That was the commitment. That you would make cells that grew in a normal fashion, that responded in a normal fashion, that looked like normal lymphocytes, that could rest like normal lymphocytes. I think we recognized very early that the closer you made a cell to the kind of cell that circulates in the blood, the more likely you were to get a cell that would function."

Easier said than done. Riddell and Watanabe concocted a liquid environment (or "medium") out of serum that the cells seemed to like better than over-the-counter elixirs. T cells need to see antigen periodically to keep in fighting trim; Riddell and Watanabe had to figure out a schedule for adding antigen to their T cell cultures. T cells need chemical signals, in the right amount at the right time, to proliferate: Riddell and Watanabe discovered that only low doses of interleukin-2, the growth factor for T lymphocytes, seemed to elicit steady, healthy growth (high doses of IL-2 actually made the T cells look and behave abnormally, forcing many into a mass cellular suicide known as apoptosis). Like any living creatures, T cells could not be relentlessly driven like livestock to divide and multiply; the researchers learned that their protocols needed to incorporate sharply defined periods of rest between stimulation. At first, all these tricks gained them double or quadruple the number of cells every two weeks; later they achieved 1,000-fold expansions per cycle, and reduced the time of each cycle to nine days. Just as important, they listened to their cells constantly—at night, on weekends and holidays—and didn't hesitate to break their own rules, or protocol, if the cells told them to. It required science, but also, for lack of a better word, intuition. "If it looks like they're not doing so well," Watanabe said, "maybe give them IL-2; if they're doing well, if they're proliferating, you don't. It's pretty obvious when cells need to be fed or when they're dividing. Beyond that, I don't know—I can't define it."

In the spring of 1991 they were ready to test the cellular therapy in humans—specifically, people who had recently received bone marrow transplants. These patients are particularly vulnerable to viral infection during a window beginning about forty days after transplant, when the cellular arm of the immune system is still rebuilding itself; about 65 percent of these patients lack the T cells that specifically kill the virus that causes CMV pneumonia. So, four to five weeks after transplantation, the Seattle group began the first of four weekly infusions of cells grown in bottles (the cells, incidentally, came not from the patients, of course, but from the bone marrow donors; just like hearts and kidneys, blood cells can be rejected if their tissues don't match the recipients' immune systems, which in this case derive from the bone marrow donor). The beauty of the system developed by Riddell and Greenberg was that they built in molecular techniques by which they could measure exactly how many of these cells persisted in the body, and for how long.

Finally, in July 1991, they had prepared 300 million T cells for "Patient 6032," a fifty-three-year-old man who had undergone a bone marrow transplant five weeks earlier. "The major issue was: Was this going to be safe?" Greenberg said. "When we gave the cells to the patient, we had both that anxiousness and enthusiasm. We were all talking and being hyperkinetic, and he just closed his eyes and nodded off."

The results in the first three patients, though limited, indicated that the transfused T cells did exactly what they were supposed to do: in patients receiving immunosuppressive drugs and whose immune responses were severely compromised, large and measurable numbers of these cells persisted for up to twelve weeks, and they retained their ability to kill CMV-infected cells. None of the first three patients developed pneumonia during the critical period up to 100 days after transplantation, although one became ill nine months after the transplant and died of a transplant-related infection. When these results were published in *Science* in July 1992, Johns Hopkins immunologist Drew Pardoll extolled the work as "among the most exciting advances in human immunotherapy since people started exploring it."

Since 1992, an additional eleven transplant patients were treated with T cell infusions, and all fourteen now have been spared CMV infection during the period of high risk; at least half would have been expected to become infected, according to statistics. The results on these first 14 patients were published in October 1995 in the *New England Journal of Medicine,* and another dozen have been treated since then. These preliminary experiments suggest two things: that the adoptive transfer of T cells can provide a significant boost to immunodeficient patients, and that the slow methodical approach can be every bit as effective in arriving at medical solutions.

The success of the CMV trial has led to two next steps, one obvious and the other audacious. If the cells worked in immunodeficient patients after bone marrow transplants, would they work in immunodeficient patients with AIDS? And could Greenberg and his colleagues do nature one better and build an improved T cell? Could they customize it, reengineer it to work better against cancer and AIDS? "I think there are lots of ways you can use these kinds of systems," Greenberg explained, "to not only have lymphocytes do what they've been programmed to do, but to do things that *you* can design them to do."

"Cells," Brad Nelson announced toward the end of a conversation, "are like Swatches. They all have the same internal mechanism, but they all have different faces." If Phil Greenberg is a throwback to the 1960s, Brad Nelson is a Gen X geneticist; the thirty-four-year-old has been known to show up at fairly prestigious scientific meetings in a T-shirt, has a serene, choirboy demeanor to match his boyishly handsome features, but can roll up his sleeves like any mechanic in a genetic chop shop and start customizing the genes of cells as if they were engine parts. In a world where cells are like Swatches, Nelson and lab colleague Mark Gilbert have begun to function as watchmaker's apprentices, creating a different cellular face for every occasion.

Nelson has focused on the cytolytic T cell, or killer T cell, which is distinguished by a surface marker known as CD8. These are known as "effector cells" because they are the agents directly responsible for destroying cancerous or virus-containing cells in the body. In order to tip the balance in favor of immune protection in immunocompromised patients, these cells need to be injected into the body in artificially high numbers similar to the quantity achieved by Stan Riddell and Kathe Watanabe in the test tube. Nelson has been trying to save his labmates the trouble of rushing like mad to prepare cells in the lab by tricking those very same cells into proliferating while they're still in the body. His strategy derives from the knowledge that T cells proliferate on the basis of two signals. T cells receive their first cue for growth when their surface tongs (the T cell receptor) encounter a foreign antigen; but they require a second signal, which is chemical and usually washes up out of the watery brine surrounding all cells in the form of the molecule interleukin-2. IL-2 is made and secreted by a kindred lymphocyte known as the "helper T cell" (distinguished by a surface marking known as CD4).

To circumvent this double signaling system, and not least to get around interleukin-2's prodigious toxicity, Nelson has genetically short-wired these growth signals. Normally the activated T cell is spurred to grow when its receptor for interleukin-2, sitting on the cell surface, detects the presence of the molecule and sends a signal to the cell nucleus to initiate cell division. Nelson has spliced two different genes together to outfit his T cells with a hybrid (or "chimeric") receptor built to receive one message from the outside, but tricked into sending a different message inside the cell. He did this by patching together different pieces of DNA and introducing the gene into lymphocytes. For the surface of the cell, he used the gene encoding instructions for building the outside-the-cell portion of the receptor for granulocyte macrophage–colony stimulating factor (or GM-CSF), a common and generally nontoxic immune system hormone that in fact has been used safely as a human drug for a number of years.

That is only half of the construct. While one part of the chimeric receptor dangles outside the cell membrane awaiting messages from GM-CSF, the other part extends inside the cell, where it initiates a process known as signal transduction; this so-called intracellular domain translates the message received by the outer antenna into a biochemical signal telling the nucleus to turn on a particular gene. At precisely the junction where the GM-CSF receptor passes through the cell membrane, Brad Nelson spliced a totally different receptor region, this one for interleukin-2. Thus a molecule of GM-CSF, which these T cells routinely make when they encounter an antigen, sends a message to the nucleus telling the T cells to multiply. By stitching these two receptor genes together, Nelson has shown that T cells exposed to GM-CSF think they are receiving the interleukin-2 signal, which is their cue to expand. "The cell thinks it's seeing interleukin-2," he explained, "and therefore by producing GM-CSF, it can promote its own growth." He has kept these self-watering T cells alive for nearly a year without ever seeing IL-2.

The Greenberg lab, however, was mindful of a potential danger. Ordinary T cells, if present in sufficient number, can cause excessive and toxic inflammatory reactions if they encounter a large amount of antigen, as might occur in an active HIV infection. Could you ever get rid of these engineered T cells? The problem in a sense has already been worked out for Nelson by his colleagues Stanley Riddell and Mark Gilbert in HIV patients. Following the success of cellular ther-

apy against cytomegalovirus (CMV) in bone marrow transplant patients, the Greenberg lab naturally wondered if a similar approach
would work against the human immunodeficiency virus (HIV) in patients with AIDS. To build in an extra safety margin, Gilbert and Riddell, working with researchers at Targeted Genetics Corp. of Seattle,
inserted a so-called suicide gene into these T cells. The gene encodes a
protein from the herpes simplex virus known as thymidine kinase
(commonly abbreviated as TK), which is precisely the vulnerable biochemical point attacked by the antiviral drug ganciclovir; inserting
the TK gene makes the T cells suddenly vulnerable to this known
drug, which can be administered safely and without toxicity for a
short period of time.

Phase I safety studies of these reengineered cells began in 1994, and
the unexpected results to date have had implications far beyond immunotherapy. The Seattle team saw "dramatic" antiviral effects in
some patients, according to Stan Riddell, but these effects were surprisingly short-lived; HIV in the blood appeared to decrease, and the
vulnerable helper T cells had increased in number following adoptive
immunotherapy. However, a puzzling and intriguing complication occurred. As Riddell and Greenberg reported in *Nature Medicine* in February 1996, the supposedly compromised immune system of HIV
patients in the Seattle trial performed only too well when it came to
reacting to the genetically altered lymphocytes. In five of six patients,
even their decimated immune systems recognized the engineered T
cells as foreign because they contained the inserted TK gene and effectively cleared them from the body within days. "What this is really
saying," Riddell said, "is that immune recognition of genetically altered cells could be a real obstacle to gene therapy. For immunotherapy, the silver lining is this. The magnitude of the immune responses
we got was encouraging, and it may be telling us that even in patients
with weakened immune systems, vaccination with gene-altered T
cells can be very powerful."

Legitimate questions remain about the ultimate utility of adoptive
immunotherapy, not least the fact that it is labor-intensive and expensive. Greenberg says that prophylactic treatment with ganciclovir to
prevent CMV infections in bone marrow transplant recipients costs
roughly $5,000 to $6,000 per patient for the entire course of therapy,
about the same as cellular therapy. However, Greenberg argues that

cells may be more economical in the long run because ganciclovir treatment, unlike T cell infusions, incurs serious side effects that often result in hospitalization, and once the drug is stopped, some patients have developed late CMV disease. Others question how practical the technology may ultimately turn out to be, since this so-called adoptive immunization essentially lasts only as long as the transfused cells. "I believe if there is one person in the world who can [make it work]," Thierry Boon has said, "it's Phil Greenberg."

Nonetheless, the Seattle group's effort has demonstrated that antigen-specific immune cells can be manipulated outside the body and administered safely, effectively, and predictably, with minimal side effects. "I think people would like it to be easier than it is," Greenberg said. "They'd like sort of a format or a recipe, but the problem is that it actually takes an enormous amount of care and effort, and just a sense of what's going on. The reality is that in the beginning, Stan spent nearly two years working on a project that had the potential of going nowhere if he couldn't do it. . . . If you're not going to be lucky and step up to the plate and hit a home run—and you know, even in the batting cage that doesn't happen real often—then you gotta be realistic that you will need to keep punching away and solving each of the small problems. So we've tried to take a very systematic approach all along. And indeed, I'm sure that makes it slower at times than it might otherwise be, but in the end, what we've done, we think, has been very definitive, and will stand up very well. And, at least for us, each observation and each solved problem has always represented a building block for the next step."

After the pessimism that suffused the field in the early 1990s, Greenberg believes there is, finally, a solid foundation to build upon for immune-based cellular medicine. In addition to the viral diseases now being tested, cellular treatments for melanoma, Hodgkin's disease, and leukemia are being developed. "It's going to be incredibly exciting in the next five or ten years," Greenberg said. "I think that for the treatment of some viral diseases, certainly viral diseases that occur in immunocompromised states, we're very close to having approaches that will work, at least first-generation approaches. I think we're even getting close to having second-generation approaches. The first-generation approach is the cells; the second-generation approach will be genetically modifying those cells. And the third-generation approach may be ways to circumvent even needing to generate the cells, ways to not have to do this in tissue culture, but to do it in people, in a sense have the human body do the tissue culture work, which would obviously be easier. And I really believe that we are at the cusp now of being able to cross over from treatment of viruses into effective specific therapies for malignancies. Not globally for all cancers—I think we are still a ways from that. But I'm confident that we're going to find that there are approachable tumors, and the combination of more hitters, in terms of cells, and better pitches to hit, in terms of target antigens in tumors, will finally start yielding some home runs."

17

"DR. LEVY'S FAVORITE GUINEA PIG"

The antibodies are magic bullets which find their
targets by themselves, so as to strike at the parasites
as hard and the body cells as lightly as possible.
—Paul Ehrlich

There are different ways to measure the unexpected cure of an incurable illness—scientific ways, to be sure, but also more subjective, occasionally even *aquatic* measures as well. In June 1981, doctors at the Stanford University Medical Center in California puzzled over some unusual changes in the blood of a terminally ill lymphoma patient named Philip Karr shortly after they began treating him with an experimental therapy known as monoclonal antibodies. Mr. Karr at the time was staying with his wife at an apartment in Palo Alto with a swimming pool, and the disease left him so weak and depleted that he could barely make it halfway across the width of the pool; ten days into the treatment, he found he could make it across a full width; and by the fourth week of his month-long treatment, he was swimming lengths. "That was when I knew," Karr said many years later of his surprisingly swift return to fitness.

Meanwhile, back at the medical center, Karr's doctors analyzed his blood after several treatments and noticed that at a microscopic level, their patient was also doing swimmingly well. His hemoglobin and platelet counts began a startling rise toward normalcy within a week or two, signs that a bone marrow clogged with cancerous white blood cells had begun to restore order and produce normal blood again. Still, doctors earn their keep by remaining less impressed than their patients with the mere appearance of change. "I didn't believe it," Ronald Levy recalled many years later. Levy, the quiet-spoken and serenely enthusiastic Stanford physician who was in charge of Karr's care, continued to disbelieve until a week later, when x rays clearly showed Karr's enlarged lymph nodes in full retreat. Only then did his doctors dare believe what Phil Karr already knew in his heart and legs: that he was on the way to a remarkable recovery.

Phil Karr's case occupies a special place in the history of immunotherapy, in its sociology and commercialization as well as its science. Fortunes have been made, huge amounts of capital have been raised, reputations burnished and rewarded, a handful of lives saved, and many hearts broken at least in part because of this one much-publicized remission. The important thing, certainly for Phil Karr, is that in 1996, fifteen years after he was given less than a year to live, Karr is alive and well and more than happy to discuss the circumstances of his case, which in a circuitous fashion has led to a promising new vaccine-based therapy for lymphoma patients.

A physicist by training and an industrial engineer and designer for most of his career, Phil Karr first found out he was deathly ill at a time when he was feeling reasonably well. The year was 1975, and at the time Karr was a sixty-two-year-old transplanted midwesterner living in Santa Barbara, California, enjoying the toys of his retirement. That autumn, he had gone into the hospital for minor hernia repair and came through the surgery beautifully. But when his doctor walked into his room the next day, he disclosed that during the operation, an enlarged lymph node had been discovered. The telltale swelling of lymph nodes, literally rock hard with densely packed, cancerous blood cells, suggested what a biopsy confirmed: a lymphoma of B cell origin.

B cell lymphomas, which include non-Hodgkin's lymphoma, are cancers that arise in lymphocytes, the very cells designed to protect the body from foreign attack. Friends of Phil Karr suggested that he go to Stanford University Medical Center, one of the leading institutions of research on the West Coast for cancers of the blood. At first, his doctors at Stanford, who included Ronald Levy and Richard Miller, did nothing. As the disease progressed over the course of a year, a Stanford researcher named Thomas Merigan tried leukocyte interferon (made in Helsinki, incidentally, according to Kari Cantell's formula), and Karr recalled that it was "dramatically effective," causing a partial remission that lasted several months. A second course of genetically engineered interferon produced a less dramatic partial response for one year. Karr's doctors bought another year with a six-month course of chemotherapy.

Lymphomas have the habit of waxing and waning; lymph nodes swell and shrink, obeying some occult biological rhythm. But physicians know that especially in the low-grade cases, the disease is chronic and comes back again and again, increasingly resistant to any form of therapy, and then essentially becomes incurable. By April 1980, Phil Karr's disease took the inevitable turn for the worse, just as his doctors had exhausted all the conventional options. The organs where lymphocytes congregate—the lymph nodes, liver, and spleen—bulged with cancerous cells; tumors cropped up on his scalp. Chemotherapy no longer halted or slowed down the progress. At one point in May 1981 the fatigue became so intense that Karr and his wife cut short a vacation in Arizona and returned to Stanford. When Levy and Miller walked into the room to examine him, Karr had no illusions. "This is it, isn't it?" he said.

His doctors had no illusions, either. "From a lot of experience that we have now with patients like this," Levy said recently, "his life expectancy was about six months."

However, that wasn't the message Levy and Miller delivered that day. "I guess we should tell him," one of them said, and Karr remembers the two doctors looking at each other "like the cat that swallowed the canary," until one of them explained, "We've got this stuff called monoclonal antibodies." They proceeded to describe this new therapy. "They asked if I would cooperate," Karr recalled, "knowing that I would, and I responded, 'Of course.' I felt I had no choice—the drumbeats were about at their end, and getting closer."

But there was a hitch. "It isn't quite ready," one of the doctors continued. "Go home, hold on, and we'll call you when it is."

Unbeknownst to Karr, Levy and Miller had given this same little "hold tight" speech to several other lymphoma patients. "We had been through the procedures on other patients and actually made products for the patients, but they didn't survive long enough to get them used," Levy admitted. "I thought we'd waited too long with Mr. Karr. He had a lot of manifestations of disease. Low reserve in his bone marrow, partly due to the chemotherapy results, and also skin sites of lymphoma in his scalp and other places. Large liver, large spleen, large lymph nodes. And so I thought we'd actually waited too long before trying the antibody that we had made for him."

During the long year of Phil Karr's decline, David Maloney and

Richard Miller, researchers in Levy's laboratory, had taken Karr's B cells and used them as the starting point in the creation of a novel form of medicine. B cells, like T cells, belong to large, extended families of related lymphocytes. Genetic shuffling produces millions of variations on this immunological theme, each subtype of which has the capacity to produce a unique, one-of-a-kind antibody. Indeed, the Greek root for "unique" is idiotype, and that is the term used to describe the specific shape and structure of antibody molecules, which sit on the surface of B cells like receptors, tasting and testing scraps of protein in the bloodstream. Those B cells are present from shortly after birth, and each idiotype may do nothing more for the lifetime of the individual than sit on the B cell like a bump on a log, waiting for a visitor who never comes. But if that receptor encounters a specific scrap of foreign protein that "fits" its idiotype, be it a bit of protein from the coat of a virus or bits of biochemical brick that form a bacterium's cell wall, the idiotype receptor makes the connection, and bells go off in the B cell. The receptor sends a signal to the cell nucleus, the B cell throbs to life, and within hours the idle B cell has transformed itself into a "plasma cell," shedding the specific idiotype protein like buzz bombs into the bloodstream, all intended to identify and tackle the protein that stirred the response in the first place.

In patients with lymphomas like Phil Karr's, however, certain B cells fail to mature properly; in fact, all it takes is one single B cell, failing to mature after its formation in the bone marrow, that becomes caught in a kind of repetitive loop of proliferation during its adolescence. Since all the deviant B cells descend from this same wayward progenitor, they are said to be a "clonal" population, and they replicate without pause until they begin to overrun the marrow and other organs of the immune system. At this point, the patient usually begins to notice symptoms: reduced ability to fight infection, fever and night sweats, incredible fatigue due in part to the fact that even the red blood cells get crowded out. The errant B cells congregate where normal B cells normally go, in the lymph nodes and spleen, and at a certain point, swelling in those organs announces the most palpable sign of an immune system gone awry.

Because of their unique, one-of-a-kind nature, however, cancerous B cells are marked cells, leaving themselves open to the same kind of immunological attack usually reserved for germs. Antibodies all share the same molecular foundation and structure, known as the constant region; it is the business end of the molecule, known as the variable region, which comes in an almost infinite variety of molecular guises.

In this sense, the antibody might be likened to an electric screw-driver—a general housing into which are fitted interchangeable tops or heads, each head (or idiotype) specific to a different job, each one "fitting" a different shape of screw (or, in an immunological setting, a different shape of antigen). There are idiotypes specific to each pathogenic occasion: idiotypes that fit, hug, and hold molecular markers that are particular to cold viruses, an idiotype that grabs parts of the measles virus, an idiotype that engages certain bacteria.

In an aberrant B cell of the sort that causes lymphoma, the variable region—the business end of the screwdriver—is unique enough, and in great enough numbers, to become a target; indeed, it appears foreign and thus vulnerable to attack by immunological agents, including other antibodies. Hence, according to then-emerging technology of monoclonal antibodies, "man-made" or "anti-idiotype" antibodies could be cultivated—farmed, literally, from the blood of mice, which had been injected with the idiotype protein that appeared on Phil Karr's malignant B cells. Especially in its infancy, the technology took the better part of a year—about six months to create cells that spewed out unlimited amounts of the aberrant idiotype, another six months to create mouse monoclonal antibodies to this protein.

Towards the end of May 1981, Maloney and Miller, assisted by a modest army of graduate students and technicians, finally succeeded. They built a custom-designed antibody, specifically tailored to the precise contours of Karr's rogue B cells; indeed, it was called an anti-idiotype antibody precisely because it targeted that portion of the cancerous B cells that was unique to the cancer cells, rather like creating a screw to fit the unique head of a screwdriver. Beginning at the end of May, Karr received two doses a week of these monoclonal antibodies over a period of four weeks. By the end of the second week, both Karr and his doctors knew something dramatic had occurred. By the third and fourth weeks, the fevers and night sweats had abated, the blood counts began to return to normal, the tumors had disappeared, and the liver and spleen had returned to normal size. Phil Karr's cancer, without a single additional treatment, had gone into complete and durable remission. "The way I put it," Karr said later, "is that I won the lottery." He immortalized the winning ticket with a vanity license plate on his car that read "MONO 1."

A report on Karr's recovery in the *New England Journal of Medicine* in March 1982 caused a sensation, and the Karr case became a cause célèbre in the field of immunotherapy. Desperate patients rang the phone off the hook in the Levy laboratory; funding poured in; Phil

Karr's story became a staple of articles, books, and documentaries, his likeness memorialized as a cancer pioneer in a pavilion of Disney's EPCOT Center in Florida. Venture capital poured into biotech companies specializing in monoclonal antibodies, in no small part, according to Levy, because of the Karr case. Unbeknownst to his doctors, Karr even sent a confidential and unsolicited letter to Armand Hammer, the wealthy industrialist and longtime patron of cancer research, describing his remarkable recovery while under the care of Levy and Miller. "Both are probably too modest . . . to go after any of your rewards," Karr wrote; Stanford subsequently received a $500,000 gift from the foundation to upgrade Levy's lab. When Hammer had occasion to meet Phil Karr at a reception several years later, he greeted him heartily as "Dr. Levy's favorite guinea pig."

"What isn't widely known is that he is still fine," said David Maloney, who spent a year handcrafting the antibodies that were used in 1981 to treat Karr and is now an oncologist at the University of Washington in Seattle. Indeed, Karr's case took several intriguing twists after its much-publicized first act, including the fact that his complete response was not permanent. In July 1987, when Karr underwent coronary bypass surgery at Stanford, Levy took advantage of the occasion to ask the surgeons to biopsy any suspicious lymph nodes as well as provide samples of bone marrow and blood; tests showed no evidence of the cancerous B cells. But Karr developed an infection on his lower leg where veins had been removed for the bypass surgery, and soon thereafter Karr noticed the first of what would turn out to be three areas of purple discoloration on his skin near the incision on his lower leg and foot, which soon developed into nodules.

"He came up and showed it to us," Levy said, "and we biopsied it, and it was the lymphoma that he'd had six years prior. It was *exactly* the same tumor, exactly the same idiotype. And apparently it had been evoked in that site by the infection—this is our theory. Presumably the release of inflammatory factors set off the growth of the tumor cells, which were obviously in his body all along in some kind of dormant state, in a subclinical form." Treated with radiation alone in 1988, the localized tumors disappeared, and Phil Karr has never been treated since. "So I take from that whole story," Levy said, "that he harbors tumor in a dormant state induced by the treatment, kicked off at least originally by the treatment, and that he holds his tumor in check. He must have it in his body, but holds it in a dormant state."

When I spoke to Karr around the beginning of 1996, he was full of octagenarian vim and vinegar. He had just undergone another, more sophisticated molecular workup at Stanford with no trace of the rogue B cells in his blood. His has been an exhilarating clinical success, and if the use of monoclonal antibodies proved as helpful to others as it did to Karr in 1981, the discovery would truly have augured, as the 1984 book *Magic Bullets* proclaimed on its cover, "the coming revolution in cancer therapy" and "the most exciting adventure in the annals of modern medicine." Those were the giddy dispatches from what one observer has dubbed "the age of optimism." The reality is somewhat more complicated.

"Most people know about his case in the sense that they know that it worked in one guy," Levy said recently in an interview. "They mostly think that it never worked again in anyone else, and they think that he failed and must have died by now. But actually none of those things are true. We have a series of patients now, six or seven of them, who have been induced into remissions that have lasted longer than the six years that the initial part of Phil Karr's remission lasted. They're still in remission now. They represent about 10 percent of all the patients that we treated this way. So we replicated that result over and over again, showing that custom-made antibody against this target can induce long-lasting complete remission of the disease. We have another series of people who have gotten shorter remissions, but not as complete. Altogether, in a series of maybe fifty patients that we treated that way, three-quarters of them had a clinically beneficial response. We would still be doing this today for *everyone,* because it worked so well, if it weren't for the difficulty and the expense."

Although Phil Karr has remained alive and healthy all these years, the field that tried to ride his coattails has never quite enjoyed such robust health. For the better part of the last twenty years, monoclonal antibodies represent biotechnology's best example of how premature claims from early clinical trials can give a good technology a bad name.

In 1975 Cesar Milstein and Georges Köhler, working at the Laboratory of Molecular Biology in Cambridge, England, stunned the world of biology by announcing the discovery of monoclonals. The technology was as breathtaking as it was simple: they created a "hybridoma," as it was called, by fusing an antibody-producing B cell from a mouse with a mouse cancer cell known as a myeloma. This hybrid cell married the exquisite specificity of an antibody-producing cell to the tireless manufacture of these critical immune proteins by immortal mouse cancer cells. Almost instantly, monoclonals promised a revolution in human diagnostics and therapy. Whether used as unadorned "naked" antibody or armed with radioisotopes or toxins, these agents tantalized biologists—and venture capitalists—with their potential to home in on tumor cells and eradicate malignancies. "Magic bullets" represented the next wave.

With a perspective of two decades, the much-touted monoclonal, naked or otherwise, has earned the reputation of a microscopic emperor without clothes, at least in the clinic. As a laboratory tool, monoclonals have been an unqualified success; to cite one of countless examples, their ability to bind and pluck the CD4 antigen on helper T cells has allowed researchers to follow the populations of these critically important immune cells in patients with AIDS, and their use as laboratory reagents to define cells and antigens has revolutionized the study of immunology. But the field has struggled mightily in recent years against a general perception that monoclonals have failed to deliver the goods as therapeutic agents—a perception that has been reinforced by an almost unrelieved succession of flops in clinical trials.

The failure of monoclonals has been the stuff of fabled Wall Street stock ascents that have the longevity and throw-weight of bottle rockets; the technology may have, until recently, enjoyed the dubious distinction of producing more millionaires than cures. One especially noteworthy crash, ably and tartly chronicled in Robert Teitelman's *Gene Dreams*, witnessed the swift rise and cruel fall of a Seattle-based company called Genetic Systems, which made its Wall Street debut in June 1981 and was swallowed, expensively, by Bristol-Myers in 1985 for $300 million. Genetic Systems merely provided a handsome model for a rash of entrepreneurial cliff divers in the 1990s. During 1993 and 1994, for example, a staple in the business press was the story about a biotech company's stock "plunging" or "plummeting" after clinical trials of a monoclonal to treat sepsis were halted; so many companies suffered this fate, including Centocor, Xoma, and Synergen, that one observer referred to antisepsis drugs as "the Bermuda triangle of the industry." As an epitaph to the "age of optimism," Eli Lilly & Co., which with great fanfare paid $350 million for the monoclonal biotech company Hybritech in 1986, dumped it in the fall of 1995, reportedly at a garage-sale price of less than $10 million. Some basic researchers privately go so far as to accuse biotech companies of fouling the nest of the fledgling technology by rushing into premature, flawed, and ultimately doomed clinical trials. "They all had to be the first to fail," groused one prominent university researcher in the field. "The Hybritechs, the Centocors, the Xomas—they just wanted to get something half-assed out there, and it's very unfortunate. It's made it very difficult for the rest of us."

Given the rocky history of monoclonals, therefore, it is not surprising that a tone of apologia and contrition has enveloped the field. At a conference held toward the end of 1995, for example, Lloyd J. Old, scientific director of the Ludwig Institute for Cancer Research, put all the past enthusiasms in sobering perspective when he stated, "The difficulties were underestimated, the timeline was unrealistic, and the claims were overstated. Add to this the pressure of keeping financial investors excited, and you have a sure formula for disappointment and disillusionment." Or, as Jean-Pierre Mach of the University of Lausanne observed more succinctly, "Some people were selling the skin of the bear before killing it."

The perceived "failure" of monoclonal antibodies, however, may be the starkest example of a cultural as well as a technological problem that plagues all of immunotherapy, and much of biotechnology, too. Success or failure of a new biomedical technology—and the timeline for measuring such results—tends to be measured against the stopwatch of financial return, while the punctuated equilibrium of scientific progress can often be clocked with a sundial. And no matter how fancy or sophisticated the work in a laboratory might be, high-tech medicine—unlike other high-tech industries such as electronics or computers—is always forced to slam on the brakes in its headlong rush to the marketplace when it hits the stage of clinical trials, because human biology is not nearly so forgiving of blunders as a computer is of a bad chip. Measured against the stern metronome of quarterly earnings reports, many new medical technologies will—unless extraordinarily lucky—be seen to fail in the short term. For all the combustive hype that attended monoclonal antibodies, ignited by legitimate scientific excitement, fueled by venture capitalists, and fanned by popular press accounts, very serious technical hurdles arose almost immediately, and as Lloyd Old has also noted, "Though you don't get any prizes for identifying the problem, you need to do so before finding the solutions."

Despite many successes in curing animals, significant problems have arisen in humans, not least the inconvenient fact that mouse-derived antibodies elicit a vigorous human immune response that in some cases limits therapy to onetime doses. Many researchers have discovered that although they found monoclonals that reacted to antigens with great affinity, the antigens themselves weren't necessarily the best ones for inducing clinical responses. And, as Ron Levy and his Stanford colleagues learned, sometimes monoclonals just aren't economical. The creation of custom-made antibodies turned out to be prohibitively expensive; the Stanford researchers estimate that the treatment that saved Phil Karr's life would, in the industrial setting, cost about $50,000 per patient. Around 1985, Levy and Richard Miller helped to form a company called IDEC Pharmaceuticals based in San Diego to commercialize the technology that saved Phil Karr's life; but, as Levy recalled, "They eventually started to make calculations about how much it was costing them to produce and how much they could charge for it in the marketplace, and came to the conclusion that it wasn't economical."

Despite all the full-dress pessimism that has dogged the technology of monoclonal antibodies, people in the field may have finally succeeded at throwing out the bathwater while hanging on to a surprisingly healthy baby. Second- and third-generation approaches to monoclonal antibodies are in full flower. These include an attempt to identify better tumor-specific antigens, efforts to attack the scaffolding of tumors, and the use of monoclonals to jam internal signals by disrupting signal transduction pathways (the signaling process that converts a message received by a receptor on the outer cell surface into a message to the nucleus) and cell cycle events. Genetic recombination technology also allows the creation of less immunogenic "humanized" antibodies designed to reduce reactions in patients, accomplished by genetically stitching the genes for the human constant region to the genes for a specific idiotype raised in mice. A team headed by Aya Jakobovits at Cell Genesys, Inc., a biotech company in Foster City, California, has engineered a strain of mouse that can create a large repertoire of "fully human" antibodies.

Moreover, several therapies appear on the verge of establishing broad clinical utility, especially against hematologic cancers such as lymphomas and leukemias, which appear more vulnerable to attack than bulky, solid tumors. The promising story of the anti-B1 antibody, first isolated in Stuart Schlossman's lab at the Dana-Farber Cancer Institute in 1980, suggests that the gestational period for a therapeutic monoclonal may be well over a decade. This antibody targets an antigen known as CD20 on B cells, and a number of researchers have been working with it to treat lymphomas. Oliver Press of the University of Washington has produced impressive results over the past eight years against non-Hodgkin's lymphoma, a B cell cancer of the immune system that will strike approximately 51,000 people in the United States this year. In the late 1980s the Seattle group began testing several different monoclonal antibodies linked to a radioactive isotope of iodine, ^{131}I, during a Phase I trial in patients whose disease had relapsed. Following preliminary biodistribution studies, a first group of 19 patients was treated with varying doses of radiolabeled anti-CD20, and every patient had a measurable response in the Seattle trial, including 16 complete responses. Eight of the 16 have remained free of disease from three and a half to eight years. In a recently completed Phase II trial of the anti-CD20 monoclonal at the maximum tolerated dose, Press said 17 of 21 patients receiving the treatment experienced complete responses.

"Using this rather radical approach," Press said, "one can obtain a high tumor response rate, with complete responses and partial responses seen in more than 90 percent of the patients treated." The Seattle group has now begun a Phase I/II trial combining the "hot" CD20 antibody with two chemotherapy drugs, again in the treatment of non-Hodgkin's lymphoma. But the bullet in this treatment, Press admitted, is more radical than magical. Patients must endure relatively high doses of radiation, must undergo a bone marrow transplant following treatment to overcome the destructiveness of that radiation, and therefore end up in the hospital for four weeks. "Our bias," Press says, "is that in order to get cures and long-lasting remissions, we're going to need high doses of radiation."

There may, however, be room for low-dose radiation approaches, as Mark S. Kaminski and coworkers at the University of Michigan Medical Center have begun to demonstrate. The Ann Arbor group has also focused on non-Hodgkin's lymphoma. In a recently completed Phase I trial in patients who had failed chemotherapy, doctors pretreated patients with unlabeled, "naked" anti-CD20 antibody to "fill sinks," such as the liver and spleen, where antibodies tend to get sequestered before they reach tumor tissue; they then administered a follow-up onetime dose of "hot" anti-CD20 antibody. Twenty-two of 28 patients responded, including 14 complete responses; the response rate in patients whose tumors had become resistant to chemotherapy was about 70 percent, including instances where the tumor burden exceeded half a kilogram. Median duration of complete responses exceeds fifteen months, and because the dosage was one-seventh the Washington regimen, patients required only a three-day hospital stay. "The main toxicity, other than reversible bone marrow suppression," Kaminski said, "was boredom." A Phase II/III multicenter trial is currently underway. Both the Press and Kaminski approaches are being supported by Coulter Pharmaceutical of Palo Alto, California.

IDEC Pharmaceuticals, the company founded by Levy and Miller, also has a CD20-based monoclonal in advanced trials. David Maloney, who headed some of those trials while still at Stanford, explained that CD20 is a good antigen because "it's only expressed on B cells. So it's not tumor-specific, but it's lineage-specific to the B cell lineage. It's not on any other cells. The advantages of this antibody-based therapy is that virtually all patients with lymphoma can be treated with it, and the response rate looks pretty good." Preliminary results suggest that complete plus partial responses total about 40 percent of those treated. Maloney is conducting a follow-up trial adding interferon to the regimen, since an earlier trial suggested that the combination improved responses.

Levy's group, too, has continued to work with a CD20 antibody experiment. The Stanford team described preliminary results at the 1995 American Society for Hematology meeting showing that a single dose of this monoclonal, linked to a radioactive chemical known as yttrium, obtained a 72 percent overall response rate in 18 patients, with 6 complete responses. "It's actually inducing remissions in about half the patients," Levy said, "and those remissions are lasting on the average eight to ten months. And again, there aren't too many side effects from it."

"I'm pretty optimistic about this form of therapy," Maloney said. "First of all, I think it's the best results that have been seen with any antibody, with any single antibody, anywhere, either published or unpublished at this point. We're getting responses in people with bulk disease. Abdominal masses. Shrinkages of enlarged spleens, livers, clearing of bone marrow. Enlarged adenopathy. I mean, multiple-centimeter-sized nodes have regressed completely. We've had some complete remissions now that seem to be pretty significant, although the most important thing will be duration. We still don't know what the real impact is."

Other monoclonal approaches that are further along in clinical testing have produced somewhat more mixed results but may nonetheless prove useful in the long run. A German group led by Gert Riethmüller of the University of Munich has shown in a large, randomized prospective trial that the use of monoclonal antibodies resulted in increased seven-year survival for patients with colorectal cancer. And Genentech, the biotech company in South San Francisco, has reported modest response rates for their HER2/neu antibody in the treatment of advanced breast cancer. These and similar results prompt Riethmüller to predict that "avoidance of overt metastatic disease may become a realistic alternative in the not-so-distant future."

As in other areas of immune-based medicine, a simplistic understanding of immune function even twenty years ago led to oversimplified hopes for a home run in the clinic with monoclonal antibodies. One of the key problems facing clinicians to date has been the selection of the right antigen to target, and understanding exactly what that antigen does. But even a decade of basic research has led to increasingly shrewd and sophisticated uses of monoclonals to target, disrupt, and otherwise confound the normal behavior of cancer cells.

If the antigen happens to be a growth factor receptor sitting on the surface of a tumor cell, for example, an antibody could be used to bind to the receptor and block the molecule that normally interacts with it, depriving cancer cells of crucial growth signals and therefore disrupting cell cycle events. That is precisely the approach pursued by John Mendelsohn and colleagues at Memorial Sloan-Kettering Cancer Center in New York. They have begun a clinical trial using what is called a chimeric mouse-human antibody, which binds to the receptor for a growth signal known as epidermal growth factor (EGF). This receptor seems especially populous on numerous tumors, and Mendelsohn is testing the hypothesis—suggested by preclinical experiments—that a combination of antibody and traditional chemotherapy will have a synergistic effect that enhances the killing of tumor cells.

A team at the biotechnology company Genentech headed by Napoleone Ferrara has described a similar effect with a similar use of monoclonals—using them to jam the biological signals that induce the formation of blood vessels feeding the tumor, known as vascular endothelial growth factor (VEGF). VEGF is present in all the cells that line blood vessels, but the level of expression found in tumors, Ferrara said, "is orders of magnitude higher than in normal tissues." It is also specific; researchers have found only two tumor cell lines in which VEGF is not expressed. In animal experiments using a cancer known as rhabdomyosarcoma, an antibody that blocks VEGF resulted "in a dramatic suppression of tumor growth," Ferrara has reported. In a subsequent experiment, the Genentech group tested the combination of antibody and the chemotherapy drug cisplatinum against the same tumor. "When we combined the chemotherapy with the monoclonal antibodies," Ferrara says, "there was really a remarkable regression." Indeed, the combination actually killed tumor cells, while each agent alone merely arrested the growth of the cancer cells. Two other promising antibodies seem to offer better specificity to solid tumors. One is called A33, which is expressed on more than 95 percent of colon cancers and is being tested by Sydney Welt and colleagues at Sloan-Kettering; another is the BR96 monoclonal, which, conjugated with the chemotherapy drug doxorubicin, has induced regressions of some spontaneous cancers in dogs, according to Ingegerd Hellström and colleagues at the Bristol Myers Squibb Research Institute in Seattle.

Several other strategies, although not so far along in the clinic, suggest how much the "naive" belief in antibodies in the early days has matured into rather ambitious biological engineering projects. One new strategy calls for shifting the focus from tumor cells per se to semi-innocent bystanders. Although the vast majority of human tumors originate in epithelial cells, Wolfgang Rettig of the German pharmaceutical company Boehringer Ingelheim has taken the approach of "looking for cells other than epithelial to attack," and has made a case for stromal cells. Stromal cells are like anatomical scaffolding—they provide support and occupy a kind of buffer zone between capillaries and the tumor tissue proper. These cells resemble fibroblasts, cells that form the connective tissue and yet are morphologically and functionally distinct from normal resting fibroblasts. They churn out growth factors, extracellular matrix proteins, and other proteinaceous excretions that suggest they have somehow been activated and recruited to the service of the neighboring tumor.

Rettig's group serendipitously discovered that a monoclonal named F19 interacts with a highly specific antigen on the surface of these tumor-activated fibroblasts. Dubbed "fibroblast activation protein" (FAP), this protein appears to be expressed normally in fetuses and newborn children, during wound healing, and in stromal cells surrounding solid tumors. In a recent biodistribution experiment done in conjunction with workers at Memorial Sloan-Kettering, 17 patients whose colon cancers had spread to the liver were infused with radiolabeled anti-FAP antibody to test its ability to traffic to tumor sites. In 14 of the 17, including 2 patients who were negative on CAT scans, the site of liver metastases was clearly identified. Moreover, FAP may offer an appealingly broad target; Rettig noted that the protein appears in about 90 percent of lung, breast, colon, and pancreatic tumors, offering a potential target in many common tumors.

Perhaps the most ambitious bit of engineering, since it literally joins the two arms of the immune response in one fanciful creation, is the "T-Body." Zelig Eshhar of the Weizmann Institute has pioneered efforts over the last several years to develop a genetic construct combining the variable region of the antibody molecule with the T cell receptor. This remarkable engineering feat aspires to take advantage of the particular strength of each immunological agent—antibodies latch with high affinity onto target antigens and, unlike T cells, do not require the elaborate presentation by MHC molecules, while T cells, once activated via their receptor, turn into especially efficient cell killers. "T cells have no problems of accessibility," Eshhar said. "They are known to reject and eliminate organs, reject and eliminate tumors." Eshhar is developing constructs using both cytolytic T cells and natural killer (NK) cells, but the agents have not yet been tried in humans.

Because of these innovations, monoclonal antibodies in cancer therapy, Old said, "has entered the age of realism." Other researchers suggested they'd been realistic all along. "Our methods are still crude," said Oliver Press. "We probably have this tool that's going to be useful, and we may not know the best way to use it yet. I think people feel apologetic that the field did not deliver on public expectations, but I always thought that it would take a long time to satisfy the expectations raised in the popular press, so I was neither surprised nor disappointed. I think we're making slow, steady progress."

There is a sense among some researchers that antibody and cellular approaches to immunotherapy are converging in the area of tumor vaccine work. Ironically, this approach might have begun more than a decade ago if Ron Levy had followed his own instincts rather than the advice of his peers; even more ironically, the story once again begins with Phil Karr.

"From the very beginning," Levy recalled, "even before we treated Mr. Karr, I remember writing a letter to George Stevenson in England." Stevenson, at Southampton University Hospital, had laid the theoretical groundwork for an anti-idiotype vaccine and performed some of the early animal experiments; he was also, to put it bluntly, a competitor with Levy in the race to be the first to try an anti-idiotype antibody in humans. At the time, Levy knew from animal experiments that he could cure lymphomas in mice by one of two routes: simply by injecting an antibody specifically raised against the B cell idiotype (a "passive" approach), or immunizing them as with a vaccine by injecting the idiotype itself—which in fact is also a protein—and letting the animals use their own immune systems to make antibody to this foreign protein.

"I wrote Stevenson telling him, 'We have a patient, we have a custom monoclonal anti-idiotype antibody, and we have the patient's protein, his idiotype protein. What do you think the best thing to do is? Should I give him the antibody, or should I immunize him with the protein?' And George wrote back and said he would give the antibody." At a major immunology meeting during that same period, several other experts in the idiotype field offered Levy the same advice. "I was suspecting that giving the protein and vaccinating would be better," Levy said recently. "But we did it with antibody, it worked, and we were in the game of giving antibody for quite some time."

In the mid-1980s, after it became clear that customized antibody therapy was prohibitively expensive, Levy and his collaborators belatedly resurrected the concept of a vaccine. As in the Phil Karr case, they began by assessing the cancerous B cell in a patient, identifying its unique idiotype, and then custom-producing the idiotype protein. But instead of going to the trouble of creating monoclonals to the protein, they would simply inject it directly into patients as a form of immunization and, it was hoped, stimulate the patient to form his or her own antibodies. In 1988 the first patients at Stanford received this form of lymphoma vaccine. In its current formulation, the cancer-specific idiotype protein is coupled to another, strongly immunogenic foreign protein from a marine gastropod called keyhole-limpet hemocyanin (or KLH). This combination is injected with an immunostimulatory potion known as an adjuvant, which typically contains an oil, a polymer, and a detergent to create an additional commotion in the blood and elicit an especially vigorous immune response.

As usual, the immune system responded with a few surprises for the doctors. All the patients treated with the vaccine had first achieved clinical remission with standard chemotherapy. A pleasant surprise has been that to date 20 of 38 patients with advanced non-Hodgkin's lymphoma mounted immune responses to the vaccines, and those same patients seemed to have longer-lasting responses; the scientific surprise has been that these remissions have been associated not just with antibodies, which was expected, but with T cells, which was not. Levy and his colleague Larry Kwak, now at the National Cancer Institute, reported the initial results in a 1992 article in the *New England Journal of Medicine,* and the Stanford laboratory updated the results in the first 38 patients at the December 1995 meeting of the American Society of Hematology in Seattle. "The results are very impressive," Levy said. "Half of them have made antibody responses against their own tumors, and half have not, and there's a big difference between their outcome. In the ones who made an antibody response, their tumors have remained in remission. In the ones who did not have an antibody response, their tumors have all come back." Levy cautioned that, short of a randomized controlled trial, these results suggest a correlation but don't prove the point. Nor is the treatment a cure, although it seems to buy a substantial amount of time for patients—on average, the responders go nearly five years before their disease begins to progress, while nonresponders have progressive disease in little more than six months.

Even more recently, Levy and coworkers have experimented with a promising vaccine against lymphoma, which combines the idiotype protein with dendritic cells. Dendritic cells are the immune sentinels, similar to macrophages and marbled throughout the body's tissues; they have enjoyed a great deal of notoriety of late because, as Ralph Steinman at Rockefeller University has convincingly shown, they are particularly powerful rabble-rousers when it comes to stirring up cellular immune responses. A number of researchers are therefore pursuing a strategy of creating vaccines using dendritic cells, including Michael Lotze and Walter Storkus at the University of Pittsburgh with melanoma and Bijay Mukherji at the University of Connecticut using Thierry Boon's MAGE-3 peptide. In the Stanford group's approach, they remove blood from lymphoma patients, separate out the dendritic cells, "pulse" them with the idiotype protein unique to each patient's cancer, and return these cells to the body. Dendritic cells loaded in such a way are believed to sound a particularly loud reveille for the immune system, and in a preliminary study with a handful of patients the Stanford group has been able to induce immune responses and tumor regressions. It took a relatively small number of dendritic cells, on the order of 10 million, to provoke a fairly strong T cell response. "We were always wondering whether T cells played a role here or not," Levy said. "It seems they do, based on what we just did with these dendritic cells. So our plan for the future is to combine the two, the dendritic cell vaccine approach and the KLH-protein approach." Larry Kwak is testing yet another promising variation on the theme; his NIH group has begun a clinical trial in which low-grade lymphoma patients, after standard chemotherapy treatment, receive a vaccine of the idiotype-KLH form plus local injections of the cytokine GM-CSF, which is emerging as a potent adjuvant in the cancer vaccine setting.

Phil Karr is living proof that immunological approaches to cancer can work. That there are not more Phil Karrs walking around these days has been seen by others as proof that such immunological interventions do not work often enough. Ronald Levy, who has lived the ups and downs of the monoclonal field as intimately as anyone, is if anything more optimistic than ever about immunological approaches to cancer. "I think that it's the only chance we really have to control systemic disease," Levy said. "And I think that it's a matter of time, although we've been saying this for a long time now. But it really seems just around the corner before we'll have a way, for certain kinds of cancer, of stimulating systemic immunity, once the tumor has been reduced to the minimum by the usual treatments. . . . I think that the science has moved forward and ideas have been perfected, and the competition hasn't done very well. The standard therapies [for cancer] have plateaued, and I think that there are a lot of different strategies that look like they might work."

18

A PIECE OF
MISLEADING NEWS

In science nothing is quite new.
—Otto Westphal,
"Hommage à Valy Menkin," 1987

๛ *Ferdy Lejeune had* what he now likes to call his "naive" idea toward the beginning of 1987. At the time, Lejeune practiced medicine at the Jules Bordet Institute in Brussels, where he had been a surgeon and oncologist for fifteen years, and if you met him, naive would probably not be the first word to come to mind. Handsome, well-dressed, garrulous, worldly (as befits someone whose primary schooling took place in colonial Leopoldville), rakishly enthusiastic, with a swashbuckling gleam in his eyes, Lejeune possesses ample amounts of the surgeon's requisite confidence, and not a little charm. A good thing, too, because he proposed a clinical experiment destined to send a shudder through the collective heart of the medical community, or at least that portion of doctors who knew a lot about cytokines. He proposed treating certain cancer patients with a drug called tumor necrosis factor, or TNF, but it was more than that. He proposed bathing the tissues of these patients in TNF—using ten times the amount that most adults could tolerate.

What made this sound positively foolhardy rather than simply naive is the fact that physicians in the United States and Europe had just spent two years testing TNF in people with advanced cancers and came away from the experience suffering from about as much shock as their patients. "There were very few responses," recalled one doctor, Douglas Fraker, "with a lot of toxicity." Of all the commotions stirred up by immune molecules, TNF instigated perhaps the rowdiest reaction in physiological terms. The side effects included a particularly brutal, life-threatening version of the "cytokine flu," with fever, chills, and perilous drops in blood pressure.

Ferdy Lejeune proposed turning up the volume on this very dangerous commotion.

Each cytokine has a history unto itself, as unique and idiosyncratic as the researchers who first hold it up to the light of the world. One of the most unusual and arguably the longest story belongs to tumor necrosis factor, a molecule that traces its lineage without exaggeration back to William Coley's cancer treatment, back to turn-of-the-century typhoid vaccines, and back to a peculiar trick of inflammation known as the Shwartzman effect, in which the immune systems of animals can be manipulated in such a way as to cause furious localized hemorrhages.

In those premolecular days it was known that bacterial endotoxins, little bits of bacterial debris from gram-negative bacteria, produced profound physiological effects. Few people have described those effects better than the essayist and researcher Lewis Thomas, who studied endotoxin for thirty-five years and aptly described it as "a sort of signal, a piece of misleading news. When injected into the bloodstream, it conveys propaganda, announcing that typhoid bacilli in great numbers (or other related bacteria) are on the scene, and a number of defense mechanisms are automatically switched on, all at once. When the dose of endotoxin is sufficiently high, these defense mechanisms, acting in concert or in sequence, launch a stereotyped set of physiological responses, including fever, malaise, hemorrhage, collapse, shock, coma, and death. It is something like an explosion in a munitions factory."

There was of course keen interest in what actually constituted the explosive material, and Murray Shear did pioneering work on bacterial endotoxin at the NIH in the 1940s to narrow the search, identifying a portion of the bacterial cell wall known as LPS that lit the fuse. But no one could identify the actual dynamite until the Memorial Sloan-Kettering Cancer Center researcher Lloyd Old had occasion to investigate further. Old, who dresses in the gray suits of a banker and speaks with the quiet authority of a minister, had been fascinated by bacterial endotoxins for many years, beginning with Coley's toxins, and performed some of the original immunostimulatory experiments with bacillus Calmette-Guérin (BCG). In 1971 Old's longtime associate Elizabeth Carswell began a seemingly innocent repetition of an experiment that had already been done by other researchers. That experiment had shown that if you gave BCG to a mouse, waited two weeks, and then injected endotoxin, the syncopated combination proved profoundly toxic to the mice—the timing, in other words, caused an especially violent explosion.

"We were very interested as to why this was," said Carswell, a personable and down-to-earth researcher whose training prior to joining Memorial Sloan-Kettering in 1957 amounted to a bachelor's degree in history at a small liberal arts college in Ohio. But she and Old collaborated well (they are still coworkers, more than thirty years later), and she had a facility for working with animals, so she tried the same two-step process—first BCG, then endotoxin—in a strain known as Swiss mice. "We noticed that these mice got very shocky after we gave them the endotoxin," Carswell recalled. "Within half an hour, they were getting diarrhea, they were getting very quiet, they were getting ruffled fur. And we knew approximately when they might die, so I bled them about an hour and a half after I gave them the endotoxin, took that serum, and spun it down." This serum was then given to leukemic AKR mice to see if it had any antitumor effects, and the animals responded very nicely.

Apprised of these results, Old suggested testing the same serum in a different animal cancer, a fast-growing sarcoma in mice called "meth A," which was known to be resistant to both chemotherapy and radiation. As Carswell later admitted, Old customarily suggested many experiments, and not all of them were done in a timely fashion. "A lot of times I'd forget or just say, 'Okay, I'll do that one of these days,'" Betsy Carswell said with a laugh, "but for some reason I listened to him this time." Carswell first injected these cancer cells into several mice and then waited seven days for "nice, small, little rosy pink tumors" typical of sarcoma to develop. Then she injected blood serum from the mice that had been treated with the BCG and endotoxin combination. "The next afternoon I went in and opened the cage and picked up the mice," Carswell said, "and here their tumors were just . . . *black!*"

Unlike so much in molecular biology, here was a reaction that didn't require a microscope or luminous radioactive chemicals to see. It was immediate, dramatic, unmistakable. Something in the serum, some factor produced in the blood, caused the tumors to hemorrhage, turn black, and dry up—to necrose, to use the technical term. Astonished by the sight, Carswell recalled being "very excited. And of course I ran right in and got Dr. Old."

Like the good and cautious scientists that they are, Carswell and Old repeated the experiments to confirm the results. What is unusual, what distinguishes those medieval days prior to genetic engineering and recombinant DNA and biotechnology patents, when the pace of modern biology had not yet reached mach speed, is that they continued to repeat their experiments and double-check their conclusions, without publishing a single word about their findings or breathing a word of it at scientific meetings, *for four years!* It is probably safe to say that in the increasingly hectic world of molecular biology, such cautious, worried husbandry of data will never happen again.

On the one hand, they appeared to have identified something in the blood that caused tumors to liquefy; on the other, the experiments smacked of dabbling in medieval humors until they could pin down the active agent. And in truth, there was a critical immunological issue to settle. What actually caused the tumors to shrivel up and die? Did the tumors in the transfused mice respond to traces of endotoxin, which would have had the same antitumor effect? Or had the endotoxin in turn provoked an immune substance—some factor, some molecule downstream in the immune response—which on its own could kill tumors? Had they, in short, simply passed along the "misleading news" of endotoxin, or had they indeed found something new?

Even by Old's exacting standards, they proceeded slowly. They repeated the experiments. Work slowed when Carswell relocated, along with all the animals, to a Sloan-Kettering lab in Rye, New York, in 1973, then resumed the work up there. They brought in a protein chemist, Saul Green, to try to purify the factor they had identified. They tried to anticipate every alternate explanation, every other feasible scenario. And the same answer kept coming back. They had found a factor, something other than endotoxin, that killed these experimental tumors in mice. Finally, in June 1975, they finished a paper, which Lewis Thomas (then head of Sloan-Kettering and a member of the National Academy of Sciences) passed on to the journal *Proceedings of the National Academy of Sciences*, where it was published the following September. They called their molecule "tumor necrosis factor," or TNF.

That at least is what they called it on the west side of York Avenue on New York's Upper East Side. Across the street, at Rockefeller University, two biochemists, Anthony Cerami and Bruce Beutler, were characterizing a molecule they called "cachectin," which seemed responsible for the severe wasting seen in patients with cancer and infectious diseases. At one point, in what one camp later called an "interesting quirk," Old's group sent a preparation of TNF across the street, where they came to the conclusion that the two factors were different. But by the mid-1980s, when the protein chemists and molecular biologists had their say, it was apparent that tumor necrosis factor and cachectin were one and the same molecule. To all but a few holdouts, TNF is the name that has stuck.

Few molecules allude by name with such vividness to their activity. Injections of tumor necrosis factor basically caused tumors to hemorrhage, soften, and turn black. And it was precisely that effect that set cloners at biotechnology companies throughout the world salivating at the prospect of treating cancer with a magical molecule that caused tumors to shrink and disappear. In 1985 an executive at the Cetus Corp. predicted annual sales of $500 million if TNF worked against cancer.

Judging from press accounts in the 1980s, tumor necrosis factor formed part of the "Big Three" of potential magic bullets, along with interferon and interleukin-2, in molecular biology's assault on cancer, and the race to clone the gene for tumor necrosis factor probably represented the last great competition of biotechnology's early gold rush years. In 1980 Charles Weissmann and colleagues cloned alpha interferon; by 1982, David Goeddel and his colleagues at Genentech had cloned eight other interferon genes, including interferon gamma; by the spring of 1983, Tadatsugu Taniguchi in Japan had cloned interleukin-2. Now came the turn of TNF. Many of the usual suspects gathered at the starting line: Goeddel's group at Genentech; Walter Fiers at the State University of Ghent in Belgium (on behalf of Biogen); a group at Cetus, the California-based biotechnology company; and researchers at the Japanese company Asahi, to name a few. Most had been on the same roller-coaster before; all knew the thrill and agony of a high-stakes cloning race. To make a swift cloning story even swifter, the Genentech team captured the flag in 1984, and despite early hints of toxicity in animal experiments, several companies moved with unusual speed into clinical trials in 1985.

The results were disastrous. Few clinical trials in biotech's admittedly short history have doused more scientists with colder water in a shorter period of time than those first early attempts to use TNF in humans. In Phase I studies to determine what is called the maximum tolerated dose (MTD), doctors found that anything more than 400 micrograms (400 millionths of a gram) per day sent patients into a dangerous tailspin: side effects included fever, rigors, headaches, and most significantly, plunging blood pressure. Critics—and there were more than a few—hastily buried the molecule in a shallow grave. Bruce Beutler and Anthony Cerami, the two biologists who had discovered "cachectin," even implied that the name "tumor necrosis factor" was false advertising; they suggested that any tumor-killing activity that had been ascribed to the molecule was an "epiphenomenon"—completely unrelated, secondary, in short a mistake.

By 1988, two short years after the gene was cloned, hardly anyone talked about TNF as a magic bullet anymore. As a *New York Times* headline put it, it was a molecule with a "dark side," a molecule with too much toxicity and too little efficacy. The disappointing first round of trials prompted two TNF experts, David R. Spriggs and Steven W. Yates, to wonder plaintively, "Will TNF ever 'necrose' tumors in humans?"

Somehow, though, Ferdinand Lejeune didn't get the message. He had followed the trials, knew how disappointing they had turned out, but thought he had a better idea. Well after TNF had become the black sheep of the cytokine family, Lejeune proposed treating certain cancer patients with *ten times* the dose of tumor necrosis factor that had scared everybody else away.

When I visited Lejeune at his office in the Fondation du Centre Pluridisciplinaire d'Oncologie in Lausanne, Switzerland, he explained how the idea had come to him. He possessed a wide, friendly face, with the hint of a cowlick at the rear of his silvery hair, and merry eyebrows that practically did jumping jacks when he wished to emphasize a point. The full mustache gave him an avuncular look; the tortoise-shell spectacles gave him a scholarly cast; and constant stolen glances at his watch made it clear he observed the harried schedule of a surgeon.

As an oncological surgeon, Lejeune had for many years performed an operation on melanoma patients that, interestingly, has never been formally approved for use in the United States, although several facilities have performed the surgery on an experimental basis for nearly forty years. The procedure is called an "isolated limb perfusion" and can provide benefit to patients with malignant melanoma if their tumors are confined to a single limb, such as a single arm or leg; indeed, one of the great merits of the procedure is that it can save a limb from amputation. Since the 1950s, when the surgery was first attempted, surgeons like Lejeune have in effect tied off the affected limb from the rest of the body with a surgical tourniquet to minimize systemwide side effects of chemotherapy and then bathed the tissues of the cancer-ridden limb with unusually high doses of an anticancer drug called melphalan. Only a limited number of melanoma patients qualify for such grueling surgery, but over the years doctors who have performed the procedure claim complete response rates of about 50 percent, meaning half of all patients undergoing the procedure see all their tumors disappear.

While performing these operations at the Jules Bordet Institute in Brussels, Lejeune wore another hat. He maintained a laboratory where he developed experimental surgical techniques with an immunological accent. He completed a thesis in 1975 on the activation of macrophages, the immune cells that respond early on during an immune crisis and are considered part of the nonspecific response. "I've been always interested in what macrophages do," Lejeune said in an interview; he knew the work of Lloyd Old and colleagues at Sloan-Kettering very well because macrophages, it turns out, begin to pump out huge amounts of tumor necrosis factor when they are recruited to the scene of an injury or infection. "And since I had been involved with isolation perfusion with high-dose chemotherapy for *years,* I thought that maybe . . ." Here Lejeune shrugged his shoulders like a raconteur. "We just have the idea. We just copied what we did with chemotherapy, which is [to give] tenfold the maximum tolerated dose [of melphalan] for system application. We naively took tenfold the MTD for TNF, you see. You know, it's like Christopher Columbus," he added with a laugh. "A ridiculous idea, but it was a good one!"

The Brussels team first tried the idea out on a fifty-three-year-old man in October 1988. Despite using ten times the safe level of TNF, they believed they could control the molecule's awesome side effects by restricting the drug to the isolated limb. The patient in question had more than forty tumors on one thigh extending down to the knee. "This was a gentleman who had been perfused before," Lejeune recalled. "Therefore the quality of the isolation was not very high, and he had a leakage during the perfusion." One of the reason that TNF is so toxic is that cells throughout the body have receptors for the molecule; if TNF is present in the blood or intracellular fluid, this piece of bad news travels very fast. These cells pick up TNF's signal, which is a cue to go into shock, the principal characteristic of which is that the patient's blood pressure suddenly nosedives (a condition known as hypotension).

"Therefore, during the operation itself, he already started to show hypotension," Lejeune said. "After half an hour or something, the anesthesiologist said, 'Well, we see like what we see when somebody goes into septic shock.'" The patient's blood pressure plummeted, the heart rate began to gallop, and fluid began to collect in the lungs, preventing good oxygen exchange; the man exhibited all the classic symptoms of septic shock, the kind of sudden and critical collapse of vital signs that occurs in patients infected with toxic shock syndrome and other acute episodes of bacterial poisoning. And so the anesthesiologist, a woman named Patricia Ewalenko, calmly said, "I will do just like I do with a genuine septic shock." She gave the patient plasma, fluids, and low doses of the drug dopamine, which enhances kidney function, and he bounced back quickly. More important, when they went to inspect the man's forty-odd melanoma nodes after the operation, they knew immediately that something extraordinary had happened. They saw what had previously been seen only in mice and the pages of journals: blackened, shriveled tumors. Necrosis. Every single tumor had in effect been strangled by the drug. I asked Lejeune what he said to this first patient. "I said, 'Good luck,'" he replied with a laugh. "You know, he experienced a complete response."

They tried the technique on a second patient, who had more than a hundred nodules. Lejeune walked over to his desk and came back with a book, which he opened to a section of color plates. "That is *two* days later!" Lejeune explained, pointing to a photograph of the woman's leg. The tumors, previously skin-colored, had turned black, not only due to the death of the cancerous tissue but because of the hemorrhage that TNF judiciously provokes in cancerous, but not normal, tissues. They were "a bit disappointed" when the third patient treated had only a minimal response. "Then," Lejeune says, "I went back to my reading."

What Lejeune and Danielle Liénard, his principal associate, did next, toward the beginning of 1989, has been repeatedly debated by researchers in the United States and elsewhere, in part because almost everyone agrees they would have been stopped cold by authorities in this country before proceeding and in part because exactly such regulatory constraints highlight a fundamental paradox between the way medical bureaucracies work and the way nature designed the immune system to work. Lejeune and Liénard hit the books and refined their recipe for melanoma with a bit more freewheeling flair than is normally the case in academic medicine, almost like chefs rescuing a passable but not entirely successful stew.

"I realized," Lejeune later said, "that a lot of good success in experimental tumors—mouse tumors and even in human tumor xenografts—was obtained when they did a combination. Because in many models, what you see is that the tumor collapses and then regrows. And especially regrows when you stop giving the TNF. . . . And they showed that with [combinations] they have much better responses and much longer-lasting responses. That was with chemotherapy, and also with interferon gamma." So Lejeune decided to pretreat his patients for two days with a low dose of gamma interferon, because this cytokine is known to stimulate cells to increase (or "up-regulate") their deployment of TNF receptors, presumably making them more responsive to the molecule. For good measure, they decided to use hyperthermia—deliberately elevating the temperature of the patient during the treatment—because this had been used for years in melphalan perfusions. Then, in addition to high doses of TNF, he would add a traditional dose of chemotherapy, attempting to improve on the 50 percent complete response rate achieved with melphalan. "I wanted to see whether, as compared to melphalan, I could by adding have a synergistic effect," he said. "So we just switched to this triple combination after the third patient."

It is not exactly clear how Lejeune's team received institutional approval to experiment with such a ragout of potent molecules on what would ultimately be twenty more patients in the study. When asked about the review mechanism in place at that time in Belgium, he admitted that it was "extremely supple." The first procedure won approval on the basis of compassionate need. When one of the melphalan perfusion patients began to relapse, Lejeune's team received permission to try TNF alone. "In that way, we were not obliged to go to the ethical committee," Lejeune explained. "Nowadays, it would not be like that. Here in Switzerland, I had to submit this protocol to the ethical committee. But in Belgium that was never the case because it was immediately taken as a great advantage, and since it was compassionate, then there was no—how do you say?—no commercial link to the use of it."

Whatever the circumstances—and even Lejeune admits that regulations were "extremely relaxed" in Belgium—the team at Jules Bordet ultimately tried this unusual triple therapy for the first time in the spring of 1989. The patient, a fifty-eight-year-old woman with melanoma tumors on her ankle and foot, qualified for all the usual reasons. In less than two weeks all nine tumors disappeared.

❧

Perhaps the most impressive response occurred in the next patient they treated. She was a seventy-six-year-old farm woman who came to Brussels in the summer of 1989; after a lifetime of working in the fields, the woman had developed a melanoma on her leg that had spread to nearby lymph nodes. "The usual rule," Liénard said, "is that with one positive lymph node, the survival at five years is 50 percent and with two or more positive lymph nodes, survival is 30 percent." Asked the prognosis in this case, Liénard later said, "She should die, probably." As her treatment history would reveal, the cancer was extremely aggressive.

Lejeune and Liénard treated the woman with a simple TNF perfusion in June 1989. She showed no response to the drug. A month later, they tried again with a combination of TNF and melphalan, and the response was dramatic. "She experienced a complete response," Liénard recalled in an interview. However, the cancer rebounded once again. "The melanoma recurred in March 1990, and she was reperfused with the triple combination of TNF, interferon gamma, and melphalan in April of that year." Six years later this patient is still alive, completely without any trace of her aggressive melanoma.

In fact, every one of the first twenty-three patients responded, and 90 percent had complete responses. Not all the patients had as durable, long-lasting responses as the elderly farm woman, but almost all of them witnessed dramatic changes in their disease, often within two or three days of treatment. It took almost a year to publish these "spectacular" results—that was the self-applauding adjective they were allowed to use in their paper in the *Journal of Clinical Oncology,* which appeared in January 1992—because, Lejeune says, peer reviewers didn't believe the results. "Why spectacular?" Liénard said. "Because we observe very acute reactions to TNF. You can touch the tumors, and they become completely liquid. In some cases, you can pump liquid out of the tumors, they have become so liquid. The tumor is soft, and this happens one day or two days after the perfusion. It's quite incredible, because we have never observed anything like this with chemotherapy or any other agents before. It's very exciting."

"I'm still enthusiastic, because of what I see," Lejeune said. "And my patients are also enthusiastic. What we see is something which is very uncommon in oncology, which is that the effect of the therapy can be seen within a *few hours!* In many patients. Not all of them, but in many you see this." He fetched a recently published book on the technique and opened it to a page showing before-and-after pictures of a leg disfigured with melanoma. "One day later, the tumor is completely collapsed, and the normal skin around it is preserved. And also the patient sees it, so that he really is very enthusiastic, because he can, with his doctor, already document that there is a response while he is still in intensive care. This is really spectacular, because when you use high-dose chemotherapy alone, although you can reach sometimes, as I said, 50 percent complete responses, those complete responses take a rather long time to establish. It takes several weeks usually. And here, the TNF—and it's specific to TNF—is just mimicking what the animal models show when you put a tumor connected to the skin, what Lloyd Old described. And that is what is spectacular to me."

While it is unclear how the tumors are actually being liquefied, previous animal studies and work by Lejeune and others suggest that tumor necrosis factor uniquely and with stunning discrimination attacks the blood vessels that feed tumors while for the most part sparing normal vessels. When high doses of TNF flow through tumors vessels, the cells lining the inside of the vessels, known as the endothelium, respond by producing what are known as "adhesion molecules"—sticky, coagulative, gummy secretions that line the vessels of tumors. These cells also secrete cytokine molecules into the blood, which recruit inflammatory and immune cells to the site, primarily macrophages and neutrophils, sometimes within fifteen minutes. The end result is something like a multicar chain reaction pileup on a fogenshrouded freeway, multiplied a million times; this mob of cells seems to pile up at the scene, leading to massive coagulation and ultimately a kind of plug that shuts off blood to the tumor. These plugs quickly cause hemorrhages and the subsequent death of the tumor. "The first time I saw this, it was incredible," said Jean Gérain, another member of Lejeune's team. "You see—Boof! Everything is disappearing."

One week after the perfusion operation, Lejeune and his colleagues have discovered that the tumors are invaded by macrophages and B lymphocytes—the kind of white blood cells responsible for making antibody. "To our surprise, we saw quite a few B cells in our patients," he remarked. "Sometimes half the population is B lymphocytes. And then full of T lymphocytes. And this would, of course, suggest that the antigenic presentation and that all the machinery for mounting the primary response would be there." More recently, they have spotted killer T cells in the vicinity of the tumors—not just any T cell, but very specific T cells that recognize the recently discovered Melan-A/MART-1 antigen (the antigen specific to many melanomas that has been identified by Thierry Boon and Steve Rosenberg, respectively). In melanoma at least, the therapeutic clues may be converging.

Lloyd Old, who directs the Ludwig Institute for Cancer Research and supports Lejeune's work, has come to refer to the triple therapy as the "Lausanne vaccine" and regards the results as "extremely interesting." Old of course may have a proprietary interest in TNF as one of its discoverers; others are more guarded in their enthusiasm, some downright pessimistic about TNF's future. "I think it's biologically interesting, but therapeutically doesn't do very much," said David Spriggs, a TNF expert at Memorial Sloan-Kettering who believes the molecule could have been a breakthrough drug if it had not become a victim of its own hype. "I think they are trying to rescue something that is probably not rescuable. I don't think it has much of a future." "I think it's wonderful for that trivial fraction of patients that has the disease distribution suitable for it," said James W. Mier of Tufts University–New England Medical Center. "I mean, we just don't see those patients. It's one in a hundred who'd come to us with a bulk primary like that with no evidence of disease elsewhere."

Whatever ultimate utility it achieves, the TNF-melphalan-interferon triple therapy is one of the more suggestive early attempts at exploring a promising approach to cancer treatment—the combination of immunotherapy and chemotherapy. In the past decade, as more cytokine molecules have become plentiful thanks to biotechnology, clinicians have been able to test them, alone at first, later paired with standard chemotherapy treatment. The key word is "begun"—these efforts are in their infancy, the results preliminary. Despite lingering skepticism about TNF, the early data from Lejeune and other European doctors suggest that it may be too soon to give up on the drug.

By the end of 1993 Ferdy Lejeune and his team had treated 160 patients with the triple therapy and claimed a complete response rate of approximately 85 percent. The follow-up data on these results are somewhat problematic, however, because, after an intramural dispute with his Belgian colleagues that prompted his move to Lausanne in 1992, Lejeune does not enjoy continuing access to his initial group of patients. "We are unable to produce a valid update on the survival of the patients that we treated in Belgium between 1988 and 1991 because half of them are lost to follow-up," Lejeune explained. As a result, critical information about duration of these responses is not as complete as one would hope.

Even before Lejeune published his results, researchers at the National Cancer Institute followed the work with keen interest, thanks to a "Deep Throat" in Europe named Alexander Eggermont. Eggermont had been a fellow at the NCI and passed along abstracts and data to his former colleagues in Bethesda. On the basis of those preliminary results, Steve Rosenberg launched several randomized trials to clear up some unanswered questions raised by Lejeune's work and to extend the applications. The NCI embarked on a series of experimental treatments to see if TNF perfusions could be applied in a broader setting— liver perfusions (of the sort described in the prologue to this book) to treat the most common metastasis of colon cancers, lung perfusions to treat lung tumors, and perfusions of the abdominal cavity to treat peritoneal cancers.

At the same time, Rosenberg, Fraker, and their colleagues at the NCI began a randomized, prospective trial of limb perfusions comparing melphalan alone against melphalan/TNF to establish the relative efficacy of each drug. The results have been mixed at best. Initial tumor regressions continue to be impressive. Fraker showed me photographs of a man with an enormous sarcoma of the wrist, as large as a bird's nest (it measured 25 centimeters by 18 centimeters), who was told at two leading medical centers that the only recourse was amputation. Because the man loved to play golf, he opted for a TNF perfusion at the NCI. The photograph of the man after treatment was remarkable: the huge tumor had completely disappeared. "It just melted away," Fraker said. But this patient has become emblematic of the Achilles' heel of the treatment, which is that in many cases the disease seems to recur; after watching the tumor on his wrist melt away, the patient developed metastases in the brain and died about a year later. Fraker, now at the University of Pennsylvania, recently reported preliminary results from this trial indicating that the TNF-melphalan combination achieved slightly better response rates than melphalan alone, especially against bulkier disease.

Meanwhile, more than 200 melanoma patients have been treated with the TNF-melphalan combination in Switzerland, Israel, and the Netherlands, and the complete response rate ranges from 60 percent to 74 percent, depending on whether interferon was included in the regimen. Lejeune's group alone recently completed a randomized trial, achieving 78 percent complete responses and 22 percent partial responses with TNF, melphalan, and interferon, and 69 and 22 percent, respectively, when interferon was dropped from the regimen.

As arduous as the treatment seems, it is already being adapted for broader application in Europe, and in a way that harks all the way back to Coley's toxins. Coley explicitly stated that sarcomas seemed particularly vulnerable to his toxins, and he advocated their use especially to avoid amputation of cancerous limbs whenever possible. Alexander Eggermont, now at University Hospital in Rotterdam, has headed a multicenter trial of triple therapy TNF perfusions in inoperable cases of soft-tissue sarcoma. These large, disfiguring tumors, measuring on average 18 centimeters (about seven inches across, bigger than a softball), usually require amputation. In the most recently reported data on 55 patients, Eggermont has claimed an 87 percent overall response rate, with 36 percent complete response. Moreover, following treatment with TNF, many tumors become operable and can be surgically removed, thus sparing limbs from amputation in 84 percent of the patients. The treatment is now being tested against squamous cell and other carcinomas.

Nonetheless, several shadows fall upon all these rosy statistics: cancers presenting this way are extremely rare, and the duration of the response leaves much to be desired. The duration of the limb perfusions for melanoma appears still to be limited, in large part because the treatment is not "systemic"—does not also root out seedlings of cancer that have spread elsewhere through the body. The median time to local progression is 495 days—about sixteen months—in patients who receive the triple therapy, 320 days in those receiving the TNF-melphalan therapy. As Lejeune has succinctly put it, "You cure a limb, but you don't cure the person."

These most recent tests bespeak the less glamorous part of academic medicine, a forced march through parameters and variables in search of the best combination—a therapy with interferon in one case, without hyperthermia in another. That is why, when describing these early days of TNF perfusions, Ferdy Lejeune can legitimately speak out of both sides of his mouth. "You know, it's fantastic!" he said. "And spectacular!" Here the eyebrows did their jumping jacks; there followed a raffish, humbling laugh. "But it's not enough!"

19

THE NEXT GREAT
MAGIC BULLET

If forced to choose, Giacometti once said, he would rescue
a cat from a burning building before a Rembrandt."
—David Markson, *Reader's Block*

&❧ *When Michael Lotze left* the National Institutes of Health in 1990 to become associate director of the Pittsburgh Cancer Institute at the University of Pittsburgh, he bore the unofficial sobriquet of "Mr. IL-2." The name adverts to Lotze's early work with interleukin-2, but he is in fact something of a maven of all the interleukins, those powerful signaling molecules used by cells within the immune apparatus to rouse, rally, and redirect an immune response. The avidity with which he administered IL-2, despite its potency and toxicity, also earned him an even more unofficial, and somewhat less flattering, nickname: "One-more-dose Lotze." To his credit, Lotze can chuckle when the name comes up.

Since arriving in Pittsburgh, Lotze has continued to use IL-2—indeed, with greater claims of success than most other practitioners in cases of melanoma and kidney cancer. But at the same time he has endeavored to reinvent himself, using a different cytokine, as "Mr. IL-12." That he is willing to attach his name to this new, barely tested molecule says a little about the personality of Dr. Lotze, who shares with many oncologists the fervent desire to try any new agent in patients as soon as possible, but even more about the intriguing immunological personality of the molecule itself. IL-12, for better or worse, has emerged as immunology's latest "magic bullet"—a mantle of expectation as unfortunate as it is inevitable. Although interleukin-12 dented the consciousness of immunologists only in the late 1980s, following its rather unexpected discovery in two different laboratories, the molecule makes its appearance on the physiological scene extremely early during immunological distress and therefore colors much of the unfolding immune response. Whatever possibilities it ultimately offers to patients with cancer, AIDS, and infectious disease (and the question remains very much uncertain as of this writing), interleukin-12 makes for an appropriate bookend to immunology's work in progress, because it clearly represents a long-lost and crucial link in explaining exactly how an immunological response unfolds.

Interleukins, as the name implies, belong to the class of molecules that carry signals between white blood cells, or leukocytes, and in order to understand why the "magic bullet" mentality still persists among people who should know better and why interleukin-12 may represent a departure from business as usual in the cancer ward in Pittsburgh and in a handful of other hospitals in the United States and Europe where it is currently being tested, it is useful first to revisit business as usual. Many case histories described so far in this narrative have, in a sense, been exceptions to the rule—the rare successes that serve as "whispers of nature." The more typical experience for oncologists is like a Monday morning not long ago in the clinic with Dr. Lotze.

On this particular January morning, back in 1993, Lotze and his team convened shortly before 8 A.M. to begin rounds. Short, broad-faced, and mustachioed, with a cheerful demeanor and almost stainless-steel sheen of confidence, Lotze manages to project more gravity and presence than his modest frame would suggest; a self-described "Doogie Howser character" who began college at age fifteen and finished medical school at Northwestern University by the time he was twenty-two, he has both the prodigy's breadth of knowledge (on this particular morning, he is quoting Hippocrates before slipping on his white coat) and, occasionally, the prodigy's habit of calling attention to that knowledge. Intellectually curious, genial as well as ambitious, candid (often to a fault, even he admits), Lotze has become inured to the process of walking into one small room after another and dispensing almost homeopathic doses of hope to people who need every kind of medicine they can get. In some cases the experimental treatments work, too, but Lotze is self-aware and savvy enough to know that the occasional clinical success can extract a high price. "We become prisoners of our anecdotes," he remarked that morning, "and then go on to torture the next ninety patients."

The first stop was to see a melanoma patient named Kathy Samuels (like all patient names in this chapter, a pseudonym); over the years, she had filled her dance card with a succession of experimental therapies, including monoclonal antibodies, alpha interferon, a crude melanoma vaccine, and interleukin-2. Out the window of her room, which overlooked the sun-drenched Oakland section of Pittsburgh, it seemed like much too beautiful a winter day to be receiving the news Lotze delivered without warning, beginning with the fact that having suffered a turn for the worse over the weekend, she now faced major abdominal surgery this very morning. "I hope it's only going to be two or three tumors, and that we'll get it all," Lotze said. "There is the *possibility* that there will be many, many tumors, in which case we won't be able to take out the whole colon. But we'll do our level-headed best to get everything, okay?" Staring up at the ceiling, each revelation bringing a wince to her young face—surgery, this very morning, a possible colostomy, possibly too many tumors—Mrs. Samuels's hollow reply of "Yeah" arrived from a distance that could be measured in light-years. Yellow flowers adorned her bedstand; her husband was in another city; and her doctor, on the way to deliver a variation on this same theme to another patient in another room, was gone as quickly as he appeared. Like a voracious fire devouring all oxygen, the news left nothing in the room but shocked silence.

And so it went in the clinic. The names of six patients had been written on a whiteboard in the seventh-floor clinic. Carl O'Connor, a man in his fifties, learned he had a tumor the size of a lemon in one kidney, with metastases to the lung; arms crossed and rigidly tucked under each other, as if this coiled posture alone might forestall further spread, O'Connor heard Lotze propose the use of interleukin-2, although the side effects were advertised as resembling "the worst cold you ever had" and required, in about 25 percent of patients, treatment in the intensive care unit. There was Carole Sims, a court stenographer, who had had a cancerous gallbladder removed the previous September; now, following radiation, Lotze was very pleased to point out on an x ray that "this little area of grave concern [near the liver] is getting smaller, and is on its way to going away" and very unhappy to learn that his patient had spent $500 on a mail-order blood test for chronic fatigue syndrome.

There was Blaine Atkinson, who'd survived the removal of most of his liver—and the cantaloupe-sized tumor bulging therein—by Lotze in 1978, and now politely declined interleukin-2 treatment for several slow-growing metastases that had appeared in the lungs. There was Mary Parker, now seventy-three years old and now seemingly healthy, and her stomach cancer. "We give you all the credit, Dr. Lotze," Parker's daughter said. "You and God," added the patient. And then there was Mary Rhoden, who had tried IL-2 and interferon, IL-2 and monoclonals, a tumor vaccine, and now faced a novel treatment awaiting NIH approval for her melanoma, which had spread to the liver and lungs. "We don't give up golf," she said, explaining her reluctance to interrupt her spring for another course of treatment with interleukin-2. "We've given up everything else, but we don't give up golf." In each room, in each way, there were small but heroic negotiations for normalcy, for routine, for the way things used to be.

Finally, there was the young man named Christopher Syzmanski, a virgin to these little negotiations. Before visiting the patient, Lotze paused in the nurses' station to look at the most recent CAT scans. "He's got a liver full of tumor, and if we don't get him in soon, he's not going to have a chance," Lotze said. "He reminds me of the first patient I treated with interleukin-2 at the NIH. Steve Rosenberg went away on vacation, and this patient had splenic, hepatic, and lung metastases, and they all went away!" That, in a sense, was the anecdote to which Lotze became a prisoner, but he has tormented "a solid 25 percent" of his melanoma and kidney cancer patients into partial or complete remissions on IL-2, which is as good or better than anyone else currently claims. In a ward crowded with tragedies, Syzmanski was the kind of case oncologists hate to see. About thirty years old, he had recently become a father for the first time and was still digesting the news that his liver was potholed with tumor. Lying on a clinic bed in blue jeans and flannel shirt, he listened with a kind of edgy anticipation while Lotze explained that, for the moment, his doctors could do nothing about it; they were still awaiting approval from the Food and Drug Administration for an experimental treatment involving natural killer cells. Then Lotze suggested starting out instead with interleukin-2 along with a drug called pentoxyphyline (which may dampen the toxic effect of another molecule stirred up by IL-2, tumor necrosis factor).

"Don't hold me to it," Lotze said, "but my intuition is that you will be a responder. . . . Want to start tomorrow?" Syzmanski had no hesitation. "Let's do it," he replied. And on the following morning, just after 8 A.M., Lotze dropped in to check on Syzmanski, who had spent an anxious, uneasy, and sleepless night. "We'll just take it dose by dose," Lotze said.

"If I have any questions," Syzmanski replied, "I'll make sure I make a nuisance of myself."

About two years later, I checked back with Lotze on the status of the patients we visited with that Monday morning in 1993, and their fates served as a powerful reminder why people like Lotze, doctors whose every workday spills over with mortal illness, push a little harder, run a little faster, and, perhaps, sometimes lean a little too aggressively to get their hands on a promising new treatment when it appears on the scene. Kathy Samuels died within six months of her operation, after battling cancer for nearly twenty years. Carl O'Connor ended up getting IL-2 after all and was still alive. Carole Sims had no recurrence and was doing fine. Blaine Atkinson remained fine, too, with no evidence of disease. Mary Parker remained fine, with no evidence of disease. Mary Rhoden died toward the end of 1994 of her metastatic melanoma. And Chris Syzmanski, the youngest of all, had a deceptively excellent response to interleukin-2; his liver tumors virtually disappeared, and he made his way into the statistics as an IL-2 responder. By the fall of 1993, Syzmanski had recovered enough to run alongside Lotze in a local 10-kilometer race in Pittsburgh. A few months later, however, he succumbed to the Achilles' heel of such remissions: metastases that spread to the brain. To Lotze he bequeathed a T-shirt he had designed and made by his own hand, which showed a Viking-like figure on the front with the legend "IL-2 WAR-RIOR." The T-shirt still hangs in Lotze's office—the uniform, unfortunately, of warriors who win a few battles but lose the war.

This is why clinicians are impatient to experiment with new treatments. This is why doctors, to paraphrase Giacometti, appreciate the Rembrandts of science, but would just like to save a few cats. And this is why, when the first whispers about interleukin-12's extraordinary powers began to circulate during the summer of 1992, Michael Lotze and just about everyone else began to jockey to get their hands on some. Since his days with Steve Rosenberg in the Surgery Branch of the NCI, Lotze has specialized in cytokines, and interleukin-12 quickly emerged as one of the most interesting such molecules. He positioned himself so well that when the time came to begin testing the drug in humans, he was the only investigator in the country to have agreements to conduct clinical trials for both companies competing to test the molecule. By that time IL-12 had become, in the words of one prominent immunologist, "hotter than a firecracker."

Just when immunologists begin to think they have roughed out the general puzzle of the immune system and need only to fit in a few missing pieces, someone discovers a new piece so large and so central that everything around it, including the general picture, must be rearranged. Something like that began to happen in the mid-1980s, when two research groups about ninety miles apart began to investigate rather minor immunological puzzles. One group was looking at aspects of the innate, nonspecific arm of early immunity while the other directed its attention at the highly specific, acquired phase of late immunity. The two projects intersected, as did these two ostensibly separate arms of immunity, on one very unique molecule.

The first group of researchers decamped from Philadelphia. Giorgio Trinchieri, an affable and thoughtful Italian-born biologist at the Wistar Institute in Philadelphia, had decided some years earlier to concentrate on the early phase of the immune response, a fairly courageous intellectual decision given the emerging importance of T cells. Indeed, when he first joined the faculty of the Basel Institute of Immunology in Switzerland in 1971, he recalls expressing an interest in studying macrophages and natural killer cells, two early and so-called nonspecific players in the immune response. But all the action in immunology was shifting toward the "specific" side of the street, represented by T cells and B cells, and nothing made his curiosity appear more out of fashion than the dismissive reaction it evoked in one of Trinchieri's mentors in Basel, the late Nobel laureate Niels Jerne. Of macrophages and NK cells, Jerne told Trinchieri, "This is not immunology."

Trinchieri begged to differ. He made the point then, as he does now, that the early, nonspecific part of the immune system is not just some biological artifact like the appendix. True, a great deal of attention had shifted to the so-called acquired immune response, but in humans this antigen-specific, adaptive aspect of immunity typically took more than a week to kick in. How did the organism, mouse or man, survive viral or bacterial assault until the specific T cells and antibodies galloped to the rescue? How did the molecular baton get passed from nonspecific cells to specific cells? In truth, Trinchieri didn't even aspire to answer those questions. In what has by now become a cliché of cytokine research, Trinchieri's group at the Wistar set out to do one thing—look at one rather narrow aspect of macrophage biology—and ended up tumbling as if through a trapdoor into an entirely different realm of immunological activity.

"We were studying the ability of lymphotoxin to induce the differentiation of human macrophage cell lines," Trinchieri explained in an interview. The cytokine lymphotoxin (also, confusingly, known as tumor necrosis factor beta) was believed to influence the maturation and activity of macrophages, cells of great interest to Trinchieri's group.

"And at that time, lymphotoxin had just been cloned, and it was very difficult to get any recombinant material. So we decided to go ahead and purify lymphotoxin ourselves. We screened several cell lines for lymphotoxin production, and RPMI was the one that produced the most." Some years earlier, researchers at Roswell Park Memorial Institute in western New York had isolated a cancerous cell line dubbed "RPMI 8866." This particular cell happened to be a B cell that had been infected by the ubiquitous Epstein-Barr virus, known most commonly as the cause of mononucleosis. The RPMI cell line was unusual in that, due to some fluke of genetic rewiring probably caused by the virus, these B cells kept dividing and dividing. Since one of the by-products of this aberrant cell was an unusually high amount of lymphotoxin, the Wistar group elected to milk it, like a microscopic cow in a test-tube barn, to gather ample amounts of the molecule. A postdoc in the lab named Michiko Kobayashi, who had come to the Wistar from a biotechnology company in Cambridge, Massachusetts, called Genetics Institute, started the thankless drudgery of growing billions upon billions of RPMI cells and then skimming off the lymphotoxin, both of which were tedious but essential preliminary tasks for the experiments Trinchieri ultimately planned to do.

Trinchieri, however, used the occasion to scratch a tiny itch of curiosity. Like Isaacs and Lindenmann's time-killing "throwaway" experiment that ultimately led them to interferon, Trinchieri did a throwaway experiment to answer a nagging, though peripheral, puzzle. For several years, his lab (and many others) had used the RPMI cell line for another purpose. For no known reason, the addition of human white blood cells, of the sort obtained from a routine blood sample, to a culture plate containing RPMI 8866 as "feeder" cells resulted in a wild proliferation of one particular species of immune cell, namely the natural killer (NK) cells. In other words, the RPMI cells seemed to send a message that caused NK cells to multiply, and multiply they did. Like foam on a pot of water in slow-motion boil, these cells bubbled up out of the flatland of cells and overgrew the cultures. "You basically waited seven to ten days," Trinchieri said, "and you'd get an incredible growth of NK cells. We had used that method for many, many years, and we knew that B cell lines, and this one in particular, had a major effect in inducing NK cell growth. So we were always curious what was responsible for that."

Kobayashi's immediate task, however, was far less exalted than scientific curiosity; her mandate was to isolate lymphotoxin, and the first step in that process required her to collect all the fluid that oozed out of the RPMI cells. This soupy extracellular fluid, known as the supernatant, might be thought of as the biochemical sweat of cells at work, and a considerable technology had matured to allow researchers to separate component molecules that made up any soup into "fractions"—rather like taking chicken soup and running it through a machine that would deconstruct its broth and separate it into component parts, or "fractions," such as onions, carrots, celery, oil, and chicken meat. As long as Kobayashi had set up the experiment to isolate the fraction that contained lymphotoxin, the Wistar group figured they might as well test some of the other fractions. They might find a factor that stimulated NK cell growth. Fortunately, they had two ways of testing this. One assay measured the ability of any given fraction to increase the activity of NK cells to kill target cells; the other measured the ability of a fraction to cause NK cells to make gamma interferon. As usual, Trinchieri discovered someone else's footprints had been down this path before; the interferonologist Jan Vilcek of New York University had noticed increased interferon production several years earlier.

"So I was running all the fractions we were getting on [these] two assays," Trinchieri recalled, "and immediately some of the fractions scored positive, quite firmly positive, for both assays. And so it was very clear immediately that there was a factor in there other than lymphotoxin that was augmenting NK cell activity and inducing interferon gamma." Those crucial, digressive experiments were performed in January 1985. Trinchieri and Kobayashi spent the next four years trying to purify the molecule and make sure it wasn't something that had been described before—an unusually long time to sit on what would turn out to be an unusually important observation. When asked if he was worried about getting scooped, Trinchieri replied, "I believe in getting it right and not making mistakes. I don't believe in rushing things not to get scooped, and we didn't!"

Like Isaacs and Lindenmann with interferon, like Morgan and Ruscetti and Gallo with interleukin-2, Trinchieri and Kobayashi had paused long enough to investigate an unexpected phenomenon in a test tube, unrelated to their main research agenda, and ended up finding a previously undiscovered and extremely important cog in the great engine of immunology. They couldn't say a great deal about it at first. In fact, the Wistar entered into an agreement with Genetics Institute in 1988 to get help in purifying the molecule, and Kobayashi returned to Cambridge, where her former employers prepared enormous industrial-sized vats of supernatant—on the order of 100 gallons—to work on. Eventually, they isolated enough of the "factor" to get a glimpse of its structure. The molecule possessed a highly unusual architecture. It came in two linked segments, the two subunits measuring (in molecular weight) 35 and 40 kiloDaltons, respectively. In April 1989, Wistar and Genetics Institute felt confident enough about the work to write it up—not simply in a paper but also in a patent application. Later that year, in September 1989, Kobayashi and her colleagues published their results in the *Journal of Experimental Medicine.* They called the molecule natural killer cell stimulatory factor (NKSF).

The other group was based a few exits up the New Jersey turnpike, at Hoffmann–La Roche, Inc. in Nutley, New Jersey, and if Trinchieri's group had scratched an itch of curiosity, the Roche group by contrast tried mightily not to rescratch an old, itchy scab that had caused plenty of corporate pain and discomfort, namely development of the molecule interleukin-2 as a potential cancer drug. Maurice Gately, an M.D. and Ph.D. who had specialized in cellular immunology during a stint at the National Institutes of Health, headed Roche's effort to get the molecule ready for clinical testing, including animal studies and toxicology tests. "We were just getting ready to go into the clinic," Gately recalled, "so there were already hints from the results that were coming out of Cetus and Steve Rosenberg's lab over toxicity issues. And following on what I had been doing at the NIH, I just wanted to see if there were other things that you might put together with IL-2 and maybe synergistically get some efficacy without the toxicity. Maybe that's naive, looking back ten years ago with the perspective we have now, but that was the rationale. So that allowed us to start working on the series of experiments that did lead to IL-12."

In August 1985, even before Steve Rosenberg's group made their big holiday splash in the *New England Journal of Medicine,* Gately's team had submitted the first fruits of their effort for publication. The paper, which ultimately appeared the following February in the *Journal of Immunology,* reported that an unidentified factor in *their* supernatant, culled from human white blood cells, turned on killer T cells. During one set of experiments, Gately and his two collaborators, Darien Wilson and Henry Wong, noticed that a test tube supernatant added to interleukin-2 synergistically enhanced the activity of T cells. Like most of their colleagues, they knew that T cells appeared to play a major (if not decisive) role in the immune defense against viral disease, and perhaps even against cancer. The results could not help but perk their interest.

Ironically, Hoffmann–La Roche ultimately decided that IL-2's toxicity made it commercially unattractive as a drug, and the company never even bothered to apply for FDA approval. But the series of experiments initiated by Gately had by that time led researchers to shift their focus to this second, and seemingly less toxic, cytokine. Pursuing this thread of unplanned research, Gately's group ended up spending several years purifying and characterizing the molecule they had identified in their test tube soup. They were still pursuing this work when the September issue of the *Journal of Experimental Medicine* arrived in the mail, with the article by Trinchieri's group describing *their* factor with 35 kD and 40 kD segments, and to quote that noted molecular immunologist Yogi Berra, it must have seemed like déjà vu all over again.

"When our paper was published," Trinchieri recalled, "Maury called me, and he said that they were also working on a factor that had some synergistic effect with IL-2. But he said it looked quite different from the one we were working on." "We immediately recognized obviously that they had something similar to ours," Gately said. "At the time, there were some differences in biologic assay results that made us think maybe it was related and not precisely the same. . . ."

Gately invited Trinchieri to give a seminar at Roche. "The Roche group did not know the molecular structure of IL-12 at that time, and thus the similarities between the two factors were not obvious," Trinchieri recalled. Within a month or two, however, they became all too obvious; Gately called to inform Trinchieri that Roche's molecule had the same unusual two-part structure and most likely they were identical. "Because of that," Trinchieri added, "the legal staff at Roche believed that my visit to Nutley could create patent complications, and they cancelled my seminar." About a year later, in 1990, Gately and colleagues published their paper announcing essentially the same discovery. They called it cytotoxic lymphocyte maturation factor (CLMF).

To most molecular immunologists, especially ones who didn't have patent attorneys peering over their shoulder (an endangered species, to be sure), "NKSF" and "CLMF" looked like one and the same molecule. For legitimate scientific reasons—but partially for legal reasons, not wanting to jeopardize their respective patent positions—both parties say now that there was no way to be sure until the actual genetic sequence of the gene was attained. About a year later, the cloners obliged with their molecular data. First at Genetics Institute, then a few months later at Hoffmann–La Roche, molecular biologists isolated and identified the genes for their respective factors, and of course the DNA sequence spelled out the unambiguous truth: both groups had discovered the same protein.

In January 1992 Genetics Institute and Hoffmann–La Roche reached agreement on two important points. First, in an exceptionally unusual (and, as it would turn out, exceptionally short-lived) arrangement, the two companies agreed to share codevelopment of the newly discovered molecule. Second, both companies agreed to give their discovery a more formal name. They decided to call it interleukin-12.

The two different routes to its discovery speak volumes about the versatility of interleukin-12. To Trinchieri's group, it revealed itself as an inducer of NK cells—that is, as a part of the early, nonspecific immune symphony. To Gately's group, it revealed itself as an inducer of T cells—that is, part of the later, specific, cell-mediated symphony. To the world of immunology, IL-12 appeared remarkably ambidextrous, a molecule able to influence both the early and late arms of the immune response. Trinchieri began to call it the "missing link," the molecule that held hands with both natural and acquired immunity.

Where did interleukin-12 come from? When researchers went looking for a natural source for the molecule in the body, they traced interleukin-12 production to a class of immune cell that had more or less been discovered exactly a century earlier in Sicily by Elie Metchnikoff—the cell that had fallen from favor during the rise of humoral immunity, and a cell that had faded further into the background as a "nonspecific" cell during the renaissance of the T cell. Macrophages made interleukin-12, and that meant its immunological signal was broadcast very close to the onset of the immune response. "After all," said Trinchieri, "IL-12 is made by the first cells to see an infectious agent." Indeed, the most remarkable thing to emerge about interleukin-12 is how decisive a role the molecule appears to play in the earliest moments of infection, and its discovery allows us to sketch out an increasingly detailed molecular anatomy of an immune response— what happens when the immune system kicks in. IL-12 forms a keystone piece in our understanding, circa 1996, of the molecular commotion that can be stirred up in the blood.

When troublemaking bacteria surmount the parapet of skin or epithelium and begin to invade the tissues, for example, the body sniffs trouble and initiates an almost instantaneous inflammatory response.

Even before the nature of the incursion is fully known, the first cells of the immune system react to the apparent distress. The table is then set for a swift cascade of cells and signals. If, as Alexander Fefer once remarked, interleukin-2 conducts the immune symphony, then interleukin-12 warms up the orchestra in a critically important way.

Two cells and three molecules, linked in a blitz of urgent biochemical messages fired back and forth, raise this first, crucial commotion in the blood, and the commotion may occur hours (if not days) before we dimly sentient humans have consciously registered, with a sneeze or fever, the fact that our body's defenses have been breached. When, for example, researchers inject into animals a large dose of bacterial endotoxin (similar to one of the ingredients, incidentally, of Coley's toxins), Trinchieri explained, "everything happens very rapidly. The first cytokine that appears is tumor necrosis factor, and it appears within one or two hours. IL-12 follows in about two to four hours. And interferon gamma, which is induced by probably both TNF and IL-12 working together, appears around five to seven hours."

Nature's mechanism for making interleukin-12 is vaguely reminiscent of the cautionary procedure by which a nuclear warhead would be activated and launched, where two keys must be turned simultaneously by two officers in two separate rooms. The two subunits of IL-12 possess their own genes, and the two genes are located at distant chromosomal locations—the gene for p40 on chromosome 5, the gene for p35 on chromosome 3. They must both be turned on simultaneously before the molecule can properly assume its active, three-dimensional shape. Stanley Wolf and his team at Genetics Institute have further shown that the p35 subunit is made by a wide variety of cells, but that the p40 subunit is produced only in macrophages and B cells, and in the macrophages only after they have become "activated"—that is, after they have bumped into an unwanted microbe. Once the two subunits are assembled, then and only then can the molecule exercise its potent effects.

In the pitched to-and-fro of the earliest moments in an immune response, cells beget molecules and molecules beckon more cells. Several hours after TNF appears, increasing numbers of inflammatory cells enter the picture. They come in stationary and stalking versions: the stalkers, or macrophages, normally patrol the lymphatics and bloodstream, chemically beckoned to an area of distress by signals from inflammatory cells, while the ambushers, or dendritic cells, essentially lie marbled and immobile through tissues like the skin until they snare a passing antigen, at which point they too begin to migrate to the lymph nodes. Both these cells amplify the cry for help into a systemwide alarm—the difference between shouting "Help!" from a window, for example, and dialing 911. First, macrophages and dendritic cells begin squirting interleukin-12 into the immediate environment, triggering what might be called a local ecology of alarm and recruiting other cells to the scene. But because they are both also known as "professional" antigen-presenting cells, they possess a special talent for swallowing, digesting, and in a sense regurgitating partially digested bits of pathogens in the form of peptides, which are packaged and shipped to the surface as antigens. Unlike the MHC Class I system in normal cells that presents antigen to killer T (CD8) cells, the professional antigen-presenting cells operate within a separate complementary MHC system known as Class II presentation and package antigens exclusively for the helper T (CD4) cells. Helper T cells are the conductors, the quarterbacks, the on-site supervisors of an unfolding response.

This later interaction and recruitment, however, may require up to seven days, and a week may be too long for an organism under siege. Hence, another key cell is beckoned to the scene, the same one whose behavior first piqued Trinchieri's curiosity: the natural killer cell. The first thing IL-12 does is yank the chain, as it were, of these NK cells. It takes about two to three days after an infection for them to respond to the tug, but once activated, the NK cells do two crucial things. They attack invaders at the scene, although somewhat indiscriminately; they cannot "see" antigens the way T cells do and thus cannot be specific about whom and what they attack. They also multiply and churn out the third important signaling molecule, interferon gamma.

Hence, in a relatively short period of time, these two cells and three molecules broadly limn in the initial immune containment strategy. Macrophages, once activated, begin preparing bits of antigen to present to the helper T cells that several days later will begin to swarm to the site of infection. Natural killer cells, more like street brawlers (to extend the metaphor) than immunological boxers schooled in the sweet science of knocking out pathogens with crisp specific punches, hold off the troublemakers until more efficient reinforcements arrive. Once macrophages and NK cells wade into action, they actively produce IL-12 and gamma interferon, and that crucial combination of molecules sends a two-part message that triggers the involvement of the specific arm of the immune response, in the form of a naive helper T cell. Once stimulated, once nudged from what immunologists refer to as a "naive" state to an activated state, these T cells oversee the essential transition of the immune response from the brawling, non-specific, stopgap activities by the rowdiest of cells in the earliest moments of immune commotion to the more specific, more efficient, and definitely more lethal cellular part of the later immune response, where highly specific killer cells or antibodies arrive to mop up. In molecular terms, these cellular reinforcements may arrive late, but they know exactly what molecular prey they are looking for (because of antigen presentation), where to find it (because of cytokine signaling), and how to eliminate it (because of the lytic ability of killer T cells). Having completed the rout, these killer T cells molecularly commit to memory the microbe or antigen that caused the problem in the first place, the better to mount a swifter response the next time around. This is why nobody gets chicken pox twice.

If that were all, the addition of IL-12 to the general picture of the immune response would represent a breathtaking advance in immunological understanding. But interleukin-12 has been shown to possess another, equally fundamental function. In the mid-1980s, several researchers pointed out that T helper (Th) cells mature from their naive virginal state, even before their initial encounter with antigen, on the basis of a kind of one-if-by-land, two-if-by-sea message. These messages, of course, arrive in the form of molecular signals and appear influenced by the nature of the threat. Bacterial infections, viruses, and intracellular parasites tend to elicit one set of cytokines, helminths (worms and nematodes), another. This is important because so-called Th1 cells nudge the developing immune response primarily down the path of cellular immunity, which calls upon killer T cells to do the dirty work, while Th2 cells direct an evolving immune response more toward humoral immunity, which favors the production of certain antibodies and also allergic responses. In other words, in response to any microbial insult, the immune system always seems to reach a point where it must choose between a Th1 or Th2 response to administer the coup de grace.

How does the immune system decide which way to go at this critical fork in the road? For reasons still being actively explored, IL-12 and a kind of alter-ego cytokine called interleukin-4 appear to function as crucial switches, sitting at precisely the junction where the immune system branches off into predominantly cellular or antibody responses. The presence of IL-12 pushes the immune system toward the

Th1 response, leading to a cellular response where killer T cells play the dominant role; its absence (and the relative dominance of IL-4) pushes the immune system toward a Th2 response.

This switch has life-or-death ramifications, certainly in animal experiments and possibly in human diseases like AIDS. Trinchieri and Phillip Scott, a colleague at the University of Pennsylvania, have illustrated just how central IL-12 is to the ultimate shape of an immune response in experimental infections of mice by the parasite *Leishmania major*. They have shown that the immune cascade in certain inbred mice takes the fork that leads to the helper T cells known as Th1; these helper cells produce a distinct set of biochemical signals (or its "cytokine profile"), including IL-2 and gamma interferon, and this chorus of signals in turn pushes the immune response toward a cell-mediated, T cell–driven reaction, with the result that the animals overcome the infection. In a different strain of inbred mice, however, a different scenario occurs. These mice are unable to make IL-12, and thus when they reach the crucial fork, the helper T cell subset known as Th2 produces molecules like interleukin-4 and interleukin-10, which drive the unfolding immune response toward a less cellular response; since this is not nearly as effective as T cells against *Leishmania major*, these animals are especially vulnerable to infection. Similar studies with other pathogens have confirmed that interleukin-12 thus plays a critical early role in setting up the Th1-like cellular response.

These insights extend well beyond animal experiments. Trinchieri's group, collaborating with Christopher Karp of Johns Hopkins University, recently demonstrated that one of the reasons that measles is such a deadly disease, killing from 1 to 2 million children each year worldwide, is that the measles virus infects macrophages and in effect knocks out the ability of those cells to make interleukin-12. It thus disables one entire arm of the immune response for several weeks, making children with an active infection especially susceptible to secondary infections that often prove fatal.

"Interleukin-12 is really important to understanding what happens in the organism immediately after infection," Trinchieri said. "The organism has the need to put it into immediate effect, and it has to happen in hours or minutes. The specific response is, first, needed to completely get rid of the pathogens, and secondly, to provide memory. But you can't wait several days for a response, or the organism will die. With interleukin-12, it is clear that what happens early in the infection really affects what kind of immunity you get."

☙

These studies began to emerge in the early 1990s. Both Genetics Institute and Hoffmann–La Roche made interleukin-12 widely available to scores of researchers for animal studies, and the results astonished even seasoned and cautious immunologists. "From the perspective of an experimentalist, it's the most potent cytokine I've ever worked with," said Alan Sher, who heads a lab at the National Institute of Allergy and Infectious Diseases. "This is like an atomic bomb. It modifies the course of one infection after another."

It became clear very quickly that such a powerful molecule could be used as a drug with a multitude of intriguing possible applications. Its early and determining role in driving the immune system toward cell-mediated immunity makes interleukin-12 particularly attractive as a potential adjuvant—an immunostimulatory agent—to be used with vaccines against infectious diseases. Scott, for example, is currently testing in monkeys a vaccine preparation including IL-12 against *Leishmania,* a parasitic disease that infects up to 12 million people in Central and South America, the Mediterranean, Africa, and Asia. Stephen Hoffman of the Armed Forces Medical Institute recently began testing IL-12 as an adjuvant in developing prophylactic protection against malaria, which causes 300 to 500 million new infections each year. And Sher's laboratory at the NIH is working on a IL-12-based vaccine for schistosomiasis, a waterborne worm infection that affects an estimated 250 million people worldwide. Genetics Institute is testing IL-12 against hepatitis, and Roche, meanwhile, has begun testing its IL-12 against hepatitis B and C in the United States and Europe.

There may even be an IL-12 connection to AIDS. Trinchieri has suggested that individuals with AIDS have an impaired ability to make IL-12, which might contribute to their vulnerability to opportunistic infections. The role of IL-12 in AIDS remains unclear, but since the summer of 1994, a handful of HIV-positive patients have received interleukin-12 in preliminary Phase I safety studies sponsored by Genetics Institute. None of those results have been publicly revealed.

To researchers interested in cancer immunotherapy (and the companies that fund them), the strong link between interleukin-12 and killer T cells did not escape notice, either. Would it have any effect against malignancies? If laboratory mice have any relevance (sometimes they do, often they don't), IL-12 may be ticketed for a different ending than interleukin-2.

In 1992 Maurice Gately and his colleague Michael Brunda at Hoffmann–La Roche began to report that IL-12 had powerful antitumor
effects against at least seventeen different mouse tumors. It worked in
experiments where the tumor was bulky, and it worked in metastatic
models where the cancer had been allowed to spread widely before
treatment was initiated. In one set of experiments, for example,
Brunda's group injected kidney cancer cells into mice, waited for two
weeks until the tumors grew up to a centimeter in diameter, and then
injected IL-12. Initially, the IL-12 merely fought the cancer to a
draw—tumor growth stalled. "But after a prolonged period of treatment," Brunda said, "the tumors go away and the animal is cured."
When they poke through the site of the tumor, they find the immunological equivalent of a field after battle: the remnants of massive inflammation, with a lot of neutrophils, a lot of macrophages, a few
lymphocytes, and a lot of scar tissue. What made the result especially
encouraging is that by waiting fourteen days for the tumors to become
established, growing to about the size of peas, the Hoffmann–La Roche
researchers crudely approximated the analogous medical scenario in
humans, who often are not diagnosed until they have already progressed to well-established, bulky disease.

In Pittsburgh, Mike Lotze and his colleagues immediately latched
onto IL-12 as a potential anticancer agent and have confirmed many of
Brunda's studies, performing some others on their own. When they
injected IL-12 into subcutaneous tumors in animals, the tumors
stopped growing. They gave these tumors as much as a two-week head
start, but even low doses of IL-12 stopped their growth. When they
implanted tumor on both sides of the animals and then injected the
molecule in just one side, both tumors disappeared, suggesting that
interleukin-12 induced systemic anticancer activity. Finally, in a
high-tech attempt to localize the molecule's effectiveness and delivery, Hideaki Taharu of Lotze's lab inserted the IL-12 gene into fibroblasts, an easily harvested cell populous in the skin, which can then
be reinjected near the site of the tumor. In animal models, the Pittsburgh team has shown that injections of these genetically altered fibroblasts ooze IL-12 around the location of the tumor and recruit T
cells to the scene.

This growing portfolio of impressive results in animal studies did
nothing to diminish the enthusiasm of researchers. In April 1995
interleukin-12 officially attained the status of an accursed cytokine
when a *Wall Street Journal* article quoted two scientists as calling it a
"magic bullet" in the experimental treatment of cancer. The comments actually referred to IL-12's role in blocking the formation of

new blood vessels of the type that feed growing tumors. The reviews served notice outside the scientific community that expectations for IL-12, building steadily since the first antitumor studies in mice several years earlier, had now reached dangerously elevated levels.

Molecules, like overnight sensations in the theater, are usually twenty years in the making. In the case of IL-12, the steps required to travel from an invisible factor in a test tube to a genetically engineered molecule abundant enough for extensive testing took a remarkably brief five years. Taking the molecule from the lab into the clinic, however, is always a more perilous and pricey journey. The trip is complicated in this instance by the weight of history (the "sordid" past of another highly touted molecule, IL-2), the unusual industrial competition (which has led to, among other things, intense scientific secrecy), and a tragic—but oddly illuminating—setback in preliminary testing in humans.

In the pharmaceutical business, no one rains on a scientific parade like someone with a sense of history, and it fell to Maurice Gately, who headed Hoffmann–La Roche's preclinical development of the ill-fated interleukin-2, to give everyone a quick splash of reality when he summarized the progress of interleukin-12 at a meeting in the spring of 1994 in New York. Tall, avuncular, cautious nearly to the point of halting speech, Gately, the codiscoverer of interleukin-12, began with an upbeat review of the molecule's potential against cancer, infectious diseases, and even allergic disorders. "All of these things sound exciting," Gately admitted, but after a telling pause, he continued, "But I came to Roche ten years ago, and sometimes I have the feeling that we're walking down a path we've walked down before, and that's IL-2. Ten years ago, another cytokine was causing a great deal of excitement. I think the New Age has not arrived yet. IL-2 has not lived up to its expectations of ten years ago, at least in part because of its considerable toxicity. Are we going to walk down the same path again? Is IL-12 another IL-2?"

A book could (and perhaps should) be written about the David-and-Goliath competition between Genetics Institute and Hoffmann–La Roche to develop interleukin-12 as a drug, because it says a great deal about the unforgiving economics of modern drug development, the sometimes uneasy relationship between clinicians and companies, and perhaps most importantly, the potential perils of entrusting decisions with potentially immense public health ramifications to the private sector, where the common weal and the bottom line may converge only by coincidence. The competition has seen bizarre moments of industrial secrecy. In January 1992, before Genetics Institute and Hoffmann–La Roche agreed to coordinate development and testing of interleukin-12, Hoffmann–La Roche's head of exploratory research, Patrick Gage, announced that he was leaving Roche to become director of research at Genetics Institute. According to Gately and others, Roche scientists who had sequenced the gene withheld the sequence information from their own lab chief. "We knew what the sequence was," Gately recalled, "and we couldn't tell him!" The secrecy proved academic, since Genetics Institute and Wistar Institute scientists ended up publishing the DNA sequence first.

Although virtually all the scientists involved speak of the collaboration with the nostalgia of former teammates reliving a championship season, the Roche–Genetics Institute handshake was always more collegial between the scientists than between managements. Corporate lawyers apparently never shared the enthusiasm of researchers, according to scientists familiar with the agreement, and strains developed in the unusual relationship relatively quickly. In truth, the two companies had very different priorities, and in both cases their objectives were captive to their corporate histories and culture. With annual global sales on the order of $12 billion and a bad aftertaste from their experience with interleukin-2, Roche viewed interleukin-12 as a peripheral product at best among dozens in the drug development pipeline, and at worst a dangerous opportunity to relive the IL-2 nightmare. For Genetics Institute, interleukin-12 surfaced in the early 1990s as one of only a handful of potential flagship products for the young biotech company. More significantly, G.I. had been roughed up badly in the 1980s by patent litigation. The company had lost a highly publicized battle with Amgen over patent rights to a blood hormone called erythropoietin, or EPO; the drug, introduced in 1989 and marketed under the name Epogen, has been a blockbuster for Amgen, and currently enjoys a global market of about $2 billion a year.

To no one's great surprise, but not without a little sadness, Hoffmann–La Roche and Genetics Institute agreed in September 1993 to go their separate ways. ("I thought it was more unusual when they were working together than now," Trinchieri later observed.) The companies agreed, however, to cross-license their patents despite the split so as not to impede each other's progress. "The scientists remain very cordial, but not as intimate as they once were," said one researcher, who like many others was tugged by divided loyalties following the split. "It *was* a business decision," explained another. "But I see it more as the big, robust, pharmaceutical house pitted against the smaller, struggling biotechnology company, although that is not entirely true because G.I. was bought out by American Home Products. [AHP owned 60 percent of Genetics Institute at the time of the split, and in December 1996 announced its intent to purchase the rest.] Genetics Institute wanted a 50–50 deal and said, 'That's our final offer,' and Hoffmann–La Roche said, 'Who do you think *you* are?' And the final decision was to not codevelop." Said another researcher familiar with the split: "At G.I., there was a feeling that Roche, because it was so much larger a company, would dictate what G.I. could do, and they would not be an equal partner. Hoffmann–La Roche took a much more conservative approach. They were looking long-term, while G.I., as a young company, wanted to be more aggressive."

With that frictive history as a backdrop, Genetics Institute played the biotech hare, racing ahead into the clinic. But the unhappy saga of IL-2 cast such a long shadow that it touched Genetics Institute too. In July 1990 an advisory panel of the Food and Drug Administration declined to recommend approval of interleukin-2 as a cancer drug, a decision that led directly to the demise of the molecule's manufacturer; Cetus Corp., which had a market capitalization of $1 billion in 1986, ceased to exist by the end of 1991 and merged into Chiron, largely because the company had gambled its entire survival on FDA approval of IL-2. The Cetus debacle put the fear of God—and of the FDA—into every biotech company in the land, and Genetics Institute took the lesson especially to heart. This hare set out in leg irons.

In May 1994 Genetics Institute sent precious vials of human IL-12 to two sites in California, UCLA and San Francisco General Hospital, for initial testing in AIDS patients. The following month, four other medical centers (Dana-Farber Cancer Institute and Tufts University–New England Medical Center in Boston, the University of Pittsburgh, and Indiana University) began Phase I safety testing in patients with advanced cancers. In the cancer studies, about ten patients were treated at each site. But the company, collaborators say, decided to take an extremely cautious approach to the testing of IL-12—too cautious, in the minds of several of the physicians participating in the trials. The company, for example, set an extremely low maximum tolerated dose for initial testing. "None of the four investigators agree that this represents a true MTD," said one doctor trying the drug.

"G.I. doesn't ever want this to be hailed as another IL-2, because they think that would be the kiss of death," said one physician participating in the trials. "So they want no toxicity, or minimal toxicity. G.I. is just scared to death of the FDA, that they'll have their study stopped or halted by the FDA."

The intense secrecy surrounding these early trials made for an excellent incubator of conflicting rumors. In the spring of 1995, at an IL-12 meeting at the NIH, the hallway buzz was that some patients had experienced complete responses, including a case of kidney cancer of some duration, while others maintained that IL-12's toxicity was high and efficacy was low. As early as December 1994, the collaborating physicians had prepared an abstract to submit for the annual meeting of the American Society of Clinical Oncology in the spring of 1995, but according to one physician, Genetics Institute discouraged, or at least delayed, publication of the results.

Without discussing its Phase I results, the company launched a quick Phase II study in 1995 against kidney cancer. And one seemingly inconsequential change in dosage schedule between the first and second trial had disastrous consequences.

In its initial safety testing, Genetics Institute's protocol called for doctors to inject patients with a single test dose of interleukin-12, wait for two weeks, and then commence daily intravenous injections of the drug. Side effects were minimal, according to several sources, and the drug seemed to be tolerated well. In planning their Phase II trial, G.I.'s clinical testing team therefore decided to forgo the priming dose. That minor change proved, most unfortunately, how much we still have to learn about the intricacies of human immune function.

Genetics Institute launched this second phase of testing of IL-12 on June 5, 1995, a Monday. Seventeen patients at four sites—Tufts University–New England Medical Center in Boston, Indiana University Medical Center in Indianapolis, Montefiore Medical Center in the Bronx, and the University of Chicago Hospitals—began to receive daily doses of the experimental drug. By Wednesday several patients—all with advanced cases of renal cell cancer—became ill, and by Thursday it had become clear that something terribly wrong had occurred. Even though they received less of a dose than the Phase I patients, fifteen of seventeen patients treated with IL-12 suffered severe, systemwide side effects, including fatigue, mouth sores, gastrointestinal bleeding, liver toxicity, and cardiac arrhythmias. Twelve patients had to be hospitalized, and two patients ultimately died—one from the Montefiore study and one from Chicago—although there were, according to the doctors involved, complicating factors in each case. "We were certainly surprised by the intensity of the reactions," said Janice Dutcher, who headed the Montefiore group. "Bewilderment" was the prevailing mood of the Tufts group.

Officials at Genetics Institute were stunned. "It was really shocking, because we thought we'd done everything right," said one company official, speaking on condition of anonymity. "This was a lot of information very quickly, and it didn't take any rocket scientists to tell that something was grossly amiss from our prior trial." On that Friday, June 9, after scrambling to understand what had gone wrong, Genetics Institute found itself making the ultimate nightmare announcement for a fledgling drug company: a brief press release stating that the trials of interleukin-12 had been halted. "Last June's tragic episode," noted *Science,* "was a serious setback for a potential wonderdrug."

The FDA asked Genetics Institute to check the formulation and potency as possible explanations, but it was the Tufts group that argued for the totally unexpected mechanism that accounted for the side effects. Several months later, company officials disclosed that tests in animals indicated an unusual pattern of toxicity—"a biologically unique phenomenon," as one put it. When animals are given a single dose of IL-12 followed by an interval of rest, they tolerate multiple high doses of the drug. But if researchers begin daily injections without the priming dose and rest period, severe toxicities develop and all the animals die. A similar phenomenon had been seen before in animal studies of interleukin-1 and tumor necrosis factor. "What makes IL-12 a little bit unique—well, actually, quite unique— is the incredible durability in this effect," says James Mier of the Tufts group, "and we're actually trying to figure that out right now."

With this sobering yet edifying knowledge in hand, Genetics Institute has had to restart from scratch. Its second Phase I safety trials against both cancer and HIV were under way by the end of 1995, and more than three dozen patients had been treated by the end of 1996 without signs of the previous toxicity. Indeed, doctors are attempting to use the crucial priming dose to their advantage, according to Nicholas Vogelzang, who heads the University of Chicago group. "You always learn with new things," he said, confirming that efforts are under way to use the priming dose as a way to ameliorate the side effects. "Everything we do in the research setting is highly educational."

Drug development is so expensive and tentative, however, that the toxicity episode may have far-reaching implications for interleukin-12's future. "It certainly made everybody more cautious, including our own management," admitted Maurice Gately of Hoffmann–La Roche, who added that Japanese medical groups declined to participate in Roche's trials of the drug after G.I.'s trials were halted. Meanwhile, Genetics Institute is proceeding so cautiously with dosages that some of the participating physicians in its trials worry that the drug is not being tested aggressively enough to reveal its potential value. "I'm disinclined to write the drug off yet," said James Mier, of the group at Tufts participating in the new Phase II study. "But I think it's unfair to push it only to the point that people are having some modest blood changes and say that that's the maximum tolerated dose. I think what needs to be done is, *somebody* has to be given permission to give this in higher doses." The Tufts group in fact has applied to the FDA and NIH for a grant that would allow it to test IL-12 at doses higher than Genetics Institute currently allows. "What'll happen," said Mier, "is either we won't get permission to do it, in which case we'll just sort of lose interest in it, or we'll get the grant. And usually, to push a Phase I to the maximum tolerated dose is certainly no more than a year."

The trials and tribulations of IL-12—probably temporary, possibly not—may illuminate a more generic problem of public concern, however, and it is that the current economics of drug development, and its enormous expense, empower the private sector to make perfectly reasonable corporate decisions, such as discontinuing a drug's development, with important public health ramifications. Animal tests, for example, suggest that IL-12 could be useful as a vaccine adjuvant in the treatment of many Third World and niche diseases, like malaria, leishmaniasis, schistosomiasis, and tuberculosis, among others. But officials at Genetics Institute have stated publicly that it could be economically disastrous for the company to test IL-12 initially against diseases where the markets are not sufficiently remunerative to offset development costs. John Ryan, G.I.'s director of clinical development, told a meeting sponsored by the NIH in 1995 that successfully testing IL-12 against leishmaniasis, for example, before it had found a commercially profitable use would be a "public relations disaster" for the young company; G.I. would be committed to scale-up and manufacturing costs running into tens of millions of dollars, yet would probably wind up distributing the drug through the World Health Organization, leaving the company with huge costs and few revenues.

From the point of view of the companies (and their stockholders), these are understandable financial postures; from the point of view of the social contract between the research community and society at large, however, public health may be ill served by the economics of drug development. "None of us are smart enough to know where IL-12 is likely to work," says Barry Bloom of the Howard Hughes Medical Institute at Albert Einstein College of Medicine. "It is likely in my judgment that it may not cure cancer and AIDS, but it may cure leishmaniasis. I don't think the companies have been very imaginative in seeking public sector funding to test these agents for tropical diseases. It's a failure of imagination." It may also be a failure of patience. The example of interferon is again illustrative: the molecule is demonstrably prolonging lives and making a difference to many patients, *and* is enormously profitable to the companies that make it. But it has taken forty years to *begin* to figure out how to use it. In the current climate, as one disgruntled clinician puts it, "Either it's a billion-dollar drug in six months, or they drop it altogether."

Despite its recent ups and downs, everyone remains upbeat about the ultimate potential of interleukin-12. "We don't know much about it yet, but we do know that it stimulates the immune system in a dramatic way," says Nicholas Vogelzang. "It stimulates NK cells and it directs the body toward cellular immunity, and cellular immunity is what we want for cancer." "If I had to bet," adds Roche's Gately, "I'd bet that IL-12 is going to be a commercial drug. I can't say at this point if it's going to be a big drug or a niche drug, but I've seen enough to think that if you can figure out how to use it, it's going to be commercial."

So, barely two years after it was declared a magic bullet and one year after the fatal trial, IL-12 has produced clinical responses against cancer, but none formally published; has been tested, but with agonizing deliberation; has revolutionized immune understanding, but may never translate into a drug of impact. Worse, it may be abandoned before its potential therapeutic role is thoroughly explored. It has, in short, become a model of all the promise and peril of developing an immunologically based medical treatment at the beginning of the twenty-first century.

∽

"The voice that always gets lost in these considerations is the patient's," says Mike Lotze. "Who speaks for the patients?" It is the clinicians on the front lines—like Lotze, like Steve Rosenberg, even like William Coley—who continue to push, to create the pressure for change. They are eyed warily by drug companies, sometimes scorned by colleagues as "cowboys," but they are beloved by patients, to whom "sooner rather than later" is not a neutral idiomatic expression but a promissory against impending death.

Lotze is hardly alone in testing IL-12, but he has ceaselessly pushed to test it more speedily. Collaborating with Genetics Institute, Lotze treated approximately a dozen cancer patients with low doses in the initial Phase I testing and noted several minor responses, but complained about having to administer "nearly homeopathic doses" (the Phase II toxicities appeared before the Pittsburgh group got under way). Collaborating with Hoffmann–La Roche on gene therapy applications of IL-12, meanwhile, Lotze and his team have managed to insert genes for the two human subunits into human fibroblasts. These are cells embedded in the skin, which as a frontline organ against the outside world is studded with immunological sentinels. In mouse experiments, the injection of these genetically altered fibroblasts just under the skin causes a local secretion of IL-12, which in turn recruits killer T cells that cure the animals of established cancers. In August 1995 Lotze treated his first patient—a fifty-five-year-old woman with breast cancer—with her own genetically altered fibroblasts. As in the mouse experiments, these cells secreted IL-12, and her immune system was incited to attack and partially reduce the tumors in her breast. Lotze has seen two other responses among the first ten patients, but these results are so preliminary and anecdotal that it is impossible to know if they have any significance.

Walter Storkus, Lotze's colleague at Pittsburgh, is taking another tack. He has devised a vaccine approach using Genetic Institute's IL-12, along with other Th1 cytokines, as ingredients for an immunologic broth in which dendritic cells are incubated with peptides known to be cancer antigens. Dendritic cells, in Lotze's opinion, "are *the* professional antigen presenting cells." They arise in the bone marrow and migrate into the tissues; when they encounter antigen, they engulf the peptide, migrate to the nearest lymph node, and "present" their catch to T cells and B cells. And they do so with unusual power—when dendritic cells present antigens to T cells, they underline the message with the costimulatory molecule B-7, a joy buzzer of a molecule that makes sure the T cell is fully aroused and alert to the antigen it is in the process of receiving. The first patients enrolled in this trial in late 1996.

Since it has taken nature millions of years to carve out a role for interleukin-12, it is not surprising that drug companies and scientists have a way to go to understand how to use it. As with interferon, as with interleukin-2, as with monoclonal antibodies and tumor vaccines, progress seems maddeningly slow to patients and doctors alike despite these encouraging signs. There is still a clinic every Monday morning at the Pittsburgh Cancer Institute, as there are at major cancer hospitals throughout the country, still a whiteboard with the names of patients waiting in little rooms, still lightboxes loaded with x-ray images where little coin-sized shadows presage lives eclipsed by disease. And still a lot of "creative tension" between clinicians and companies. "My goal is to make it work, and try it in different ways to get it to work," Lotze said not long ago. "And their goal is to go step . . . by step . . . by step . . ."

The last time I visited Lotze, a different T-shirt hung in his office; it bore the logo of Genetics Institute, and it showed four little cartoon gremlins, each emblematic of an immune function: one held up a Th1 cell, another pushed a cytotoxic T cell, two more lugged around an NK cell. It hung right next to the "IL-2 WARRIOR" T-shirt that Lotze inherited from Chris Syzmanski just before the young man died. One could read it as a sign that "kinder, gentler" cytokine drugs, as Lotze likes to put it, are on the way to the clinic. But one can also read it as a sobering reminder that while doctors are in the process of trying new drugs and new immunotherapeutic techniques, sometimes as if they were articles of fashion, the cancer patient has only one life to give to the cause. If IL-12, or any of the other intriguing immunotherapeutic approaches currently under study, delivers on its initial promise, it will mark a breakthrough a century in the making. But as that same century of dogged research reminds us, some promises take a very long time to keep.

Epilogue

METAPHORS,
MANIC DEPRESSION,
AND THE "C" WORD

We know what we know, and we don't have
the foggiest notion about what we don't yet know.
—Peter Lengyel, Yale University, 1994

ॐ *On January 26, 1993,* almost one hundred years to the day after William Coley inaugurated the first human use of his cancer vaccine by injecting sixteen-year-old John Ficken with the mixed bacterial toxins, the scientific resurrection of Coley's legacy enjoyed a particularly sweet moment. Lloyd J. Old—medical director of the Cancer Research Institute, current scientific director of the Ludwig Institute for Cancer Research in New York, longtime researcher at the Memorial Sloan-Kettering Cancer Center, and one of the most esteemed scholars in the field of tumor immunology—opened a symposium in New York by reminding participants that "many of the current approaches to cancer immunotherapy represent variations on the theme Coley initiated a century ago."

Old has always been fascinated by Coley, more so than many of his contemporaries, but what gave the statement its power was the range of modern immunological expertise in the audience deemed worthy of consideration as heirs to that legacy. Sitting in the room were Thierry Boon, whose work on tumor antigens has laid the scientific groundwork for specific anticancer vaccines; Philippa Marrack and John Kappler, who have exquisitely laid bare the structure of that marvelous sleuth and slayer, the T cell receptor; Pamela Bjorkman, Don Wiley, and Jack Strominger, who were the first to produce an atomic snapshot of the immune system at work, a three-dimensional structure of an antigen nestled in the groove of the immune MHC molecule. There was Philip Greenberg, who had managed to turn the laboratory farming of T cells into predictable medicine, and leading figures in research on cytokines and tumor antigens, pioneers in gene therapy and the adoptive transfer of immune cells, T cell transplanters and antibody experts—many of the key players whose clinical work accounts for renewed excitement in a field that has seen its share of ups and downs—were also there. For Old to trace their intellectual genealogy all the way back to Coley was a subtle, indirect, but eloquent way of answering the question that always hovers over this early vaccine. Yes, it did work.

It is significant that every immunological mechanism discussed in part IV of this narrative, with the possible exception of monoclonal antibodies, probably formed part of the complex immunological response triggered by Coley's toxins. Tumor necrosis factor was certainly a component; interleukin-12, animal experiments have shown, is induced by endotoxin as well, along with a host of other cytokines, some of which may not yet have even been identified; and T cells probably form part of the reaction as well, although the exact mechanism is not known. That it has taken the better part of a century to explain results that, if believed, compare favorably to the best modern therapies against advanced metastatic disease in certain cancers is reason to celebrate the extraordinary power of the immune system and at the same time to concede how much we still need to learn about it.

We still don't know how Coley's toxins worked—or, for that matter, why they so often *didn't* work—but the wealth of immunological and molecular evidence accumulated over the past half-century, especially since the advent of genetic engineering, provides a reasonable foundation for speculating on how Coley's unusual vaccine may have functioned. Indeed, it is possible not only to translate Coley's crude approach into today's molecular idiom but to see many of today's current immunotherapeutic approaches as quoting (usually without attribution) from aspects of the Coley phenomenon. And although immunotherapists have made, and broken, more promises than politicians over the past century, there is very good reason to speak of promise again. But promise, perhaps, within a different, more realistic context.

❧

Dr. Oliver Wendell Holmes once remarked that the key to longevity is to have a chronic disease and take very good care of it. The "C" word in cancer research has always been "cure," but the experience of A. J. Goertz may offer a different route to longevity. Goertz, you may recall, was diagnosed with hairy cell leukemia in 1977 and began taking interferon in 1982. It is now twenty years since he was diagnosed with a disease that at the time was uniformly fatal; in the span of a generation, the disease has become treatable and chronic. Goertz has never been cured of his leukemia, but whenever it begins to recur, he injects himself with interferon—just like a diabetic with insulin, he pointed out, right down to the tiny needles—and the disease is driven into remission. He has lived long enough to see his three children grow up, his business prosper, and for an even better medicine than interferon to reach the market for the treatment of hairy cell leukemia. When cancers can be turned into manageable chronic diseases, a great battle has been won. And with the reprieve won for Goertz by interferon, clinicians have had the opportunity to show that it can extend the survival of patients with more common malignancies. "You get the drug approved for the two hundred people in the world with hairy cell leukemia," said oncologist Peter Wiernik, "and then you find out what it's really good for. And it's really good for chronic myelogenous leukemia."

A number of immunotherapeutic interventions currently undergoing trials offer the legitimate hope of turning cancers like lymphoma, malignant melanoma, and leukemias in at least some instances into chronic ailments, extending survival, and doing so in a less toxic way. Examples include the use of low-dose radioactive antibodies against non-Hodgkin's lymphoma of the sort being tested by Mark Kaminski of the University of Michigan, the lymphoma vaccine being tested by Ronald Levy at Stanford, and preliminary work by Gert Riethmüller of the University of Munich with an antibody against colorectal cancer. None of these approaches represents "the" cure of cancer. But one clear message of twentieth century research is that, as Peter Medawar stated a quarter of a century ago, "Cancer is not one disease and it will not have one cure." Medawar also predicted that "the way things are going the treatment of a cancer patient is going to acquire more and more the characteristics of a research problem in which the patient, after scrupulously careful biochemical, pathological, and immunological assessments, will have a treatment or a system of treatments exactly tailored to suit his condition, by a physician competent to appraise and give due weight to all the evidence that will come before him of the tumor's whereabouts, its degree of malignancy, and the patient's natural power to combat it—which is likely to depend, at least in part, on the soundness of the immunological response system." With customized antibodies and MHC-specific vaccines, Medawar's prophecy is becoming a reality.

The use of the immune system to treat disease—applied immunology, if you will—has made steady rather than spectacular headway over the past century. By the measure of basic discoveries in immunology, progress has been astonishing; by the measure of benefit to patients, progress has been rather more modest. What distinguishes much of the work in the recent past, however, is a profound transition to molecular-based immunotherapies. Although the pieces for such a transition have been assembled over decades, only since the advent of recombinant DNA, monoclonal antibodies, and techniques like T cell cloning have immunologists moved beyond the general commotion described by Hans Sloane to the more precise, almost pinpoint commotions being attempted at the bedside today. Major obstacles remain, not least the cleverness of tumors in evading the immune system, but the optimism that pervades the field of tumor immunology today rests on a solid, molecular foundation.

Perhaps the most hopeful aspect of this campaign is that it is still in its infancy. Countless combinations await testing, and if the history of immunotherapy to date is any indication, there will be many surprises—including, inevitably, unhappy ones—in the tedious process of finding the right combination of medicines for the right disease. As Medawar suggests, disease is very personal, tied up in genetics and lifestyle and life experience, and so too is any response that depends on the patient's own immune system.

The immune system has customarily been described, by physicians and popularizers alike, in military terms. Cells (killer T cells, natural killer cells) attack, kill, lyse, eliminate, assassinate, terminate, destroy, and repulse invaders, interlopers, the enemy. Like weeds, the militant idiom crops up everywhere, including in this very same narrative.

But perhaps the time has come to mount a disarmament campaign against the usual immunological metaphors. "Commotion in the blood," as Hans Sloane termed it, is a lovely metaphor for a kind of healthy, restorative, biological cacophony, but we now know that the immune system is much more sophisticated and supple than random noisemaking. Given the coordinated and syncopated nature of this response, perhaps the better analogy is to a beautifully scored piece of music. Indeed, one could view the immunological research during this century as a kind of molecular and cellular archaeology, digging up the instruments of the immune symphony, learning what they do and how they might be used, the concert enrichening and developing new subtleties of tone with each new discovery. The composition has always been rich; we are just acquiring the ears to hear its full complexities.

To extend this musical metaphor, a major emphasis of future immunological investigations may well be to move beyond the science of solo instrumentation and begin to sort out the music these instruments make in ensemble. In truth, this effort has been under way for some time, but there may well be some fascinating disclosures in what might, for lack of better terms, be called chrono-immunology and contextual immunology. How do the instruments play together? Which lead and which play counterpoint? How does the score for such impromptu compositions unfold in the very real auditorium of human infection and disease? What accounts for variations on the usual theme?

Nature has left us abundant clues that timing and context play important roles in the immune response, and thus the dosage and regimen of cytokines and cells—how often they are given, how much, by what route, with what other therapies, and into what kind of local immunological milieu—probably play critically important roles in immune modulation. The unfortunate deaths in the interleukin-12 clinical trials shed light on a highly unusual aspect of timing. During Phase I trials, in which a minimum of toxicity was observed, patients received a priming dose of IL-12; doctors then waited two weeks before beginning daily high-dose therapy, which was well tolerated. In the second phase, the priming dose was omitted, and because of this single and seemingly trivial change in the protocol, even lower doses than the Phase I trial plunged many patients into life-threatening toxicities.

Another intriguing clue to the importance of timing—and patience—has occurred in the initial Phase I testing of the MAGE-3 peptide as a cancer vaccine. Thierry Boon and Alex Knuth decided to administer their vaccine once a month and planned only three injections; in the complete and partial responses seen so far, clinical and diagnostic evidence of remission has not even become apparent until after the treatment has ceased—in other words, more than three months after treatment has begun. "The thing I find so surprising," Boon remarked recently, "is how slow it is."

In this context, finally, it is worth repeating a possibly apocryphal story related by Michael Lotze of the Pittsburgh Cancer Institute, who once asked a representative of a pharmaceutical company why the recommended dosage for a form of interferon was three times a week. "Because," he was told, "that's what the marketing department decided would be best." As our knowledge of these new medicines matures, we can hopefully all agree that the immune system will ultimately provide better clues to the healthy orchestration of immune modulation than the marketing departments of major drug companies.

Finally, a few words about the term "magic bullet," whose meaning has passed down to us, tattered, like a game of telephone played over a century. When Paul Ehrlich coined the term in 1908, he intended to suggest a single chemotherapeutic drug that would knock out a disease, precisely and efficiently. In the subsequent decades, "magic bullet" has come to imply a fabulously precise agent—drug or cytokine or cell—that will miraculously target cancer cells. Biomedical research has persistently dedicated itself to the identification of magic bullets, but this is an especially foolhardy pursuit in immunology. If we have learned anything in the last twenty years, it is that the immune system does not like lone gunmen! Indeed, the consistent message—holistically true since the time of Coley, molecularly true since the advent of genetic engineering—is that the immune system works by coordination, by combination, by timing, by redundancy, and by gentle nudges, not sledgehammer blows. Perhaps it is time to retire all of those militaristic metaphors, and begin to think in more exalted, uplifting terms, appropriate to the wonder of such a fabulous machine. Think of a symphony, our understanding of it still unfinished. When it is complete, Old has been heard to say, "I'm sure it will be Mahlerian."

We live in an age of scientific marketing. As long as the stock market tracks biotechnology companies and as long as scientists rely on grants to do their research, academic and industrial researchers alike will be hustling a product, and the sound will remind no one of Mahler. This has been true at least since the end of World War II; consider the zeal of Britain's Medical Research Council to patent, test, and publicize the "antiviral penicillin" in the 1950s and 1960s. However, commerce has changed biology significantly in the last fifteen years, and we are entering a period of potential crisis in drug development, in the sense that the temptation to oversell the virtues of a medicine or molecule may preclude a fair trial of its usefulness.

The story of tumor necrosis factor is a cautionary tale, according to David R. Spriggs of Memorial Sloan-Kettering Cancer Center, who has reviewed the extensive clinical literature on TNF. "It could have been an approvable drug," Spriggs said in an interview, "but this was less about making a drug than about making money. For biotechnology companies, it was really about selling stock on hype. TNF was never as good as it was said to be during the hype, and never as bad as it was said to be later." How good it might be, Spriggs added, is something that we probably will never know because of the reluctance of pharmaceutical companies to manufacture the molecule now.

The cycle of boom and bust, of promise and disappointment, has become a staple of biomedicine these days. If we value the free flow of information (and undoubtedly we should), it is difficult to see an alternate to the kind of rowdy, competitive media free-for-all that accompanied the first reports about interleukin-2, for example. But society should be prepared to pay a price for these manic-depressive episodes, including the distinct possibility that we will give up too soon on useful, perhaps dramatically beneficial, drugs because of the aura of failure that attaches to premature, ill-conceived, or abbreviated testing of new compounds.

The flip side of the depression is the mania. Now that biologists have become entrepreneurs, the danger of economically motivated hype can ultimately jeopardize an entire technology. One of the great (although perhaps unintentional) virtues of Barry Werth's entertaining *The Billion-Dollar Molecule* is the way the book illuminates a drastic evolution in the concept of the scientific "story." "Story" in scientific circles used to denote a thread of distinguished research over the course of many years that opened up and then developed an area of biological importance; that story was always told in the past tense, as something that had already been accomplished. In Werth's telling, the "story" now refers to the pitch that entrepreneurial scientists make about the work they *will* do, the story they tell in the future tense in order to raise money. In a culture that seems to forgive hype or, to put it kindly, misleading statements, if there is a pot of gold at the end of a fictive tale, it is understood that the story merely needs to sound plausible and good, not necessarily be true.

But the story, as it has come to be defined in contemporary science, sets up an ecology of expectation and disappointment, and this can have harmful long-term ramifications when those expectations intersect with the health and, often, survival of desperately ill human beings. Unrealistic expectations can poison an entire field, as occurred with monoclonal antibodies and may be happening again with gene therapy; unrealistic pessimism can deprive a field of funding, talent, and the ability to make the slow, steady progress that is the real hallmark of drug development. Interleukin-12 is on the way to becoming a negative object lesson; it may offer great promise as a vaccine adjuvant against infectious diseases like malaria, tuberculosis, leishmaniasis, and schistosomiasis. But because those diseases are endemic in less remunerative Third World markets, the realities of market economics force drug companies to test the cytokine first against diseases like cancer. Interferon—a genuine billion-dollar molecule—may be a positive history lesson. It has been hailed as a miracle drug and then declared dead as many times as Rasputin. Even though no one fully understands its activity yet, it has proven to be useful in a variety of settings.

Anecdote" has become a dirty word in modern research, code for a tantalizing case history that cannot be extrapolated to larger biological truths. The recent history of immunotherapy shows, to the contrary, just what kinds of lessons can be learned by the enlightened study of a single anecdotal case—when the right scientific tools are brought to bear.

The melanoma patient Frau H, in particular, has provided bountiful information about tumor antigens and cell-mediated immunity, with repercussions far beyond her unusual remission; indeed, this patient, studied over fifteen years, has inspired the cancer vaccines that have already, even in the earliest testing, aided other patients. With today's molecular biology, the anecdote can be deconstructed in such a way as to provide truly groundbreaking information.

Beyond that, the anecdote can be just as inspiring to researchers as a clean column of data, and perhaps more influential. Edward Jenner had his milkmaids, William Coley had Fred Stein and Mr. Zola. Jorge Quesada listened to A. J. Goertz, Ronald Levy listened to Phil Karr, and Steve Rosenberg listened to "James DeAngelo" and later Linda

Taylor. In each case the anecdote led to a new insight, a new molecule, a new therapy. Patients play a much larger role than they are credited for in the progress of biomedical research, and the unsung heroes of immunotherapy are the patients who have opened doors to further biological information.

Enthusiasm," too, has been considered a dirty word, especially when applied to a discipline that has often promised so much and delivered so little. As in any field, the older voices are more tempered and cautious, the younger voices more enthusiastic. But there is a sense among some levelheaded scientists that the field of immunotherapy not only has arrived but is poised to assume a major role in medicine.

"I personally am *quite* optimistic," Drew Pardoll said, shortly before submitting a paper reporting the first results of a vaccine against kidney cancer. "And not because of the clinical results we've had so far, although that certainly bolsters it. The reasons are twofold. One, tumor immunology has finally—and Thierry Boon's work has been a very important element of this—entered the realm of hard-core molecular science. It's not black box, chicken soup anymore. Two, the most astounding thing about our clinical trial is that the [results] were identical to what were produced in mice. What that tells us is that the mouse is a good model for humans. The final point is, we in the field are hopefully more intelligent about designing good clinical trials. There were literally thousands of patients vaccinated with BCG and tumors, and not a lot came out of that. The key is that the scientists have to take over this field, and do the clinical trials. That is happening now, and in ten years, cancer vaccines will be players."

Our understanding of the immune system—fabulously advanced over the last century and significantly burnished in just the last decade—is still far from complete, and it is difficult to know how limitless or how marginal our manipulations of the immune system will ultimately turn out to be. Shortly before his death in 1976, the Chinese prime minister Chou En-lai was asked if he thought the French Revolution had been, on balance, a good thing. "Surely," he replied, "it is far too soon to tell." So, too, it is far too soon to tell what role immune modulation will ultimately play in the treatment of human diseases. But unlike in earlier eras, there is good reason—good *scientific* reason—to be optimistic.

NOTES

A GENERAL NOTE

Several general sources, cited at various points in the notes, provide an especially good overview of the two main fields described in the text, namely immunology and cancer.

On immunology, good accounts for general readers include William Clark, *At War Within: The Double-Edged Sword of Immunity*, Oxford University Press, New York, 1995; Steven B. Mizel and Peter Jaret, *In Self Defense: The Human Immune System—The New Frontier in Medicine*, Harcourt Brace Jovanovich, San Diego, 1985; and Robert S. Desowitz, *The Starfish and the Thorn: The Immune System and How It Works*, Norton, New York, 1987. Somewhat more technical but excellent texts include Debra Jan Bibel, *Milestones in Immunology: A Historical Exploration*, Springer Verlag/Science Tech, Madison, Wis., 1988; Arthur M. Silverstein, *A History of Immunology*, Academic Press, New York, 1989; and Richard B. Gallagher et al., eds., *Immunology: The Making of a Modern Science*, Academic Press, San Diego, 1995. For a more advanced textbook, consult Charles A. Janeway Jr. and Paul Travers, *Immunobiology: The Immune System in Health and Disease*, 2d ed., Current Biology/Garland Publishing Co., New York, 1996.

For a good cultural history of cancer, see James T. Patterson, *The Dread Disease: Cancer and Modern American Culture*, Harvard University Press, Cambridge, 1987. For a good popular discussion of the work that has led to the current thinking about the genetic basis of cancer, see Robert A. Weinberg's *Racing to the Beginning of the Road: The Search for the Origins of Cancer*, Harmony, New York, 1996. The history and current practice of biological therapy is summarized in V. T. DeVita, Jr., S. Hellman, and S. A. Rosenberg, *Biologic Therapy of Cancer*, Lippincott, Philadelphia, 1991.

In addition, two special issues of *Scientific American* merit mention: "Life, Death and the Immune System," which appeared in September 1993, and "What You Need to Know About Cancer," which appeared in September 1996.

Prologue: "Shooting Rubber Bands at the Stars"

3 "Okay, we're going to start": interview, Douglas Fraker, National Cancer Institute, Bethesda, Md., August 4, 1993; rounds, Aug. 5, 1993, Sept. 16, 1993, March 29, 1995, and March 22, 1996; D. Fraker, letter to author, Aug. 27, 1996. Special thanks to Rich Alexander, Arleen Thom, the nurses, and others who accommodated my operating-room visit at the Clinical Center.

5 Background on Raiford illness: Claude Raiford, telephone interview, August 19, 1995, and Douglas Fraker. See also *Statesville Record & Landmark*, Nov. 14, 1993, p. C-1 ("Raiford Observes Another Milestone—Survival") and *Charlotte News and Observer*, April 10, 1994, p. 21 ("12 Months Go By, Giving Cancer Patients a Reason to Celebrate").

9 As an example of animal studies as front-page news: L. Altman, "Scientists Report Finding a Way to Shrink Tumors," *New York Times*, Dec. 30, 1994, p. A-1. Altman addressed the issue of hype several weeks later in a thoughtful column, making the point that, in contrast to the claims that scientists make in the literature, "When speaking to the public, some scientists, in conjunction with their institution's press office, are willing to make much bolder claims for their work" (L. Altman, "Promises of Miracles: News Releases Go Where Journals Fear to Tread," *New York Times*, Jan. 10, 1995, p. C-3). Altman described a number of factors that contribute to public expectations, but might also have included the placement of such preliminary results on the front page of the country's most influential newspaper.

10 "miserable failures": Douglas Fraker, interview, Bethesda, Md., Dec. 9, 1992.

10 Background on TNF perfusion studies in Europe: D. Liénard et al., "High-Dose Recombinant Tumor Necrosis Factor Alpha in Combination with Interferon Gamma and Melphalan in Isolation Perfusion of the Limbs for Melanoma and Sarcoma," *Journal of Clinical Oncology*, 10:52–60 (January 1992).

16 on results of TNF liver perfusions: Douglas L. Fraker, "Isolated Hepatic Perfusion (IHP) with TNF," abstract for "TNF and Related Cytokines: Clinical Utility and Biological Action," symposium at Hilton Head Island, SC, March 10–16, 1996.

16 Raiford obituary: *Charlotte News and Observer*, Nov. 26, 1995, p. B-8.

I. THE OCCASIONAL MIRACLE

19 Epigraph: "When his old pupil" W. B. Coley, "The Idea of Progress," *Transactions of the New York and New England Association of Railway Surgeons*, 1920, p. 10.

Chapter 1: Laudable Pus

21 Epigraph: "The first hope," Pearce A. Gould, "The Bradshaw Lecture on Cancer," *Lancet*, 2:1665–1673 (1910).

21 Blood test report, James Ewing to William Coley, dated December 7, 1903, archives, Cancer Research Institute, New York, N.Y. This and all other CRI archival materials are not officially cataloged and therefore cannot be cited with the same precision as more formal archival collections.

21 Details of the relationship between Bessie Dashiell and John D. Rockefeller, Jr. are contained in *The Legacy of Bessie Dashiell*, a privately printed com-

memorative volume issued by the Woodstock Foundation, Inc., Woodstock, Vermont, 1978, in a limited edition of 50 copies; the foundation is associated with Laurence Rockefeller, one of John D. Rockefeller Jr.'s sons. This volume describes itself as "an informal account of nearly eighty years of dedicated interest, effort, and continuous financial support by three generations of the Rockefeller family to assist Memorial Hospital in the improvement of the care and treatment of victims of cancer, and in searching for causes, prevention, and cure of the dreaded disease." On Rockefeller as a youth, see also Allan Nevins, *John D. Rockefeller: The Heroic Age of American Enterprise*, Scribner's, 1941, Volume II, pp. 164–174, 286.

23 The exact nature of Dashiell's injury, involving the Pullman seats, is reported by Dr. Coley's daughter, Helen Coley Nauts. That physical trauma—a blow or bruise—can sometimes play a role in the development of bone cancers is not a commonly held belief.

24 "slight blow" and "the usual local applications": and the subsequent description of Dashiell's symptoms are most fully described in William B. Coley, "Contribution to the Knowledge of Sarcoma," *Annals of Surgery*, 14:199–220 (1891); "The pain soon became" and "The appetite": W. B. Coley, ibid., p. 202.

25 "I cannot tell you": J. D. Rockefeller Jr. to Elizabeth Dashiell, Oct. 19, 1890, Record Group III 2 Z, Rockefeller Family Archives, Rockefeller Archive Center (RAC), Pocantico Hills, North Tarrytown, N.Y.

26 "one of the most malignant": W. B. Coley, *Glasgow Medical Journal*, 126:49–86, 128–170 (1936), p. 82.

28 "I cannot tell you": J. D. Rockefeller Jr. to Mary Dashiell, Feb. 15, 1891, Record Group III 2 Z, Rockefeller Family Archives, RAC, North Tarrytown, N.Y.; "I think it goes back": quoted in *Legacy* (1978), frontispiece.

29 "deep impression": W. B. Coley, 1936, p. 82.

29 "A disease that": W. B. Coley, 1891, p. 205; "Nature often gives": Coley, ibid., p. 210.

30 Details of William Coley's early life have been pieced together through interviews with his daughter, Helen Coley Nauts; inspection of diaries, letters, and other archival material in the possession of the family. I am also grateful to Mrs. Nauts for providing a copy of the manuscript of an unpublished biography of her father, which was commissioned by the Cancer Research Institute and reflected the efforts of Wendy Murphy, James Gollin, and Carol Kay Taylor. Ms. Taylor in addition provided especially valuable assistance in locating archival material and bringing new information to my attention. A final version, completed by Mrs. Nauts and Brian Quinn, is scheduled for publication in 1997.

   "Played cards": and other remarks by Coley during his Yale and Oregon years come from these diaries.

33 "I have been asked": J. Marion Sims, *The Story of My Life*, Da Capo, New York, 1968 (reprint of 1883 edition), pp. 114–15.

34 For background on the state of medicine in the late nineteenth century, see Paul Starr, *The Social Transformation of American Medicine*, Basic Books, New York, 1982, and Sherwin Nuland, *The Doctors: The Biography of Medicine*, Vintage, New York, 1988.

35 "one of the three greatest": Walter Graeme Eliot, *Portraits of Noted Physicians of New York, 1750–1900*, American Biographical Society, New York, 1900; for a biography of Bull, see W. B. Coley, "William Tillinghast Bull, M.D.," *Transactions of the American Surgical Association* 27:xxix–xxxiii

(1909). According to Coley, Bull "was the first American to devote himself to surgery alone, from the start." On Bull's illness, see "Dr. Bull Takes the Air" (New York Times, Jan. 8, 1909, p. 6) and "Dr. Bull Out in the Park" (New York Times, Jan. 9, 1909, p. 5).

36 W. B. Coley, "Treatment of Penetrating Shot Wounds of the Abdomen," Boston Medical and Surgical Journal, Vol. 119, pp. 373–380 (Oct. 18, 1888).

36 "from places of dreaded": Starr, Social Transformation, p. 145.

36 "at the most opportune": W. B. Coley, Yale Alumni Weekly, Jan. 20, 1933, p. 345.

38 "The one quality of mind": W. B. Coley, "Some Unsolved Problems of Medicine and Surgery," Therapeutic Gazette, February 1902, p. 26.

38 "some little light": Coley 1891, p. 205.

38 "Clinical Records, Dr. Robert Weir": dated Oct. 23, 1886, to Feb. 22, 1887, Rare Books Room, New York Academy of Medicine Library, New York. All quotations from this volume, unpaginated. See also Reports of Cases from Surgical Clinics at the New York Hospital, given by Dr. Robert F. Weir and Dr. William T. Bull, Dec. 1889, Jan., Feb., March 1890. This latter also contains a fascinating description of the conditions of asepsis then employed at New York Hospital (and which Coley therefore learned as an intern):

The "seat of operation" was shaved, scrubbed with soap and water, alcohol, and ether, then washed with an antiseptic solution and covered with antiseptic towels, which also surrounded the field of operation. The surgeon (or "operator") and assistants all washed with soap and water as well as antiseptics, "paying special attention to nails and joints." Instruments were cleaned and boiled prior to the procedure, then handed to the operator from a carbolic solution. Iodoform, either in powder or crystal form, might be sprinkled on the wound, and the wound would be packed with iodoform gauze, then bichloride gauze (pp. 226–27).

39 On the issue of numbers not adding up: This appears to be a flaw that occurred not infrequently in Coley papers and has been noted by other researchers. In addition to the cited example (see p. 205), other examples in the 1891 paper occur on page 207, in a discussion of eight cases of Samuel Gross (Coley appears to describe nine), and on page 211, where his discussion of Bruns's cases is similarly confusing. Scholars sifting the record will have to come to terms with the fact that in many instances, dates, ages, and descriptions vary from paper to paper; when facts are in conflict, an effort here has been made to select what seems like the most authoritative source material, usually the closest chronologically to the event described.

One possible explanation for this may come from Coley's daughter, Helen Coley Nauts. In reviewing documents from the Bone Sarcoma Registry, she noted, "Father was obviously far from methodical in his paperwork. His desk, while he was writing, would be an indescribable chaos of papers, books, reprints, etc., and of course his writing was so illegible that even he had trouble reading it sometimes." Nauts insists that the desk in question was at Coley's country home only.

39 "subsequent history": Coley 1891, p. 205; "the most intensely": Mansell Moulin, quoted in ibid., p. 206; "the large proportion": ibid., p. 210.

40 Details of Fred K. Stein case: Sources include Coley 1891, pp. 211–12; case report "FKS," archives, Cancer Research Institute; W. B. Coley, "The Treatment of Malignant Tumors by Repeated Inoculations of Erysipelas: With a

Report of Ten Original Cases," *American Journal of the Medical Sciences,* May 1893, pp. 487–511; and W. B. Coley, "A Preliminary Note on the Treatment of Inoperable Sarcoma by the Toxic Products of Erysipelas," *Post-Graduate* 8:278–86 (1893).

40 "absolutely hopeless": W. B. Coley, *Practitioner* 83:589–613 (1909), p. 590.

41 "spread a furious": Nuland, *The Doctors,* p. 344; Shelby Foote, *The Civil War: A Narrative,* vol. 1, *Fort Sumter to Perryville,* Vintage, New York, 1986, for example, p. 381.

41 "after each, the cicatrization": FSK case history based on transcript of hospital records, archives, Cancer Research Institute.

41 "That was the last note": Coley 1936, p. 49.

42 Details of the search: This account was pieced together from interviews with Helen Nauts and Coley's own remarks in Coley 1893; Coley 1909; Coley 1912; and Coley 1936. In prepared remarks for an address to the Royal Society of Medicine in the fall of 1935 (archives, Cancer Research Institute), Coley said of the Stein case: "I felt that it was of great importance to learn the end-result, so I started out with the man's address of seven years' previously, and after tracing him from one tenement-house to another, finally, in the spring of 1891, I located him."

42 "If erysipelas": Coley 1909, p. 591.

43 "laudable pus": Arthur M. Silverstein, *A History of Immunology,* Academic Press, New York, 1989, p. 41.

43 "A moment's thought": William Boyd, "The Meaning of Spontaneous Regression," *Journal of the Canadian Association of Radiologists* 8:63 (1957); see also Boyd, "Spontaneous Regression," p. 1. These two lucid, almost lyrical papers confront the paradox of spontaneous regression. Apropos of the current subject matter, Boyd writes, "It is a curious and perhaps significant fact that in many of the recorded cases some such partial and half-hearted interference has been attempted. It would almost appear as if such procedures had triggered a mechanism or started a chain reaction which cut short the neoplastic process and resulted in the downfall of the tumor. Severe febrile attacks such as erysipelas and the use of Coley's fluid may be included in this category. Perhaps in such cases we are justified in speaking of induced regression."

43 "balsamic blood": A. Silverstein, *History of Immunology,* p. 10; "This experiment involved": p. 12; "raising such a commotion": p. 32.

44 Background on simultaneous infection and regression: Boyd, "Spontaneous Regression"; H. C. Nauts, "The Beneficial Effects of Bacterial Infections on Host Resistance to Cancer: End Results in 449 Cases," Cancer Research Institute Monograph 8 (2d ed.), New York, 1980, pp. 147–55; and H. C. Nauts, "Bacteria and Cancer—Antagonisms and Benefits," *Cancer Surveys* 8:713–23 (1989); L. Old and H. Oettgen, in *Biologic Therapy of Cancer,* Lippincott, Philadelphia, 1991, pp. 87–119. See also A. Nowotny, "Antitumor Effects of Endotoxins," in L. J. Berry, ed., *Handbook of Endotoxin,* vol. 3, *Cellular Biology of Endotoxin,* Elsevier, Amsterdam, 1985, pp. 389–448.

45 "He who escapes": quoted in Nauts, "Beneficial Effects," p. 8.

45 On malariotherapy: see Henry J. Heimlich, "Try Malariotherapy," letter to the *New York Times,* Feb. 21, 1995, p. A-18.

46 "Here, gangrene seems": Tanchou, quoted in Nauts, "Bacteria and Cancer," p. 714.

46 On "suppuration" as an adjuvant to sarcoma surgery, see J. Da Costa, *Clinical Hematology: A Practical Guide to the Examination of the Blood, With Reference to Diagnosis,* P. Blakiston's Sons & Co., Philadelphia, 1901, p. 263.

46 "a partial or complete": Tilden C. Everson and Warren H. Cole, *Spontaneous Regression of Cancer: A Study and Abstract of Reports in the World Medical Literature and of Personal Communications Concerning Spontaneous Regression of Malignant Disease,* W. B. Saunders Co., Philadelphia, 1966.

47 "whispers of nature" and "I'm going to put": see "Spontaneous Cancer Regression," *Medical World News,* June 7, 1974, pp. 13–15; see also S. Hall, "Cheating Fate," *Health,* April 1991, pp. 38–46.

47 Coley's familiarity with spontaneous regression: We know that Coley familiarized himself with the literature of spontaneous regression early in his career, because he investigated it to refute criticism that his claimed successes could have been spontaneous remissions or errors in diagnosis. According to an account of a talk he gave to the Johns Hopkins Medical Society on April 6, 1896, a chronicler present stated, "Such an explanation [spontaneous regression] might be entitled to some consideration were a single case only involved, but those who would seriously propose it as a satisfactory explanation in view of the results in more than 20 cases, could not claim to be guided by scientific principles. The writer [Coley] stated that he had carefully examined the literature of the subject of spontaneous disappearance of tumors supposed to be malignant, but had failed to find a single instance in which the diagnosis had been confirmed by the microscope. It would appear remarkable that these cases should be the first on record with a clinically and microscopically confirmed diagnosis to disappear spontaneously, and it would seem more remarkable still that this disappearance should be coincident with the beginning of the treatment with the toxins" (Coley 1896, p. 160).

47 "That erysipelas has": Coley 1891, p. 210.

47 Busch case described in Nowotny, "Antitumor Effects," pp. 390–91.

48 F. Fehleisen, "Über die Züchtung der Erysipelkokken," *Deutsche Medizinische Wochenschrift* 8:553–54 (1882). Also Otto Westphal, professor emeritus, University of Freiburg, letter to author, July 8, 1996.

49 P. Bruns, *Beitrage f. Klinische Chirurgie,* 1888, p. 443; "He has collected": Coley 1891, p. 211.

49 "entire disappearance": W. B. Coley, "The Treatment of Cancer," *Guy's Hospital Gazette,* Jan. 6, 1912, pp. 7–14.

50 "Having satisfied myself": Coley 1893, p. 489.

Chapter 2: "The Man Who Does the Most Work Does the Best Work"

51 Epigraph: B. Waisbren Sr., *Journal of Biological Response Modifiers* 6:12 (1987).

51 Coley discussed the Zola case several times, first in W. B. Coley, "Contribution to the Knowledge of Sarcoma," *Annals of Surgery* 14:199–220 (1891); several more rounded versions appear in W. B. Coley, "The Treatment of Malignant Tumors by Repeated Inoculations of Erysipelas: With a Report of Ten Original Cases," *American Journal of the Medical Sciences* 10:487–511

(1893) and W. B. Coley, "The Treatment of Inoperable Sarcoma by Bacterial Toxins," *Proceedings of the Royal Society of Medicine, Surgical Section* 3:1–48 (1909–10). The present account has also benefited from the Zola case notes, archives, Cancer Research Institute. There is good reason to believe that the Zola diagnosis was correct: it was based on microscopic analysis of a biopsy sample, and perhaps more tellingly, the disease recurred within 12 months of excision, a trademark of sarcoma.

51   "about the size of": The entire pathology report on Zola is included in W. B. Coley, *Johns Hopkins Hospital Bulletin,* Aug. 1896.

52   "Apparently he had": Coley, "Treatment of Cancer," p. 7.

53   The erysipelas paradox: Charlie O. Starnes and Bernadette Wiemann have written, "One of the disconcerting aspects of the history of this subject is that so much attention has been apparently rightfully centered upon the evaluation of the antitumor effects of endotoxin, yet in nature it was primarily erysipelas, a gram-positive bacterial infection (hence, devoid of endotoxin) that provided the original observations upon which all subsequent work was supposedly based." They have advanced an interesting thesis to provide a possible explanation for this paradox. During normal metabolism, endotoxin (LPS) from the gastrointestinal tract routinely leaks into the portal circulation of the bloodstream. This naturally occurring LPS drains directly into the liver, where it is immediately cleared from the body. However, patients suffering from group A streptococcus infections (a group that includes erysipelas) are continuously exposed to streptococcal exotoxins, which have the capacity to compromise liver function. Wiemann and Starnes suggest that during an erysipelas infection, the body might therefore fail to clear the LPS as it normally leaks from the G.I. tract, causing it instead to spill over into the general circulation, resulting in a slow, constant infusion of LPS. For a more elaborate discussion, see B. Wiemann and C. O. Starnes, "Coley's Toxins, Tumor Necrosis Factor, and Cancer Research: A Historical Perspective," *Pharmacology & Therapeutics* 64:529–64 (1994).

54, 55   "intense local redness" and "the tumor": Coley 1891, p. 214.

55   "In both cases": ibid., p. 217.

56   "I worked continuously": Coley 1909, p. 591.

56   "The tumor of the neck": ibid.

57   "The patient's general condition": Coley 1893, p. 490.

57   "Golden Age of Quackeries": in James T. Patterson, *The Dread Disease,* Harvard University Press, Cambridge, 1987, pp. 39–41.

57   Chelidonium and other purported cures: see W. B. Coley, "The Treatment of Inoperable Cancer," *Practitioner* 9:497–517 (1899).

58   "the latch-string": letter to Coley from John D. Rockefeller, Feb. 16, 1926; "those little devices": letter to Coley from John D. Rockefeller Jr., May 3, 1912, both from CRI archives.

58   The hospital building where Coley first administered his biological treatment has had a checkered history. Built in 1884 as the New York Cancer Hospital, it was renamed the General Memorial Hospital in 1899; thus it may be considered the original site of what has ultimately become known as Memorial Sloan-Kettering Cancer Center, which opened at its present site on the Upper East Side of Manhattan in 1939. The "Bastille" later served as the Towers Nursing Home until 1974, when its owner was arrested and later convicted of Medicaid fraud; it was declared a landmark building in 1976 and

remains vacant to this day, although the property was purchased in 1988 by Ian Schrager, once part-owner of Studio 54. See "After 6 Years, Landmark Eyesore Is Still an Eyesore," *New York Times*, Dec. 4, 1994, p. 6/City.

59 "many attempts": Coley 1909, p. 592.

59 "From the beginning": Coley 1893, p. 490; "On November 14" and "probably metastatic": Coley 1909, p. 593.

59 Summation of first twelve cases: W. B. Coley, "A Preliminary Note on the Treatment of Inoperable Sarcoma by the Toxic Products of Erysipelas," *Post-Graduate* 8:278 (1893). "The uncertainty": ibid., p. 279.

60 "The clinical and experimental": ibid., p. 278. As an aside, several scholars of Coley's work, including Starnes, suggest that what Coley calls "sarcoma" or "lymphosarcoma" probably refers to what doctors now call lymphoma.

60 "Shortly afterward": Coley 1909, p. 594; "absolutely confirming": ibid., p. 594. "practical difficulties": ibid., p. 594. On Fehleisen's leaving the faculty of Würzburg: Otto Westphal, telephone interview, June 18, 1996, and letter to author, July 8, 1996.

61 "And, most important": Coley 1909, p. 594.

61 "a portion at least": ibid., p. 594.

62 "Roger had never": ibid., p. 595; G. H. Roger, "Contribution a l'Etude Expérimentale du Streptocoque de l'Erysipèle," *Revue de Mèdecin*, 1892, p. 12. Coley must have assiduously read European medical journals, because Roger's article came out shortly before he treated John Ficken.

63 "perhaps the most sensitive": Lewis Thomas, *The Youngest Science: Notes of a Medicine Watcher*, Viking, New York, 1983, p. 151.

63 Coley discusses Spronck's work extensively in the *Post-Graduate* paper (1893). C. H. H. Spronck, "Tumeurs Malignes et Maladies Infectieuses," *Annales de L'Institut Pasteur* 6:683–707 (October 1892).

64 "You should embrace": W. B. Coley, "Some Unsolved Problems in Medicine and Surgery," *Therapeutic Gazette*, Oct. 15, 1901, pp. 11–30. This paper, reprinted from a talk Coley gave to students at Thomas Jefferson Medical College in Philadelphia, is probably the best single compendium of his medical philosophy and beliefs. Coley's medical heroes are all mentioned, including Oliver Wendell Holmes, Charles Darwin, Huxley, Lister, Jenner, and Benjamin Rush, but what emerges most strongly is his concern for the patient. "If one were to attempt to give the qualifications of the ideal surgeon of today," he writes, "he would place high up in the list: skill in diagnosis, manual dexterity, and a thorough knowledge of aseptic technique, but above all these he would place the knowledge of when not to operate."

66 "The chill and tremblings" and other details of the John Ficken case come from a Cancer Research Institute file based upon Memorial Hospital records. The Ficken case was also discussed in Coley 1893; 1896; 1909; 1911.

66 "I want to see you": Letter from John F. Ficken to William Coley, September 17, 1910, archives, Cancer Research Institute.

67 "sufficiently encouraging" and "While the treatment": Coley 1893, pp. 283, 286.

67 "There is no longer much question": "The Failure of the Erysipelas Toxins," *Journal of the American Medical Association* 23, no. 24 (Dec. 15, 1894), p. 919. The editorial also noted: "The plain duty of the profession is clear. We should patiently wait until those having the facilities for investigation complete their work, and bearing in mind the infirmities of poor, weak, human

nature, we should not accept as conclusive, reported cures by any single investigator no matter how prominent. We should demand that these cures be corroborated by other observers, making investigations independently."

68 "That a few physicians": W. B. Coley, *Journal of the American Medical Association*, Jan. 5, 1895, p. 28. "To hold out": in "The Erysipelas Toxins," *Medical Record*, 50:557 (1896). This editorial came on the heels of a report by a committee of the New York Surgical Society, which concluded that "the alleged successes are so few and doubtful in character that the most that can be fairly alleged for the treatment by toxins is that it may offer a very slight chance of amelioration."

68 "enough to cause": Sir Rickman Godlee, *Lord Lister*, Macmillan, London, 1917, p. 532.

69 "These results are indisputable": W. B. Coley, 1909; "if we claim": W. B. Coley, "The Influence of the Roentgen Ray upon the Different Varieties of Sarcoma," *Transactions of the American Surgical Association* 20:308–34 (1902).

70 "What renders": E. Jenner, quoted in Robert Desowitz, *The Thorn in the Starfish*, Norton, New York, 1987, p. 23; "What he did": ibid. See also William R. Clark, *At War Within: The Double-edged Sword of Immunity*, Oxford University Press, New York, 1995, pp. 20–26.

71 On the schism between cellular and humoral immunology: Silverstein devotes an entire chapter (pp. 38–58) to the "epic" nineteenth-century debate between the humoralists and cellularists—"battles," he writes, "that saw opposing schools engage in passionate debate and a degree of vilification almost unknown in present-day science." (Silverstein, *History of Immunology*, p. 38). Silverstein describes how Elie Metchnikoff proposed the phagocytic (or cellular) theory of inflammation in 1884, how the theory came under "severe and protracted attack" (47), and how the tide began to swing inexorably toward the seeming primacy of antibodies following the discovery that immune serum conferred protection against diphtheria and tetanus in the early 1890s. This debate, he points out, was not without nationalist overtones: the cellularists, rallying around Metchnikoff at his base at the Pasteur Institute in Paris, tended to be French, while his fiercest detractors tended to be German (Prussian especially), and tensions simmering since the Franco-Prussian War of 1870 led to vitriolic exchanges between Louis Pasteur and Robert Koch. (Paul de Kruif even suggested in *The Microbe Hunters* that the debate between the cellularists and the humoralists contributed to the outbreak of World War I). Silverstein concludes that "the fall from favor of Metchnikoff's cellular (phagocytic) theory of immunity carried with it profound implications for future developments in the young discipline of immunology" and that "nascent immunology more and more turned away from medicine and biology and toward chemistry" (54). For a good survey of the early history of cancer immunotherapy, see Ilana Löwy, "Experimental Systems and Clinical Practices: Tumor Immunology and Cancer Immunotherapy, 1895–1980," *Journal of the History of Biology* 27:403–35 (1994).

71 Coley on immunology: Coley never had much to say about immunology. In a talk in 1901, he stated, "Of all the problems that are before the physician and surgeon today, there are few more interesting or important than immunity and serum therapy, for the two are so closely related that they should be considered together" (p. 18). Later, in 1909, he spoke of immunity in the

context of his belief in the microbial origins of cancer: "Assuming such origin, we have but to follow the analogy of other diseases of known germ origin. We know that in all such diseases there is a natural immunity and an acquired immunity. In the case of malignant tumors there is probably a natural immunity which is very great, but in certain cases it is absent or becomes lowered and the germ finds a favorable site and here starts the primary malignant tumor. . . . Assuming such extrinsic origin, the action of the toxins appears to me to produce certain changes in the blood serum that restore the weakened or lost immunity, or natural resisting power of the tissues, and the sarcoma cell, no longer finding conditions favorable for further growth and development, undergoes a process of degeneration with absorption in some cases, and the formation of slough in others. . . . Many and repeated blood examinations of sarcoma patients treated with the toxins show almost universally a marked leucocytosis as a result of the treatment."

72 "nearly one-half" and "showed more or less": Coley, "Further Observations Upon the Treatment of Malignant Tumors with the Toxins of Erysipelas and Bacillus Prodigiosus, With a Report of 160 Cases," *Johns Hopkins Hospital Bulletin*, August 1896, pp. 157–62; "great value" and "I have been": Finney and Welch remarks, recorded in the discussion section of the paper, p. 162.

73 "The treatment of inoperable": N. Senn, "The Treatment of Malignant Tumors by the Toxins of the Streptococcus of Erysipelas," *Journal of the American Medical Association* 25:131–34 (1895).

73 "This simply shows": Roswell Park, cited in "Discussion" section of W. B. Coley, "Treatment of Inoperable Malignant Tumors with the Toxines of Erysipelas and the Bacillus Prodigiosus," *Transactions of the American Surgical Association*, 12:183–212 (1894), p. 26. Roswell Park actually spent "a day or two" with Coley studying the method in New York. "In no case have I done harm. In every case," he said, "I have seen more or less benefit, but I have not had a positive cure." For follow-up editorial, see "Erysipelas and Prodigiosus Toxins" (Coley), *JAMA*, 103:1067 (1934).

74 "It is natural": Coley 1909, p. 612.

74 "I have never": Coley 1912, p. 14.

Chapter 3: Bleak House

76 Epigraph: H. Cushing to W. B. Coley, May 1921, archives, CRI.

76 For early x-ray history, see Stanley Joel Reiser, *Medicine and the Reign of Technology*, Cambridge University Press, Cambridge, 1978, pp. 58–68. The Radithor history is described in Roger M. Macklis, "The Great Radium Scandal," *Scientific American*, August 1993, pp. 94–99 (William Bailey, who "invented" Radithor, also manufactured a gold-plated radium-containing harness called the Radioendocrinator that, according to Macklis, could be worn in a special jockstrap by fatigued males). See also Ron Winslow, "The Radium Water Worked Fine Until His Jaw Came Off," *Wall Street Journal*, Aug. 1, 1990, p. A-1; the practice of using radiation as a home remedy persists, as in Eric Morgenthaler, "For a Healthy Glow, Some Folks Try a Dose of Radon," *Wall Street Journal*, Oct. 12, 1990, p. A-1. Codman remarks: E. A. Codman, "No Practical Danger from X Ray," *Boston Medical and Surgical Journal* 144:197 (1901); this letter, however, primarily discusses diagnostic uses of x rays.

77   "We find abundant": W. B. Coley, "The Present Status of the X-ray Treatment of Malignant Tumors," *Medical Record* 63, no. 12:441–51 (1903).

79   "He had few patients": Fred W. Stewart, "James Ewing" (obituary notice), *Archives of Pathology*, 36:325–30 (Sept. 1943). Other details of Ewing's career and background can be found in Guy F. Robbins, "James Ewing—The Man," *Clinical Bulletin* (a publication of Memorial Sloan-Kettering Cancer Center) 8, no. 1 (1978); Arthur I. Holleb, "James Ewing: The Man," in "A History of The Society of Surgical Oncology," 1995; and obituary notice, *New York Times*, May 17, 1943, p. 15. See also "Cancer Crusade," *Time*, Jan. 12, 1931, pp. 24–28, with "Cancer Man Ewing" on the cover.

80   "After the death": Stewart, "James Ewing," p. 326.

80   On Ewing's distaste for surgeons: "the chief difficulties in the way of cancer research in hospitals lie in the obstructions placed in the way by the attending surgeons." (J. Ewing to W. B. Coley, Oct. 1910; archives, Cancer Research Institute).

81   Ewing's "foot-hold": W. B. Coley to J. Ewing, April 14, 1916; archives, CRI.

81   "by far the best": W. B. Coley, letter to George Crocker, Oct. 20, 1908; archives, Cancer Research Institute. The Coley correspondence cited here belongs to the papers of William Bradley Coley in the archives of the Cancer Research Institute. This is an incomplete record of Coley's correspondence and must perforce be treated with caution. A further difficulty for scholars is that the collection has not been cataloged and organized for easy consultation; some of the letters are also in extremely fragile condition. The portions cited here represent but a small sample of the exchanges between Coley and Ewing, many of which are devoted to the arcana of pathological analysis and a benumbing amount of bureaucratic detail about the assignment and disposition of hospital cases of bone cancer. In the tiny smattering of remarks quoted in the main text, I have attempted to give a flavor of some of the recurring themes of the correspondence. These include the close early collaboration between Ewing and Coley to shape the mission of Memorial Hospital as a cancer research facility; Ewing's sharp criticism of Coley on the issues of diagnosis, writing papers, and self-promotion; Coley's sharp criticism of Ewing on the issues of his hostility toward the toxins and his clear enthusiasm for radiation in the treatment of bone sarcomas; and perhaps most important, the fact that both men could be, and were, wrong about important medical matters.

That there was friction, however, is undeniable. As Coley wrote (WBC to J. Ewing, Feb. 3, 1925): "This is the first letter that I have received from you in a good many years which, I think, has been written in a friendly spirit, and showing a desire to cooperate. I am only too happy to reply in like spirit. As you say, we are the oldest men on the Board; we have a longer and a deeper interest in the Hospital than any of the other men have; and we ought not to let any of our personal animosities interfere with the harmony which is necessary for the welfare of the Hospital. I think that the great trouble has been that neither of us has really understood the other's point of view. Whenever we attempted to discuss things personal, one or the other has said something that gave offense and made calm discussion impossible."

81   "the hope of determining": S. P. Beebe and M. Tracy, "The Treatment of Experimental Tumors with Bacterial Toxins," *Journal of the American Medical Association* 49:1493–98 (1907). The Beebe and Tracy experiments are

ultimately difficult to interpret because these transplanted tumors were not inserted into inbred animals. Thus the transplanted tumors, like a mismatched organ transplant, probably incited an immune response not simply to the cancer but rather to the foreignness of the tissue.

In an accompanying article, "The Growth of Lymphosarcoma in Dogs" (*JAMA* 49:1492), Beebe describes experiments in transplanting the form of cancer known as lymphosarcoma from dog to dog, and the paper notes that James Ewing had ruled out an infectious organism. The two men may have concluded that something in the blood serum prevented tumors from taking hold on occasion, because they suggest that "the blood contains immune factors which confer a passive immunity on a susceptible animal." An interesting sidelight to this work is that during the highly publicized illness of William Bull in 1909, Beebe and Ewing may have attempted to treat Bull's neck and facial cancer with a form of passive immunotherapy; Ewing was quoted in the *New York Times* (Jan. 29, 1909, p. 1) as saying, "One of the several methods for the treatment of cancer which have been experimentally tested in the Cornell Medical College has been used in Dr. Bull's case. As to the results we are unable to form any opinion." This account reported that Ewing and Beebe applied the treatment; a story several days earlier suggested the experimental treatment took the form of injections of lymph from an animal.

83 James Ewing, "The Treatment of Cancer on Biological Principles," *New York Medical Journal* 96, no. 16 (October 19, 1912), pp. 773–79. One of the most remarkable aspects about this paper is how actively, during the first decade of this century, researchers pursued immune modulation therapies as a legitimate medical approach to cancer and other diseases.

84 "quiet, introspective": from *Legacy*, pp. 35 and 46.

84 "I see in this": *New York Times*, June 15, 1939, p. 48; "If I had a million": Ewing, quoted in *New York Times*, Jan. 9, 1938, p. II-8; "the pathologist's pathologist" and "sound, thorough, and virile": Dr. Jacob M. Ravid, letter to editor, *New York Herald Tribune*, May 13, 1944.

85 "extremely": J. Ewing to W. Coley, March 9, 1910, archives, Cancer Research Institute. The debate over the so-called parasitic origin of cancer is an interesting one. The leading cancer authorities in the U.S., Ewing and Francis Carter Wood of Columbia University, thought such a theory bordered on quackery, but Coley had rather distinguished company. Several prominent scientists also shared this belief, including the British physician William Gye, director of the Imperial Cancer Research Fund, and Peyton Rous of the Rockefeller Institute, who in fact won a Nobel Prize (in 1966) for his discovery (in 1911) of the Rous sarcoma virus, which causes a form of cancer in chickens. Late in Coley's life, he exchanged letters with Rous in which both reiterated their firm belief in a microbial origin of cancer. "Like yourself," Rous wrote, "I firmly believe that viruses cause malignant tumors. There is more factual ground for that belief than for any other about cancer causation" (PR to WBC, Jan. 15, 1936, archives, Cancer Research Institute).

Ironically, after many decades in eclipse, this view has been at least partially revived by recent research, although neither Gye nor Coley deserve any credit for holding it prior to proof. Among instances of cancers currently believed to have a microbial component are the association of HTLV-1 virus to a form of leukemia, Epstein-Barr virus to Burkitt's lymphoma, human papilloma virus to cervical cancer, the bacterium *Helicobacter pylori* to gastric cancer, and a recently discovered herpes virus to Kaposi's sarcoma. In addition, research in the 1960s and 1970s saw a revival in the belief that viruses might cause cancer, and this eventually led, by a circuitous route, to the discovery of oncogenes. (For a brief, excellent account of this research, see Daniel J. Kevles, "Pursuing the Unpopular: A History of Courage, Viruses, and Cancer," in Robert B. Silvers, ed., *Hidden Histories of Science*, New York Review Books, New York, 1995; for a longer treatment, see Robert Weinberg, *Racing to the Beginning of the Road*, Harmony, New York, 1996.)

86 "of pernicious anemia": from Bob Considine, *That Many May Live: Memorial Center's 75 Year Fight Against Cancer*, Memorial Center for Cancer and Allied Diseases, New York, 1959, p. 41.

86 "a man of ideas": James Ewing, *Neoplastic Diseases: A Treatise on Tumors*, W. B. Saunders Co., Philadelphia, 1940 (4th ed.), frontispiece.

86 "go out of their way": R. Weil to Coley, April 9, 1917, archives, Cancer Research Institute; "I often wish": Ewing to Coley, July 18, 1917.

87 "You are quite right": Ewing to Coley, Feb. 16, 1924.

87 "these are three well recognized," Ewing to Coley, July 18, 1917; "I was never more": Coley to Ewing, April 24, 1917, archives, Cancer Research Institute.

87 "Your letters are hardly": Ewing to Coley, Jan. 15, 1917, archives, Cancer Research Institute.

88 "It is the first time": Coley to Ewing, Jan. 16, 1917, archives, Cancer Research Institute. In this same letter, Coley writes, "You openly made the statement at the conference a few weeks ago that you have never seen any other type of tumor yield to the toxins except lymphosarcoma, apparently forgetting that you have seen the patients, personally examined the slides, and given a written diagnosis of nearly every other type except the melanotic, which have been treated with the toxins."

88 On the issue of accurate diagnosis: Many histological slides documenting Coley's early cases were irreparably damaged in a hansom cab accident when they were being transferred from the Loomis Laboratory in downtown New York to the new pathology department at Memorial Hospital on the upper West Side (source: Helen Coley Nauts).

88 On the Memorial Hospital policy on ward cases of bone cancer: see W. B. Coley, "The Treatment of Sarcoma of the Long Bones," *Transactions of the American Surgical Association*, 50:383–417 (1932). Coley writes, "Between 1915 and 1928, practically all the service cases, including those of giant-cell tumor, at the Memorial Hospital were treated by primary irradiation. Having a large amount of radium at our disposal, at first four grams, and later eight grams, a considerable number of cases were treated with the radium pack; this was sometimes supplemented by bare tubes of radon or gold seeds inserted into the tumor. The majority of cases, however, were treated with rontgen rays." (p. 392). A review in 1928 of more than 140 cases of operable malignant bone sarcomas treated by radiation at Memorial, Coley wrote, showed only 4 patients alive and well beyond five years.

88 "You do not hesitate": Coley to Ewing, Feb. 27, 1922, archives, Cancer Research Institute; "I feel we": Ewing to Coley, March 5, 1922, archives, Cancer Research Institute.

89 "I am not a mere": Ewing to Coley, March 5, 1922.

89 "quite incompetent" and "Janeway and I feel": Ewing to Coley, July 18, 1917, archives, Cancer Research Institute.

89 For Codman background, see L. Altman, "A Reformer's Battles," *New York Times*, June 12, 1984, p. C-5; "probably more living": E. A. Codman to W. B. Coley, July 22, 1920, archives, Cancer Research Institute. Coley's daughter, Helen Coley Nauts, has done a yeoman's job of cataloging and summarizing the extensive correspondence surrounding the Bone Sarcoma Registry, from which many of these citations come.

91 "when we come": J. Bloodgood to Coley, Nov. 6, 1921. Of particular contention between Bloodgood and Coley was the diagnosis of giant cell sarcoma. Bloodgood insisted that these were invariably benign tumors, that they could be treated by curettage and radiation alone, and that their cure could not be claimed by Coley with the toxins. Coley believed some cases of giant cell sarcomas had the clinical behavior of malignancy and that pathologists could not distinguish between benign and malignant cases by examining tissue samples alone. It is now recognized that a small portion of giant cell sarcomas do indeed become malignant.

91 "I must admit": E. Codman to WBC, July 22, 1920, archives, CRI. Codman's sentiments were quite clear: to another physician who had apparently cured a case of bone sarcoma with bacterial toxins, he spoke of the patient "as having gotten well in spite of Coley's toxins instead of with the help of them." Coley refers to the remark in a letter, WBC to Codman, February 7, 1922; Codman apparently made the remark to Dr. George Packard of Denver, the attending physician in the case of a nine-year-old boy diagnosed with Ewing's sarcoma in 1920 who recovered with prolonged toxin treatment.

91 "There is only one way": WBC to J. Bloodgood, Nov. 16, 1921.

91 "I think Ewing": Charles Mayo to Coley, Jan. 19, 1923. William Mayo expands on this same sentiment in a November 19, 1924, letter to Coley. "Codman is hopeless. He is honest, but he was born in the days when witches were burned in Salem. He means to do what is right, but he hears the voice of the Lord in the Wilderness telling him that he must do certain things regardless, and I fear that if we traced that voice we should find out what Balaam did. We don't send our material to him for that reason."

91 "*Confidentially,* Ewing": WBC to W. Welch, Feb. 23, 1922.

91 "I think these men": W. H. Welch to Coley, May 22, 1922, archives, CRI.

92 "It is true": WBC to E. Codman, February 7, 1922, archives, CRI.

92 Leonard Portal Mark, *Acromegaly: A Personal Experience,* Bailliere, Tindall, and Cox, London, 1912.

93 "In the journal": Helen Coley Nauts, interview, Sept. 10, 1992, New York; "a cruel shock": Helen Coley Nauts, "Coley's Toxins: The First Century," paper read at International Clinical Hyperthermia Society, Rome, Italy, May, 1989. Apropos of the benign tumor he knew to be pressing on the pituitary gland in his brain, Coley asked the neurosurgeon Harvey Cushing, who ironically treated Ewing for his nerve disorder, to examine his brain tissue after his death, which he did.

94 "I have not registered": E. Codman to WBC, Feb. 6, 1922.

94 "As a matter of fact": WBC to Codman, Feb. 7, 1922.

94 "Codman's letter was": Helen Coley Nauts, "Coley's Toxins."

95 "self-appointed triumvirate": Coley to Welch, Feb. 23, 1922.

95 "Do not trouble": Codman to WBC, Dec. 6, 1921.

95 "steel barriers" and "I am 60 years": Coley to Welch, Feb. 23, 1922. In this same letter, Coley tells Welch, "If the method has been able to save the lives of a considerable number of hitherto-regarded-as-hopeless cases, it naturally follows that a greater knowledge of the method and a wider application will save a good many more lives. It is possible for men of such high position in the scientific world, as Ewing, Bloodgood, and Codman, to take away this chance of life from a great many people who might otherwise receive it; they might, perhaps, honestly persuade themselves that their hypocritical attitude was an example of a high scientific standard (although of this, I personally have some doubts)."

Chapter 4: The Method of Choice

96 Epigraph: W. B. Coley, "Some Thoughts on the Problem of Cancer Control," *American Journal of Surgery* 14:618 (1931).

96 "are credited with": *New York Times*, May 7, 1934, p. 34; other news from the *New York Post*, May 25, 1934.

97 "And most baffling": in "Cancer Crusade," *Time*, Jan. 12, 1931, p. 24.

97 "miraculous cures": James Ewing, as quoted in the *New York Times*, Jan. 13, 1933, p. 10.

98 "increasingly grotesque appearance": Helen Coley Nauts, letter to author, March 25, 1996.

98 "The work of convincing": W. B. Coley, "Some Unsolved Problems in Medicine and Surgery," *Therapeutic Gazette*, Feb. 1902, p. 28.

98 "Back Ewing": *Legacy*, p. 33.

99 "were of similar nature": ibid., p. 46.

99 1934 bone cancer symposium: The papers presented at the 1934 Memorial Hospital Symposium were gathered and published the following January in the *American Journal of Surgery* 27:3–49 (1935). In the same volume, regarding the radiation results alone, Bradley L. Coley, "The Treatment of Osteogenic Sarcoma by Irradiation," ibid., pp. 43–47.

100 "While many of the earlier": W. B. Coley, "Endothelial Myeloma or Ewing's Sarcoma," *American Journal of Surgery* 27:7–17 (1935).

101 The most comprehensive discussion of the Christian and Palmer case can be found in S. L. Christian and L. A. Palmer, "An Apparent Recovery from Multiple Sarcoma," *American Journal of Surgery*, n.s., 4, no. 2:188–197 (1928).

102 Effect of radiation on immune system: see Hartmann Stahelin, "The Development of Immunosuppressive Agents from X-rays to Cyclosporin," in P. M. H. Mazumdar, ed., *Immunology, 1930–1980: Essays on the History of Immunology*, Walls & Thompson, Toronto, 1989, pp. 185–201.

103 "The results obtained": W. B. Coley, *AJS*, 27:7–17 (1935), p. 16; "willing to have": p. 17.

104 "I venture to urge": Ewing, "Place of the Biopsy," pp. 26–28. Bloodgood remarks: ibid., p. 41; "remains to be proved"; C. Simmons, ibid., p. 25.

105 "occasional miracles": E. A. Codman, "Symposium on the Treatment of Primary Malignant Bone Tumors," ibid., pp. 3–6.

106 "In every case": H. W. Meyerding, "Surgical Treatment of Osteogenic Sarcoma," ibid., pp. 29–34. According to Helen Nauts, the toxins cost about $2 per bottle.

106 Current regimens for osteosarcoma treatment include amputation of the limb followed by high-dose methotrexate, alone or in combination with adriamycin; in limb-sparing surgery, radiation treatment may precede surgery, and chemotherapy follows it. Current regimens for Ewing's sarcoma call for long-term (up to 18 months) systemic chemotherapy; if the disease is limited at diagnosis, remission rates of 80 to 90 percent can be achieved.

106 "auxiliary cells": in H. N. Claman, E. A. Chaperon, and R. F. Triplett, "Thymus-marrow Cell Combinations: Synergism in Antibody Production," *Proceedings of the Society for Experimental Biology and Medicine* 122:1167–71 (1966). This paper actually argued for a more important role for T cells, but the word "auxiliary" still suggests that T cells played a secondary role to antibody.

107 On the role of lymphocytes in transplant and tumor rejection, see James B. Murphy, "The Lymphocyte in Resistance to Tissue Grafting, Malignant Disease, and Tuberculosis Infection: An Experimental Study," *Monograph of the Rockefeller Institute for Medical Research,* No. 21, Sept. 20, 1926, pp. 1–168. Murphy had worked with Peyton Rous and, later, in an elegant series of experiments summarized here, demonstrated that lymphoid cells played a decisive role in rejecting foreign tissue transplants and tumor transplants. "We have been led to conclude that mice resistance to malignant tumors, whether transplanted or spontaneous, is closely associated with the lymphoid tissue and there are indications that the same is true of other species including man," Murphy wrote (p. 163). He was also among the first to appreciate the odd gestalt of the lymph system: "The lymphoid tissue considered as an organ has been particularly difficult to investigate. It not only composes largely the spleen and numerous lymph nodes but small deposits scattered through most of the organs and tissues of the body and in addition large numbers of the cells are to be found in the circulation. If all this tissue could be brought together in a single mass it would represent an organ of considerable size" (p. 155). But one of the most remarkable ironies of this work is that Murphy offered the activity of lymphoid cells as an *alternative* to an immune response. "Having implicated the lymphocyte in the rejection of tumor grafts," Silverstein writes, "little more could be said on this subject at that time" (Silverstein, *History of Immunology,* pp. 281–82).

107 "The pioneering cell": ibid., p. 55.

107 "I will leave": Coley 1909, p. 612; "unscientific": Almroth Wright, quoted in I. Löwy, "Experimental Systems and Clinical Practices: Tumor Immunology and Cancer Immunotherapy, 1895–1980," *Journal of the History of Biology* 27 (1994), p. 411.

108 "What is the explanation": Coley 1909, p. 599. Apropos of quality control, Paul Starr observes, "There was hardly an advance of medical science whose introduction into medical practice was not initially marred by uncertainty and disillusionment because of errors in application or failures of quality control" (*Social Transformation,* p. 138).

108 "crucial decades" and following: P. Starr, *Social Transformation,* p. 121. Starr also writes that "the assimilation of medical education into the universities drew academic medicine away from private practice" and that "in the twentieth century, academic and private physicians began to diverge and represent distinctive interests and values" (p. 122), trends that may explain

why Coley's work was viewed in such poor light by his contemporaries in academic medicine.

108 "not of such promise" and "that kind of study": quoted in *Legacy*, pp. 27, 25.

109 "Whatever the future of radium": Fred W. Stewart, obituary, *Archives of Pathology* 36:328 (1943).

109 On Pasteur and Koch, Silverstein recounts the story of how Pasteur returned an honorary degree to the University of Bonn and requested that his name be "effaced" from the university archives following the Franco-Prussian war; Koch, for his part, dismissed as "meaningless" many of Pasteur's experiments (Silverstein, *History of Immunology*, pp. 42–46). For recent scholarship on Pasteur, see Gerald L. Geison, *The Private Science of Louis Pasteur*, Princeton University Press, 1995, and also L. Altman, "Revisionist History Sees Pasteur as Liar Who Stole Rival's Ideas," *New York Times*, May 16, 1995, p. C-1.

110 "I have had no difficulty": W. B. Coley, *AJS*, 1935, p. 17.

112 "radiation alone": J. Ewing, p. 314. Ewing apparently had conceded that the toxins played some role in remissions, but declined to publicly acknowledge that fact for political reasons. In a Nov. 25, 1922, letter to Coley, Ewing writes: "I am not by any means disinclined to believe that the administration of the toxins may act as a sensitizing agent in radiation therapy. When I suggested this matter to some of your colleagues they advised me not to do so, saying that you would quote the statement on all occasions, interpret it as justification for the indiscriminate use of toxins and make an unwise use of the suggestion. On the whole I have decided to follow this line of advice and to withhold any recommendation of the use of the toxins with x-ray until you have succeeded in demonstrating the wisdom of the combination." Ironically, Coley's son Bradley, no great fan of the toxins, once observed that radiation probably inhibited the activity of the vaccine, because it damaged the very blood vessels that carried the salutary effects of the toxins into tumors.

112 "It appears": Council on Pharmacy and Chemistry, "Erysipelas and Prodigiosus Toxins (Coley)," *JAMA* 103:1067–69 (1934).

113 "I was pleased to see": W. Mayo to WBC, Oct. 25, 1934, archives, Cancer Research Institute.

113 On the nonuse of the toxins for Mrs. Coley: Helen Coley Nauts maintains that her mother's advanced age (69 years) and medical history (a recent heart attack) precluded the rigorous side effects of toxin treatment, and adds that "many cases" of colon cancer were successfully treated with the toxins (H. C. Nauts to author, March 25, 1996).

113, 114 Details of Coley's final hours: interview with Helen Coley Nauts and Archibald Young, "Postscriptum: Dr. William B. Coley," *Glasgow Medical Journal*, August 1936, pp. 165–70. Young quotes directly from a letter sent by Coley's secretary on April 28, 1936.

114 "leading American exponent": *New York Post*, April 16, 1936, p. 12.

Chapter 5: The Coley Phenomenon

115 Epigraph: Seneca, quoted in W. B. Coley, "The Idea of Progress," *Transactions of the New York and New England Association of Railway Surgeons*, 1920, p. 6.

115 "evangelist": in James T. Patterson, *The Dread Disease: Cancer and Modern American Culture*, Harvard University Press, Cambridge, 1987, p. 195.

115 "an outstanding symbol": in "Frontal Attack," *Time*, June 27, 1949, pp. 66–75 (cover story on "Cancer Fighter Cornelius P. Rhoads").

115 "proceed immediately": C. P. Rhoads to Helen Coley Nauts, Oct. 2, 1942, archives, Cancer Research Institute. In this letter, Rhoads outlined four specific activities: "1. To make tests of the toxin on experimental lymphosarcoma rats. 2. To make similar tests on similar cases in mice with spontaneous breast cancer. 3. To obtain from Dr. [Walker] Swift data on the case you referred to. 4. To provide a suitable patient with either lymphatic leukemia or mammary cancer for clinical test."

Apropos of this moment, Coley's son Bradley, head of the Bone Service at Memorial, sent a November 1, 1943, letter to the hospital's chief pathologist, Fred W. Stewart, implying that the hospital was giving his sister the runaround. Brad Coley had never been a strong proponent of the toxins during his father's lifetime, but following his sister's analysis of the cases, he wrote, "I feel that until some really controlled scientific evaluation of the method has been made, it is unjustifiable to let it slip quietly into the discard." In an unsent letter to Stewart composed around the same time, Brad Coley was more emphatic: "I believe that, despite the war, the time has come for the Memorial to quietly study the method in a scientific manner." He also indicated that Shwartzman at Mt. Sinai would test titers of tumor-hemorrhage toxin "without expense to the Memorial." (Source: CRI archives).

116 The source for Rhoads's contacts with Parke-Davis is H. C. Nauts, personal interview, Aug. 9, 1993; also referred to in H. C. Nauts, "Coley Toxins—The First Century," read at the International Clinical Hyperthermia Society, Rome, May 1989.

116 no source in America: Unbeknownst to Coley or anyone else in New York, a German company based in Munich called Südmedica had since 1914 manufactured a product called "Vaccineurin," which was expressly used to induce fevers and was composed of extracts of the two bacterial species used in Coley's toxins (it sold for 5 German Marks, less than $4, over the counter until discontinued when Südmedica went out of business in 1991). See letter from Dr. Bader of Südmedica to Helen Coley Nauts, Feb. 19, 1979, archives, Cancer Research Institute. In his letter, Bader states, "we were totally unaware of Dr. Coley's vaccine therapy."

116 For *Unproven Methods of Cancer Management*, see Ralph W. Moss, *The Cancer Industry: The Classic Exposé on the Cancer Establishment*, Paragon House, New York, 1989, pp. 97–118. Moss provides a good, often acidic survey of mainstream medicine's response to alternative approaches to cancer therapy.

117 "He'd never thrown away": Helen Coley Nauts, personal interview, New York, N.Y., Sept. 9, 1992.

118 "It's always been my": William Curtis, personal interview, Seattle, Wash., Dec. 27, 1994.

118 For a comprehensive and critical history of the Cancer Research Institute, see Ilana Löwy, "Innovation and Legitimation Strategies: The Story of the New York Cancer Research Institute," in *Medicine and Change: Historical and Sociological Studies of Medical Innovation*, edited by

O. Amsterdamska, J. Pickstone, and Patrice Pinell, Colloque Inserm, vol. 220, 1993, pp. 337–358.

118 See H. C. Nauts, W. E. Swift, and B. L. Coley, "The Treatment of Malignant Tumors by Bacterial Toxins as Developed by the Late William B. Coley, M.D., Reviewed in the Light of Modern Research," *Cancer Research* 6:205–16 (1946). The key finding was that "at least fifteen different preparations of Coley toxins have been used since the method was introduced in 1892, of which three were considerably more potent than the rest. It was further observed that the technique of administration has varied considerably as regards site, dosage, frequency, and duration of treatment." See also H. C. Nauts, G. A. Fowler, and F. H. Bogatko, "A Review of the Influence of Bacterial Infection and of Bacterial Products (Coley's Toxins) on Malignant Tumors in Man," *Acta Medica Scandinavica* 145 (suppl. 276), 1953, p. 11.

118 "Father didn't have": H. C. Nauts, personal interview.

119 On the role of fever in cancer regression, see K. F. Kölmel, O. Gefeller, and B. Haferkamp, "Febrile Infections and Malignant Melanoma: Results of a Case Control Study," *Melanoma Research*, 2:207–11 (1992).

119 Follow-up studies with Coley's toxins: Since Coley's death in 1936, there have been sporadic attempts to experiment further with the mixed toxins. In every case, the results have been inconclusive; but in every case, distinct clinical responses have been noted. Unfortunately, these studies have never been designed in such a way to categorically answer important questions about the vaccine's efficacy. These attempts are briefly discussed in order:

1) In the 1950s, a group at Temple University led by H. Francis Havas conducted tests on no less than 10,000 mice, trying to figure out which component of Coley's toxins destroyed tumors; they concluded that most of the activity could be attributed to *Serratia marcescens*, although they found that the toxins of the two mixed bacteria had greater effect on tumors than either bacterium used singly. Then, in a series of papers published in the journal *Cancer Research* between 1958 to 1961, they showed that this formulation of bacterial toxins reliably killed transplanted tumors in mice. "If there exists any host resistance to tumor development and growth," the Temple researchers wrote, "it may be mediated through a non-specific natural resistance, a more or less specific immunological reaction (antigen-antibody phenomenon), or both" (in A. J. Donnelly, H. F. Havas, and M. E. Groesbeck, "Mixed Bacterial Toxins in the Treatment of Tumors, II; Gross and Microscopic Changes Produced in Sarcoma 37 and in Mouse Tissues," *Cancer Research* 18:149–54 [1958]).

2) In 1962, a physician at Bellevue Hospital in New York named Barbara J. Johnston published two studies in *Cancer Chemotherapy Reports*. In the first, a controlled study comparing 34 patients treated with the toxins versus 37 controls who received a typhoid vaccine (as a general immunological stimulant), Johnston claimed 9 "objective" responses (26 percent) with the toxins versus one response in 37 controls. In the second, uncontrolled study, Johnston claimed 30 objective responses in 93 patients (32 percent). Many of these responses were short-lived, but there were several striking regressions, supported by photographic evidence of tumor shrinkage, including a man whose neck tumor almost totally disappeared. All too typical of the ill fortune attached to the toxins, this case ended in tragedy. "Two months after the toxin therapy was begun the patient died," Johnston noted. "While in an

inebriated state, he fell down a flight of stairs and fractured his cervical vertebrae. An autopsy failed to reveal the presence of a gross or microscopic tumor" (B. J. Johnston, "Clinical Effect of Coley's Toxins, I: A Controlled Study," *Cancer Chemotherapy Reports*, no. 21 [August, 1962], pp. 19–41; and B. J. Johnston and E. T. Novales, "Clinical Effect of Coley's Toxin, II: A Seven-Year Study," ibid., pp. 43–68).

Johnston went on to become head of oncology at St. Vincent's Hospital in New York, but her studies are considered flawed. It is part of the folklore of Coley's toxins that Johnston was dismissed under suspicious circumstances by Lewis Thomas, then medical director of New York University Hospital; Ralph Moss quotes Johnston in *The Cancer Industry* as saying, "They let us finish so as to prove that it was wrong. But it didn't work out that way." But there can be no doubt that Thomas had a long-standing and committed interest to immunological approaches to disease, and when specifically asked if Thomas was justified in letting Johnston go, Helen Nauts agreed. "She really wasn't proving much," Nauts said. "She wasn't very good or imaginative."

3) In 1987 Rita S. Axelrod, heading a group at Temple, published a study in the journal *Cancer* reporting on thirteen patients treated with a "completely modified" version of the vaccine, and although none of the patients showed "dramatic" tumor responses, "many had stabilization of disease and subjective improvement," including a remarkable instance of a man receiving long-term therapy lasting at least 650 days despite extensive metastatic disease spread to the liver and kidney. Echoing almost the same sentiments as Coley a century earlier, the Temple group concluded that the vaccine "is worthy of further investigation, which should focus on the optimal immunomodulating dose in patients with earlier disease" (Rita S. Axelrod et al., "Effect of Mixed Bacterial Vaccine on the Immune Response of Patients with Non-Small Cell Lung Cancer and Refractory Malignancies," *Cancer* 61:2219–30 [June 1, 1988]).

4) Sanford J. Kempin et al., "Improved Remission Rate and Duration in Nodular Non-Hodgkin Lymphoma (NNHL) With the Use of Mixed Bacterial Vaccines (MBV)," abstract, *Proceedings of the American Society for Clinical Oncology* 22:514 (1981). In this study, conducted at Memorial Sloan-Kettering Cancer Center, patients with advanced non-Hodgkin's lymphoma were divided into two groups. One received two regimens of standard chemotherapy; the other received chemotherapy plus a set of inoculations of a mixed bacterial vaccine made according to the Coley formula. The complete response rate in the vaccinated patients was 85 percent compared to 44 percent with chemotherapy alone. "However, the survival curves of the two populations eventually met," said one of the participating researchers, Lloyd Old, "raising the question of whether longer treatment with the vaccine would have provided more persistent benefit." An attempt is currently underway to do follow-up studies on patients who participated in this trial. One physician who participated in the trial, Dr. Herbert Oettgen, said in an interview that physician compliance in a second, unpublished sarcoma study was poor. Some patients who were supposed to receive the mixed bacterial vaccine did not. (H. Oettgen, personal interview, Feb. 15, 1996.)

5) Burton A. Waisbren, Sr., "Observations on the Combined Systemic Administration of Mixed Bacterial Vaccine, Bacillus Calmette-Guérin, Transfer Factor, and Lymphoblastoid Lymphocytes to Patients with Cancer, 1974–

1985," *Journal of Biological Response Modifiers* 6:1–19 (1987). This paper describes the treatment of 139 cancer patients in Wisconsin with a minestrone of immune modifiers—BCG, a Coley-like vaccine, immune cells, and so on. It was not a controlled study, and 85 of 139 patients had died by the time of publication. There were several notable regressions, however, and Waisbren argued that the therapy was safe, well tolerated, at the very least spared patients the rigors of chemotherapy, and "had a salutary effect on the courses of a number of the patients treated."

   6) Finally, Klaus F. Kölmel, a dermatologist at the University of Göttingen Klinik in Germany, treated 15 patients with advanced melanoma with a commercial preparation of bacterial lysates intended to induce fever ("Vaccineurin"). Kölmel reported that three cases resulted in a "total and long lasting remission." The most obvious shortcoming of this report, in addition to the small sample in a notoriously variable disease, is that it is not randomized. Kölmel has launched a large, multicenter European study of the effects of fever on the course of cancer; results are not anticipated until 1997 at the earliest. Source: K. F. Kölmel et al., "Treatment of Advanced Malignant Melanoma by a Pyrogenic Bacterial Lysate: A Pilot Study," *Onkologie* 14:411 (1991).

120   "were abandoned when": Steven A. Rosenberg, *The Transformed Cell*, Putnam, New York, 1992, p. 59.

120   "Those who have scrutinized": L. Old and H. Oettgen, "The History of Cancer Immunotherapy," in *Biologic Therapy of Cancer*, edited by V. DeVita, S. Hellman, and S. Rosenberg, Lippincott, Philadelphia, 1991, p. 97.

121   "one cannot fail": Frances Balkwill, *Cytokines in Cancer Therapy*, Oxford University Press, New York, 1989, p. 82.

121   "I can't explain": Frances Balkwill, personal interview, Imperial Cancer Research Fund, London, December 7, 1993.

121   "Coley's toxin was": Michael Osband, personal interview, Pittsburgh, June 3, 1992.

122   "Although certainly considerably": C. O. Starnes, "Coley's Toxins in Perspective," *Nature* 357:11 (1992). This article generated considerable correspondence, most suggesting that Starnes had been too conservative in discussing the potential utility of Coley's toxins or tumor necrosis factor.

123   "Given these striking": Lloyd Old, opening remarks, "Frontiers of Immunology and Cancer Immunology," sponsored by the Cancer Research Institute, New York, January 1993.

123   Sequence of TNF and BCG work: Murray Shear et al., "Chemical Treatment of Tumors, V: Isolation of the Hemorrhage-Producing Fraction from Serratia Marcescens (Bacillus prodigiosus) Culture Filtrate," *Journal of the National Cancer Institute* 4:81 (1943); L. J. Old, D. A. Clarke, and B. Benacerraf, "Effect of Bacillus Calmette-Guérin Infection on Transplanted Tumors in the Mouse," *Nature* 184:291–92 (1959); Georges Mathé, J. A. Amiel, L. Schwartzenberg, et al., "Active Immunotherapy of Acute Immunoblastic Leukemia," *Lancet* 1:697 (1969); A. Morales, D. Eidinger, and A. W. Bruce, "Intracavity Bacillus Calmette-Guérin for Transitional-Cell Carcinoma of the Bladder," *Journal of Urology*, 116:183 (1976); Elizabeth Carswell et al., "An Endotoxin-induced Serum Factor That Causes Necrosis of Tumors," *Proceedings of the National Academy of Sciences* 72:3666 (1975); M. J. Berendt, R. J. North, and D. P. Kirsten, "The Immunological Basis of Endo-

toxin-induced Regression. Requirement for T Cell–mediated Immunity," *Journal of Experimental Medicine,* 148:1550–59 (1978); Michiko Kobayashi et al., "Identification and Purification of Natural Killer Cell Stimulatory Factor (NKSF), A Cytokine with Multiple Biologic Effects on Human Lymphocytes," *Journal of Experimental Medicine* 170:827–45 (September 1989); and A. S. Stern et al., "Purification to Homogeneity and Partial Characterization of Cytotoxic Lymphocyte Maturation Factor from Human B-lymphoblastoid Cells," *Proceedings of the National Academy of Sciences* 87:6808–12 (Sept. 1990). For an extended review of Coley's work, see B. Wiemann and C. O. Starnes, "Coley's Toxins, Tumor Necrosis Factor, and Cancer Research: A Historical Perspective," *Pharmacology & Therapeutics,* 64:529–64 (1994).

125 One paradox of great interest to scholars of Coley's toxins is: Why did so many independent researchers, including Coley, report that tumors regressed during an erysipelas infection, and how to explain Coley's first puzzling case of Zola? The conundrum is this: as research by Havas suggested and has been confirmed since, including by Starnes at Amgen, the antitumor effect of Coley's toxins has little or nothing to do with *Streptococcus pyogenes,* the gram-positive bacterium that causes erysipelas and launched Coley on his lifelong vaccine work. It is the other component—known as *Bacillus prodigiosus* to Coley, as *Serratia marcescens* to modern microbiologists—that seemed to contain the active principle. As noted previously, Starnes and Bernadette Wiemann have speculated that certain gram-positive bacteria can, during active infection, block the liver's ability to clear lipopolysaccharide (LPS), which normally enters the bloodstream from the gastrointestinal tract.

126 "One of the things": Drew Pardoll, remarks at "Cancer Research and Treatment Beyond the Year 2000: Harnessing the Immune System," science writers briefing sponsored by Cancer Research Institute, New York, January 12, 1995.

126 "Today, over a hundred": C. O. Starnes and B. Wiemann, 1994, pp. 554–55.

127 Statistics on cancer incidence: see S. L. Parker et al., "Cancer Statistics, 1996," *CA—A Cancer Journal for Clinicians,* 46:5–27 (1996), p. 8.

II. THE PATRON SAINT OF CYTOKINES

129 Alexis Carrel and Albert H. Ebeling, "Leucocytic Secretions," *Journal of Experimental Medicine,* 36:645–59 (December 1922).

Chapter 6: In Search of an Interferon

131 Epigraph: Peter Medawar, in "Induction and Intuition in Scientific Thought," *Pluto's Republic,* Oxford University Press, New York, 1984, p. 101.

131 "I myself was extremely": Jean Lindenmann, personal interview, Nov. 1, 1993, Zurich, Switzerland. All remarks attributed to Dr. Lindenmann come from this interview unless otherwise noted. I am deeply indebted to Dr. Lindenmann for his patience in answering a long series of follow-up questions with a succession of gracious, informative, good-humored letters.

General sources on interferon include: Sandra Panem, *The Interferon Crusade,* Brookings Institution, Washington, D.C., 1984; Mike Edelhart with Jean Lindenmann, *Interferon: The New Hope for Cancer,* Addison-Wesley,

New York, 1981; W. E. Stewart II, *The Interferon System*, Springer Verlag, New York, 1979; S. Baron et al., "The Interferons: Mechanisms of Action and Clinical Applications," *Journal of the American Medical Association* 266:1375–83 (1991); and Alfons Billiau, ed., *Interferon: General and Applied Aspects*, Elsevier, Amsterdam, 1984.

132 "Viruses know more": Joost J. Oppenheim, "Cytokines: Past, Present, and Future," in S. Kasakura, ed., *Cytokine 1994*, Japan Publishing, Tokyo, 1994, pp. 11–18.

132 On interference: the original observation was G. W. M. Findlay and F. O. MacCallum, "An interference phenomenon in relation to yellow fever and other viruses," *Journal of Pathology and Bacteriology*, 44:405–24 (1937). At the time of Isaacs and Lindenmann's collaboration, the leading researchers into the phenomenon were Werner and Gertrude Henle, a husband and wife team at Children's Hospital in Philadelphia, who attempted the first clinical trial in 1946 (see W. Henle et al., "Experimental Exposure of Human Subjects to the Viruses of Influenza," *Journal of Immunology* 52:145–65 (1946).

133 Technique to inactivate enzymes: A. Isaacs and M. Edney, *Australian Journal of Experimental Biology and Medical Science*, 28:219 (1950).

136 "Alick was very amusing": Susanna Isaacs Elmhirst, personal interview, Dec. 5, 1993, London, England, and April 16, 1994, New York. For background on Alick Isaacs, I am grateful to Dr. Elmhirst for also sharing recollections, photographs, and family memories of her late husband (including "Origins," by Alick's brother, Bernard Isaacs), as well as a scrapbook of his *Lancet* contributions. See also C. H. Andrewes, "Alick Isaacs, 1921–1967," obituary notice in *Biographical Memoirs of Fellows of the Royal Society*, vol. 13, pp. 205–221 (1967). In addition, a number of former friends and colleagues provided recollections and insights, including Anthony Allison, Samuel Baron, Leslie Brent, Derek Burke, Donna Chaproniere, Margaret Edney, Norman Finter, Robert Friedman, Ion Gresser, Ian Kerr, James Porterfield, Helio Pereira, Joseph Sonnabend, Joyce Taylor-Papadimitriou, and David Tyrrell. I am also grateful to Robert Moore of the National Institute for Medical Research, Mill Hill, for sharing material related to Isaacs in the Mill Hill archives.

136 "His sense of humor": David Tyrrell, personal communication to author, March 3, 1994; "Most scientists are": Anthony Allison, telephone interview, May 13, 1994.

137 "He enjoyed school": from "Origins," by Dr. Bernard Isaacs, a brother of Alick, who emigrated to Israel; copy provided by S. I. Elmhirst.

138 "All assistance short of": S. I. Elmhirst, interview. Under Scottish law at the time, according to Dr. Elmhirst, it was not legal for a family to single out a child for disinheritance; hence, Louis Isaacs disinherited all four of his children as a result of his falling out with Alick.

138 clonal selection theory: see F. M. Burnet, *The Clonal Selection Theory of Acquired Immunity*, Vanderbilt University Press, 1959. In the 1950s, Burnet developed this theory, which built upon a hypothesis advanced in 1955 by Niels Jerne. Burnet correctly surmised that the formation of antibodies is governed by cells—in this case the lymphocytes known as B cells—and that random mutation had generated enormous genetic variation in the "business end" of receptors on the surface of these cells, the receptors in essence identical to the antibodies that would be made by that particular cell. Thus,

humans were born with the entire repertoire of possible antibodies (now known to number about 10 million variations), and it was only when the appropriate B cell encountered its complementary antigen that the cell would be "selected" both to proliferate and to manufacture and release into the bloodstream its particular variant of antibody, which of course was specific to the antigen. The theory, originally published in 1957 in the *Australian Journal of Science,* was confirmed two years later in experiments by Joshua Lederberg (see Gordon L. Ada and Gustav Nossal, "The Clonal Selection Theory," *Scientific American,* August 1987, pp. 62–69).

Isaacs, interestingly, never wrote a commentary on the theory for the *Lancet*—a conspicuous omission. When asked about this, Sue Isaacs confirmed that he hadn't, adding a surprising explanation: "I don't think Alick wrote anything about it, and I don't know why. And I remember thinking at the time, because he was slightly dubious about it, why he wasn't writing about it. I felt he wasn't sure if FMB was right about this. Burnet had come to England, a bit earlier than that, and went through a phase where he was very pro-apartheid. And it shocked us rigid. And I wonder if that colored Alick's reaction to the clonal theory" (Sue Isaacs, personal interview, April 16, 1994, New York).

139 "is one who has": A. Isaacs, unsigned book review, *The Lancet,* Feb. 7, 1953.

139 "In the course of": J. Lindenmann to author, March 25, 1994.

140 The first Isaacs-Lindenmann encounter and background on Mill Hill experiments: J. Lindenmann, "From Interference to Interferon: A Brief Historical Introduction," *Philosophical Transactions of the Royal Society of London, B,* 299:3–6 (1982); J. Lindenmann, "The First Stirrings of Interferon," *Current Contents* 17:46 (1989); and J. Lindenmann, "Induction of Chick Interferon: Procedures of the Original Experiments," *Methods in Enzymology,* Vol. 78, Academic Press, New York, 1978, pp. 181–88. In addition, Dr. Lindenmann provided much additional detail in a series of letters to the author.

142 "a crucial, *crucial* manipulation": Derek Burke, personal interview, Norwich–London train, Dec. 6, 1993.

143 "could be demonstrated" and "I believe that": Lindenmann to author, March 25, 1994.

144 "In Search of an Interferon": Alick Isaacs, laboratory notebook, in the collection of the National Library of Medicine, Bethesda, Maryland. These notebooks ended up in the United States, according to Sue Isaacs, because no one in England wanted them.

144 "an almost idiotically": Robert Friedman, personal interview, Bethesda, Maryland, Dec. 9, 1992.

145 "seemed very impressed": A. Isaacs to J. Lindenmann, June 25, 1957, courtesy of J. Lindenmann.

146 Predecessor papers: A. R. Rich and M. R. Lewis, "Mechanisms of Allergy in Tuberculosis," *Proceedings of the Society of Experimental Medicine* 25:596–98 (1928); V. Menkin, "Chemical Basis of Fever," *Science* 100:337–38 (1944); I. L. Bennett Jr. and P. B. Beeson, "Studies on the Pathogenesis of Fever, II: Characterization of Fever-producing Substances from Polymorphonuclear Leukocytes and from the Fluid of Sterile Exudates," *Journal of Experimental Medicine* 98:493–508 (1953); and R. Levi-Montalcini and V. Hamburger, "A Diffusible Agent of Mouse Sarcoma Producing Hyperplasia of Sympathetic

Ganglia and Hyperneurotization of the Chick Embryo," *Journal of Experimental Zoology*, 123:233–388 (1953).

146 "Calamity Joe": Lindenmann says Isaacs coined the name because "Whenever he handled a test tube, there was a 50-50 chance that it would break."

147 Burke's collaboration: "Alick came along": Burke interview and personal communication, July 18, 1994.

147 The two papers published together were A. Isaacs and J. Lindenmann, "Virus Interference, I: The Interferon," *Proceedings of the Royal Society of London, B*, 147:258–67 [1957]), and A. Isaacs, J. Lindenmann, and R. C. Valentine, "Virus interference, II: Some Properties of Interferon," ibid., pp. 268–73 (1957). In a letter, Lindenmann noted that he and Isaacs had a minor dispute over the order of authorship; Lindenmann was to be first author on the second of the two original papers, but editors at the Royal Society insisted that the names of the authors appear in alphabetical order. As a result, Lindenmann was the first author on the third paper in the series, on the biochemistry (J. Lindenmann, D. C. Burke, and A. Isaacs, "Studies on the Production, Mode of Action, and Properties of Interferon," *British Journal of Experimental Pathology* 38:551 [1957]), which in turn upset Burke.

147 "must be accounted": C. H. Andrewes, *Biographical Memoirs* 1967, p. 210.

147 "Alick was very enthusiastic": J. Lindenmann, letter to author, May 4, 1994.

149 Early follow-up work: Isaacs and Burke published A. Isaacs and D. Burke, "Mode of Action of Interferon," *Nature* 182:1073–74 (Oct. 18, 1958) and D. C. Burke and A. Isaacs, "Some Factors Affecting the Production of Interferon," *British Journal of Experimental Pathology*, 39:452–58 (1958).

149 Mooser episode: J. Lindenmann described this in a letter to the author, June 10, 1994. When Mooser learned that Isaacs and Lindenmann would publish their work on interferon, he decided to publish Lindenmann's previous work in the journal *Experientia*, where he had influence with the editors. "The sad fact," Lindenmann writes, "is that later, when it became obvious that interferon was here to stay, Mooser became more and more convinced that he had made a major contribution, and that he had been cheated out of a due recognition. When I returned to Zurich in mid-1957 he was still quite friendly, and I had hopes that he held no grudges, but his feeling of having been unfairly treated festered. . . . The situation in Zurich became more and more untenable, Mooser's resentment grew so much that in the fall of 1959 he asked me to resign, which I did in a letter dated Sept. 10, 1959." Isaacs also acknowledged the priority dispute in a letter, A. Isaacs to H. Mooser, letter dated Nov. 18, 1958, courtesy of J. Lindenmann.

150 "Mooser died": J. Lindenmann, *Methods in Enzymology* (1981), p. 185.

150 "It is my opinion": K. Cantell, unpublished manuscript, p. 47.

150 "Alick is ill": C. H. Andrewes to J. Lindenmann, Nov. 10, 1958, courtesy of J. Lindenmann.

150 Public disclosure of interferon: The interest of the British press in the original publication is described in A. Isaacs to J. Lindenmann, October 23, 1957, courtesy of Lindenmann. Isaacs received a call from the science correspondent of the *Daily Express*, who had seen the original publications and sought further information. "I told him very firmly," Isaacs wrote, "that we couldn't give out information which was not published in the scientific literature first."

151 "Alick didn't shrink": D. Burke interview.

152 first clinical trial: Scientific Committee on Interferon, "Effect of Interferon on Vaccination in Volunteers," *Lancet*, pp. 873–75 (April 28, 1962). A Dutch historian of science, Toine Pieters, has provided a richly detailed account of this first clinical trial in T. Pieters, "Interferon and Its First Clinical Trial: Looking Behind the Scenes," *Medical History* 37:270–95 (1993).

152 "Work on interferon has": in A. Isaacs, "Interferon Tried in Man," *New Scientist*, May 3, 1962, pp. 213–14.

153 "When I was there": Leslie Brent, personal interview, St. Mary's Hospital, London, England, Nov. 18, 1994.

153 "Interferon can be thought": in Isaacs, "Interferon Tried," p. 214.

154 "complained bitterly that": N. Finter, "The Wellferon Story: From 'Virus Interference' to Namalwa," *Wellcome World*, Sept/Oct. 1993, pp. 7–11. Finter has written a four-part history of Wellcome's involvement in the commercial production of human interferon.

154 On diminishing work on interferon at Mill Hill: annual reports of the Division of Bacteriology and Virus Research, National Institute for Medical Research at Mill Hill, years 1962–67, courtesy of archives, Mill Hill.

154 Burnet remark: M. Edney to author, September 27, 1994.

154 Follow-up work to Isaacs and Lindenmann: I. Gresser, "Production of Interferon by Suspensions of Human Leukocytes," *Proceedings of the Society for Experimental Biology and Medicine* 108:799–803 (1961); K. Paucker, K. Cantell, and W. Henle, "Quantitative Studies on Viral Interference in Suspended L Cells, III: Effect of Interfering Viruses and Interferon on the Growth Rate of Cells," *Virology* 17:324–34 (1962); J. Taylor, "Inhibition of Interferon Action by Actinomycin D," *Biochem. Biophys. Res. Commun.* 14:447–53 (1964); E. F. Wheelock, "Interferon-like Virus Inhibitor Induced in Human Leukocytes by Phytohemagglutinin," *Science* 149:310–11 (1965); I. M. Kerr, R. A. Brown, and L. A. Ball, "Increased Sensitivity of Cell-free Protein Synthesis to Double-stranded RNA after Interferon Treatment," *Nature*, 250:57–59 (1974); and I. Gresser et al., "Action inhibitrice de l'interferon brut sur le développement de la leucémie de Friend chez la souris," *C. R. Acad. Sci. Paris*, 263:586–89 (1966). Details of this last paper also from Ion Gresser, personal interview, Dec. 29, 1993, New York, and telephone interview, Aug. 6, 1994, and S. Panem, *Interferon Crusade*, 1984, pp. 12–13.

156 Isaacs's illness: Sue Elmhirst openly provided many details; "After this": C. H. Andrewes, *Biographical Memoirs*, p. 215.

157 "The chief loss": R. Friedman, interview. Helio Pereira, who visited Isaacs when he was hospitalized, said in an interview, "In his depressive states, [Alick] was afraid that the whole thing was nothing. That there was nothing to it at all, that interferon didn't even exist" (H. Pereira, personal interview, Dec. 4, 1993, Twickenham, England).

158 "round unvarnish'd tale" and "The glimpses": in A. Isaacs, "Interferon: A Round Unvarnish'd Tale," *Journal of Pharmacy and Pharmacology Supplement* 13:57T–61T (1961).

Chapter 7: Lymphodrek

159 Epigraph: Piero Camporesi, *Juice of Life: The Symbolic and Magic Significance of Blood*, Continuum, New York, 1995, p. 14.

160 "by nature": Cantell manuscript.

160 "typically nordic": Cantell manuscript.

160 On Vainio's death and the Cantell-Strander collaboration: Kari Cantell, personal interview, Nov. 8, 1994, Helsinki; Hans Strander, personal interview, Nov. 11, 1994, Stockholm, Sweden; also, Kari Cantell, *The Interferon Story: The Ups and Downs of the Life of a Scientist,* unpublished English translation of Finnish edition, 1994, and K. Cantell, personal communication, Oct. 10, 1996.

162 Interferon's thousand-mile journey: On the early clinical results using interferon against cancer, see H. Strander and K. Cantell, "Studies on Antiviral and Antitumor Effects of Human Leukocyte Interferon In Vitro and In Vivo," in *The Production and Use of Interferon for Treatment and Prevention of Human Virus Infection: Proceedings of a Tissue Culture Workshop,* Lake Placid, N.Y., 1973, published by Tissue Culture Association, Rockville, Md.; and H. Strander, K. Cantell, G. Carlstrom, and P. A. Jakobsson, "Clinical and Laboratory Investigations on Man: Systemic Administration of Potent Interferon to Man," *Journal of the National Cancer Institute* 51:733–42 (1973).

162 "action-oriented empiricism": G. Nossal, "Summary and Prospects for Future Research," in *The Role of Non-Specific Immunity in the Prevention and Treatment of Cancer,* edited by Michael Sela, proceedings of a meeting at the Vatican, Oct. 17–21, 1977, Pontificia Academia Scientiarum, Vatican, 1979, p. 582.

163 The Krim meeting has been well documented. See, for example, S. Panem, *The Interferon Crusade,* Brookings Institution, Washington, D.C., 1984, pp. 16–24.

163 "Our shortage of interferon": K. Cantell to Gerhard Bodo, Oct. 13, 1975, courtesy of author.

163 "Das Blut ist": Goethe, quoted in Cantell manuscript, 1994.

164 From the very first application: For an excellent collection and discussion of milestone research in immunology during the past century, from which this discussion has benefited greatly, see Debra Jan Bibel, *Milestones in Immunology: A Historical Exploration,* Springer Verlag/Science Tech, Madison, Wis., 1988. Bibel reprints excerpts of landmark papers, accompanied by a lively, intelligent, and entertaining commentary. Another key resource is R. B. Gallagher, J. Gilder, G. J. V. Nossal, and G. Salvatore, eds., *Immunology: The Making of a Modern Science,* Academic Press, U.K., 1995, a collection of somewhat more personal essays from key immunological researchers on their pioneering work.

164 For Landsteiner's epochal work in immunochemistry, see Karl Landsteiner, *The Specificity of Serological Reactions,* Dover, New York, 1962 (reprint; the original version was published in 1936).

166 "some hysterical character": Paul de Kruif, *The Microbe Hunters,* Pocket Books, New York, 1940 (original published by Harcourt Brace in 1926), p. 196. This is one of the most influential books ever written about science and inspired a generation of young people to pursue biology as a career, but the writing may strike modern readers as rather exaggerated.

166 "The discovery of antibodies": Bibel, *Milestones,* p. 117.

166 "the complete ignorance": Arnold Rich, quoted in J. L. Gowans, "The Mysterious Lymphocyte," in Gallagher et al., *Immunology,* p. 66.

166 the rehabilitation of the cellular arm: K. Landsteiner and M. W. Chase,

"Experiments on Transfer of Cutaneous Sensitivity to Simple Compounds," *Proceedings of the Society for Experimental Biology and Medicine,* 49:688–90 (1942); and M. W. Chase, ibid., 59:134 (1945).

167 Perhaps the most startling: B. Glick, T. S. Chang, and R. G. Jaap, "The Bursa of Fabricius and Antibody Production," *Poultry Science* 35:224–25 (1956).

169 "of no general interest": quoted in Bibel, *Milestones* p. 129.

169 "we shall come to": P. Medawar, quoted in Jacques F. A. P. Miller, "The Discovery of Thymus Function," in Gallagher et al., *Immunology,* p. 78.

169 One of the earliest: J. L. Gowans, "The Recirculation of Lymphocytes from Blood to Lymph in the Rat," *Journal of Physiology* 146:54–69 (1959); "effete small lymphocytes": ibid., p. 67. In a letter, Gowans explained that he meant "effete" to mean worn-out. "I was referring to the disposal in the body of lymphocytes at the end of their life span—just small lymphocytes generally; we did not know about the distinction between T and B lymphocytes in those days." J. Gowans to author, Aug. 15, 1996.

170 In the early 1960s: J. F. A. P. Miller, "Immunological Function of the Thymus," *Lancet* 2:748–49 (1961); "were essential to allow": J. F. A. P. Miller in Gallagher et al., *Immunology,* p. 81.

172 "the lymphocyte is perhaps": Gustav J. V. Nossal, "Life, Death, and the Immune System," *Scientific American,* September 1993, p. 62.

173 Lymphocytes in tissue culture: Peter C. Nowell, "Phytohemagglutinin: An Initiator of Mitosis in Cultures of Normal Human Leucocytes," *Cancer Research* 20:462–66 (1960). This, too, was an accidental observation. Nowell writes that PHA was "employed originally as a means of separating the leukocytes from whole blood in preparing the cultures" and by chance was "found to be a specific initiator of mitotic activity."

173 They reported the existence: Two good general overviews can be found in Steven B. Mizel and Peter Jaret, *In Self-Defense,* Harcourt Brace Jovanovich, San Diego, 1985; and Frances Balkwill, *Cytokines in Cancer Therapy,* Oxford University Press, Oxford, 1989. In addition, see also Joost J. Oppenheim and Igal Gery, "From Lymphodrek to Interleukin 1 (IL1)," *Immunology Today* 14:232–34 (1993); Joost J. Oppenheim, "Cytokines: Past, Present, and Future," in S. Kasakura, ed., *Cytokine 1994,* Japan Publishing, Tokyo, 1994, pp. 11–18; and Byron H. Waksman and Joost J. Oppenheim, "The Contribution of the Cytokine Concept to Immunology," in Gallagher et al., *Immunology,* pp. 133–43.

173 "for a considerable period": Waksman and Oppenheim, "Contribution," p. 133.

174 There were, as always, early hints: Alexis Carrel, "Leukocyte Trephones," *Journal of the American Medical Association* 82:255–58 (Jan. 26, 1924). "The growth of tissues," this paper begins, "probably depends on the presence in the pericellular fluid of substances necessary for the synthesis of protoplasm, and of hormones of the type called by Gley harmozones." Carrel conceived of "trephones" as "principles" made by certain cells and used by others to build up protoplasm, or to feed cells. However, Carrel envisioned lymphocytes and macrophages as "mobile unicellular glands" that converted substances in the serum into food that could be delivered to fixed cells. This sounds tantalizingly close to lymphokines, but Gley's concept of "harmozones" (as described by Carrel) may even be closer; they were "secreted by certain cells, and their function consists in arousing other cells to

activity. Their nature is utterly unknown." Carrel unfortunately did not cite a source for Gley's work in the paper. I am grateful to Michael Lotze for bringing this paper to my attention. In addition, see Alexis Carrel and Albert H. Ebeling, "Leucocytic Secretions," *Journal of Experimental Medicine*, 36:645–59 (December 1922).

174 "phenomenological detection," J. Oppenheim, "Cytokines."

174 "Instead of doing biochemistry": Steven Mizel, Bowman-Gray School of Medicine, telephone interview, Dec. 11, 1992.

174 On the discovery of specific factors:

LMF: S. Kasakura and L. Lowenstein, "A Factor Stimulating DNA Synthesis Derived from the Medium of Leukocyte Cultures," *Nature* 208:794–95 (Nov. 20, 1965); and J. Gordon and L. D. MacLean, "A Lymphocyte-stimulating Factor Produced *In Vitro*," ibid., pp. 795–96.

MIF: J. R. David et al., "Delayed Hypersensitivity *In Vitro*, I: The Specificity of Inhibition of Cell Migration by Antigens," *Journal of Immunology* 93:264–73 (1964); J. R. David, "Delayed Hypersensitivity In Vitro: Its Mediation by Cell-free Substances Formed by Lymphoid Cell—Antigen Interaction," *Proceedings of the National Academy of Sciences* 56:73–77 (1966); and Barry R. Bloom and Boyce Bennett, "Mechanism of a Reaction in Vitro Associated with Delayed-Type Hypersensitivity," *Science* 153:80–82 (July 1, 1966).

MAF: C. F. Nathan, M. L. Karnofsky, and J. R. David, "Alterations of Macrophage Functions by Mediators from Lymphocytes," *Journal of Experimental Medicine* 133:1356–76 (1971).

LAF: I. Gery, R. K. Gershon, and B. H. Waksman, "Potentiation of Cultured Mouse Thymocyte Responses by Factors Released by Peripheral Leucocytes," *Journal of Immunology* 107:1778–80 (1971).

TNFs: N. H. Ruddle and B. H. Waksman, "Cytotoxic Effect of Lymphocyte-Antigen Interaction in Delayed Hypersensitivity," *Science* 157:1060–62 (1967); E. A. Carswell et al., "An Endotoxin Induced Serum Factor That Causes Necrosis of Tumors," *Proceedings of the National Academy of Sciences* 72:3666–70 (1975); and M. Kawakami and A. Cerami, "Studies of Endotoxin-induced Decrease in Lipoprotein Lipase Activity," *Journal of Experimental Medicine* 154:631–39 (1981).

175 "The uncharacterized lymphocyte-derived": J. Oppenheim and I. Gery, "From Lymphodrek," p. 232. Oppenheim claims paternity for the term "lymphodrek," using it for the first time at a meeting in 1967.

176 "cytokine": see Stanley Cohen, Pierluigi E. Bigazzi, and Takeshi Yoshida, "Similarities of T Cell Function in Cell-Mediated Immunity and Antibody Production," *Cellular Immunology* 12:150–59 (1974). In this "commentary," Cohen et al. also proposed distinguishing between lymphokines, or molecules that traffic solely between lymphocytes, and cytokines, or molecules that traffic not only between lymphocytes but also from lymphocytes to nonimmune cells. In practical usage, this distinction often gets blurred.

176 The debate reached a climax: For an account of the Ermatingen meeting, see Mizel and Jaret, pp. 75–76. Also S. Mizel, telephone interview; Werner Paetkau, telephone interview, Dec. 10, 1992; and Joost Oppenheim, telephone interview, Nov. 24, 1992. According to several participants, a cohort of young researchers—Mizel, Paetkau, Kendall Smith (then of Dartmouth), Steve Gillis (then of the University of Washington), and a token elder (Op-

penheim]—retired to a bar. In the ensuing discussion, they all agreed that there were not hundreds of different factors, as was being described in the meeting, but dozens of names for a much smaller number of distinct entities, so they proposed the radically different nomenclature.

"As we got drunker and drunker," Mizel admitted, "we came up with all sorts of names." They were written down, Paetkau added, on a paper towel. Paetkau claims credit for coining the term "interleukin." He said, "It came to me in a flash. These messages go between leukocytes so, okay, it's 'interleukin.' That mixes Greek and Latin roots, but I thought, what the hell, we're scientists. We're not classical scholars. It doesn't matter." There is apparently no truth to a story occasionally floated by Oppenheim that one of the terms under consideration was "Heidikines," in honor of the barmaid who kept the committee members free of dehydration during their deliberations.

Chapter 8: "The Cloning of Interferon and Other Mistakes"

178 Epigraph: Albert Rosenfeld, "Provocative Talk at an Interferon Workshop," *Life*, July 1979, p. 62.
178 Background on the Weissmann-Cantell collaboration: Kari Cantell, personal interview, Nov. 7, 1994, Helsinki, Finland; Charles Weissmann, personal interview, Nov. 3, 1993, Zurich, Switzerland; Peter Lengyel, telephone interview, April 29, 1994, and personal communication, May 13, 1994. Two other invaluable sources on this work are Charles Weissmann's superb account of the cloning experiment (C. Weissmann, "The Cloning of Interferon and Other Mistakes," *Interferon* 3:101–34 [1981]); and Kari Cantell's charming autobiography, *The Interferon Story: The Ups and Downs of the Life of a Scientist*, unpublished English translation of Finnish edition, 1994. I am grateful to Dr. Cantell for permission to quote from this manuscript.
179 "a name then unknown": K. Cantell, unpublished manuscript.
180 "He just looked": C. Weissmann, interview; "It looked so fantastic": K. Cantell, interview; "During his visit": K. Cantell, unpublished manuscript; "I gained the impression": Weissmann, "Cloning," p. 105.
182 "I was not interested": P. Lengyel, interview.
182 Even as Lengyel and Weissmann strolled: For background on the early work in recombinant DNA, especially by Genentech, see S. S. Hall, *Invisible Frontiers*, Atlantic Monthly Press, New York, 1987. On the early history of recombinant DNA, including the moratorium, see James D. Watson and John Tooze, *The DNA Story: A Documentary History of Gene Cloning*, W. H. Freeman, San Francisco, 1981; Michael Rogers, *Biohazard*, Knopf, New York, 1977; and John Lear, *Recombinant DNA: The Untold Story*, Crown, New York, 1978. For a later and more critical assessment of biotechnology, see Robert Teitelman's *Gene Dreams: Wall Street, Academia, and the Rise of Biotechnology*, Basic Books, New York, 1989.
183 "it became evident": Weissmann, "Cloning," p. 101; "I actually considered": Weissmann, interview.
184 "I think every time": Weissmann, interview.
185 On the Mary Lasker modus operandi, see James T. Patterson, *Dread Disease*, pp. 173–74; Richard Rettig, *Cancer Crusade*, pp. 18–41; Sandra Panem, *Interferon Crusade*, pp. 21–22; Ronald Cape, former chairman, Cetus Corp., personal interview, Emeryville, Calif., Sept. 29, 1992.
185 "Maybe at the back": Weissmann, interview.

186 "To the delight": Weissmann, "Cloning," p. 103. The projections mentioned by Weissmann at the Biogen meeting were saved in a scrapbook that he shared with the author.

187 "I had to find": Weissmann, "Cloning," p. 105.

190 "We had made": Weissmann, interview.

191 "The decision to do": David Goeddel, senior scientist, Genentech, Inc., personal interview, South San Francisco, California, June 16, 1992.

192 "I felt like a passenger": Edgar Pick, "The Fox and the Raven," in J. Oppenheim et al., eds., *Interleukins, Lymphokines, and Cytokines: Proceedings of the Third International Lymphokine Workshop,* Academic Press, New York, 1983, p. 765.

192 "I knew that Pestka": Weissmann, interview.

192 "one of the most gifted": Weissmann, "Cloning," p. 106.

192 "They were like animals": Cantell, interview.

193 "Although I was advised": T. Taniguchi, "B.C., D.C., and A.C. of an Interferon Gene," *Journal of Interferon Research* 7:481–85 (1987), a lovely memoir for a *festschrift* volume dedicated to Jean Lindenmann; "I was very shocked" and "He was very embarrassed": Weissmann, interview; "We never discussed": Tadatsugu Taniguchi, personal interview, October 18, 1994, Washington, D.C.

194 "exposure to the vicissitudes": Weissmann, "Cloning," p. 108.

196 "The assay was plagued": Weissmann, ibid.

196 On the false positives, Weissmann said, "That was really bugging us, because what it started to look like is that there were too *many* clones coming out. And then, combined with the fact that there was this business that the assays were not very reproducible, we thought that we were actually getting false positives, whereas most likely we were getting true positives which sometimes were negative. That was very confusing. Nobody *dreamed* of the fact that there would be anything like twelve [interferon] genes or whatever around, or that the messenger would be that much more frequent than we had estimated, by about a factor of ten."

    For his part, Cantell realized in retrospect that the bacteria had been making interferon all along; Weissmann had sent him bacterial extracts to test, but the extracts were difficult to work with and toxic in his assay system. "There seemed to be some activity," Cantell said in an interview, "but I did not believe it. Which I regretted bitterly afterwards, because surely there was active interferon and we would have saved one year if I would have had more confidence."

196 In February 1979: Weissmann had preserved a copy of the telegram; "technically bankrupt": Weissmann, "Cloning," p. 110.

197 "Wally was just talking": Weissmann, interview.

198 "There is this squidlike": Weissmann, interview.

198 "To be honest": D. Goeddel, personal interview. Weissmann, as part of a lawsuit between Biogen (with its marketing partner Schering-Plough Corp.) and Genentech (with its partner, Hoffmann-La Roche, Inc.), had the opportunity to inspect the laboratory notebooks of both Pestka and Goeddel, and said in an interview, "It was very interesting to see how Pestka was operating and the problems they had. They didn't want to send this clone, this partial clone they had, to Genentech. He wanted to do it himself, you see. . . . Then only—and that we found out later on—only under the pressure of the events and the announcement of Biogen did [Roche] management force Pestka then to send out stuff and collaborate."

198 "No, it wasn't smooth": S. Pestka, chairman, molecular genetics and microbiology, Robert Wood Johnson School of Medicine, Newark, N.J., telephone interview, Nov. 8, 1996. Pestka stresses that the most valuable patents to emerge from the work on alpha interferon were obtained by Genentech and Roche on the basis of his protein data and not on the nucleotide sequence data produced by Weissmann's group. In 1996, the federal Patent and Trademark Office upheld Genentech and Roche's claim in a longstanding dispute with Biogen and Schering over which group was the first to engineer the recombinant protein (see "Genentech, Partner Win Patent Ruling on Alpha Interferon," *Wall Street Journal*, Jan. 18, 1996, p. B-10).

199 T. Taniguchi et al., "Construction and Identification of a Bacterial Plasmid Containing the Human Fibroblast Interferon Gene Sequences," *Proceedings of the Japanese Academy* 55B:464–69 (1979).

199 "On the way": Taniguchi, 1987, p. 483.

199 "I was disinclined": Weissmann, "Cloning," p. 110; "No one was clear": Goeddel, interview; "That was the right": Tanaguchi, interview.

200 "My thought was" and "After that party": Weissmann, interview.

201 "Indeed, the assay": Weissmann, "Cloning," p. 113.

201 "Everything was done": Weissmann, interview.

202 "the date on which": Nicholas Wade, "Cloning Gold Rush Turns Basic Biology into Big Business," *Science* 208:688–92 (May 16, 1980). On the press conference and aftermath, see H. Schmeck, "Natural Virus-Fighting Substance Is Reported Made by Gene Splicing," *New York Times*, Jan. 17, 1980, p. A-1; and V. Cohn, "Scientists Making a Copy of Disease-Fighting Gene," *Washington Post*, Jan. 17, 1980. p. A-7. Of the press coverage, Weissmann remarked, "Instead of concentrating on what I thought was an interesting piece, an *inventive* piece, of research, they were trying to get me to say that the reason this is being reported is to boost Biogen. Which was true! I didn't want to do it, really, but there was a lot of—well, I won't say pressure, but a lot of arguments that this was *the* occasion for Biogen to make its mark and so on. They were also worried that some other company might come out with the announcement before."

202 "The cloning of the interferon gene was a technological tour de force": The Weissmann lab's work appeared in March 1980 as S. Nagata et al., "Synthesis in *E. coli* of a Polypeptide with Human Leukocyte Interferon Activity," *Nature* 284:316–20 (1980); and Genentech-Roche team published six months later as D. V. Goeddel et al., "Human Leukocyte Interferon Produced by *E. coli* is Biologically Active," *Nature* 287:411–17 (1980). After their brief period of "discomfort," Taniguchi and Weissmann resumed healthy scientific relations with a collaboration comparing the DNA sequences of the two interferon genes they discovered, published as T. Taniguchi et al., "Human Leukocyte and Fibroblast Interferons are Structurally Related," *Nature* 285:547–49 (1980).

203 "He's about the only person": Arthur Goertz, telephone interview, Bastrup, Texas, April 12, 1996; Jorge Quesada, telephone interview, May 12, 1994, and personal communication, Feb. 21 and 23, 1996. See also Jorge R. Quesada, "Alpha Interferons in Hairy Cell Leukemia: A Clinical Model of Biological Therapy for Cancer, *Interferon* 8:111 (1987); J. R. Quesada et al., "Alpha Interferon for Induction of Remission in Hairy-cell Leukemia," *New England Journal of Medicine* 310:15 (1984); and J. R. Quesada and J. U. Gutterman,

"Interferons in the Treatment of Human Neoplasms," *Journal of Interferon Research* 7:575 (1987).

205 "There's no question": Kari Cantell, personal interview, Nov. 7, 1994, Helsinki, Finland.

206 From the point of view of basic science: For subsequent history of interferon, see H. M. Johnson, F. W. Bazer, B. E. Szente, and M. A. Jarpe, "How Interferons Fight Disease," *Scientific American*, May 1994, pp. 68–75. For signal transduction, see J. E. Darnell, I. M. Kerr, and G. R. Stark, *Science*, 264:1415 (1994) and Tony Hunter, "Cytokine Connections," *Nature* 366:114–16 (1993). For interaction with killer T cells, see D. F. Tough, P. Borrow, and J. Sprent, "Induction of Bystander T Cell Proliferation by Viruses and Type I Interferon in Vivo," *Science* 272:1947–50 (1996).

206 "On the learning curve": John Kirkwood, University of Pittsburgh Medical Center, personal interview, October 9, 1996, New York.

206 On Namalwa cell line: Norman Finter, "The Wellferon Story," *Wellcome World*, 1993–1994 (four-part series on development of lymphoblastoid interferon).

207 From the point of view of business: Market for interferons: Edmund A. Debler, financial analyst, Mehta and Isaly, New York, telephone interview, October 29, 1996.

207 "During evolution": K. Cantell interview.

Chapter 9: "One of My Best Known Accidents"

209 Epigraph: Herodotus, quoted in A. Cambrosio and P. Keating, *Exquisite Specificity*, Oxford University Press, New York, 1995, p. v.

209 "I don't bear": Doris Morgan, personal interview, Hahnemann University, Philadelphia, Jan. 25, 1994. In addition, Morgan has written an account of her work in the Gallo lab in Doris Morgan, "Foreword," in Kendall A. Smith, ed., *Interleukin-2*, Academic Press, New York, 1988, pp. xvii–xxi.

209–210 "two of the most": Robert E. Gallagher, personal interview, Montefiore Hospital, Aug. 25, 1994, Bronx, New York. I am particularly grateful to Dr. Gallagher for consulting daily logs from the period in question to establish a reliable chronology of events.

210 "because of the reduced": Phillip Markham to Doris Morgan, undated (but shortly after January 23, 1978).

211 Robert Gallo, *Virus Hunting: AIDS, Cancer, and the Human Retrovirus: A Story of Scientific Discovery*, Basic Books, New York, 1991. Of Gallo's account, Morgan said in an interview, "That was, without a doubt, not an accurate recollection of those events."

A phenomenal amount has been written about Robert Gallo and his laboratory because of the controversy surrounding the discovery of the first human immunodeficiency virus, or HIV. For several accounts, see John Crewdson, "The Great AIDS Quest: Science under the Microscope," *Chicago Tribune*, Nov. 19, 1989; and Randy Shilts, *And the Band Played On*, St. Martin's Press, New York, 1987. A special panel of the National Academy of Sciences later made note of "a pattern of behavior on Dr. Gallo's part that repeatedly misrepresents, suppresses, and distorts data and their interpretation in such a way as to enhance Dr. Gallo's claim to priority" (see

J. Crewdson, "Scientific Panel Accuses Gallo of 'Recklessness,'" *Chicago Tribune,* March 27, 1992, p. 1).

211 "They almost never": Francis Ruscetti, personal interview, National Cancer Institute, Sept. 30, 1994, Frederick, Md.

213 On the first reported discovery of a human cancer retrovirus, see R. E. Gallagher and R. C. Gallo, "Type C RNA Tumor Virus Isolated from Cultured Human Acute Myelogenous Leukemia Cells," *Science* 187:350–53 (Jan. 31, 1975). About midway through 1975, according to Robert Gallagher, serious doubts about contamination arose about this work, and it was in response to this development that other researchers attacked the Gallo lab at the Virus Cancer Program meeting in Hershey, Pa., in October and reported that the putative human virus was in fact contamination by primate viruses. In his book, Gallo writes, "What surprised me were not the findings—as I say, I was already developing my own doubts—but the vehemence with which they were delivered . . . nothing compared with the feelings that passed over me as I sat that day in Hershey, Pennsylvania, hearing not just HL-23 but much of my life's work—the search for tumor-causing RNA viruses in humans— systematically and disdainfully dismissed" (p. 85). In an interview, however, Gallo said, "I'm confident that there was a virus. I'm sure that wasn't a contaminant."

214 The first public acknowledgment of the freezer accident appeared in R. E. Gallagher et al., "Growth and Differentiation in Culture of Leukemic Leukocytes from a Patient with Acute Myelogenous Leukemia and Re-identification of Type-C Virus," *Proceedings of the National Academy of Sciences* 72:4137–41 (October 1975). The paper noted that "Seeds of these [WHE] cells were stored in a freezer which thawed." In terms of chronology, the freezer accident occurred at the end of September or early October 1974, according to Frank Ruscetti, who investigated the incident. The *Science* paper was published four months later, but did not mention the loss of the WHE factor because, Gallagher explained, the ramifications of the loss were still not clear at that point.

214 "So we published": Robert Gallo, personal interview, National Institutes of Health, Jan. 28, 1994, Bethesda, Md. All other Gallo remarks from this interview.

214 On the search for a replacement for WHE: see R. Gallagher, F. Ruscetti, S. Collins, and R. Gallo, "Growth and Differentiation of Human Myelogenous Leukemia Cells in Conditioned Medium from Human Embryo Cultured Cells," in P. Benjvelzen and J. Hilgers, ed., *Advances in Comparative Leukemia Research 1977,* Elsevier, Amsterdam, 1978, pp. 303–6.

214 "one of my best known": Morgan, "Foreword," p. xvii. Interestingly, in a manuscript copy of this essay provided by Morgan, she initially refers to it as "one of my best-known failures."

217 "Just for no other reason": Gallo's group had previously identified a growth factor in stimulated cells; see J. T. Prival, M. Paran, R. C. Gallo, and A. M. Wu, "Colony-Stimulating Factors in Cultures of Human Peripheral Blood Cells," *Journal of the National Cancer Institute* 53:1583–88 (December 1974).

219 "an incredible wave": Morgan, "Foreword."

222 The abstract of the presentation at the 1975 American Society of Hematology appeared as D. Morgan, F. Ruscetti, R. Gallagher, A. Wu, and R. Gallo,

"Selective In Vitro Growth of Human Hemopoietic Cells by PHA-stimulated Lymphocyte Conditioned Medium (LY-CM)," *Blood*, 46(6), 1975.

223 D. A. Morgan, F. W. Ruscetti, and R. Gallo, "Selective in Vitro Growth of T Lymphocytes from Normal Human Bone Marrows," *Science* 193:1007–8 (Sept. 10, 1976).

223 F. W. Ruscetti, D. Morgan, and R. Gallo, "Functional and Morphological Characteristics of Human T Cells Continuously Grown in Vitro," *Journal of Immunology* 119:131 (1977); purification was reported in J. W. Mier and R. C. Gallo, "Purification and Some Characteristics of Human T-cell Growth Factor from Phytohemagglutinin-stimulated Lymphocyte-conditioned Media," *Proceedings of the National Academy of Sciences* 77:6134–38 (1980).

223 Gallo's decision to share information with Ron Herberman and Steve Rosenberg may actually have been in response to a collaboration initiated by Ruscetti, who had told Kendall Smith at Dartmouth about the discovery of T cell growth factor in March 1976. Smith proposed the obvious next step—to see if antigen-specific T cells could be grown—but Gallo terminated this collaboration before it got off the ground, according to both Ruscetti and Smith.

224 "was discovered in": D. Morgan, "Foreword," p. xxi.

224 "She was keenly feeling": Richard Smith, Lofstrand Laboratories, Gaithersburg, Md., telephone interview, July 17, 1996.

226 "Given a four- or five-year reprieve": see K. A. Smith, "Interleukin-2: Inception, Impact, and Implications," *Science*, 240:1169–76 (May 27, 1988); and K. A. Smith, "Interleukin-2: The First Hormone of the Immune System," *Scientific American* 262:50–57 (1990).

227 "have had an extraordinary": K. A. Smith, "Interleukin-2: Inception," p. 1169.

227 Role of IL-2 in discovery of HTLV-1 and HIV: For historical purposes, it is worth recording here Frank Ruscetti's recollections about the discovery of the first human retrovirus, HTLV-1, and the role of IL-2 in it. "He [Gallo] was disinterested in it [IL-2] and didn't see the significance of it. But when Bernie Poiesz was working under me in Gallo's lab and we discovered a bonafide human retrovirus that was associated with the disease Adult T Cell Leukemia, *then* Gallo's story was he directed us to grow T cells to find the virus. Then it became important for him and his story to incorporate IL-2. And I can tell you quite honestly that he said to me, 'Don't work on T cells. You'll never find a virus there.' And that's a direct quote. And I'll testify to that under grounds of perjury if I have to. So that's when the story changed. Early on, he's right; he didn't see the significance because he was distracted. The minute a virus came out of it, the story changed and the importance changed and his position in the story changed."

227 "We knew it was important": R. Gallo, quoted in *Wall Street Journal*, Dec. 10, 1983, p. 33.

III. THE RISE OF THE T CELL CHAUVINISTS

229 Epigraph: Steven A. Rosenberg book (as below), p. 57.

Chapter 10: The Silk Purse Years

231 Epigraph: Michael Lotze, personal interview, June 2, 1992.

232 "We all reconstruct history": Norman Wolmark, interview, January 26,

1993, Montefiore Hospital, Pittsburgh, Pa. All remarks by Wolmark come from this interview. For a discussion of the pig lymphocyte work, see also Steven A. Rosenberg with John Barry, *The Transformed Cell: Unlocking the Mysteries of Cancer,* G. P. Putnam and Sons, New York, 1992, pp. 63–75. Wolmark's experiments were not, however, the first attempt at transplanting animal tissues into humans; a team at the Tulane University Medical Center headed by Keith Reemtsma transplanted chimpanzee kidneys into six patients in 1963.

233 For background on transplantation immunology, see "Transplantation and Immunogenetics" chapter in A. M. Silverstein, *A History of Immunology,* Academic Press, New York, 1989, p. 275. See also Joseph E. Murray, "Organ Transplantation and the Revitalization of Immunology," in R. B. Gallagher et al., *Immunology: The Making of a Modern Science,* Academic Press, New York, 1995, pp. 179–89.

234 "I went way back": Steven Rosenberg, interview, Dec. 7, 1992, Surgery Branch, National Cancer Institute, Bethesda, Md. Interviews with Dr. Rosenberg took place on Dec. 7, 1992; Dec. 9, 1992; and Sept. 26, 1994; remarks are identified by the date. Other remarks come from his previously cited book, *The Transformed Cell,* and are designated TTC, followed by the page number.

234 Georges Mathé's group published the BCG work as G. Mathé et al., "Active Immunotherapy for Acute Lymphoblastic Leukemia," *Lancet* 1:697–99 (1969). He referred to the book dedication in a personal interview, Nov. 16, 1994, Hôpital Suisse de Paris, Paris, France. Directly or indirectly, the Mathé work inspired a great deal of nonspecific immunotherapy. According to an international registry of immunotherapy sponsored by the National Cancer Institute, 50 immunotherapy protocols were registered in 1973, but by 1976 the number had climbed to 347 (Source: I. Löwy, "Experimental Systems," p. 429).

234 "To Bill": G. Mathé, *L'Homme qui voulait être guéri,* Anthropos, Paris, 1989, p. 11.

235 Details of the Symes research: M. O. Symes and A. G. Riddell, *British Journal of Surgery,* 60:176–80 (1973). The overall review can be found in Steven A. Rosenberg and William D. Terry, "Passive Immunotherapy of Cancer in Animals and Man," *Advances in Cancer Research,* vol. 25, edited by Georg Klein and Sidney Weinhouse, Academic Press, New York, 1977, pp. 323–88.

235 "excitable": Rosenberg, *TTC,* p. 66; one NIH scientist who attended the seminar referred to Symes's presentation as a "disaster."

238 "there is something else": Rosenberg, *TTC,* p. 72.

239 "Of course I knew": Rosenberg, interview, Sept. 26, 1994; "It was considered": Rosenberg, ibid. Repeated requests were made to see the manuscripts of the pig cell transfer experiments, along with the comments of reviewers, but Wolmark's office said it was unable to locate them.

240 On xenotransplantation: see S. S. Hall, "A First in Cell Transplantation: Researchers Organize, Meet," *Science* 256:1522 (June 12, 1992). Baboon bone marrow experiment: Results of the experiment have not yet been published, but press accounts have noted both the failure of the baboon cells to take and the patient's improvement (see L. Altman, "Baboon Cells Fail to Thrive, but AIDS Patient Improves," *New York Times,* Feb. 9, 1996, p. A-14 and L. Altman, "Baboon-Cell Transplant Failed, but AIDS Patient Is Improved,"

New York Times, Dec. 16, 1996, p. A-12.] For the biotech boon in xenotransplant technology, see R. Nowak, "Xenotransplants Set to Resume," *Science* 266:1148–51 (Nov. 18, 1994); and J. Bishop, "Using Pigs for Transplants Shows Promise," *Wall Street Journal*, May 1, 1995, p. B-4. Several biotech companies, including Nextran, Alexion Pharmaceuticals, Inc., and Imutran, Ltd. in Britain, have experimentally introduced human genes into pigs that seem to reduce (but not entirely blunt) an immune rejection of the pig tissue. In July 1995, the FDA approved an experimental procedure by doctors at Duke University Medical Center to transplant pig livers into humans to keep them alive until human donor organs became available; as of December 1996, the procedure had not yet been attempted.

241   "You know, I've always felt": S. Rosenberg, interview, Sept. 26, 1994.

Chapter Eleven: The Rise of the T Cell Chauvinists

242   Epigraph: Steven A. Rosenberg, "The War on Cancer: Stalking the Killer," The Learning Channel, Nov. 18, 1996.

242   "I got real excited": Rosenberg, interview, Sept. 26, 1994, Surgery Branch, National Cancer Institute, Bethesda, Md.; Richard Hodes, telephone interview, Sept. 23, 1994, National Institute on Aging, Bethesda, Md. Hodes didn't remember the turbulence, but did remember the conversation. "It was right around the time when people began to know enough about cellular immunity to begin to make a transition to it as a separate arm of immunity. . . . We batted ideas back and forth about cellular immunology, which was just then coming to the consciousness of tumor immunologists."

243   "very lower middle class": Rosenberg, interview.

243   "Indirectly, my family": Steven A. Rosenberg with John Barry, *The Transformed Cell*, Putnam, New York, 1992, p. 34; Rosenberg described his family history in the book (pp. 31–34), and expanded on that material in interviews with the author. "I think that": Rosenberg, interview; "I recall my": *TTC*, p. 33; "allow nothing to deflect": Rosenberg, ibid., p. 25; "You're getting in": ibid., p. 27.

245   "more than 500 papers": S. Rosenberg, interview, Dec. 7, 1992. The egocentrism has not diminished over time; when this project was described to Rosenberg in an early interview, he confided, "Since the overwhelming amount of immunotherapy that's been done recently has been done by my group, my guess is it's going to have to play a very prominent part about anything that you write about immunotherapy." (S. A. Rosenberg, personal interview, December 7, 1992, Surgery Branch, National Cancer Institute, Bethesda, Md.)

246   The "DeAngelo" case of spontaneous regression: See S. A. Rosenberg, E. Fox, and W. H. Churchill, "Spontaneous Regression of Hepatic Metastases from Gastric Carcinoma," *Cancer* 29:472–74 (Feb. 19, 1972); S. Rosenberg, *TTC*, pp. 11–23; and personal interviews with Rosenberg.

246   "No diagnostic or therapeutic" and "The cause of such regression": Rosenberg et al., 1972, p. 474; "In retrospect": Rosenberg, *TTC*, p. 22.

248   Background on National Cancer Act: see James T. Patterson, *The Dread Disease: Cancer and Modern American Culture*, Harvard University Press, Cambridge, 1987, pp. 171–254; Richard A. Rettig, *Cancer Crusade: The Story of the National Cancer Act of 1971*, Princeton University Press, Princeton, N.J., 1977; Natalie Davis Spingarn, *Heartbeat: The Politics of Health Re-*

search, Robert B. Luce, Washington, D.C., 1976; and Stephen P. Strickland, *Politics, Science and Dread Disease: A Short History of United States Medical Research Policy*, Harvard University Press, Cambridge, 1972. See also Tim Beardsley, "A War Not Won," *Scientific American*, January, 1994, pp. 130–138. Beardsley quotes John C. Bailar III, a respected cancer epidemiologist, as stating that death rates from cancer in the U.S. continued to rise between 1950 and 1990, with increasing death rates for non-Hodgkin's lymphoma, multiple myeloma, and cancers of the prostate, brain, kidney, esophagus, and breast. The NCI estimated that the overall improvement in five-year survival since the War on Cancer began is about 4 percent. Bailar is quoted as saying "our decades of war against cancer have been a qualified failure." In November 1996, the National Cancer Institute and the American Cancer Society reported the first sustained decline in age-adjusted mortality rates for cancer in this century, with an annual 0.6 percent drop since 1990. Total number of deaths, however, continued to increase (American Cancer Society, "Cancer Facts & Figures—1997").

248   "best funded cancer researcher": Jeffrey Lyon and Pete Gorner, *Altered Fates*, Norton, New York, 1995, p. 142.

249   Statistics on cancer: American Cancer Society, "Cancer Facts and Figures—1996," New York, 1996. The percentages of cures per treatment modality came from a slide presented by George Canellos, Harvard Medical School, at a Cancer Research Institute symposium, New York, January 12, 1995; Canellos was unable to cite a source for these figures when asked, and sources at the American Cancer Society and the National Institutes of Health were unable to substantiate them, although an NIH spokesman said they were in line with other statistics. For an interpretation of survival and cure rates, see Howard P. Greenwald, *Who Survives Cancer?*, University of California Press, Berkeley, 1992.

249   "The big lie": Michael Lotze, telephone interview, Dec. 5, 1992.

250   "The reality is": Phillip Frost, telephone interview, Jan. 3, 1994.

250, 251   "That, it seemed to": S. Rosenberg, interview, Sept. 26, 1994; "That was something": S. Rosenberg, interview, Dec. 7, 1992.

251   "when interest in BCG": "Topical BCG for recurrent superficial bladder cancer," *Lancet*, 337:821–22 (April 6, 1991). On BCG: the original observation of BCG's effectiveness was A. Morales, D. Eidinger, and A. W. Bruce, "Intracavity Bacillus Calmette-Guérin in the Treatment of Superficial Bladder Tumors," *Journal of Urology*, 116:183 (1976), with followup confirmed in D. L. Lamm et al., "A Randomized Trial of Intravesical Doxorubicin and Immunotherapy with Bacille Calmette-Guérin for Transitional-Cell Carcinoma of the Bladder," *New England Journal of Medicine*, 325:1205–09 (Oct. 24, 1991). On ganglioside vaccines, see P. O. Livingston et al., "Improved Survival in Stage III Melanoma Patients with GM2 Antibodies: A Randomized Trial of Adjuvant Vaccination with GM2 Ganglioside," *Journal of Clinical Oncology*, 12:1036–44 (May 1994).

252   "What I decided": S. Rosenberg, interview, Dec. 7, 1992; "It was just a": S. Rosenberg, ibid.

252, 253   The overall review can be found in Steven A. Rosenberg and William D. Terry, "Passive Immunotherapy of Cancer in Animals and Man," *Advances in Cancer Research*, Vol. 25, edited by Georg Klein and Sidney Weinhouse,

Academic Press, New York, 1977, pp. 323–388. "Cells had a lot": S. Rosenberg, interview, Dec. 7, 1992; "a few examples": S. Rosenberg, ibid.

253 "various and sundry places": Alexander Fefer, personal interview, December 27, 1994, University of Washington Medical School, Seattle, Wash.

255 For Fefer's early work, see A. Fefer, "Immunotherapy and Chemotherapy of Moloney Sarcoma Virus-induced Tumors in Mice," *Cancer Research*, 29:2177–83 (December, 1969); L. Fass and A. Fefer, "Studies of Adoptive Chemoimmunotherapy of a Friend Virus-induced Lymphoma," *Cancer Research*, 32:997–1001 (May 1972); and L. Fass and A. Fefer, "Factors Related to Therapeutic Efficacy in Adoptive Chemoimmunotherapy of a Friend Virus-induced Lymphoma," *Cancer Research*, 32:2427–31 (November 1972).

Chapter 12: To Be in Motion . . .

257 Epigraph: "In research, the usefulness": L. Thomas, *The Youngest Science: Notes of a Medicine Watcher*, Viking, New York, 1983, p. 82.

258 "We're always working": S. Rosenberg, Sept. 26, 1994.

259 "Every day I take care": S. Rosenberg, Dec. 7, 1992; interviews with other lab members include: Michael Lotze, Dec. 5, 1992, Jan. 24, 1993, Sept. 24, 1994; Jim Mulé, personal interview, Dec. 7, 1992; Douglas Fraker, personal interview, Dec. 7, 1992; Suzanne Topalian, Sept. 26, 1994; and John Yannelli, Sept. 26, 1994.

260 "We didn't know if": S. Rosenberg, Dec. 7, 1992.

261 "seemed significant": S. Rosenberg, *TTC*, p. 80. There is some disagreement over the original reception of IL-2's discovery by the immunologic community. Rosenberg suggested in an interview that when Doris Morgan's paper appeared in *Science*, its importance was immediately appreciated, but Kendall Smith later noted, also in *Science*, "This report went unnoticed by most immunologists because it dealt with a way to culture T cells from bone marrow, a site not ordinarily viewed as a source of mature T cells. Moreover, even though the cells continued to proliferate in conditioned medium, they appeared to be immunologically immature, as they could not be shown to perform any antigen-specific functions. Consequently, the phenomenon was dismissed as merely another aspect of the antigen-nonspecific nature of mitogenic factors in conditioned media; indeed this alone was more than enough reason to assure that it would be ignored by the immunology community." (K. A. Smith, "Interleukin-2: Inception, Impact, and Implications," *Science*, 240:1169–76, 1988, p. 1170).

261 "The ability to propagate": in S. Gillis and K. A. Smith, "Long Term Culture of Tumor-specific Cytotoxic T Cells," *Nature* 268:154 (1977).

261 "registered strongly": S. Rosenberg, *TTC*, p. 81; "We were looking for": S. Rosenberg, Sept. 26, 1994.

262 Meeting in Oxford: According to Mitchison, "Bill Terry suggested to me that now would be a good time to rethink cancer immunology" (Avrion Mitchison, telephone interview, Oct. 10, 1996).

262 "It was in the course," "to shift some," and "I came back": S. Rosenberg, Sept. 26, 1994.

263 M. Rosenstein et al., "In Vitro Growth of Murine T Cells, VI: Accelerated Skin Graft Rejection Caused by Adoptively Transferred Cells Expanded in T Cell Growth Factor," *Journal of Immunology* 127:566–71 (1981). Additional

details from Rosenstein, personal interview, Jan. 28, 1993, Pittsburgh Cancer Institute, Pittsburgh, Pa.

263 "is about twenty-five years old": Gustav J. V. Nossal, "The Case History of Mr. T. I.: Terminal Patient or Still Curable?" *Immunology Today* 1:5–9 (1980). For historical overviews of tumor immunology, see Lloyd J. Old, "Cancer Immunology," *Scientific American*, May 1977, pp. 62–79; and "Tumor Immunology: The First Century," *Current Opinion in Immunology* 4:603–7 (1992); Georg Klein and Eva Klein, "Tumor Immunology," in R. B. Gallagher et al., eds., *Immunology: The Making of a Modern Science*, Academic Press, New York, 1995, pp. 203–21; and Ilana Löwy, "Experimental Systems and Clinical Practices: Tumor Immunotherapy and Cancer Immunotherapy, 1895–1980," *Journal of the History of Biology* 27:403–35 (1994).

264 "This single devastating fact": For original work on histocompatibility, see Peter A. Gorer, "The Genetic and Antigenic Basis of Tumor Transplantation," *Journal of Pathology and Bacteriology* 44:691–97 (1937). Klein and Klein ("Tumor Immunology," pp. 204–5) make the point that very few scientists understood the work of Gorer and George Snell, who reached similar findings while working at the Jackson Memorial Laboratory in Maine, and that Gorer and Snell did not make much effort to make their work more understandable. Gorer, they write, "was known by very few. In his seminars and conference lectures, he did nothing to explain the complicated jargon. We used some of his papers as the acid test to distinguish between truly motivated and less-interested students."

267 For background on the development of inbred mice, see Michael Potter, "Some Reminiscences on the Contributions of Cancer Research to Immunology in the 1950s," in P. M. H. Mazumdar, ed., *Immunology, 1930–1980: Essays on the History of Immunology*, Walls & Thompson, Toronto, 1989, pp. 95–105 and A. Silverstein, *History of Immunology*, pp. 282–83.

267 "conflicting to a surprising": T. Lumsden, "Tumor Immunity," *American Journal of Cancer* 15:563–640 (1931).

267 It began with a landmark paper: Ludwik Gross, "Intradermal Immunization of C3H Mice against a Sarcoma That Originated in an Animal of the Same Line," *Cancer Research* 3:326 (May 1943); "humoral antibodies are not": in L. Gross, ibid., p. 332. For background on Gross, see also Daniel J. Kevles, "Pursuing the Unpopular: A History of Courage, Viruses, and Cancer," in Robert B. Silvers, ed., *Hidden Histories of Science*, New York Review Book, New York, 1995, pp. 69–112. As Kevles recounts, Gross also showed in the early 1950s that mouse leukemias were caused by a virus, a finding that revived the field of viral oncogenesis and led indirectly to the discovery of oncogenes.

268 "The history of attempts": Richmond T. Prehn and Joan M. Main, "Immunity to Methylcholanthrene-Induced Sarcomas," *Journal of the National Cancer Institute* 18:769–78 (June 1957).

268 E. J. Foley, "Antigenic Properties of Methylcholanthrene-induced Tumors in Mice of the Strain of Origin," *Cancer Research* 13:835–37 (1953); "put tumor immunology": H. Oettgen, personal interview, Feb. 15, 1996, New York.

268 G. Klein et al., "Demonstration of Resistance against Methylcholanthrene-induced Sarcomas in the Primary Autochthonous Host," *Cancer Research* 20:1561–72 (Dec. 1960).

269   "These findings placed": Lloyd Old, remarks at "Frontiers of Immunology and Cancer Immunology," symposium on Jan. 25–26, 1993, New York.
269   "There had been this": S. Rosenberg, interview, Sept. 26, 1994.
269   On lack of antigens in spontaneous tumors: H. B. Hewitt et al., "A Critique of the Evidence for Active Host Defence against Cancer, Based on Personal Studies of 27 Murine Tumors of Spontaneous Origin," *British Journal of Cancer* 33:241 (1976).
270   the Endangered Self: In 1996, Polly Matzinger's group at the NIH proposed a "danger model" of immunology, in which the immune system does not so much perceive self versus non-self as perceive antigens associated with physiological harm. See Elizabeth Pennisi, "Teetering on the Brink of Danger," *Science,* 271:1665–67 (1996).
271   The first hint of an answer: For the role of MHC molecules in antigen presentation, see R. M. Zinkernagel and P. C. Doherty, "Restriction of In Vitro T Cell-mediated Cytotoxicity in Lymphocytic Choriomeningitis within a Syngeneic and Semiallogeneic System," *Nature* 248:701–2 (1974); for the role of MHC plus peptides, see A. R. M. Townsend et al., "The Epitopes of Influenza Nucleoprotein Recognized by Cytotoxic T Lymphocytes Can Be Defined with Short Synthetic Peptides," *Cell* 44:959–68 (1986).
273   For the identification of the costimulatory molecule B-7, see S. E. Townsend and J. P. Allison,"Tumor Rejection After Direct Costimulation of CD8t T Cells by B-7-Transfected Melanoma Cells," *Science* 259:368–70 (1993). P. J. Bjorkman et al., "Structure of the Human Class I Histocompatibility Antigen, HLA-A2," *Nature* 329:506–512 (1987).
274   "T. I. is not": Klein and Klein, *Tumor Immunology,* p. 204.
274   "A lot of people": M. Lotze, personal interview, June 2, 1992.
274   For the Seattle work, see M. A. Cheever et al., "Augmentation of the Antitumor Therapeutic Efficacy of Long-term Cultured T Lymphocytes by *in Vivo* Administration of Purified Interleukin-2," *Journal of Experimental Medicine* 155:968–80 (1982).
275   "He goes ahead": A. Fefer, personal interview, Dec. 27, 1994, Seattle, Wash. This opinion was seconded by Fefer's colleague Phil Greenberg. "I think Alex really was the first person to convincingly demonstrate that he could take an animal with a disseminated tumor and develop a reproducible model where he could treat that animal with cells specifically reactive to that tumor and eliminate it, and cure mice. I mean, we've actually told Steve [Rosenberg] this. Steve often tries to take credit for that, which is remarkable in that after I had been here probably five years, Steve actually called up Alex and asked him to send him our FBL tumor so he could develop the model. And he makes some very curious distinctions about why his model is different, and his model *is* different. But what wasn't different is that for years before that—*years* before that!—Alex had already demonstrated that he could treat disseminated disease with intravenous administration of lymphocytes. So Steve himself was part of the reason, I think, that Alex doesn't receive the credit" (P. Greenberg, personal interview, Sept. 21, 1993, Seattle, Wash.).
275   "Cinderella experiment": S. Rosenberg, *TTC,* p. 131.
276   E. A. Grimm et al., "Lymphokine-activated Killer Cell Phenomenon," *Journal of Experimental Medicine* 155:1823–41 (June 1982); and E. A. Grimm et al., "The Lymphokine-Activated Killer Cell Phenomenon: In Vitro and In Vivo Studies," in J. Oppenheim and S. Cohen, eds., *Interleukins,*

Lymphokines, and Cytokines: Proceedings of the Third International Lymphokine Workshop, Academic Press, New York, 1983, pp. 739–48.

276 "We tried to throw": M. Lotze, interview, June 2, 1992.

277 "The original papers": Jerome Ritz, personal interview, Dana-Farber Cancer Institute, Boston, Mass., Feb. 8, 1994; "Lymphokine-activated killer cells": S. Rosenberg, Sept. 26, 1994.

278 The first use of lectin-activated LAK cells in 10 patients was completed by 1982 and published in 1984 as A. Mazumder et al., "Phase I Study of the Adoptive Immunotherapy of Human Cancer with Lectin Activated Autologous Mononuclear Cells," *Cancer* 53:896–905 (1984).

278 "Failures in lab experiments": S. Rosenberg, *TTC,* p. 138.

279 "had the ability to": ibid., p. 145; "When I first went": M. Lotze, telephone interview, Sept. 24, 1994; "Michael would always set": M. Rosenstein interview; "In very Mike Lotze": M. Lotze, interview, Sept. 24, 1994; "I had huge vats": M. Lotze, June 2, 1992.

280 On Du Pont involvement in interleukin-2 production: see S. Rosenberg, *TTC,* pp. 146–49 and Richard Robb, telephone interview, Feb. 20, 1996.

281 "The side effects noted": in C. Bindon et al., "Clearance Rates and Systemic Effects of Intravenously Administered Interleukin 2 (IL-2) Containing Preparations in Human Subjects," *British Journal of Cancer* 47:123 (1983). P. Hersey to author, Aug. 8, 1994.

281 "There were no": Rosenberg, *TTC,* p. 161.

282 "Thunderbolt moments": ibid., p. 166; "Perhaps for the first": ibid., p. 183; "I had put": ibid., p. 193; "flailing and futile": ibid., p. 193.

283 "You know, we'd been": S. Rosenberg, interview, Sept. 26, 1994.

Chapter 13: The End of the Beginning?

285 Epigraph: Talmud, cited by J. Groopman, *New York Times,* Aug. 3, 1996.

285 Background on the Linda Taylor case: S. A. Rosenberg, *The Transformed Cell,* Putnam, New York, 1992, pp. 199–213, and S. A. Rosenberg et al., "Special Report: Observations on the Systemic Administration of Autologous Lymphokine-activated Killer Cells and Recombinant Interleukin-2 to Patients with Metastatic Cancer," *New England Journal of Medicine* 313:1485–92 (Dec. 5, 1985). This patient, identified by the pseudonym "Linda Granger" in Rosenberg's book, was interviewed using her real name on "Primetime Live" (ABC), Oct. 30, 1996. Also personal interviews, S. A. Rosenberg and Michael Lotze.

286 "Dose escalation": In the mid-1980s, clinicians were just beginning to stop dose escalation in individual patients, according to Michael Lotze. However, William Coley, for example, did exactly that with his toxins, starting with a minimal dose, waiting a few days, and then increasing the amount of toxins with each subsequent injection until he reached a maximum tolerated dose in each patient. Although a nonstandardized seat-of-the-pants approach, it also bespeaks a certain clinical wisdom; people are biologically different, and not all patients tolerate the same "maximum" dose.

286 "experiments do not work": S. Rosenberg, *TTC,* p. 246; "Everything was on": ibid., p. 207; "I was made": ibid., p. 207; "I think it would": S. A. Rosenberg, Sept. 26, 1994, Bethesda, Md.

288 "James Jensen" case described in *TTC*, pp. 208–13; "And I felt desperate": Rosenberg, *TTC*, p. 210.

288 "My God!": Rosenberg, *TTC*, p. 211; "That for me was": S. Rosenberg, personal interview, Sept. 26, 1994.

289 "It's always a wonder": S. A. Rosenberg, personal interview, Dec. 7, 1992.

289 "Cancer researchers smelled a storm": James W. Mier, Tufts University–New England Medical Center, Boston, Mass., telephone interview, August 1, 1996. All other remarks from Mier from this interview.

290 "It was very straightforward": Vincent DeVita, telephone interview, January 10, 1995. Despite the thrust of DeVita's remarks, there was only one complete response in the first 25 patients.

291 "totally off the cuff": Rosenberg, interview, Sept. 26, 1994; "The President has cancer": Rosenberg, *TTC*, p. 225. For Rosenberg's discussion of the Reagan case, see *TTC*, pp. 222–26.

291 "a lot of trouble": DeVita, interview. DeVita further revealed in an interview that Rosenberg was "the first candidate to replace me as director, and Mrs. Reagan herself took his name off the list because he got up and said in public, 'The President has cancer.' "

291 "It was the first": Rosenberg, interview, Sept. 26, 1994.

291 "He was one cell-doubling": M. Lotze, personal interview, Sept. 24, 1994, and personal communication, April 4, 1996. At the time, IL-2 therapy alone was also being studied elsewhere; an Italian group had published on the use of IL-2 in bladder cancer, and there were reports of responses at Memorial Sloan-Kettering Cancer Center and M. D. Anderson Hospital.

292 "achievable": V. DeVita, quoted in T. Beardsley, "A War Not Won," *Scientific American*, Jan. 1994, p. 135.

292 "It was clear": Arnold Relman, telephone interview, Oct. 13, 1994. As to the rationale for publishing such a preliminary study, Relman added: "While I was well aware that the Rosenberg study was preliminary and in no sense definitive, at the time I thought that as a preliminary report on what might prove to be a worthy treatment for cancer, it was worth publishing. I think there is room within the *New England Journal of Medicine* for articles which are identified as being preliminary if it is very interesting, novel, and opens up new pathways for treatment."

292 The anecdote about the General Motors Award is related in Robert Teitelman, *Gene Dreams*, Basic Books, New York, 1989, pp. 188–89.

293 "many clinicians believe": in G. Bylinsky, "Science Scores a Cancer Breakthrough," *Fortune*, Nov. 25, 1985, pp. 16–21.

293 "unheard of in cancer": ibid., p. 16. Among the assertions in this influential story was that physicians "have now discovered how to use a small group of substances produced by the body's own immune system to control all cancers." "Totally blew me away": Rosenberg, interview, Sept. 26, 1994.

294 "When the article came": Rosenberg, interview, Sept. 26, 1994.

294 "articulate and reassuring": *Newsweek*, Dec. 16, 1985, p. 61; "First new kind of approach": S. A. Rosenberg, quoted in ibid., p. 64.

294 "It's not a cancer cure": S. A. Rosenberg, quoted in *People*, Dec. 23, 1985, pp. 46–49; "a researcher of near": "New Weapon in the Cancer War?" *Time*, Sept. 22, 1986, p. 58. "The whole thing": S. Rosenberg, interview, Sept. 26, 1994.

295 "A lot of people": Mier interview; "It was the height": A. Fefer, interview,

Dec. 27, 1994. Estimates of the cost of the confirmatory studies vary; DeVita suggested it cost about $6 million (personal communication, Nov. 7, 1996); Mier suggests that, with direct and indirect costs figured over two years, the actual cost may have been between $12 and $15 million. "People thought": interview; "I think Steve was": M. Lotze, interview, Sept. 24, 1994.

296 "disturbing, even frightening": Rosenberg, *TTC*, p. 235; "I am really anxious": S. A. Rosenberg, quoted in C. Wallis, "Arming Cancer's Natural Enemies," *Time*, Dec. 16, 1985, p. 58; "In Steve's case": DeVita interview.

297 "There's a general sense": M. Lotze, interview, Dec. 5, 1992.

299 "Olympic athletes": Robert Oldham, remarks at Society of Biological Therapy annual meeting, Oct. 30, 1992, Williamsburg, Va.

300 On first successful use of IL-2 alone: M. Lotze et al., "High-Dose Recombinant Interleukin 2 in the Treatment of Patients with Disseminated Cancer," *Journal of the American Medical Association* 256:3117–24 (Dec. 12, 1986). See also S. Rosenberg, *TTC*, p. 250.

301 "She had a huge": M. Lotze, interview, Sept. 24, 1994; "This was right before": ibid.

302 "one of these fishy": Peter Wiernik, personal interview, Aug. 23, 1994, Montefiore Medical Center, Bronx, N.Y.

302 "a very cautious" and "One of the things": M. Lotze, interview, June 2, 1992.

303 "Following their first": C. Moertel, "On Cytokines, Lymphokines, and Breakthroughs," *JAMA* 256:314 (Dec. 12, 1986). For press coverage, see H. Schmeck Jr., "Medical Journal Assails the Use of an Experimental Cancer Drug," *New York Times*, Dec. 12, 1986, p. A-1 and M. Bloom, "Cancer M.D.'s Clash Over Interleukin Therapy," *Science* 235:154–55 (Jan. 9, 1987). "first new kind of approach": S. A. Rosenberg, quoted in *Newsweek*, Dec. 16, 1985, p. 64.

304 "See, the problem": M. Lotze, interview, June 2, 1992; "There have been three": Nicholas Vogelzang, University of Chicago, telephone interview, Nov. 1, 1996. According to Ronald Herberman—former head of the NIH's Biological Response Modifiers Program and present director of the Pittsburgh Cancer Institute, whose group discovered the NK cell in 1975—the confusion between LAK cells and NK cells had unhappy ramifications for the entire field. "The biggest problem, I think, that occurred with the LAK cell area was that the studies were done with really just a mixed population of white blood cells that were cultured in IL-2," Herberman said in an interview. "And putting the name of 'LAK cell' on it in a way was perhaps the most unfortunate or misleading aspect of it, because it led most people to think of it as kind of a single population of cells, these IL-2 cultured lymphocytes, that all in some way might have antitumor effects. Where it's turned out that only a very small proportion of the cells seem to be important in therapy, and most of those are the natural killer cells."

Even a former member of the Rosenberg lab who participated intimately in the work suggests that the story was always more complex than it seemed. "LAK activity is the ability to kill fresh tumor cells without previous [exposure] to that tumor antigen—in other words, the lack of previous sensitization, which is nominally an important aspect of immunization," said Michael Lotze. "The cells that we give to patients are largely IL-2 activated cells from the peripheral blood [meaning blood drawn from a vein] that have been cultured in IL-2 for three to five days but include T cells, NK cells, macrophages, a variety of different cell types, *some of which*, not all of which, have this ability to kill fresh tumor. It's primarily the NK cells. What we really give to patients is not LAK cells, although that's been the term that's used. We give them IL-2-activated peripheral blood mononuclear cells. Now the reason why that's an important distinction to make is that some of the antitumor effects, if any, may come from other cells, other than the ones that mediate the LAK activity."

305 "This immunotherapeutic approach": S. A. Rosenberg et al., "A Progress Report on the Treatment of 157 Patients with Advanced Cancer Using Lymphocyte-activated Killer Cells and Interleukin-2 or High-Dose Interleukin-2 Alone," *New England Journal of Medicine* 316:889–97 (April 9, 1987).

305 W. H. West et al., "Constant-infusion Recombinant Interleukin-2 in Adoptive Immunotherapy of Advanced Cancer," ibid., 316:898–905 (April 9, 1987).

306 "an oncologic perspective" and "If LAK cells": J. Durant, "Immunotherapy of Cancer: The End of the Beginning?" ibid., pp. 939–941.

306 "To me, that was": Rosenberg, interview, Sept. 26, 1994.

307 Their moment came: R. I. Fisher et al., "Metastatic Renal Cancer Tested with Interleukin-2 and Lymphokine-Activated Killer Cells: A Phase II Clinical Trial," *Annals of Internal Medicine* 108:518–23 (1988).

307 Reaction to the LAK/IL-2 studies: "No one has reported": J. Mier, interview; "represents a significant": S. Hellman, "Immunotherapy for Metastatic Cancer: Establishing a 'Proof of Principle,' " *Journal of the American Medical Association* 271:945 (March 23/30, 1994); "In some cases": J. Mier interview. "Whatever else you can": A. Fefer, interview, Dec. 27, 1994;

308 Expanded applications of IL-2 include: J. A. Kovacs et al., "Increases in CD4 T Lymphocytes with Intermittent Courses of Interleukin-2 in Patients with Human Immunodeficiency Virus Infection," *New England Journal of Medicine* 332:567–75 (1995); J. A. Kovacs et al., "Controlled Trial of Interleukin-2 Infusions in Patients Infected with the Human Immunodeficiency Virus," ibid., 335:1350–56 (1996); and E. L. Jacobson et al., "Rational Interleukin 2 Therapy for HIV Positive Individuals: Daily Low Doses Enhance Immune Function without Toxicity," *Proceedings of the National Academy of Sciences* 93:10405–10 (September 1996).

309 "fiasco": Kendall A. Smith, Division of Immunology, New York Hospital–Cornell Medical Center, personal interview, New York, Dec. 30, 1993.

309 "Because of either his": Ronald Herberman, director, Pittsburgh Cancer Institute, personal interview, Jan. 25, 1993, Pittsburgh, Pa.; "That's actually been one": P. Greenberg, Fred Hutchinson Cancer Research Center, personal interview, Sept. 21, 1993. Seattle, Wash.

311 "a new approach": in S. A. Rosenberg, P. Spiess, and R. Lafreniere, "A New Approach to the Adoptive Immunotherapy of Cancer with Tumor-Infiltrating Lymphocytes," *Science* 233:1318–21 (Sept. 19, 1986).

312 "extraordinarily excited": Rosenberg, *TTC*, p. 254; "substantial tumor regression": Rosenberg quoted in S. Okie, "Biological Cancer Treatment Improved," *Washington Post*, March 30, 1988.

312 For first clinical application, see S. A. Rosenberg et al., "Use of Tumor-Infiltrating Lymphocytes and Interleukin-2 in the Immunotherapy of Patients with Metastatic Melanoma," *New England Journal of Medicine*, Dec. 22, 1988, p. 1676.

312 "The pressure to improve": Rosenberg, *TTC*, p. 254; "incremental improvements" and "to leap a chasm": *TTC*, p. 255.

313 "About that time": *TTC*, p. 261. The idea of inserting new genes into TIL cells had occurred to Rosenberg as soon as he realized that TIL cells sought

out and homed in on tumors. He asked Werner Green, then an immunologist at the NIH, to try to insert the gene for the interleukin-2 receptor into TIL cells, the idea being that this would make them more receptive to the IL-2 he was pumping into patients.

Chapter 14: "There's Just So Much You Can Learn from a Mouse"

314 Epigraph: Chester Southam, "Applications of Immunology to Clinical Cancer: Past Attempts and Future Possibilities," *Cancer Research* 21:1302 (1961).

314 "Ever since the 1960s": The birth of gene therapy has been one of the most closely chronicled and exhaustively documented developments in biomedical science since the end of World War II. Even before data from the first human ADA gene transfer experiment were published in the literature, two book-length popular histories had already been issued: Larry Thompson, *Correcting the Code: Inventing the Genetic Cure for the Human Body*, Simon and Schuster, New York, 1994; and Jeffrey Lyon and Pete Gorner, *Altered Fates: Gene Therapy and the Retooling of Human Life*, Norton, New York, 1995. In addition, meetings of the Recombinant DNA Advisory Committee and its human gene therapy subcommittee have been documented in government publications as well as by extensive coverage in the lay and scientific press. The National Institutes of Health has documented public meetings and correspondence in Office of Recombinant DNA Activities, *Recombinant DNA Research*, NIH Publications; in particular, see "Recombinant DNA Research, Volume 13: Documents Relating to 'NIH Guidelines for Research Involving Recombinant DNA Molecules,' May 1987–December 1988," U.S. Department of Health and Human Services, NIH Publication No. 91-3205, Bethesda, Md., Dec. 1990.

315 For background on the Martin Cline case, see Lyon and Gorner, *Altered Fates*, pp. 68–77.

316 Background on French Anderson: L. Thompson, "French Anderson's Genetic Destiny," *Washington Post Magazine*, Jan. 20, 1991; Robin Marantz Henig, "Dr. Anderson's Gene Machine," *New York Times Magazine*, March 31, 1991. Lyon and Gorner write of Anderson's scientific contribution: "The discoveries made by [others] have been more crucial to gene therapy's embryonic development than those of W. French Anderson. In truth, Anderson's ideas have never been the most seminal of those seeking the grail of therapeutic gene replacement" *(Altered Fates*, p. 257). Richard Mulligan, quoted in the Thompson story, said, "The key question is looking at the track record and the contributions to the field. If you look at his [Anderson's] contributions, in technological development, there is nothing there" (p. 36). Anderson said in an interview that these and other disparaging comments "were so obviously false that I didn't bother to respond. But all one has to do is go back and look at the published papers. If you look at what was published in the literature between 1986 and 1989, almost every significant contribution came out of my lab." (F. Anderson, telephone interview, Feb. 21, 1997).

316 "I think French saw": Ken Culver, telephone interview, Aug. 8, 1994; this and all other remarks by Culver come from this interview unless otherwise indicated.

316 "wonderfully wily": Michael Lotze, telephone interview, Dec. 5, 1992. Most of Lotze's remarks in this chapter come from this interview, although there

were also conversations on June 2, 1992, January 25, 1993, January 26, 1993, Sept. 24, 1994, and Nov. 23, 1996.

316 "French has had a history": R. Michael Blaese, personal interview, June 3, 1994, Cold Spring Harbor Laboratory, New York. This and all other remarks by Blaese came from this interview unless otherwise noted.

317 "I think when": N. Wivel, former executive director, Subcommittee on Gene Therapy, telephone interview, Aug. 3, 1996.

318 Donald Kohn experiments: P. W. Kantoff et al., "Correction of Adenosine Deaminase Deficiency in Human T and B Cells Using Retroviral-mediated Gene Transfer," *PNAS* 83:6563–67 (1986).

320 "like trying to send": Lyon and Gorner, *Altered Fates*, p. 114.

322 "early rock 'em, sock 'em": W. F. Anderson, "Musings on the Struggle, Part I—The Phonebook," *Human Gene Therapy* 3:251–52 (1992).

322 "the most formidable bureaucracy": Lyon and Gorner, *Altered Fates*, p. 153.

322 "Phonebook": see W. F. Anderson, "Musings—Part I," Reading between the lines, part of Anderson's strategy may have been the old bureaucratic ploy of burying the opposition in paper. According to the minutes of the Dec. 7, 1987, meeting of the human gene therapy subcommittee, Charles Epstein, a member, "stated that the structure of the protocol had been criticized. Because it contains a tremendous amount of data, the protocol is very difficult to read. He hoped that in future edits, it would become easier to read" (Office of Recombinant DNA Research, vol. 13, 1990, pp. 93–94).

322 "We had to provide": Anderson, "Musings Part I," p. 252.

323 Background on the Dec. 7, 1987, meeting of the human gene therapy subcommittee: ORDA, vol. 13, 1990, pp. 91–98; Thompson, "Genetic Destiny," pp. 319–26.

324 "I have always had": Anderson, "Musings—Part I," p. 251.

326 "About that time": Steve Rosenberg with John Barry, *The Transformed Cell*, Putnam, 1992, p. 261; gene therapy described, ibid., pp. 259–336.

326 "certainly wasn't one of": S. Rosenberg, telephone interview, Nov. 14, 1996. Apprised of Rosenberg's remark, Anderson said, "All I can say is that he [Lotze] was intimately involved in the beginning, and not involved later on." (F. Anderson, telephone interview, Nov. 27, 1996).

326 On March 17, 1988, meeting: interviews and K. Culver, personal communication, Nov. 19, 1996. "In a little over": Rosenberg, *TTC*, p. 266.

328 "Steve had the persona": M. Lotze, interview, Sept. 24, 1994.

329 "hot": On internal scientific meetings of the Rosenberg-Anderson-Blaese groups, see Rosenberg, *TTC*, p. 272.

330 On the July 1988 and October 1988 meeting of the Recombinant DNA Advisory Committee, see "Recombinant DNA Research, Volume 13," "The Subcommittee wrestled": ibid., p. 183. On October meeting: ibid., pp. 225–65. Anderson also discussed this meeting in Anderson, "Musings," and "Musings on the Struggle, Part III—The October 3 RAC Meeting," *Human Gene Therapy*, 4:401–02 (1993).

330 "That's a pretty stiff": R. Scott McIvor, University of Minnesota, telephone interview, Nov. 15, 1996.

331 "disingenuous and intended": Robert Cook-Deegan, Office of Technology Assessment observer at RAC meetings, personal communication, Aug. 26, 1996.

331 On the subcommittee "bypass": McIvor recalled in an interview that on the night prior to the Oct. 3 meeting of the RAC, he joined several other committee members for dinner, including Richard Mulligan, and that the main

topic of conversation was "we don't want French to come in with a pile of new data, and that's exactly what he did."

331 "We felt that logic": W. F. Anderson, "Musings—Part III," p. 401.

332 "deleterious effects": Recombinant DNA Research, vol. 13, p. 225.

332 "I was asked at": W. F. Anderson, "Musings—Part III," p. 402. "From the look": Robert Cook-Deegan, personal communication.

332 "It is virtually not possible": B. Davis, quoted in Lyon and Gorner, p. 159.

334 "Well, I haven't": ibid., p. 168.

334 "would open a door": Rosenberg, TTC, p. 289.

336 "It is true that": W. F. Anderson, telephone interview, Nov. 27, 1996.

336 "It was the smoothest": Rosenberg, telephone interview, Nov. 14, 1996.

337 Both the Blaese-Anderson team: Lyon and Gorner have a particularly good account of the July 1990 meeting in which both Rosenberg and Anderson-Blaese won approval for therapeutic gene therapy protocols, pp. 152–161. "the day there was a crack": M. Lotze, telephone interview, Nov. 23, 1996. This remark was in part occasioned by the fact that that same day, in the afternoon, an advisory committee of the Food and Drug Administration declined to approve the drug interleukin-2 for clinical use.

337 Ashanti DeSilva experiment: Described in much detail in the press. The results of this treatment were finally published recently as R. Michael Blaese et al., "T Lymphocyte-Directed Gene Therapy for ADA-SCID: Initial Trial Results After 4 Years," Science 270:475–80 (1995).

337 Rosenberg did not get: S. A. Rosenberg et al., "Gene Transfer into Humans— Immunotherapy of Patients with Advanced Melanoma, Using Tumor-infiltrating Lymphocytes Modified by Retroviral Gene Transduction," New England Journal of Medicine 323:570–78 (Aug. 30, 1990); S. A. Rosenberg, "The Immunotherapy and Gene Therapy of Cancer," Journal of Clinical Oncology 10:180–99 (Feb. 1992); and S. A. Rosenberg, "Gene Therapy for Cancer," Journal of the American Medical Association 268:2416–19 (Nov. 4, 1992).

338 In March 1995, Genetic Therapy; "amounted to a major": Lyon and Gorner, Altered Fates, pp. 245–46. Anderson stated in an interview that the Lasker incident was in part triggered by NIH politics; the National Heart, Lung and Blood Institute, to which he belonged, felt the National Cancer Institute was publicizing gene therapy in part to get more funding from Congress. "The Cancer Institute had been issuing press releases about gene therapy mentioning Rosenberg and not Blaese and I, and holding press conferences with Rosenberg that we didn't even know about (F. Anderson, telephone interview, Feb. 21, 1997).

339 On patent dispute, see R. Nowak, "Patent Award Stirs a Controversy," Science 267:1899 (March 31, 1995). Culver also provided a copy of a deposition he gave on March 26, 1996, in defense of his claims. In addition, he confirmed that the NIH team used a vector provided by University of Washington researchers and adapted by Gene Therapy Inc. because otherwise "it would have taken a lot longer" to get a human experiment underway (K. Culver, telephone interview, Nov. 15, 1996).

339 Nonetheless, a number of: Gary Nabel, presentation at Cold Spring Harbor Symposium, June 3, 1994; Drew Pardoll, telephone interview, Jan. 24, 1996; Michael Lotze, telephone interview, Jan. 26, 1996. See also L. Thompson, "At Age 2, Gene Therapy Enters a Growth Phase," Science 258:744–46 (Oct. 30, 1992).

340 T. Friedmann, "Human Gene Therapy—An Immature Genie, but Certainly Out of the Bottle," *Nature Medicine* 2:144–147 (Feb. 1996).

340 On the hype surrounding gene therapy: see Eliot Marshall, "Gene Therapy's Growing Pains," *Science* 269:1050–55 (August 25, 1995) and E. Marshall, "Less Hype, More Biology Needed for Gene Therapy," *Science* 270:1751 (Dec. 15, 1995), the latter reporting on the "Report and Recommendations of the Panel to Assess the NIH Investment in Research on Gene Therapy," known as the Orkin-Motulsky report. This report, which criticizes gene therapists for weak study designs and inadequate data, is available on the Internet at http://www.nih.gov/news/panelrep.html.

342 Christopher Anderson, "Gene Therapy Researcher Under Fire over Controversial Cancer Trials," *Nature* 360:399 (Dec. 3, 1992). *Science* later covered the controversy as well in "A Speeding Ticket for NIH's Controversial Cancer Star" (C. Anderson, *Science* 259:1391–92 [March 5, 1993]). Rosenberg's funding was restored in June 1994 (see "Rosenberg Regains Funds," *Science* 264:1850 [1994], and "NCI Restores Funding to Gene Therapist Seen as 'Too Hasty,'" *Nature* 369:598 [1994]).

343 "It was pretty tough": Steven Rosenberg, personal interview, Dec. 9, 1992, Bethesda, Md.

343 "See, you have to": S. Rosenberg, interview, Dec. 9, 1992.

343 "Steve has never published": N. Wivel, telephone interview, Aug. 3, 1996.

344 "You know, these are": Steven Rosenberg, personal interview, Sept. 26, 1994, Bethesda, Md.

345 "The unforgiving yardstick": L. Old, closing remarks, "Cancer Vaccines, 1994" symposium, New York, October 1994.

IV. IN VIVO VERITAS

347 Steve Rosenberg, public remarks, "Cancer Vaccines 1996," New York, October 9, 1996. Lloyd Old used the term "In vivo Veritas" at the 1994 meeting.

Chapter 15: "One Plus One Equals Ten"

349 Epigraph: Drew Pardoll, "Cancer Vaccines: A Road Map for the Next Decade," *Current Opinion in Immunology* 8:619 (1996).

349 Frau H case: Alexander Knuth, personal interview, Nov. 14, 1994, Krankenhaus Nordwest, Frankfurt Germany; "Frau H," personal interview, Nov. 14, 1994, Frankfurt; Thierry Boon, personal interviews, Brussels, Dec. 9, 1993; Cold Spring Harbor Laboratory, New York, June 6, 1994; New York, December 6, 1994; and Jan. 10, 1996 (telephone). Records and accounts of Frau H's hospitalizations: A. Knuth, personal communication, Jan. 2, 1996.

351 For patient "A. V.," see P. O. Livingston et al., "Cell-mediated Cytotoxicity for Cultured Autologous Melanoma Cells," *International Journal of Cancer* 24:34 (1979); and A. Knuth, "T-cell-mediated Cytotoxicity against Autologous Malignant Melanoma: Analysis with Interleukin-2-dependent T-cell Cultures," *Proceedings of the National Academy of Sciences* 81:3511 (June 1984).

354 For general background on Boon work, see T. Boon, "Teaching the Immune System to Fight Cancer," *Scientific American,* March 1993, pp. 82–89; and T. Boon, "Toward a Genetic Analysis of Tumor Rejection Antigens," in *Ad-*

vances in Cancer Research, 1992, vol. 58, edited by G. F. Vande Woude and G. Klein, Academic Press, San Diego, 1992, pp. 177–210.

356 "Before he worked": Norton Zinder, Rockefeller University, New York, telephone interview, Dec. 5, 1994.

359 H. B. Hewitt et al., "A Critique of the Evidence for Active Host Defence against Cancer, Based on Personal Studies of 27 Murine Tumors of Spontaneous Origin," *British Journal of Cancer* 33:241 (1976).

360 Some American scientists point out ruefully: Researchers who admire Boon's tenacity and patience make the point that it is precisely the kind of research against which the current culture of funding in the U.S. militates, partly because the grants administered by the National Institutes of Health usually last only three years. "It's not the type of work that likely would have succeeded in the milieu of NIH funding," explained one American researcher. "If he had to have had progress reports and recompeted [for grants] every three years, he would have had a lot of trouble." Indeed, Thierry Boon had nothing to show for the better part of a decade; in Germany, there was pressure to cut off Alex Knuth's portion of the funding for the same perceived lack of progress. What kept both of them in business, they are quick to point out, was the support of the Ludwig Institute. Such long-term funding, said an American immunologist, not without a hint of envy, permitted good researchers to attempt what he called "adventuresome science."

361 The Boon group's "refutation" of the Hewitt paper appears in A. Van Pel, F. Vessiere, and T. Boon, "Protection against Two Spontaneous Mouse Leukemias Conferred by Immunogenic Variants Obtained by Mutagenesis," *Journal of Experimental Medicine* 157:1992–2001 (June 1983).

363 On technology of taking "surface cuttings": A. L. Cox et al., "Identification of a Peptide Recognized by Five Melanoma-Specific Human Cytotoxic T Cell Lines," *Science* 264:716–19 (1994).

364 P. van der Bruggen et al., "A Gene Encoding an Antigen Recognized by Cytolytic T Lymphocytes on a Human Melanoma," *Science* 254:1643 (Dec. 13, 1991). "Human tumor immunology": Drew Pardoll, telephone interview, Jan. 26, 1996.

366 Since the discovery of MAGE-1: For a general review of tumor antigens discovered in the wake of the *Science* paper, see Drew M. Pardoll, "Tumor Antigens: A New Look for the 1990s," *Nature* 369:357 (June 2, 1994).

366 "inaugurated a new era": L. Old, "Cancer Vaccines 1994" meeting, New York, October 1994. For the Self vs. Non-Self argument, see Alan N. Houghton, "Cancer Antigens: Immune Recognition of Self and Altered Self," *Journal of Experimental Medicine* 180:1 (July 1994).

367 W. F. Bodmer et al., "Tumor Escape from Immune Response by Variation in HLA Expression and Other Mechanisms," *Specific Immunotherapy of Cancer with Vaccines*, edited by J-C. Bystryn, S. Ferrone and P. Livingston, New York Academy of Sciences, New York, 1993, pp. 42–49.

369 "for the first time": T. Boon, *Scientific American*, 1993, p. 89.

369 For a review of nonspecific cancer vaccines, see also J. Cohen, "Cancer Vaccines Get a Shot in the Arm," *Science* 262:841–43 (Nov. 5, 1993) and S. Hall, "The Cancer Vaccine," *Health*, March/April 1994, pp. 77–86.

370 The first results of the MAGE-3 vaccine appeared in M. Marchand et al., "Tumor Regression Responses in Melanoma Patients Treated with a Peptide Encoded by Gene *MAGE-3*," *International Journal of Cancer* 63:883 (December 1995). Alex Knuth, Drew Pardoll, Steven Rosenberg, and Pramod Srivastava described preliminary vaccine results at a session at the annual meeting

of the American Association for the Advancement of Science, Baltimore, Maryland, Feb. 10, 1996, and all four presented updated results at "Cancer Vaccines 1996," Oct. 7–9, 1996, New York.

370 "We often have to": T. Boon, presentation, Cancer Vaccines 1996, Oct. 7, 1996, New York.

370 The preparation of tumor vaccines has become a hot area in immunology, with a number of book-length proceedings of meetings. One of the best is *Specific Immunotherapy of Cancer*, 1993.

372 "I think we're at": Victor Engelhard, University of Virginia School of Medicine, telephone interview, Jan. 26, 1996.

Chapter 16: "Beautiful Living Things"

375 Epigraph: P. B. Medawar and J. S. Medawar, *The Life Sciences: Current Ideas of Biology*, Harper & Row, New York, 1977, p. 137.

375 Discussion of cell preparation and background: personal interviews with Kathe Watanabe, Stanley Riddell, and Philip Greenberg, Fred Hutchinson Cancer Research Center, Seattle, Wash., Sept. 21 and 22, 1993. Additional interviews with Dr. Greenberg, Dec. 17, 1994 and April 12, 1995.

385 The results in the first three: S. R. Riddell et al., "Restoration of Viral Immunity in Immunodeficient Humans by the Adoptive Transfer of T Cell Clones," ibid., *Science* 257:238 (1992).

385 "among the most": Drew Pardoll quoted in M. Hoffman, "Transferred Immune Cells May Help Fight Viral Infection," ibid., p. 166.

385 For a review of the use of cellular therapies against viral disease, see Stanley R. Riddell and Philip D. Greenberg, "Principles for Adoptive T Cell Therapy of Human Viral Diseases," *Annual Review of Immunology 1995*, 13:545–86 (1995).

385 E. A. Walter et al., "Reconstitution of Cellular Immunity against CMV in Recipients of Allogeneic Bone Marrow by Adoptive Transfer of T Cell Clones From the Donor," *New England Journal of Medicine* 333:1038–44 (1995). Another example of using adoptive T cell therapy in the bone marrow transplant setting is the Sloan-Kettering group (see E. B. Papadopoulos et al., "Infusions of Donor Leukocytes to Treat Epstein-Barr Virus-Associated Lymphoproliferative Disorders after Allogeneic Bone Marrow Transplantation," *New England Journal of Medicine* 330:1185–91 (April 28, 1994).

386 "Cells are like Swatches": Brad Nelson, personal interview, Seattle, Wash., Sept. 21, 1993.

388 "Phase I safety studies": S. R. Riddell et al., "T-cell Mediated Rejection of Gene-modified HIV-specific Cytotoxic T Lymphocytes in HIV-infected patients," *Nature Medicine* 2:216–23 (February 1996).

388 "What this is really saying": Stanley Riddell, telephone interview, Jan. 25, 1996.

389 "I believe": Thierry Boon, Ludwig Institute for Cancer Research, personal interview, Brussels, Belgium, Dec. 9, 1993.

Chapter 17: Dr. Levy's Favorite Guinea Pig

391 Epigraph: Paul Ehrlich, as quoted in Grant Fjermedal, *Magic Bullets*, Macmillan, New York, 1984, p. 6.

391 Phil Karr case: discussed in ibid., pp. 224–38. See also Robert Teitelman, *Gene Dreams: Wall Street, Academia, and the Rise of Biotechnology*, Basic

Books, New York, 1989; Ronald Levy, personal interview, Jan. 11, 1993, Stanford, California, and telephone interview, Dec. 22, 1995; David Maloney, personal interview, Dec. 27, 1994, Seattle; Phil Karr, telephone interviews, Nov. 14, 1995 and Dec. 6, 1995; and Phil Karr, letter to author, May 14, 1996.

The Karr case was originally described in Richard A. Miller et al., "Treatment of B-cell Lymphoma with Monoclonal Anti-idiotype Antibody," *New England Journal of Medicine* 306:517 (March 4, 1982); see also R. Levy, "B Cell Lymphomas," in *Biologic Therapy of Cancer*, edited by V. DeVita, Jr., S. Hellman, and S. Rosenberg, J. B. Lippincott, Philadelphia, 1991. A summary of the entire Stanford program with customized anti-idiotype antibodies is included in R. Levy, "Monoclonal Antibody Therapy," in *The Molecular Basis of Cancer*, edited by J. Mendelsohn et al., W. B. Saunders Co., Philadelphia, 1995, pp. 467–73.

396 "Both are probably too": P. Karr to Armand Hammer, Jan. 7, 1982, courtesy of Mr. Karr.

397 "the age of optimism": Lloyd Old, Ludwig Institute for Cancer Research, personal interview, New York, Feb. 15, 1996. For an example of biotechnology's enthusiasm for monoclonals, see G. Bylinsky, "Coming: Star Wars Medicine," *Fortune*, April 27, 1987.

397 In 1975 Cesar Milstein: G. Köhler and C. Milstein, "Continuous Cultures of Fused Cells Secreting Antibodies of Predefined Specificity," *Nature* 256:495–97 (1975). See also Alberto Cambrosio and Peter Keating, *Exquisite Specificity: The Monoclonal Antibody Revolution*, Oxford University Press, New York, 1995. In an informed obituary of Georges Köhler that appeared in the *New York Times*, Nicholas Wade described how Köhler was the one to have the idea and get the experiment to work. Milstein and Köhler won Nobel Prizes in 1984. N. Wade, "Georges Köhler, 48, Medicine Nobel Winner," *New York Times*, March 4, 1995, p. 26.

Apropos of the unrealized expectations for monoclonals in the clinic, Cesar Milstein in 1993 remarked, "Is this a failure of the ideas, or an oversimplification of the difficulties?" (Quoted by Rakesh Jain, "Monoclonals 1995").

398 During 1993 and 1994: On the failure of antisepsis drugs, see Richard Stone, "Search for Sepsis Drugs Goes On Despite Past Failures," *Science* 264:365–67 (April 15, 1996). The problems of Centocor, Xoma, and Synergen were well documented in the business press, but as an indication that the tide may be turning, see Laura Johannes, "Magic Bullet Drugs Try for a Comeback," *Wall Street Journal*, Jan. 31, 1996, p. B-1. "the Bermuda triangle": quoted in *Wall Street Journal*, July 19, 1994, p. A-3.

398 "They all had to": Stuart Schlossman, personal interview, New York, Oct. 16, 1996. Schlossman made the additional point that when researchers established a leukocyte antigen workshop to systematize the existing monoclonals in 1993, fully 25 percent had been uncorrectly characterized, including some that were being tested in humans.

399 "The difficulties were underestimated": L. Old, opening remarks, "Monoclonal Antibodies and Cancer Therapy: The Next Decade," Oct. 16–18, 1995, New York, sponsored by the Cancer Research Institute. "Some people were selling": Jean-Pierre Mach, University of Lausanne, remarks at the same meeting. For an account of the meeting, see S. Hall, "Monoclonal Antibodies at Age 20: Promise at Last?" *Science* 270:915–16 (Nov. 10, 1995).

401 "Using this rather radical": Oliver Press, University of Washington, personal interview, Oct. 17, 1995, New York. For clinical results, see O. W. Press et

al., "Radiolabelled-antibody Therapy of B Cell Lymphoma with Autologous Bone Marrow Support," *New England Journal of Medicine* 329:1219–24 (1993).

401 low-dose radiation approaches: M. S. Kaminski et al., "Iodine-131—Anti-B1 Radioimmunotherapy for B-Cell Lymphoma," *Journal of Clinical Oncology* 14:1974–81 (July 1996); and M. Kaminski, University of Michigan Medical Center, telephone interview, Nov. 22, 1996.

403 "avoidance of overt metastatic": G. Riethmüller, abstract, "Monoclonal Antibodies and Cancer Therapy," October 1996. For trials in colorectal cancer, see G. Riethmüller and J. P. Johnson, "Monoclonal Antibodies in the Detection and Therapy of Micrometastatic Epithelial Cancers," *Current Opinion in Immunology,* 4:647–55 (1992).

403 For brief discussion of work by Mendelsohn, Ferrara, Welt, Hellström, Rettig, and Eshhbar, see S. Hall, "Monoclonals at Age 20", (1995).

406 On peptide vaccine work: L. Kwak et al., "Induction of Immune Responses in Patients with B-cell Lymphoma against the Surface Immunoglobulin Idiotype Expressed by Their Tumors," *New England Journal of Medicine* 327:1209 (1992); F. J. Hsu et al., "Results of a Clinical Trial of Idiotype Specific Vaccine Therapy for B Cell Lymphoma (Abstract)," *Blood* 86, no. 10:1077 (suppl.; Nov. 15, 1995). Idiotype plus GM-CSF trial: Larry Kwak, National Cancer Institute, Frederick, Md., telephone interview, Oct. 30, 1996.

407 On the use of dendritic cells in vaccines: F. J. Hsu et al., "Vaccination of Patients with B-cell Lymphoma Using Autologous Antigen Pulsed Dendritic Cells," *Nature Medicine* 2:1–9 (January 1996). See also B. Mukherji et al., "Induction of Antigen-specific Cytolytic T Cells *in Situ* in Human Melanoma by Immunization with Synthetic Peptide-pulsed Autologous Antigen Presenting Cells," *Proceedings of the National Academy of Sciences* 92:8078–82 (August 1995).

Chapter 18: A Piece of Misleading News

409 Epigraph: Otto Westphal, "Hommage à Valy Menkin," opening remarks in Bonavida, edited by Gifford, Kirchner, Old, *Tumor Necrosis Factor/Cachectin and Related Cytokines,* proceedings of an international meeting in Heidelberg, 1987, Karger, Basel, 1988, p. 1.

409 "naive" idea and origin of TNF perfusions: F. Lejeune, personal interview, Centre Pluridisciplinaire d'Oncologie, Lausanne, Switzerland, Nov. 2, 1993; and F. Lejeune, personal communication, Feb. 28, 1996.

409 "There were very few": Douglas Fraker, personal interview, National Cancer Institute, Bethesda, Md., Dec. 7, 1992.

410 "a sort of signal": from "Endotoxin," in Lewis Thomas, *The Youngest Science: Notes of a Medicine Watcher,* Viking, New York, 1983, p. 151.

410 On discovery of TNF and overviews of its activity: E. A. Carswell et al., "An Endotoxin-induced Serum Factor That Causes Necrosis of Tumors," *Proceedings of the National Academy of Sciences* 72:3666–70 (September 1975). One of the noteworthy aspects of this paper is that Carswell, a technician, was first author, an unusually gracious gesture by the lab chief, Lloyd Old, for a paper destined to attract a lot of attention. "There are a lot of people who would never allow technical people in their labs to be first author—even appear on papers, let alone be first author," Carswell acknowledged in an interview. See also Lloyd J. Old, "Tumor Necrosis Factor," *Scientific Ameri-*

can, May 1988, pp. 59–75. "The story of the discovery of tumor necrosis factor," Old notes in this article, "properly begins with William B. Coley. . . ."

Otto Westphal, a retired expert in endotoxin, has made the case that the Russian-born Valy Menkin may have discovered tumor necrosis factor even earlier, when he described in 1956 a "factor" released by injured cells that he called "necrosin." "Today it appears—and seems more and more accepted— *that TNF is identical with Menkin's necrosin,*" Westphal writes (see O. Westphal, "Hommage à Valy Menkin," p. 5).

It has been suggested that a paper by O'Malley in 1962 also may have made early note of tumor necrosis factor, and that Carswell et al. merely transferred it. "I don't think we even knew about this paper at the time," Carswell said in an interview, "but ironically, other people have referred to this as being the first TNF. And they are absolutely wrong because we now know, and we have tested this, that what he found was that the earlier you took the serum out of the mice after they were injected with the endotoxin, the more active this serum was in making the tumor-bearing animals you injected, had more antitumor effect. And the longer you waited to take the serum, the less effect there was. Well, of course, it makes a lot of sense that the longer you leave the endotoxin floating around in the serum of mice, it's just not going to be there." Carswell said she repeated O'Malley's experiments and found that the antitumor effects were tracable to endotoxin, not TNF, which did not appear for 25 to 30 minutes after animals were injected with endotoxin.

411 "We were very interested": Elizabeth Carswell, personal interview, Memorial Sloan-Kettering Cancer Center, New York, Aug. 4, 1994.

413 Few molecules allude: Frances Balkwill, "Tumor Necrosis Factor and Lymphotoxin," in *Cytokines in Cancer Therapy*, Oxford University Press, 1989, pp. 54–98; David R. Spriggs, "Tumor Necrosis Factor: Basic Principles and Preclinical Studies," in V. DeVita, S. Hellman, and S. Rosenberg, eds., *Biologic Therapy of Cancer*, Lippincott, Philadelphia, 1991, pp. 354–77; and David R. Spriggs and Steven W. Yates, "Cancer Chemotherapy: Experiences with TNF Administration in Humans," in *Tumor Necrosis Factors: The Molecules and Their Emerging Role in Medicine*, Bruce Beutler, ed., Raven Press, New York, 1992, pp. 383–405.

413 "interesting quirk": B. Beutler and A. Cerami, "Introduction," in ibid., p. 4.

413 Judging from press accounts: J. Bishop, "Tests of Potential Cancer Weapon TNF Indicate Possible Serious Side Effects," *Wall Street Journal*, Sept. 9, 1985, p. 12; and H. Schmeck, "Anticancer Substance Reveals Its Dark Side," *New York Times*, Dec. 9, 1986, p. C-1.

414 epiphenomenon: Beutler and Cerami, "Introduction," p. 3.

414 "Will TNF ever": Spriggs and Yates, "Cancer Chemotherapy," p. 402.

414 On original limb perfusion: O. J. Creech et al., "Chemotherapy of Cancer: Regional Perfusion Utilizing an Extracorporeal Circuit," *Annals of Surgery* 148:616–32 (1958).

418 "The usual rule": Dr. Danielle Liénard, personal interview, Centre Pluridisciplinaire d'Oncologie, Lausanne, Switzerland, Nov. 2, 1993; Jean Gérain, personal interview, Nov. 2, 1993.

418 In fact, every one: Triple therapy results: The first results were reported in D. Liénard et al., "High-dose Recombinant Tumor Necrosis Factor Alpha in

Combination with Interferon Gamma and Melphalan in Isolation Perfusion of the Limbs for Melanoma and Sarcoma," *Journal of Clinical Oncology* 10:52–60 (January 1992). Follow-up studies include D. Liénard et al., *World Journal of Surgery* 16:234–40 (1992); D. Liénard et al., *Melanoma Research 4*, suppl. 1:21–26 (1994); and F. Lejeune et al., "Rationale for Using TNF Alpha and Chemotherapy in Regional Therapy of Melanoma," *Journal of Cellular Biochemistry* 56:52–61 (1994).

420 "extremely interesting": Lloyd Old, personal interview, March 5, 1993, Ludwig Institute for Cancer Research, New York; "I think it's biologically": David Spriggs, telephone interview, Memorial Sloan-Kettering Cancer Center, New York, July 31, 1996. "I think it's wonderful": James Mier, telephone interview, Aug. 1, 1996.

421 Updated clinical results were published as part of the meeting abstracts for "TNF and Related Cytokines: Clinical Utility and Biology of Action," Hilton Head Island, South Carolina, March 10–16, 1996. For a summary of this meeting, see F. Lejeune, "TNF Alpha: Shock-free Cancer Remission," *Biotechnology* 14:706–8 (1996).

422 In sarcoma: A. M. M. Eggermont et al., "Isolated Limb Perfusion with High-Dose Tumor Necrosis Factor-Alpha in Combination with Interferon Gamma and Melphalan for Nonresectable Extremity Soft Tissue Sarcomas: A Multicenter Trial," *Journal of Clinical Oncology* 14:2653–65 (October 1996).

Chapter 19: The Next Great Magic Bullet

423 Epigraph: David Markson, *Reader's Block,* Dalkey Archive Press, Normal, Ill., 1996, p. 10.

424 On Pittsburgh Cancer Institute patients: M. Lotze, Jan. 26, 1993, Pittsburgh; follow-up status, telephone interview, April 22, 1995. I am especially grateful to Dr. Lotze and his patients for allowing me to sit in on these consultations.

428 "This is not immunology": Giorgio Trinchieri, Wistar Institute, telephone interview, April 11, 1995; personal interview, New York City, March 11, 1994.

431 Original IL-12 papers: Michiko Kobayashi et al., "Identification and Purification of Natural Killer Cell Stimulatory Factor (NKSF), a Cytokine with Multiple Biologic Effects on Human Lymphocytes," *Journal of Experimental Medicine* 170:827–45 (Sept. 1989); Maurice Gately et al., "Synergy between Recombinant Interleukin 2 (rIL 2) and IL 2-Depleted Lymphokine-containing Supernatants in Facilitating Allogeneic Human Cytolytic T Lymphocyte Responses in Vitro," *Journal of Immunology* 136:1274–82 (Feb. 15, 1986); and A. S. Stern et al., "Purification to Homogeneity and Partial Characterization of Cytotoxic Lymphocyte Maturation Factor from Human B-Lymphoblastoid Cells," *Proceedings of the National Academy of Sciences* 87:6808–12 (Sept. 1990).

431 "We were just getting": Maurice Gately, Hoffmann–La Roche, personal interview, Bethesda, Md., May 16, 1995. Apropos of putative differences beween the two molecules at first, Gately said, "And also, in their final [purification] step, their NKSF was killed by the reverse-phase HPLC, and we were doing reverse-phase HPLC also as our final step and it was maintaining activity."

433 Date of agreement: "Timeline of Events, rhIL-12," public relations document, Genetics Institute, Cambridge, Mass.

436 For mechanism of action and effects in infectious and malignant disease, see Phillip Scott, "IL-12: Initiation Cytokine for Cell-mediated Immunity," *Science* 260:496–97 (April 23, 1993); for work on measles, see C. L. Karp et al., "Mechanism of Suppression of Cell-mediated Immunity by Measles Virus," *Science* 273:228–31 (1996).

438 "From the perspective of": Alan Sher, National Institute of Allergy and Infectious Diseases, Bethesda, Md., telephone interview, May 5, 1995. For anticancer studies in animals, see Michael J. Brunda et al., "Antitumor and Antimetastatic Activity of Interleukin-12 against Murine Tumors," *Journal of Experimental Medicine* 178:1223–30 (October 1993). For general review of IL-12 effects, see S. S. Hall, "IL-12 at the Crossroads," *Science* 268:1432–34 (June 9, 1995).

439 "magic bullet": in J. E. Bishop, "Cancer Drug Surprises Researchers With 'Double-Barrel' Attack on Tumors," *Wall Street Journal*, April 19, 1995, p. B-8 (the story referred to comments by R. S. Kerbel and R. G. Hawley in "Interleukin 12: Newest Member of the Antiangiogenesis Club," *Journal of the National Cancer Institute*, 87:557–59, April 19, 1995).

440 "All of these things": M. Gately, remarks at "From Coley's Toxins to Recombinant Cytokines: The Role of Interleukin-12 in the Immune Response, Disease, and Therapy," workshop sponsored by the Cancer Research Institute, New York, March 10–11, 1994.

442 On demise of Cetus, see Paul Rabinow, *Making PCR*, University of Chicago Press, Chicago, 1995, pp. 134–58.

442 On conflict between the doctors and Genetics Institute: according to sources, tension existed about the protocol for the testing. The company, fearful that any negative news about side effects might compromise the molecule's reception both by the public and by regulatory agencies like the FDA, set a rather conservative Maximum Tolerated Dose (MTD) for IL-12 of 500 nanograms per kilogram of body weight, or roughly 34 *millionths* of a gram in a patient weighing 150 pounds. The doctors believed this dosage level is too low.

443 Details of Genetics Institute's Phase I results were presented by Matthew Sherman, Department of Clinical Development, at "Interleukin-12: Cellular and Molecular Immunology of an Important Regulatory Cytokine," meeting sponsored by the New York Academy of Sciences, New York, Nov. 12, 1995.

443 The toxicities that resulted in a halt of the trial were most closely reported in the *Boston Globe* (see R. Rosenberg, "2 More Hospitalized in Drug Test," *Boston Globe*, June 13, 1995, p. 45, and R. Rosenberg, "Changes Made Before Cancer Test Went Awry," *Boston Globe*, June 14, 1995, p. 1 [Metro]); see also L. Seachrist, "Death Halts Trial of Kidney Cancer Drug," *Science News*, June 17, 1995, p. 375. In the case of the two fatalities, the Chicago patient was said to be "seriously ill from other things" at the time the toxicities appeared, according to his physician; in the case of the New York fatality, this patient apparently declined to return to Montefiore Medical Center when the side effects became severe and later died at North Shore University Hospital on Long Island.

444 "We were certainly": Janice Dutcher, Montefiore Medical Center, Bronx, New York, telephone interview, Nov. 6, 1996; "Bewilderment": James Mier, Tufts University–New England Medical Center, Boston, telephone inter-

view, August 1, 1996; Oct. 29. 1996. An employee of Genetics Institute also discussed the company's view of the episode on condition of anonymity.

444 "Last June's tragic": in J. Cohen, "IL-12 Deaths: Explanation and a Puzzle," *Science* 270:908 (Nov. 10, 1995).

444-45 "You always learn": Nicholas Vogelzang, University of Chicago, telephone interview, Nov. 1, 1996.

445 "public relations disaster": in S. S. Hall, "IL-12 at Crossroads", (1995).

446 "None of us are": Barry Bloom, Howard Hughes Medical Institute, Albert Einstein College of Medicine, telephone interview, May 25, 1995.

Epilogue: Metaphors, Manic Depression, and the "C" Word

449 Epigraph: Peter Lengyel, telephone interview, April 29, 1994.

451 "You get the drug": P. Wiernik, Montefiore Medical Center, Bronx, N.Y., personal interview, August 25, 1994.

451 "Cancer is not one": Peter Medawar, *Pluto's Republic*, Oxford University Press, New York, 1982, p. 157.

455 "It could have been": D. Spriggs, telephone interview, July 31, 1996.

455 Barry Werth, *The Billion-Dollar Molecule*, Simon & Schuster, New York, 1994.

457 The Chou En-lai remark is noted in Gore Vidal, *Palimpsest: A Memoir*, Random House, New York, 1995.

ACKNOWLEDGMENTS

∂ *I have sometimes been accused* of possessing an unusually liberal interpretation of the word "acknowledgment," bordering on confession, but books—like scientific discoveries—often begin with a mistaken premise pursued in earnest. In that spirit, I would like to thank Dr. Steven Rosenberg of the National Cancer Institute for inadvertently luring me into the topic of immunotherapy. Dr. Rosenberg wrote an article called "Adoptive Immunotherapy for Cancer" in the May 1990 issue of *Scientific American,* and by the time I finished reading it, I had the impression that immunotherapy had arrived as a full-fledged arm of clinical medicine. That impression, of course, was not quite true in 1990, but it launched me into the fascinating history of immunology, basic and applied, where I made the acquaintance of William Coley, Alick Isaacs, Dr. Rosenberg himself, and a host of other intriguing figures, all of whom have worked toward translating basic knowledge in the laboratory to useful therapies in the clinic. I am happy to report that immunotherapy is more established in 1996 than it was in 1990, and happier still to predict that, in the same slow and stepwise fashion, it is destined to assume even greater importance in the years to come.

This project would never have been undertaken without the generous support and enthusiasm of the Alfred P. Sloan Foundation, so I would like to thank Victor McElheny first and foremost for doggedly soliciting a proposal from me; Arthur B. Singer Jr. and the rest of the book committee for their willingness to support a rather more speculative member of the Technology Book family; and others associated with the foundation's Technology Book Series who have lent helpful hands along the way, including Richard Rhodes, Elting Morison, and Samuel Gibbons.

Many researchers have assisted in this project, as I will acknowledge below, but I would especially like to thank Lloyd J. Old of the Ludwig Institute for Cancer Research, whose historical memory of immunological inquiry is prodigious, whose scholarship is impeccable, and whose well-intentioned caution will never go out of fashion in a field in where discretion and restraint have not always been in conspicuous abundance. Several others offered assistance beyond mere sufferance of an interview or two, so it is a special pleasure to cite Mrs. Helen Coley Nauts, Dr. Susannah Isaacs Elmhirst, Jedd Levine, Jean Lindenmann, Philip Greenberg, Alexander Knuth, Joost Oppenheim, and Michael Lotze for special thanks.

For the Coley section, I would in addition to Mrs. Nauts like to thank her present and former staff members, including Kidgie Williams, Margaret Evans, and Carol Taylor, and the staff of the Cancer Research Institute, including Jill O'Donnell-Tormey, Rob Durkee, and Brian Quinn. In addition to Dr. Old, much useful information and perspective was provided by Frances Balkwill, Lent Johnson, Sanford Kempin, Klaus Kölmel, Ilana Löwy, Robert North, Herbert Oettgen, Michael Osband, Charlie Starnes, and Otto Westphal. Peter Wolfe and Sandra Horner of Warner-Lambert responded to queries about the Parke-Davis involvement, and Michael O'Hara handled queries at the Mayo Clinic.

For the section on cytokines, I wish to thank a great many researchers in the interferon field who consented to interviews, first and foremost Jean Lindenmann. Others included Anthony Allison, Samuel Baron, Alfons Billiau, Baruch Blumberg, Leslie Brent, Derek Burke, Kari Cantell, Donna Chaproniere, Edmund A. Debler, Walter Fiers, Norman Finter, Robert Friedman, Walter Gilbert, David Goeddel, Ion Gresser, Monto Ho, Hanna Kauppinen, Ian Kerr, John Kirkwood, Mathilde Krim, Peter Lengyel, Hilton Levy, Thomas Merigan, Michael Oxman, Helio Pereira, Sidney Pestka, Toine Pieters, James Porterfield, Jorge Quesada, Margaret Edney Sabine, Joseph Sonnabend, Hans Strander, Michel Streuli, Tadatsugu Taniguchi, Joyce Taylor-Papadimitriou, David Tyrrell, Jan Vilcek, and Charles Weissmann. For other aspects of the cytokine story, I am indebted to John Crewdson, Robert Gallagher, Robert Gallo, Steve Gillis, James Gowans, Steven Mizel, Doris Morgan, Joost Oppenheim, Werner Paetkau, Bernard

Poiesz, Frank Ruscetti, Ethan Shevach, Kendall Smith, and Richard Smith.

For the section on the National Cancer Institute, I would especially like to thank Steven Rosenberg for accommodating several visits, including one on rounds at the Surgery Branch, as well as Douglas Fraker, Richard Hodes, Mike Lotze, Jim Mulé, Maury Rosenstein, Suzanne Topalian, Norman Wolmark, and John Yannelli. Others who consented to interviews, offered insights, or were otherwise helpful included W. French Anderson, Michael Blaese, Robert Cook-Deegan, Ken Culver, Vincent DeVita, Alexander Fefer, Ronald Herberman, Peter Hersey, Georges Mathé, Scott McIvor, James Mier, Avrion Mitchison, Robertson Parker, Arnold Relman, Jerome Ritz, Richard Robb, Kendall Smith, Peter Wiernik, and Nelson Wivel.

At one point in this project, I developed a considerable amount of information about the discovery of interleukin-2, its development by the Cetus Corp., and its consideration by the Food and Drug Administration as a potential drug. Unfortunately, space considerations did not allow this material to be included, although it may yet see the light of day in another form. I nonetheless wish to thank the many researchers and sources who were so generous with their time, including Frederick Appelbaum, Judy Blakemore, J. Yule Bogue, Edward Bradley, Stephen Burrill, Ronald Cape, Budd Colby, Wanda de Vlaminck, Ross Di Leo, Edward Engleman, Peter Farley, Robert Fildes, Art Frankel, Gwen Fyfe, David Gelfand, Michael Innes, Michael Konrad, Kirston Koths, Frank Kung, Cory Levenson, David Mark, Hugh McDevitt, Kary Mullis, Saul Neidelman, Paul Rabinow, Hollings Renton, James Rurka, Jay Siegel, Peter Wiernik, and Thomas White.

For the final section, I would like to thank Richard Alexander, Michael Atkins, Christine Biron, Barry Bloom, Sir Walter Bodmer, Thierry Boon, Pierre van der Bruggen, Michael Brunda, Elizabeth Carswell, Mario Colombo, Janice Dutcher, Alexander Eggermont, Victor Englehard, Zelig Eshhar, Napoleone Ferrara, Olivera Finn, Greg Folkes, Maurice Gately, Jean Gerain, Philip Greenberg, Lee Hall, Dennis Harp, Stephen Hoffman, Rakesh Jain, Mark Kaminski, Martin Kast, George Klein, Joseph Kovacs, Alexander Knuth, Larry Kwak, H. Clifford Lane, Ferdinand Lejeune, Ronald Levy, Danielle Liénard, Philip Livingston, James Mier, David Maloney, Philippa Marrack, Cas Melief, Malcolm Mitchell, Robert Modlin, Bijay Mukherji, Kenneth Murphy, Nadia Noguiera-Cross, Drew Pardoll, Oliver Press, Wolfgang Rettig, Stanley Riddell, Jerome Ritz, John Ryan, Stuart Schlossman, Phillip Scott, Gene Shearer, Alan Sher, Craig Slinguff, David Spriggs,

Pramod Srivastava, Walter Storkus, Hideaki Tahara, Arleen Thom, Giorgio Trinchieri, Nicholas Vogelzang, Sidney Welt, Don Wiley, Darien Wilson, and Stanley Wolf.

Since *A Commotion in the Blood* is a story that belongs as much to the patients as the doctors, I wish to thank the "anecdotes" who consented to interviews and in most cases allowed their real names to be used: William Curtis, A. J. Goertz, "Frau H," Phil Karr, Claude Raiford, and "Chris Syzmanski," and to the many others who, although not represented in the final text, spoke with candor, courage, wisdom, and dignity.

Every chapter in this book has been vetted by a qualified immunologist conversant with the scientific work under discussion. I wish to thank my many conscientious reviewers, who performed this task in confidence in part so that they cannot share any responsibility for any errors or inaccuracies that may remain.

In terms of library and archives, I would like to thank Thomas Rosenbaum of the Rockefeller Archives, Pocantico Hills, New York; Shaw Kinsley of the Rare Book Room at the New York Academy of Medicine Library, as well as the nameless research assistants at the library itself who fetched many a dusty, century-old tome from the stacks; Jeanne Becker at Memorial Sloan-Kettering Cancer Center in New York; Robert Moore of the National Institute for Medical Research (Mill Hill) in London; and Helen Nauts, Carol Taylor, and Brian Quinn at the Cancer Research Institute in New York.

I wish to thank my colleagues at *Science,* especially John Benditt and Colin Norman, for allowing me to cover developments in the field as they unfolded. In addition, Anton Regalado helped out with some of the research, and I am grateful to Robin Iida, Jessica Nicoll, and Wendy Scheir for transcribing interviews.

As usual in a lengthy book project, I turned up on many doorsteps, sang for many suppers, and crawled gratefully into many spare beds over the course of the research; it is always special pleasure to thank friends simply for doing what friends do, which is help you get through an ordeal with offers of sustenance, whether in the form of a meal or well-timed words of encouragement. So, thanks to: Roger and Chris Blumberg; Patrick Cooke; Barbara Ellmann and Joe McElroy; Anne Friedberg and Howard Rodman; Bill Grueskin; Tom O'Neill; Mary Hawthorne and Richard Rothman; Richard Klug and Kate Stearns; Charles Mann; Wallace Matson; Terry and Josephine Monmaney; Peggy Northrop and Sean Elder; Delores and Robert Hall; Jerry and Linda Roberts; Steve Rubenstein and Caroline Grannan;

Russ Rymer; Margareta Schiappa; Lisa Shea; John Simmons; Nelson Smith; Virginia Stern; Larry Sulkis; and Gary Taubes.

I would like to thank all the people at Henry Holt for making this an exceptionally pleasant editorial experience. It begins with Bill Strachan, my editor, who has been enthusiastic about this book from day one (but hasn't let that enthusiasm cloud sound editorial judgment), but thanks also go to John Candell for the cover design; Miranda Ottewell for copyediting; Diane Tong for proofreading; and to Darcy Tromanhauser and Mary Ellen Burd for assistance. As always, my agent, Melanie Jackson, knows when to nudge and when to nurture.

Finally, it has been a special pleasure and challenge to bring a project of such complexity to completion while in the midst of starting an even more daunting endeavor: a family. After 180,000 words or so, I can't quite come up with the well-turned phrase that will adequately express my gratitude to my wife Mindy for her unconditional support, boundless patience, good judgment (editorial and otherwise), and all-around love in getting me over the hump. And I cannot thank Micaela Lucia enough for learning how to knock on her old man's office door at so precocious an age, reminding him to get on with the rest of his life.

INDEX

544 · *Index*